Inhibitors of Cyclin-dependent Kinases as Anti-tumor Agents

CRC Enzyme Inhibitors Series

Series Editors
H. John Smith and Claire Simons
Cardiff University
Cardiff, UK

Carbonic Anhydrase: Its Inhibitors and Activators
Edited by Claudiu T. Supuran, Andrea Scozzafava and Janet Conway

Enzymes and Their Inhibition: Drug Development
Edited by H. John Smith and Claire Simons

Inhibitors of Cyclin-dependent Kinases as Anti-tumor Agents
Edited by Paul J. Smith and Eddy W. Yue

CRC Enzyme Inhibitors Series

Inhibitors of Cyclin-dependent Kinases as Anti-tumor Agents

Edited by

Paul J. Smith
Eddy W. Yue

Taylor & Francis
Taylor & Francis Group
Boca Raton   London   New York

CRC is an imprint of the Taylor & Francis Group,
an informa business

CRC Press
Taylor & Francis Group
6000 Broken Sound Parkway NW, Suite 300
Boca Raton, FL 33487-2742

International Standard Book Number-10: 0-8493-3774-7 (Hardcover)
International Standard Book Number-13: 978-0-8493-3774-1 (Hardcover)

Library of Congress Cataloging-in-Publication Data

Inhibitors of cyclin-dependent kinases as anti-tumor agents / edited by Paul J. Smith, Eddy W. Yue.
    p.cm. -- (Enzyme inhibitors ; 3)
      Includes bibliographical references (p.   ).
    ISBN 0-8493-3774-7 -- ISBN 1-4200-0540-5
    1. Cyclin-dependent kinases--Inhibitors. 2. Cyclin-dependent kinases--Inhibitors--Therapeutic use--Testing. 3. Cancer--Chemotherapy. I. Smith, Paul J. (Paul James), 1953- II. Yue, Eddy W. III. Series.

QP609.C93I54 2006
616.99'407--dc22   2006045588

Visit the Taylor & Francis Web site at
http://www.taylorandfrancis.com

and the CRC Press Web site at
http://www.crcpress.com

Series Preface

One approach to the development of drugs as medicines, which has gained considerable success over the past two decades, involves inhibition of the activity of a target enzyme in the body or invading parasite by a small molecule inhibitor, leading to a useful clinical effect.

The CRC Enzyme Inhibitor Series consists of an expanding series of monographs on this aspect of drug development, providing timely and in-depth accounts of developing and future targets that collectively embrace the contributions of medicinal chemistry (synthesis, design), pharmacology and toxicology, biochemisty, physiology, and biopharmaceutics necessary in the development of novel pharmaceutics.

H. John Smith
Claire Simons

Preface

The mechanisms for controlling when and how a eukaryotic cell divides are fundamental to the biology of multicellular organisms. Abnormal regulation can provide a driver for disease processes, not the least, cancer. The aim of this volume is to provide an overview of the key regulator molecules — the cyclin-dependent kinases (CDKs) — as molecular and functional entities, together with both their involvement in different disease processes and their potential for pharmacological modulation.

The cell cycle is at the heart of the rationale for this area of intensive endeavor in basic research and drug development. The cell cycle comprises the transitions that a cell undergoes to achieve proliferation. The central importance of the cell cycle regulation machinery was recognized in the awarding of the Nobel Prize in Physiology or Medicine for 2001 to Leland Hartwell, Paul Nurse, and Timothy Hunt for their discoveries in the fields of cell cycle genes, checkpoint controls, the CDKs, and the cyclins. In the mid-1970s, Paul Nurse discovered the archetypal CDK gene (CDC2) in the *Schizosaccharomyces pombe* yeast, later isolating the corresponding gene in humans (CDK1), the function of this CDK being highly conserved during evolution. The chapters in this volume touch upon the cell cycle regulatory genes and their physiological inhibitors, the cyclin-dependent kinase inhibitors or CDKIs, distinct from the small molecule inhibitors that by chance or design have been found to impact upon CDK function. This volume was prompted by the realization that we are at a critical point in time for our understanding of the availability, advantages, and drawbacks of the CDKs as therapeutic targets for small molecules. Accordingly, the volume offers a current perspective on where we stand in terms of the success of the agents developed so far, the possibility of new routes to more selective inhibitors and our growing appreciation of the critical issues for therapy.

This book partitions its chapters into four sections, addressing in sequence the integration of cell cycle control pathways and the opportunities for targeting, the targets of inhibitors and their evaluation, the chemistry of the pharmacological CDK inhibitors, and finally a therapeutic perspective and horizon scan.

The first seven chapters, covering two sections, set the scene for the chemistry focus. The opening chapter by Paul Smith, Emeline Furon, and Rachel Errington provides an overview of cell cycle dynamics and the challenges for CDK targeting, highlighting issues raised in more detail in other chapters with a critical discussion on the problem of metastatic disease control. Functional regulation of the physiological CIP/KIP CDK inhibitors is reviewed by Mong-Hong Lee and Ruiying Zhao, revealing the recently discovered molecular regulation of the CKI family members. The use of mouse models to study the *in vivo* function of CDKs in normal homeostasis and tumor development is described by Marcos Malumbres, Pierre Dubus, and Sagrario Ortega. The elegant studies overviewed reveal how the absence of what are often regarded as key regulators can still be compatible with cellular survival, underlining the functional buffering that is programmed into mammalian cells. This redundancy has clear implications for therapeutic intervention in cancer.

Moving the focus away from cancer, Martin Crook and Manfred Boehm make the case for understanding the molecular mechanisms of growth control in cardiovascular cells and how this can lead to the development of exciting new molecular therapies for heart and vascular diseases. Stéphane Bach, Marc Blondel, and Laurent Meijer contribute a chapter on the evaluation of CDK inhibitor selectivity, with an insightful review of methodologies from affinity chromatography to yeast genetics. Sachin Mahale and Bhabatosh Chaudhuri bring a focus to bear on cyclin-dependent kinase 4, exploring its selective inhibition *in vivo* and *in vitro* by designed small molecules. Ursula Schulze-Gahman and Sung-Hou Kim provide an excellent overview of the structural biology of the CDKs. The structures of CDKs are covered from their apoenzymes to activated CDK/cyclin complexes to CDK/inhibitor complexes. Their analysis of the structural basis for CDK specificity provides an important foundation for the chemistry section.

The focus of the third section is the design, development, and chemistry of small molecule CDK inhibitors. The chapters in this section range from overviews of late-stage compounds that are already in clinical development to the early-stage discovery of new chemical entities with a capacity to inhibit the CDKs. At the beginning of this section, Adrian Senderowicz reviews two of the better known CDK inhibitors, flavopiridol (a flavonoid) and UCN-01 (a staurosporine analog), with an emphasis on their mechanism of action and results in clinical trials. The following two chapters also focus on well-known inhibitors of CDKs. Laurent Meijer and co-authors provide a comprehensive review of roscovitine, a purine that was discovered using a classical medicinal chemistry approach, whereas Conrad Kunick and co-authors describe the discovery of the novel paullones found through the Antitumor Drug Screening Program at the National Cancer Institute. In the first of several reviews by groups in the pharmaceutical industry, John Hunt addresses the discovery and development of BMS-387032, a potent and novel aminothiazole CDK inhibitor currently in clinical study. The next chapter focuses on the oxindole class of CDK inhibitors reviewed by Philip Harris who notes the existence of this pharmacophore has been studied by a number of research groups in both pharmaceutical companies and academia. The indenopyrazoles, reviewed by David Carini, Catherine Burton, and Steven Seitz, is another potent and novel class of CDK inhibitors discovered from high-throughput screening and optimized to compounds with a good biological profile.

The remaining chapters in the chemistry section illustrate the extensive SAR knowledge that is generated from groups in the pharmaceutical industry intent on discovering inhibitors of CDKs with the right balance of activity, selectivity, and pharmacological properties. Guoxin Zhu describes the development of indolocarbazoles which are in the same structural class as UCN-01 and staurosporine. Paolo Pevarello and Anna Vulpetti review the use of pyrazoles as potent and selective inhibitors of CDKs. Jay Markwalder and Steven Seitz end this section with a review of the pyrazolopyrimidines, another unique class of inhibitors also discovered from high-throughput screening. The diversity of the compounds and pharmacophores explored in this section highlights the opportunities for discovery of novel inhibitors.

As a number of CDK inhibitors have now entered clinical trials, a report on their current status is provided by Lloyd Kelland, with an evaluation of their potential use in combination with various anticancer cytotoxic agents. Chapter 17 reveals the challenges faced by the therapeutic deployment of small molecule inhibitors, resonating with the issues raised in other chapters. Finally, in Chapter 18, Richard Pestell and co-authors provide a perspective on the future of developing CDK inhibitors as anticancer agents. The design and discovery of small molecule inhibitors of CDK has focused on targeting the ATP-binding site and finding compounds that compete with ATP. The authors point to the use of detailed structural information of CDK/cyclin complexes as one way to develop better inhibitors. The book affirms the importance of selectivity with respect to inhibitor design, a matter made more critical by the increasing number of functions ascribed to CDK/cyclin complexes. An intriguing outlook is the potential for CDK inhibitors to regulate the local tumor microenvironment, diverting therapeutic attention away from direct targeting of the cancer cell, with the prospect of reducing microenvironmental signals that promote tumor progression. New concepts promise solutions to the therapeutic challenges and conflicts considered in the opening chapter of the volume.

Our primary aim as editors was to provide readers — from research chemists to clinicians — with a contemporary position to view the importance of rational drug design and knowledge-based therapeutics in relation to these exceptional kinase targets. The extensively referenced chapters draw upon perspectives and insights from expert groups in academia, government institutions, and the pharmaceutical industry to meet this brief.

Paul J. Smith
Cardiff, Wales, United Kingdom

Eddy W. Yue
Wilmington, Delaware

The Editors

Paul J. Smith has been active in the fields of DNA repair, drug development, cytometry, and imaging technologies for more than 20 years. His current academic research focuses on the cell cycle and anticancer drugs, imaging technologies, and mathematical modeling of complex biological systems. He received his Ph.D. in cellular radiobiology from Manchester University. After fellowship positions at the Atomic Energy of Canada Chalk River Nuclear Laboratories in Ontario, including support by the U.S. National Cancer Institute, he returned to the United Kingdom, becoming a senior scientist with the Medical Research Council at the MRC Centre in Cambridge. There he pursued basic research in a  clinical oncology unit. In 1995 he was appointed to the Chair of Cancer Biology at the University of Wales College of Medicine in Cardiff, incorporated in 2004 into the merged Cardiff University. Professor Smith is author of numerous research papers and is the senior editor of the multiauthor book *DNA Recombination and Repair.* He is cofounder and director of a spin-out company serving the biotechnology, drug screening, and healthcare market sectors. As the principal investigator in the U.K. Optical Biochips Consortium, he also leads an interdisciplinary effort to develop microtechnologies in the fields of drug screening and diagnostics. He is a member of U.K. Research Council's review and expert panels and a member of journal editorial boards. He has represented national bioscience interests in the United States and the Far East, and has promoted the public understanding of science in open lectures and the broadcast media.

Eddy W. Yue attended the University of California, Berkeley, as an undergraduate where he quickly developed a fondness for synthetic organic chemistry. During his time at UC Berkeley, he performed independent research at Lawrence Berkeley National Laboratory in the Division of Nuclear Medicine where he synthesized compounds for positron emission tomography studies. Dr. Yue received his B.S. in chemistry in 1991 and then entered the graduate program at The Scripps Research Institute in La Jolla, California, where he worked with Prof. K. C. Nicolaou. He was involved in the total synthesis of the  natural product Zaragozic Acid A, a squalene synthase inhibitor that contains a fascinating highly oxygenated core structure. He also contributed to the development of novel reactions in the areas of enediyne chemistry and polyether macrocyclic natural products. In 1996 he received his Ph.D. degree and then joined the medicinal chemistry department at The DuPont Merck

Pharmaceutical Company (which eventually became The DuPont Pharmaceuticals Company and then Bristol-Myers Squibb Company). He has conducted research in many therapeutic areas including inflammation, metabolic diseases, CNS disorders, and cancer. His work in the CDK area has led to several manuscripts and patents. He is currently a senior principal investigator in medicinal chemistry at Incyte Corporation where he continues his quest to find novel small molecule anticancer medicines.

Contributors

Stéphane Bach
C.N.R.S.
Amyloids and Cell Division Cycle
  Group
Station Biologique
Bretagne, France

Karima Bettayeb
C.N.R.S.
Amyloids and Cell Division Cycle
  Group
Station Biologique
Bretagne, France

Marc Blondel
C.N.R.S.
Amyloids and Cell Division
  Cycle Group
Station Biologique
Bretagne, France

Manfred Boehm
Cardiovascular Branch
National Heart, Lung,
  and Blood Institute
National Institutes of Health
Bethesda, Maryland

Catherine R. Burton
Bristol-Myers Squibb Pharmaceutical
  Research Institute
Wallingford, Connecticut

David J. Carini
Bristol-Myers Squibb Pharmaceutical
  Research Institute
Wallingford, Connecticut

Bhabatosh Chaudhuri
Leicester School of Pharmacy
De Montfort University
Leicester, England, United Kingdom

Martin F. Crook
Cardiovascular Branch
National Heart, Lung,
  and Blood Institute
National Institutes of Health
Bethesda, Maryland

Pierre Dubus
Histologie et Pathologie
  Moléculaire
University of Bordeaux
Bordeaux, France

Rachel J. Errington
Department of Medical Biochemistry
  and Immunology
Wales College of Medicine
Cardiff University
Cardiff, Wales, United Kingdom

Emeline Furon
Department of Pathology
Wales College of Medicine
Cardiff University
Cardiff, Wales, United Kingdom

Hervé Galons
Laboratoire de Chimie Organique 2
Université René Descartes
Paris, France

Philip A. Harris
GlaxoSmithKline
Research & Development
Collegeville, Pennsylvania

John T. Hunt
Bristol-Myers Squibb Pharmaceutical
   Research Institute
Princeton, New Jersey

Lloyd R. Kelland
Department of Basic Medical Sciences
St. Georges Hospital Medical School
London, England, United Kingdom

Sung-Hou Kim
Calvin Laboratory
University of California-Berkeley
Berkeley, California

Conrad Kunick
Institut fuer Pharmazeutische Chemie
   Technische
Universitaet Braunschweig
Braunschweig, Germany

Mong-Hong Lee
Department of Molecular and
   Cellular Oncology
The University of Texas
M.D. Anderson Cancer Center
Houston, Texas

Thomas Lemcke
University of Hamburg
Institute of Pharmacy
Hamburg, Germany

Sachin Mahale
Leicester School of Pharmacy
De Montfort University
Leicester, England, United Kingdom

Marcos Malumbres
Molecular Oncology Program
Centro Nacional de Investigaciones
   Oncológicas (CNIO)
Madrid, Spain

Jay A. Markwalder
Bristol-Myers Squibb Pharmaceutical
   Research Institute
Princeton, New Jersey

Laurent Meijer
C.N.R.S.
Amyloids and Cell Division Cycle
   Group
Station Biologique
Bretagne, France

Sagrario Ortega
Biotechnology Program
Centro Nacional de Investigaciones
   Oncológicas (CNIO)
Madrid, Spain

Nagarajan Pattabiraman
Departments of Cancer Biology
   and Medical Oncology
Thomas Jefferson University
Philadelphia, Pennsylvania

Richard G. Pestell
Departments of Cancer Biology
   and Medical Oncology
Thomas Jefferson University
Philadelphia, Pennsylvania

Paolo Pevarello
Medicinal Chemistry Department
Experimental Therapeutics Programme
Centro Nacional de Investigacones
   Oncologicas
Madrid, Spain

Eliot M. Rosen
Departments of Cancer Biology
   and Medical Oncology
Thomas Jefferson University
Philadelphia, Pennsylvania

Ursula Schulze-Gahmen
Physical Sciences Division
Lawrence Berkeley National
   Laboratory
Berkeley, California

Steven P. Seitz
Bristol-Myers Squibb Pharmaceutical
  Research Institute
Princeton, New Jersey

Adrian M Senderowicz
Radiation Biology Branch
National Cancer Institute
National Institutes of Health
Bethesda, Maryland

Jayalakshmi Sridhar
Departments of Cancer Biology
  and Medical Oncology
Thomas Jefferson University
Philadelphia, Pennsylvania

Anna Vulpetti
CADD
Novartis Institutes for BioMedical
  Research
Basel, Switzerland

Ruiying Zhao
Department of Molecular
  and Cellular Oncology
The University of Texas
M.D. Anderson Cancer Center
Houston, Texas

Guoxin Zhu
Lilly Research Laboratories
Indianapolis, Indiana

Table of Contents

SECTION III CDK Inhibitors: Chemistry Focus

SECTION IV   Therapeutic Perspective

Section I

Integration of Cell Cycle Control Pathways and the Opportunities for Targeting

1 Cell Cycle Dynamics and the Challenges for CDK Targeting

Paul J. Smith, Emeline Furon,
and Rachel J. Errington

CONTENTS

1.1 INTRODUCTION

The cell cycle comprises the temporal sequence of structural and functional transitions that a cell undergoes during processes of cell growth, replication of the genetic material, nuclear division, and cytoplasmic separation. It can be perceived at the level of observable landmarks (e.g., metaphase plate formation) or as complex shifts in the status of the cell cycle engine — an integrated network of cell cycle proteins showing evidence of considerable conservation from yeast to man. The cell cycle regulatory machinery is vital in controlling the balance between proliferation, growth and, indeed, cell death. The universal regulators comprise a set of unstable regulatory subunits called *cyclins* and a set of inactive catalytic subunits called *cyclin-dependent kinases* (CDKs). The regulators function to permit the cell to maintain the timing of both the onset and the fidelity of DNA replication and mitosis (Georgi et al., 2002). These key players determine the cell cycle transitions in concert with the influence of external and internal control signals. CDKs as targets for small molecules can be viewed from multiple standpoints providing pharmaceutical, therapeutic, and mechanistic perspectives (Shah and Schwartz, 2005; Blagden and de Bono, 2005; Eisenbrand et al., 2004; Huwe et al., 2003; Hirai et al., 2005; Pei and Xiong, 2005; Fischer 2004). The aim of this chapter is to provide an overview from those different perspectives, particularly with respect to the challenges faced in oncotherapeutics such as dealing with the problem of metastatic disease.

1.2 THE TARGET AND THE CHALLENGE

The concept of exploiting the pharmacological sensitivity of the cell cycle regulation machinery via the CDK components, as a means of controlling cellular proliferation or behavior, is highly attractive — particularly within an anticancer strategy. However, this concept is immediately challenged by our lack of understanding of how complex systems respond to perturbation and our relative inability to predict the likely impact of proliferation-controlling measures at different stages of the neoplastic process. The problem is further compounded by the inevitable need to combine CDK targeting therapies with other regimens. However, CDKs remain both important and novel targets despite the emerging complexities of their cellular physiology and their sometimes surprising redundancy as revealed in knockout studies (Dai and Grant, 2003). Although we may consider CDKs as legitimate targets in terms of proposed mechanisms of action in neoplasia, the reality is that the usefulness of pharmacological inhibitors as selective therapeutic agents will be determined by factors that confront many anticancer agents. These include:

1. The relative inability to monitor critical higher-order behavioral responses of targeted cell populations, e.g., intratumoral mobility, induction of angiogenesis, intravasation, extravasation, tissue-homing, and so on
2. The influence of macro- and micropharmacokinetic barriers
3. The treatment-limiting responses of collaterally exposed tissue systems
4. The disrupting influence of both phenotypic plasticity and genomic instability in maximizing Darwinian fitness and the development of acquired resistance
5. The effect of inherent heterogeneity giving rise to innate resistance
6. The formidable problem of quiescent micrometastases
7. The eradication of metastatic disease at disparate organ locations

1.3 THE CELL CYCLE IN PERSPECTIVE

We are aware that in multicellular organisms, as development proceeds, cell populations and lineages impose both spatial organization and temporal control over growth and proliferation. Here, the regulation of the cell cycle is increasingly subjected to signaling via intercellular communication rather than that arising from cell-autonomous mechanisms. There is strong evidence for cell size checkpoints in single-cell organisms (Polymenis and Schmidt, 1999) although it is likely that mammalian cells do not need, and probably do not have, such cell size checkpoints to coordinate their growth and division (Conlon and Raff, 2003) — reflecting the impact of signals from other cells in tissues. Internal signals arising from the imposition of genomic stress, for example through DNA damage, can still act to rapidly override the normal cell cycle program (Vogelstein et al., 2000). This override provides checkpoint-mediated opportunities for repair, recovery, and event completion without the need for an irreversible commitment to programmed cell death (apoptosis). The emergence of neoplasia with the loss of spatial control of growth and the temporal control of proliferation invariably reflects a dysregulation of the cell cycle.

1.3.1 KEY CELL CYCLE COMPONENTS

The cyclin family of proteins, structurally identified by their conserved "cyclin box" regions, are the regulatory subunits of the holoenzyme CDK complexes — responsible for guiding cell cycle traverse. In general schemes, holoenzyme function and longevity is modulated by kinases, phosphatases, and cyclin dynamics. Thus, CDK activity *per se* is controlled by: (1) the expression levels balanced by directed destruction via ubiquitination pathways and the subcellular location of cyclins, (2) the association with members of two families of cyclin-dependent kinase inhibitors (CDKIs), including p16 (INK4a), p15 (INK4b), p21 (CIP1), p27 (KIP1), and p57 (KIP2), (3) the addition of inhibitory phosphates, and (4) the dephosphorylation of a conserved threonine residue in the T-loop (Lee and Yang, 2001). Understanding the interplay of phosphorylation will continue to be enhanced by the introduction of approaches such as phosphoproteomic fingerprinting (Lim et al., 2003) and system-level approaches for the quantification of protein phosphorylation stoichiometry (Steen et al., 2005).

The broad-acting CDK inhibitor p21(WAF1) is an example of a naturally occurring braking system for the cell cycle, playing a physiological role in cell cycle regulation for self-renewing tissues such as oral mucosa and skin (Weinberg and Denning, 2002). p21(WAF1) also occupies a central position in regulating the cell cycle under stress, integrating genotoxic insults into competing growth arrest and apoptotic signals and providing a means of determining cell fate within cell populations (Perkins, 2002). Significant advances are being made at a systems level, in understanding how stress signals are integrated with the proliferation regulation, by the use of microarray approaches to track the genomewide responses to DNA-damaging agents in bacteria, yeast, and mammalian systems (Fry et al., 2005). Importantly, not all CDK family members are obligate direct regulators of the cell cycle; rather, they may act by influencing transcription (e.g., CDK7, CDK8, and CDK9) and their roles are not yet fully understood, for example, in neuronal and secretory cell function (CDK5) (Sausville, 2002).

Deregulation of CDK activity is linked to changes in cell proliferation and the emergence of cancer. Overexpression of positive regulators (e.g., cyclins) or the underexpression of negative regulators (e.g., CKIs) can subvert the cell cycle. It is clear that cell cycle regulatory proteins can act as oncogenes or tumor suppressor genes, or are closely associated with the transformation process. Functional defects in cell cycle transitions, as highlighted by the G_1/S abnormalities in breast cancer (Landberg, 2002), are readily demonstrable in cancer cell lines *in vitro*. A recent census of the complete human gene set, as revealed by the human genome sequence, indicates that mutations in more than 1% of genes contribute to human cancer (Futreal et al., 2004). Of the 291 cancer genes reported in that census, about 90% show somatic mutations in cancer, 20% show germline mutations, and 10% show both (Futreal et al., 2004). The identified cancer genes were often found to encode protein kinase, DNA binding, or transcriptional regulation domains. As new relationships are established between cancer genes, opportunities for cancer-targeting agents increase while the difficulty in understanding their logical application becomes compounded.

1.3.2 CDKs AND CDK-LINKED TARGETS

Many human cancers sustain mutations that alter the function of the retinoblastoma tumor suppressor protein (Rb) or the p53 transcription factor by direct mutation of gene sequences or by targeting genes that act epistatically to prevent their normal function (Nabel, 2002). A central role for a compromised Rb pathway in many human cancers has underpinned the concept of *CDK hyperactivation* (Senderowicz, 2005) and has driven the search for novel direct and indirect CDK inhibitors, with the first two direct CDK inhibitors tested in clinical trials being flavopiridol and UCN-01 (Senderowicz, 2002). X-ray crystal structures of the catalytically active and inactive forms of cdk2 and the crystal structures of a number of key inhibitors bound to their CDK targets have been used to inform structure–activity relationships and to further guide the design of more potent and selective inhibitors (Hardcastle et al., 2002; Furet, 2003).

Currently, there is an expanding therapeutic horizon for small molecule inhibitors of CDKs beyond that readily envisaged for neoplasia (see review, see Knockaert et al., 2002). For example, degenerative, nonneoplastic disease processes may also call upon the dysregulated CDK function. In Alzheimer's disease (AD) brain, the microtubule-associated protein tau is hyperphosphorylated at specific epitopes and abnormally aggregates into filamentous structures. A CDK inhibitor was found to reduce cell death induced by the topoisomerase I inhibitor camptothecin in a differentiated cell system, with evidence that increased tau phosphorylation is a generalized outcome of the apoptotic processes in neuron-related cells, and that CDKs probably play a role in this process (Mookherjee and Johnson, 2001). A specific role for cyclin-dependent kinase-5 (CDK5) in tau phosphorylation has also been suggested as a contributing factor to the pathogenesis of AD. In AD brains CDK5 and the activator p25 are colocalized with neurofibrillary tangles, driving a search for brain-permeable CDK5 inhibitors as therapeutic agents (Tsai et al., 2004).

At the physiological level the process of tissue remodeling is essential to the survival of adult organs. The treatment of "remodeling" dysfunctions may also offer opportunities for molecules that target CDKs or their normal endogenous inhibitors (CKIs) (Nabel, 2002). Furthermore, because CDKs are required for replication of viruses that do so only in dividing cells, such as adeno- and papillomaviruses, they are legitimate targets for antiviral drugs that have activity *in vivo* at nontoxic doses (Schlang, 2002). Accordingly, agents are currently being evaluated not only in the management of cancer but also for alopecia, neurodegenerative disorders (e.g., amyotrophic lateral sclerosis, stroke, and AD), cardiovascular disorders (e.g., atherosclerosis and restenosis), glomerulonephritis, viral infections (e.g., HCMV, HIV, and HSV), and diseases caused by parasitic protozoa (*Plasmodium* sp. and *Leishmania* sp.) (Knockaert et al., 2002).

Many biological pathways demand the controlled destruction of members as part of regulatory loops and the ubiquitin ligases serve a vital role in this process. The ligases target specific proteins for timely destruction and are tightly regulated by extracellular stimuli (Gao and Karin, 2005), as seen in MHC Class I antigen processing. Cell cycle transitions are driven by waves of ubiquitin-dependent degradation of the key cell cycle regulators. A number of oncoproteins and tumor suppressor proteins, including MDM2, Skp2, pVHL, APC, and Cdc4, are components of ubiquitin ligase complexes (Harris and Levine, 2005). Polyubiquitinated target proteins are recognized and processed by the 26S proteosome organelle, resulting in protein turnover and ubiquitin recycling (Mani and Gelmann, 2005). Recognition involves diverse ubiquitin receptors (Elsasser and Finley, 2005). Histone ubiquitination promotes chromatin relaxation and transcription complex assembly (Dhananjayan et al., 2005) and therefore impacts chromatin remodeling (Kinyamu et al., 2005). The profile of the ubiquitin–proteasome system in cancer is increasing (Devoy et al., 2005), and its role in cell cycle control is central to this mounting interest (Hershko, 2005; Ciechanover, 2005).

Two major classes of ubiquitin ligases are the metaphase/anaphase cell-cycle-transition-linked anaphase-promoting complexes (APCs) and the G1/S-transition-linked SCF (Skp1–Cullin–F-box protein) complexes (Vanoosthuyse and Hardwick, 2005; Fung and Poon, 2005; Ang and Wade Harper, 2005; Nakayama and Nakayama, 2005). The principal targets for the APC are involved in chromatid separation (securin) and exit from mitosis (cyclin B). Although this role of the APC in mitosis is relatively clear, there is mounting evidence that APCs containing Cdh1 (APC(CDH1)) may also have a function in the G1 phase of the cell cycle (Wei et al., 2004). The major target of the Skp1/Cul1/Skp2 (SCF(SKP2)) complex is thought to be the CKI p27 during S phase. Importantly, proteosome inhibitors have therapeutic potential (Zavrski et al., 2005; Mitsiades et al., 2005; Chauhan et al., 2005). A recently developed agent, bortezomib (Velcade; Millennium Pharmaceuticals Inc.), has been found to interfere with ubiquitin–proteasome complex dynamics, causing breakdown of cell cycle regulators and cell cycle arrest (Dubey and Schiller, 2005).

1.3.3 CYCLE ARREST AND CELL DEATH

In terms of tumor growth arrest and ablation, the ability of a CDK inhibitor to induce cycle arrest or cell death is critical. The discrete mechanisms by which these effects are engaged are still not clear and depend on the tumor cell type under study. Conceptually, direct CDK inhibitors target the ATP-binding site of the kinase, whereas indirect inhibitors interfere with the upstream pathways required for CDK activation (Senderowicz, 2003). The direct acting agent flavopiridol has a wide spectrum of activity against CDKs and is capable of inducing cell cycle arrest and apoptosis in various tumor cells *in vitro* and *in vivo* (Schrump et al., 1998; Shapiro et al., 1999). Flavopiridol's capacity to halt transcriptional elongation and cause more subtle shifts in cellular behavior, such as new blood vessel generation and differentiation induction, complicates its inclusion in coherent strategies for CDK modulation (Senderowicz, 2003).

The linkage between the cell-cycle-arresting properties of flavopiridol and its capacity to induce cell death is not obligatory (Schwartz, 2002; Mayer et al., 2005). For example, there is evidence that flavopiridol induction of cell death in small cell lung cancer appears to target S phase without a requirement for a severe arrest of cell cycle traverse (Litz et al., 2003). The cell cycle effects of flavopiridol may have an impact on responses to other agents. The S phase depletion in concert with an accumulation of cells in G_1 and G_2 for treated ovarian carcinoma cells (24 h exposure to 300 nM flavopiridol) has been proposed as a contributing mechanism in flavopiridol-induced cell radiosensitivity (Raju et al., 2003). As new agents are developed, there will be a constant need to explore the cycle-modulating activities of a candidate agent in relationship with the triggering of cell death pathways. CYC202 (R-roscovitine) is an example of a more recently investigated CDK inhibitor, showing a potent inhibition of recombinant CDK2/cyclin E kinase activity ($IC_{50} = 0.10$ μM) (McClue et al., 2002). Against a panel of human tumor cell lines, CYC202 shows an average cytotoxicity of 15.2 μM (IC_{50}) (McClue et al., 2002). CYC202 causes a substantial increase in p53 levels in competent cells, with a reciprocal effect of an endogenous caspase inhibitor (Mohapatra et al., 2005). The agent induces apoptosis in tumor cells (Hahntow et al., 2004) and normal cells in a proliferation-dependent manner (Atanasova et al., 2005). However, the pan-cell cycle capacity of CYC202 to induce cell death does not correspond with a predictive targeting of one part of the cycle (McClue et al., 2002), reenforcing ambivalence in the relationship between cycle arrest and the triggering of cell death (Schwartz, 2002).

1.4 CHALLENGES FOR TRACKING CELL CYCLE PROGRESSION AND KEY COMPONENT ACTIVITY

Accurate, noninvasive dynamic monitoring of cell cycle position of individual live cells is very valuable in the exploration of novel and potentially multiple pharmaco-dynamic actions of CDK inhibitors for *in vitro* drug-screening programs. The ability to dynamically monitor single-cell responses addresses the confounding effects of asynchronous populations and heterogeneity in response pathways. These responses may be invisible to methods that report population "snapshots" of the very limited number of informative cells exhibiting transient expression at any one time. Kinetic cell cycle tracking provides a method of connecting cellular events, in which a checkpoint history can be linked with delayed downstream outcomes. This type of analysis also helps the identification of the operation, abrogation, or bypass of a checkpoint. The retention of live cell function optimizes the potential for multiplexing with other biochemical assays to determine the effect of cell cycle status on putatively disparate cellular processes. In the following text we describe a conventional approach, as well as new ones, to the tracking of CDK-related events in live cells *in vitro* and *in vivo*.

1.4.1 S Phase Tracking *in vitro* and *in vivo*

Functional analyses for cell cycle progression, such as bromodeoxyuridine incorporation, provide important tools in the study of CDK inhibitors (Senderowicz, 2004). Bromodeoxyuridine incorporation for S phase tracking is a standard experimental approach permitting the retrospective analysis of cells labeled *in vitro* for cell cycle distribution and checkpoint function (Heinrich et al., 2006). The effects of the CDK inhibitors olomoucine and roscovitine on cell kinetics have been analyzed using pulse/chase labeling with bromodeoxyuridine, revealing dose-dependent inhibitions of G1/S phase progression and the exit from G_2 (Schutte et al., 1997). The ability to *in vivo* label cellular subpopulations with bromodeoxyuridine has been used to investigate tumor cell kinetics (for example, Zackrisson et al., 2002) and more recently stem cell behavior (Pang et al., 2003).

1.4.2 Live Cell *in vitro* Cell Cycle Tracking

The proliferative status of cells can be tracked by observing cyclin profiles (Widrow et al., 1997). It is generally viewed that cyclin B1 levels accumulate as cells progress through S phase and reach their maximal levels at mitosis but are effectively absent in G1 phase cells. Overexpression of cyclin B1 in tumor cells may be an indicator of risk of locoregional recurrence and metastasis (Hassan et al., 2002), and unscheduled expression has been noted in the G1 phase in breast cancer cells from patient tissues and in lymphocytes from patients with leukemia (Shen et al., 2004).

Figure 1.1 shows the visualization of cellular progression through the cell cycle as tracked by the expression of a fluorescently tagged cyclin reporter. This nontoxic molecular probe changes its expression in specific temporal windows, because it comprises a green fluorescent protein (GFP) (Tsien, 1998; Steff et al., 2001) under the control of a key cell cycle regulator (Thomas and Goodyer, 2003). The reporter system depends on the control of the expression levels and location of GFP as a cell progresses to the later cell cycle stages and negotiates mitotic entry and exit. This is achieved by using the functional components from cyclin B1 (Clute and Pines, 1999; Hagting et al., 1998, 1999) that regulate reporter expression (promoter region), removal via the destruction box (D-box), and translocation from the cytoplasm to the nucleus compartment via the cytoplasmic retention signal (CRS). Cyclin B1 expression is tightly regulated and acts as a major control switch suitable for following the transition from S phase through the G_2 phase into mitosis (Takizawa and Morgan, 2000). During interphase, cyclin B1 shuttles in between the nucleus and the cytoplasm because constitutive nuclear import is counteracted by rapid nuclear export. During prophase, five serine residues (predominantly S126 in late G2/M) within the CRS are phosphorylated. Phosphorylation of regions within the CRS decreases nuclear export and increases the rate of nuclear import, leading to accumulation of cyclin B1 in the nucleus (Hagting et al., 1999). The accumulation of cyclin B1 in the nucleus occurs approximately 10 min before breakdown of the nuclear envelope. The D-box represents the N-terminal domain of mitotic cyclins containing a conserved 9 amino

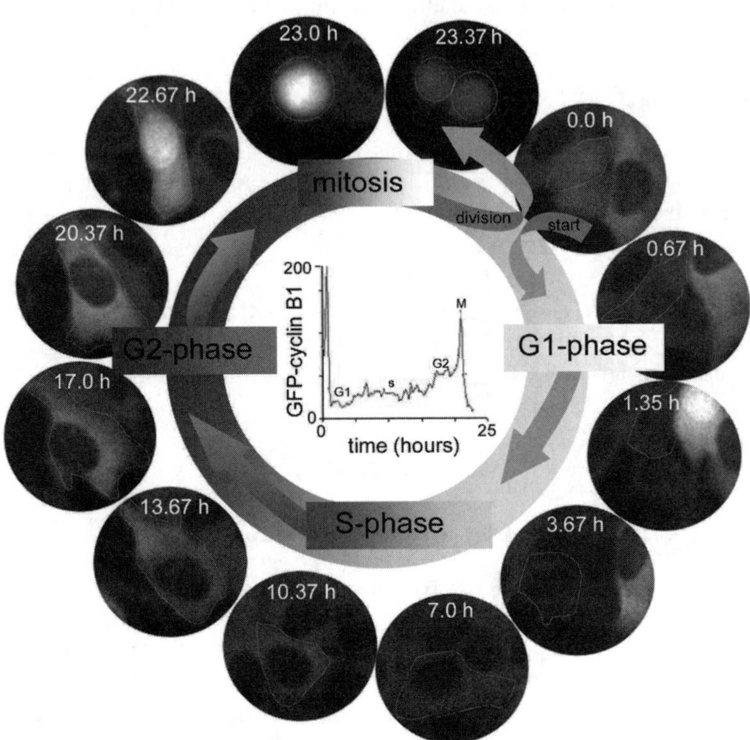

FIGURE 1.1 Diagrammatic representation of the mammalian mitotic cell cycle indicating the relative expression of a GFP/cyclin B1 fluorescent reporter stably transfected and expressed into the human osteosarcoma cell line U-2 OS. GFP/cyclin B1 expression increases during cell cycle traverse from G1 phase (0.0 h) to mitosis (23.0 h); upon commitment to mitosis the cyclin B1 reporter translocates to the nucleus (at 22.67 h). The cyclin expression is tracked for the outlined cell and measured at a given time point from the start of the cell cycle (0h). The inset trace shows the mean fluorescence intensity of the GFP/cyclin B1 reporter for a complete cell cycle traverse.

acid motif (RTALGDIGN) and is necessary for cyclin ubiquitination and subsequent degradation. Although the deletion of the N-terminal region does not interfere with the capacity of mitotic cyclins to activate CDC2 and drive the cells into mitosis, these mutations dominantly arrest cell division in telophase. Grafting of the cyclin B D-box onto otherwise stable proteins results in fused proteins becoming unstable in mitosis (i.e., the D-box is portable). However, Yamano et al. (1998) have demonstrated that insertion of the D-box itself was not sufficient for degradation of cyclin B and that inclusion of lysines flanking the D-box is necessary for efficient degradation. Importantly, because the cyclin B1/CDK1 binding site is absent, this reporter does not interfere with or perturb cell cycle progress; thus, it is thought that this is a nonperturbing cell cycle sensor (Thomas et al., 2005).

The reporter can be stably expressed through many generations and obviates the need for cell synchronization, which was previously mandatory for analysis of heterogeneous responses within a complex population. Earlier (Marin and Bender, 1966; Hurwitz and Tolmach, 1969a, 1969b; Thompson and Suit, 1969; Thompson and Suit, 1967; Hopwood and Tolmach, 1971) and more recent (Forrester et al., 1999; Forrester et al., 2000; Chu et al., 2002; Chu et al., 2004) studies have revealed the value of individual cell tracking by the use of video time lapse to reveal the complex time course of cell cycle delay, arrest, mitotic catastrophe, apoptosis, and clonogenic potential. Premitotic cell cycle events can now be tracked in parallel with morphologically recognizable changes associated with mitosis in live cells using transmission and fluorescence time-lapse microscopy of cells stably transfected with the cyclin B1/GFP reporter construct (Thomas and Goodyer, 2003; Thomas, 2003; Thomas et al., 2005). Global time-lapse analyses can be performed on U-2 OS cells ex to candidate agents to track lineages of up to five generations, permitting early event linkage with conventional clonogenic assay results.

Figure 1.1 shows the changes in expression levels of cyclin B1-GFP stably expressed in human osteosarcoma cells (U-2 OS cell line) as an untreated control cell negotiates cell cycle traverse. Unlike most human carcinoma cell lines, U-2 OS expresses wild-type p53 (Landers et al., 1997). The U-2OS cancer cell line is aberrant for p16 (INK4A) but is wild type for the retinoblastoma gene product (pRb), in keeping with MCF-7 breast cancer cells (Craig et al., 1998). Similar to MCF-7 cells, U-2 OS osteosarcoma cells are known to be functional for DNA-damage-induced transcriptional activation of p53 through phosphorylation at Ser15 (Florenes et al., 1994). Previous work has also shown that U-2 OS cells express the expected mitotic arrest responses and coordinated changes in cyclin B1/CDC2 kinase activity when exposed to a mitotic spindle checkpoint (MSC) activator (Raj et al., 2001). The MSC is therefore functional in U-2 OS cells (Seong et al., 2002), suggesting that rate of mitotic traverse could also report checkpoint activation. Recent work has however underlined the potential delay in the engagement of apoptosis after cells exit from a compromised mitosis (Allman et al., 2003), indicating the need to track outcome of mitotic events over extended periods (Marquez et al., 2003).

1.4.3 *In vivo* Cell-Cycle-Tracking Technologies

Combining the GFP reporting technology with simultaneous live cell DNA content analysis or nuclear discrimination provides a flexible, extended map for cell cycle characterization. This includes the marking of the nuclear compartment, which is useful in the identification of cyclin B1 translocation as an indicator of mitotic commitment. The characteristics of GFP demand the application of a red-shifted fluorescent cell-permeable DNA/nuclear reporter to minimize spectral overlap. We have developed a solution using a novel far-red DNA dye DRAQ5 (Smith et al., 2000). The dye can rapidly report cellular DNA content of live cells at a resolution adequate for cell cycle analysis. The cell cycle can be segmented into G1, S phase, and G2 or their subcompartments according to relative GFP expression using the correlated DRAQ5 signal as analyzed by multiparameter flow cytometry (Figure 1.2). Combining GFP reporting technology and flow cytometry with an *in vivo* hollow fiber containment of defined tumor and tumor/stroma populations (Shnyder et al., 2005)

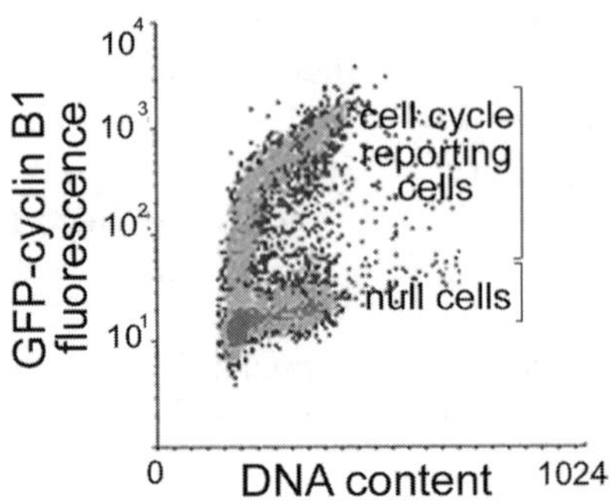

FIGURE 1.2 A multiparameter FACS analysis of DNA content (DRAQ5 fluorescent dye labeling) versus the expression of the GFP/cyclin B1 reporter, revealing the pattern of increase in reporter expression as DNA content, and hence cell cycle position, increases.

would provide a route for the rapid *ex vivo* analysis of cell cycle perturbation for cells treated with CDK inhibitors *in vivo*.

Understanding the complex dynamics of inhibitor targeting *in vivo* requires new solutions. Recently a p27Luciferase fusion protein has been used to monitor CDK2 activity both *in vitro* and *in vivo* (Zhang et al., 2004) with options for reporter validation in cell culture using siRNA against CDK2 (or its partner cyclin A). A significant advance is the ability to image cells producing p27-Luc grown in transparent hollow fibers in nude mice after treatment with CDK2 inhibitory drugs *in vivo* (Zhang and Kaelin, 2005). However, although the tracking of the hollow fiber/*in vivo* action of a candidate agent is an important step, the response at critical tumor cell "homing" sites remains a significant challenge. The assessment of the chemosensitivity of micrometastases represents an important hurdle for the development of new therapeutic approaches. Protocols have been described for the use of GFP-tagged micrometastasis tumor models combined with noninvasive detection devices (Nakanishi et al., 2005). Xenograft mouse models of human breast cancer based on luciferase-expressing MDA-MB-231 tumor cells that show metastasis to clinically relevant tissues, have been used to provide *in vivo* monitoring of both primary and secondary tumor sites using bioluminescence imaging (Jenkins et al., 2005), overcoming the considerable problems of excitation-dependent fluorescence methods.

1.5 METASTASIS

As primary and secondary tumor sites become more accessible to detection and analysis, it will become increasingly important to understand the contribution of CDK inhibitors in the control of early phase and ongoing spread of neoplastic disease, that is, metastasis. We continue to be disadvantaged in this because the underlying processes governing the various patterns of metastasis observed in different tumor types remain unclear.

1.5.1 METASTASIS: A MULTISTEP CASCADE

Metastasis can be considered as a multistep cascade. Early changes in specific integrin signals appear to enable cancer cells to detach from neighboring cells, facilitate matrix remodeling, and reorient their polarity during migration (Guo and Giancotti, 2004). As the metastatic process continues, integrin-mediated formation of microemboli — composed of tumor cells, platelets, and leukocytes — enable the docking of tumor cells to the endothelium and extravasation at a metastatic site (Guo and Giancotti, 2004). Studies on disseminated tumor cells found in the bone marrow of patients with various types of solid tumors, including lung tumors, suggest that the cells seem to first disseminate from the early primary lesions and then acquire additional genetic defects (Pantel and Brakenhoff, 2004). The amalgamation of intravital imaging and gene expression profiling methods has started to place into context behavioral processes occurring at early invasion stages (Wang et al., 2005). Such approaches have highlighted the role of enhanced motility for successful tumor cell invasion of the microenvironment (Condeelis and Segall, 2003).

Cytoplasmic localization of p21 (CIP1) and p27 (KIP1) is associated with high tumor grade, tumor cell invasiveness, and metastasis; the function of CKIs of the CIP/KIP family, for example, is potentially modulated by cytoplasmic relocation rather than gene mutation *per se*. At this location they may act to regulate cytoskeletal functions rather than the cell cycle *per se* (see review: Besson et al., 2004). These alternative CKI functions, acting in the cytoplasm to regulate small GTPase (e.g., Rho) signaling, provide a new therapeutic target for the modulation of changes in cytoskeletal organization and the accompanying changes of cell migration (Besson et al., 2004). Recent work has revealed that a kinase-independent function of Raf-1 may be to act as a spatial regulator of Rho downstream signaling during migration (Ehrenreiter et al., 2005).

The opportunities for the rational development of antimetastasis drugs essentially began with the description of a metastasis suppressor gene *nm23* (Steeg et al., 1988). Confirmed metastasis suppressors (see Steeg et al., 2003; Steeg, 2004) include: Nm23, differentiation-related gene (Drg-1), Src-suppressed C Kinase substrate (SSeCKs), Vitamin D_3 upregulated protein 1 (VDUP), the CRSP3 transcriptional coactivator, mitogen-activated protein kinase kinase 4, Raf kinase inhibitor Rkip, RhoGDI2, Brms1, Kiss-1 (metastin), Claudin-4, and Kai1. Most of these metastasis suppressor genes encode cell signaling products and are expressed during embryonic development, yet appear to have no impact on *in vitro* proliferation or primary tumor size. Novel mechanisms by which suppressors may act include transcriptional repression through an impact on chromatin remodeling. For example, the breast cancer metastasis suppressor 1 (BRMS1) suppresses metastasis of multiple human and murine cancer cells without inhibiting tumorigenicity, potentially operating through histone deacetylase complexes (Meehan et al., 2004).

1.5.2 METASTASIS AND PHENOTYPIC PLASTICITY

The importance of genetic changes in neoplastic progression is well recognized and has been highlighted by the drive for the development of animal models for the reiteration of critical steps. For example, the inactivation of both Rb1 and p53 were found to be prerequisites for the pathogenesis and high incidence of aggressive lung tumors with striking morphologic and immunophenotypic similarities to SCLC (small cell lung cancer): diffuse spread through the lung and propensity for extra-pulmonary metastasis (Meuwissen et al., 2003; Meuwissen and Berns, 2005). The proposal has been made that gain of cytoplasmic CKI function might also be involved in the process of tumor invasion and metastasis (Besson et al., 2004).

Ramaswamy et al. (2003; see also Pan et al., 2005) have suggested that metastatic potential of human tumors appears to be encoded in the bulk of the primary tumor, challenging the concept that metastases arise from rare cells at a primary site. Despite the consensus that malignant tumor progression represents a ratchet of genetic alterations, it is often found that metastases of carcinomas recapitulate the organi-zation of their primary tumors, exhibiting a degree of plasticity. This has led to the suggestion that human tumor progression and metastasis may be underpinned by the behavior of migrating and phenotypically plastic cancer stem cells (Brabletz et al., 2005). Epithelial mesenchymal transitions occur during embryo development, although restrictive mechanisms that prevent such cellular transitions are thought to operate in adult tissues (Prindull and Zipori, 2004). It has been suggested that reversible transitions may not only enable the development of embryonic stem (ES) cells to adult stem cells (hematopoietic stem cells) but also, through the fundamental property of plasticity, facilitate metastatic spread and the progression of microme-tastases when restrictive mechanisms fail (Prindull, 2005).

Behavioral transformations of cells, reflecting phenotypic plasticity, characterize embryogenesis, wound healing, physiological adaptation, and neoplasia (Eyden, 2004; Jechlinger et al., 2003; Grünert et al., 2003). Limiting the plasticity envelope, and thereby the potential for variant generation, using containment therapies is intuitively a route for controlling neoplastic progression but demands favorable cell population dynamics and in any case would need to be integrated with surgical, cytotoxic, or novel tumor eradication regimens. However, for effective therapeutic intervention, attention is also being drawn to the later stages of the metastatic cascade in which the location of micrometastases is not merely determined by blood flow patterns, but also by the active participation of the tissues to which cancer cells appear to be drawn. This view has been strengthened by the assignment of new biological roles for chemokine receptors in assisting the spread of primary tumors to distant secondary sites.

The dynamic relationship between the G-protein-coupled seven-span transmem-brane receptor CXCR4 and the alpha-chemokine stromal-derived factor (SDF-1/CXCL12) is thought to be an important regulator of trafficking of both normal and cancer stem cells (Kucia et al., 2005). For example, there is growing evidence that the CXCR4 extravasation response to CXCL12 regulates migration and homing of detached cells in a variety of cancers including SCLC-stimulating cell motility, invasion, survival, and proliferation at metastatic sites (Marchesi et al., 2004). Using *in vivo* biolumines-cence imaging of metastasis in a mouse lung model, inhibition of CXCR4 has revealed a role for the chemokine receptor in regulating the growth of both primary and metastatic breast cancer (Smith et al., 2004). Interestingly, the chemokine receptors CCR5 and CXCR4 serve as coreceptors for the human immunodeficiency virus 1 (HIV-1) and

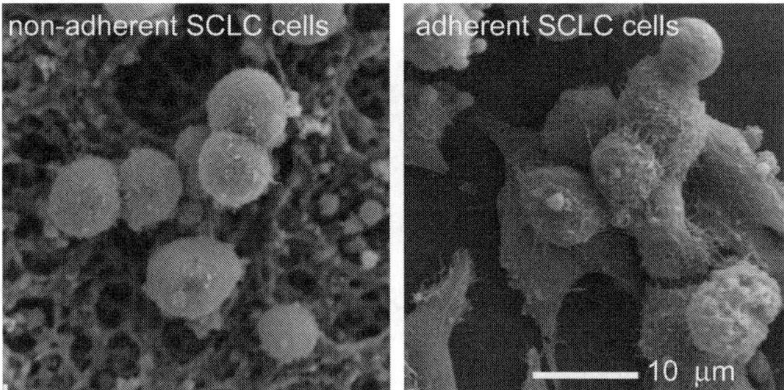

FIGURE 1.3 Scanning electron microscopy images of small cell lung carcinoma NCI-H69 cells demonstrating the adherent and nonadherent (prometastatic) *in vitro* phenotypes.

thus are also important cellular components during HIV-1 cell entry (Bachmann and Gamper, 2005); they have therefore become foci for antiviral therapies.

1.5.3 METASTASIS AND CELL CYCLE REGULATION: SCLC

The balance between drivers for proliferation and metastasis are critical for the potential effectiveness of CDK targeting. For example, SCLCs show early spread but rarely have ras mutations, suggesting that ras activation may not confer a growth advantage. SCLC cell matrix interactions play an important role in defining potential responsiveness to therapy through integrin-driven effects on cell cycle regulators (Hodkinson et al., 2006). The *in vitro* selection of anchorage-independent growth of tumor cells has been used previously to enhance the invasive potential of SCLC as well as the metastatic potential of the NCI-H69 cell line (Onn et al., 2003). Adhesion-variant phenotypes can be generated in culture (Figure 1.3), where the loss of anchorage to a substrate gives rise to prometastatic behavior. Here we briefly explore the SCLC cell system to profile the impact selectable phenotypic shifts in cell behaviour on cell cycle targets. Comparing the adherent and nonadherent phenotypes under conditions of optimal *in vitro* cell proliferation for gene expression provides an insight into the potential compensation by the cell cycle regulatory machinery, including CDK expression, as cells explore prometastatic behavior.

Microarray analysis (Figure 1.4) reveals that there is a significant modulation of only a small fraction of cell-cycle-related genes, cyclin A1 gene expression being reduced in adherent cells (Figure 1.4). However, there is a relatively unperturbed expression of the partner CDK2 expression and unchanged expression of other G1/S transition CDK2 partners, including: cyclin A2 (Nigg, 1995), cyclin E1 (Sgambato et al., 2003), and cyclin E2 (Zariwala et al., 1998). Similarly, no significant changes were found in the gene expression for the G1/S transition-related CDK4 and partner cyclin D3 (Bartkova et al., 1998), or CDK7/cyclin H, a regulator of transcription and CDK-activating kinase (CAK) (Fisher and Morgan, 1994). CDK9 and its binding partner cyclin T1, comprising the positive elongation factor b (P-TEFb) (Peng et al., 1998), were unchanged, this complex being highly regulated during differentiation (Napolitano et al., 2003).

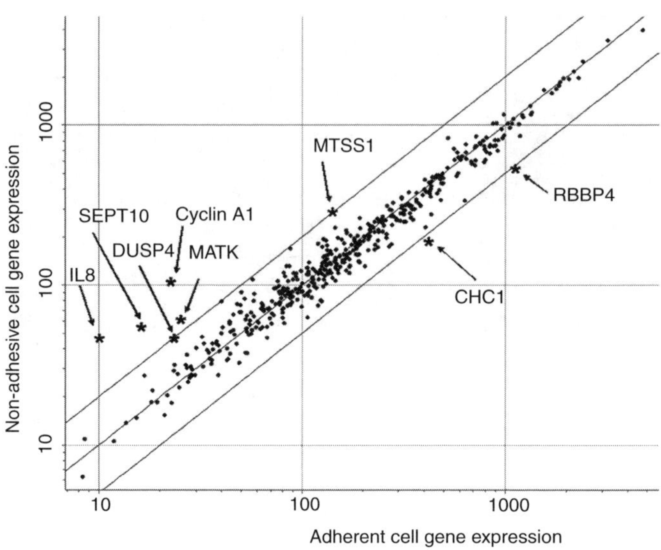

FIGURE 1.4 Relative expression profiles for cell cycle genes comparing adherent and non-adherent (prometastatic) phenotypes of the small cell lung carcinoma cell line NCI-H69 grown *in vitro*. RNA was prepared from exponentially growing culture and gene expression analyzed using HGU133A2 Affymetrix microarrays. Data were normalized using Agilent Genespring software. The lines indicate the twofold change threshold. Annotated genes are the significantly up- or downregulated genes in adherent cells, as determined by a two-way ANOVA (p-value .05) and a fold change > 2.

Active cyclin A/CDK2 complexes are thought to be a critical factor for cell proliferation in human primary lung carcinomas (Shimizu et al., 1997; Dobashi et al., 1998), with cyclin A and CDK2 being frequently expressed in SCLC with a link between cyclin A/CDK2 complex expression and higher CDK2 kinase activity (Dobashi et al., 1998). The activity of CDK4 and CDK2 is reduced in SCLC cells undergoing cycle arrest (Ravi et al., 1999). Cultures of both SCLC phenotypes showed similar cell cycle distributions and proliferation rates, and therefore the changes in cyclin A1 expression do not appear to limit proliferation. A fall in expression of IL8 was detected, interleukin-8/CXCL8 being a growth factor for human lung cancer cells (Figure 1.4; Zhu et al., 2004), suggesting that the adherent cells may be under reduced, but not limiting, drivers for proliferation. A recent study has indicated that cyclin-A1-deficient murine cells accumulate in G_1 and G_2/M phase and that siRNA-mediated silencing of cyclin A1 in highly expressing leukemic cells significantly slowed S phase entry (Ji et al., 2005). Ji et al. (2005) have suggested that cyclin A1 might be a therapeutic target because its silencing inhibited leukemia cell growth. Small molecule targeting of CDK2 could act to reduce proliferative potential and may provide a means of reducing proliferation in cells with critical levels of partner cyclin.

1.5.4 Therapeutic Intent and the Complexity of Drug Resistance

The potential for monotherapy or combinations of CDK-targeting agents with other anticancer agents is clear (Fischer and Gianella-Borradori, 2003) but raises issues of the recruitment of collateral targets (Kelland, 2000) and drug resistance interactions. A potential for nonspecific action of inhibitors is suggested by the discovery of the direct binding of flavopiridol to DNA (Bible et al., 2000). Reverse-phase high-performance liquid chromatography has demonstrated that flavopiridol binds to genomic DNA to a similar extent as the intercalator ethidium bromide and the minor groove binding ligand Hoechst 33258 (Bible et al., 2000). Resistance to flavopiridol has also been noted (Incles et al., 2003). The Goldie–Coldman model (Goldie and Coldman, 1979) predicts the presence of drug-resistant mutants even within the 10^9 neoplastic cells nominally present at diagnosis for treatment-naïve primary tumor sites. This has informed a chemotherapeutic approach of rapidly alternating the administration of equally effective drugs to prevent clonal resistance (Goldie et al., 1982; Panetta, 1998). Clonal resistance can occur within a background of plasticity, enhancing the probability of development of the metastatic phenotype.

Disappointingly, it is general experience that tumors that respond rapidly to cytotoxic drugs also relapse rapidly (Epstein, 2005; Sierocki et al., 1979). It is likely that optimal adjuvant drug therapy will include both induction ("cytotoxic") and maintenance components (Epstein, 2005), the latter comprising metastasis-suppressive maintenance drugs such as histone deacetylase inhibitors (Chiba et al., 2004). One of the early models that attempted to describe effective adjuvant chemotherapy is that of Norton and Simon (Norton and Simon, 1986, 1977). They assumed that all tumor growth, tumor regression, and tumor regrowth is Gompertzian. The Gompertz model, originally a demographic description of age-dependent mortality rate, has been the conceptual basis for models of solid tumors (Norton, 1988; Speer et al., 1984; Norton and Simon, 1977, 1986; Spratt et al., 1993).

According to Gompertzian kinetics, as the tumor becomes smaller its growth fraction increases and it regrows at a faster rate. Accordingly there is a role for modified CDK dynamics during the regrowth phase. The Norton–Simon hypothesis, at variance with the earlier log-cell kill hypothesis (Skipper et al., 1964), suggested that tumors given less time to regrow between treatments are more likely to be destroyed if therapeutic intent is to effect cell kill or depopulation. Chemoresistance due to dormancy or transient cell cycle arrest may be reduced by increasing the frequency of cytotoxic drug treatment (Norton and Simon, 1977). Thus, a cure will never be effected if regrowth rate is not matched or exceeded by the rate of cell kill. The implication for treatment is that the only way to overcome this "kinetic resistance" barrier is the absolute eradication of every viable micrometastatic cell. Thus, the intervals between chemotherapy treatment are critical (Clare et al., 2000). Reducing any ongoing supply of micrometastatic cells would be advantageous while adjuvant therapies that induced dormancy would act to spare sources of chemoresistance. A critical question for the use of CDK inhibitors is, therefore, therapeutic intent. It is likely that CDK inhibitors used in the adjuvant setting could address the attainment of induced and maintained dormancy as a positive therapeutic endpoint (Epstein, 2005).

1.5.5 MATHEMATICAL MODELS

It is likely that intervention is never without unseen and unforeseen consequences. Indeed, relapse in breast cancer has been attributed to the possible postsurgical stimulation, via angiogenesis, of distant dormant micrometastases (Retsky et al.,1997; Retsky et al., 2004). Here the hypothesis is that a burst of angiogenesis of distant dormant micrometastases after surgery provides for increased aggressiveness of the tumor and, paradoxically, a recruitment into a more chemosensitive state (Retsky et al., 2004). Epstein (2005) has recently highlighted the experience (Braud et al., 1999) that "… the relationship between nonresponse to chemotherapy and adverse outcome is far stronger than that between response and favorable outcome." One implication is that the metrics for detection of response, usually favoring the more "active" cytotoxic induction modes of therapy, may be missing potentially important anticancer effects achieved by non-cytotoxic maintenance modes of action (Epstein, 2005). How to evolve that view into coherent strategies for drug screening and development is a current challenge. Understanding in this area, in terms of prediction of the dynamics of complex systems, is being advanced using systems biology approaches to the perturbation of the mammalian cell cycle and the dynamic responses of targeted populations.

The potential for successful therapeutic intervention in cancer by using CDK inhibitors — whether acting as antiproliferative agents, engaging novel pathways for limiting the generation of highly metastatic cells or the maintenance of dormancy — is likely to depend on our understanding of the initiation time and how candidate metastatic cells accrue at primary tumor sites and/or continue to evolve their metastatic potential following intravasation and extravasation. Developing a better understanding of the natural history of cancer via mathematical models can suggest more effective methods of screening and treatment. Although the use of mathematical models to determine relapse and to predict response to chemotherapy cannot replace experimental and clinical results, they can both eliminate and suggest strategies (review: Clare et al., 2000). There have been attempts to model the dynamics of metastatic cell population expansion and inform a rationale for screening to capture patients with premetastatic lesions (Koscielny et al., 1985). Modified Gompertzian models allow for stepwise growth patterns with the possibility of dormant phases associated with limiting processes such as the development of a tumor microvasculature. However, model selection can have a significant effect on the estimation of the preclinical growth phase (Norton, 1988). Different CDK inhibitor treatment strategies would be pertinent if a model outcome indicated that there was a high probability that metastasis has already occurred by primary diagnosis, rather than if the metastasis were unlikely to have been initiated (Clare et al., 2000).

1.6 CONCLUSIONS

There are new opportunities for combining CDK inhibitors with metastasis-limiting therapies that do not compromise cell proliferation signals *per se*. Here, therapeutic intent is critical in the application of CDK inhibitors, with the likely rise of dormancy-inducing strategies for tumors not susceptible to outright eradication regimens (Uhr et al., 1997; Hedley et al., 2004; Epstein, 2005). A refined model of primary tumor and metastases growth dynamics, rejecting an exponential growth assumption, has revealed that metastatic initiation could occur at a tumor volume of the primary lesion that was only slightly smaller than that at the time of tumor diagnosis (Koscielny et al., 1985). One implication is that metastasis treatment and containment therapies will become increasingly important as screening program evolve and attempt to inform early intervention. New *in vivo* tracking technologies for monitoring the action of CDK targeting agents offer solutions in the drug discovery and development process. It will be exciting to see the application of such tracking methods to *in vivo* models of tumors in early phases of progression. CDK inhibitors offer therapeutic promise at multiple levels as notions of intent and target are translated into medicines designed for purpose.

ACKNOWLEDGMENTS

The authors acknowledge grant support from the Research Councils U.K. Basic Technology Research Programme (GR/S23483) and the Biotechnology and Biological Sciences Research Council (75/E19292). We thank the Medical Microscopy Unit for the electron microscopy study.

REFERENCES

Allman, R., Errington, R.J., and Smith, P.J. (2003). Delayed expression of apoptosis in human lymphoma cells undergoing low-dose taxol-induced mitotic stress. *Br. J. Cancer* **88**, 1649–1658.

Ang, X.L. and Wade Harper, J. (2005), SCF-mediated protein degradation and cell cycle control, *Oncogene* **24**, 2860–2870.

Atanasova, G., Jans, R., Zhelev, N., Mitev, V., and Poumay, Y. (2005). Effects of the cyclin-dependent kinase inhibitor CYC202 (R-roscovitine) on the physiology of cultured human keratinocytes. *Biochem. Pharmacol.* **70**, 824–836.

Bachmann, A.S. and Gamper, I. (2005), Killing two birds with one drug: a new application for HIV-1 cell entry inhibitors in the treatment of metastatic cancer. *Hawaii Med. J.* **64**, 220–223.

Bartkova, J., Lukas, J., Strauss, M., and Bartek, J. (1998). Cyclin D3: requirement for G1/S transition and high abundance in quiescent tissues suggest a dual role in proliferation and differentiation. *Oncogene* **17**, 1027–1037.

Besson, A., Assoian, R.K., and Roberts, J.M. (2004). Regulation of the cytoskeleton: an oncogenic function for CDK inhibitors? *Nat. Rev. Cancer* **4**, 948–955.

Bible, K.C., Bible, R.H., Jr., Kottke, T.J., Svingen, P.A., Xu, K., Pang, Y.P., Hajdu, E., and Kaufmann, S.H. (2000). Flavopiridol binds to duplex DNA. *Cancer Res.* **60**, 2419–2428.

Blagden, S. and de Bono, J. (2005). Drugging cell cycle kinases in cancer therapy. *Curr. Drug Targets* **6**, 325–335.

Brabletz, T., Jung, A., Spaderna, S., Hlubek, F., and Kirchner, T. (2005). Opinion: migrating cancer stem cells — an integrated concept of malignant tumour progression. *Nat. Rev. Cancer* **5**, 744–749.

Braud, A.C., Asselain, B., Scholl, S., De La Rochefordiere, A., Palangie, T., Dieras, V., Pierga, J.Y., Dorval, T., Jouve, M., Beuzeboc, P., and Pouillart, P. (1999). Neoadjuvant chemotherapy in young breast cancer patients: correlation between response and relapse? *Eur J. Cancer* **35**, 392–397.

Chauhan, D., Hideshima, T., and Anderson, K.C. (2005). Proteasome inhibition in multiple myeloma: therapeutic implication. *Annu. Rev. Pharmacol. Toxicol.* **45**, 465–476.

Chiba, T., Yokosuka, O., Fukai, K., Kojima, H., Tada, M., Arai, M., Imazeki, F., and Saisho, H. (2004). Cell growth inhibition and gene expression induced by the histone deacety-lase inhibitor trichostatin A on human hepatoma cells. *Oncology* **66**, 481–491.

Chu, K., Leonhardt, E.A., Trinh, M., Prieur-Carrillo, G., Lindqvist, J., Albright, N., Ling, C.C., and Dewey, W.C. (2002). Computerized video time-lapse (CVTL) analysis of cell death kinetics in human bladder carcinoma cells (EJ30) X-irradiated in different phases of the cell cycle. *Radiat. Res.* **158**, 667–677.

Chu, K., Teele, N., Dewey, M.W., Albright, N., and Dewey, W.C. (2004). Computerized video time lapse study of cell cycle delay and arrest, mitotic catastrophe, apoptosis and clonogenic survival in irradiated 14-3-3sigma and CDKN1A (p21) knockout cell lines. *Radiat. Res.* **162**, 270–286.

Ciechanover, A. (2005). Intracellular protein degradation: from a vague idea thru the lysosome and the ubiquitin-proteasome system and onto human diseases and drug targeting. *Cell Death Differ.* **12**, 1178–1190.

Clare, S.E., Nakhlis, F., and Panetta JC. (2000). Molecular biology of breast cancer metastasis: the use of mathematical models to determine relapse and to predict response to chemotherapy in breast cancer. *Breast Cancer Res.* **2**, 430–435.

Clute, P. and Pines, J., (1999). Temporal and spatial control of cyclin B1 destruction in metaphase, *Nat. Cell Biol.* **1**, 82–87.

Condeelis, J. and Segall, J.E. (2003). Intravital imaging of cell movement in tumours. *Nat. Rev. Cancer* **3**, 921–930.

Conlon, I. and Raff, M. (2003). Differences in the way a mammalian cell and yeast cells coordinate cell growth and cell-cycle progression. *J. Biol.* (published online) 2(1), 7.

Craig, C., Kim, M., Ohri, E., Wersto, R., Katayose, D., Li, Z., Choi, Y.H., Mudahar, B., Srivastava, S., Seth, P., and Cowan, K. (1998). Effects of adenovirus-mediated p16INK4A expression on cell cycle arrest are determined by endogenous p16 and Rb status in human cancer cells. *Oncogene* **16**, 265–272.

Dai, Y. and Grant, S. (2003). Cyclin-dependent kinase inhibitors. *Curr. Opin. Pharmacol.* **3**, 362–370.

Devoy, A., Soane, T., Welchman, R., and Mayer, R.J. (2005). The ubiquitin-proteasome system and cancer. *Essays Biochem.* **41**, 187–203.

Dhananjayan, S.C., Ismail, A., and Nawaz, Z. (2005). Ubiquitin and control of transcription. *Essays Biochem.* **41**, 69–80.

Dobashi, Y., Shoji, M., Jiang, S.X., Kobayashi, M., Kawakubo, Y., and Kameya, T. (1998). Active cyclin A-CDK2 complex, a possible critical factor for cell proliferation in human primary lung carcinomas. *Am. J. Pathol.* **153**, 963–72.

Dubey, S. and Schiller, J.H. (2005). Three emerging new drugs for NSCLC: pemetrexed, bortezomib, and cetuximab. *Oncologist* **10**, 282–291.

Ehrenreiter, K., Piazzolla, D., Velamoor, V., Sobczak, I., Small, J.V., Takeda, J., Leung, T., and Baccarini, M. (2005). Raf-1 regulates Rho signaling and cell migration. *J. Cell Biol.* **168**, 955–964.

Eisenbrand, G., Hippe, F., Jakobs, S., and Muehlbeyer, S. (2004). Molecular mechanisms of indirubin and its derivatives: novel anticancer molecules with their origin in traditional Chinese phytomedicine. *J. Cancer Res. Clin. Oncol.* **130**, 627–635.

Elsasser, S. and Finley, D. (2005). Delivery of ubiquitinated substrates to protein-unfolding machines. *Nat. Cell Biol.* **7**, 742–749.

Epstein, R.J. (2005). Maintenance therapy to suppress micrometastasis: the new challenge for adjuvant cancer treatment. *Clin. Cancer Res.* **11**, 5337–5341.

Eyden, B. (2004). Fibroblast phenotype plasticity: relevance for understanding heterogeneity in "fibroblastic" tumors. *Ultrastruct. Pathol.* **28**, 307–319.

Fischer, P.M. (2004). The use of CDK inhibitors in oncology: a pharmaceutical perspective. *Cell Cycle* **3**, 742–746.

Fischer, P.M. and Gianella-Borradori, A. (2003). CDK inhibitors in clinical development for the treatment of cancer. *Expert Opinion Invest. Drugs* **12**, 955–970.

Fisher, R.P. and Morgan, D.O. (1994). A novel cyclin associates with MO15/CDK7 to form the CDK-activating kinase. *Cell* **78**, 713–724.

Florenes, V.A., Maelandsmo, G.M., Forus, A., Andreassen, A., Myklebost, O., and Fodstad, O. (1994). MDM2 gene amplification and transcript levels in human sarcomas: relationship to TP53 gene status. *J. Natl. Cancer Inst.* **86**, 1297–1302.

Forrester, H.B., Albright, N., Ling, C.C., and Dewey, W.C. (2000). Computerized video time-lapse analysis of apoptosis of REC:Myc cells X-irradiated in different phases of the cell cycle. *Radiat. Res.* **154**, 625–639.

Forrester, H.B., Vidair, C.A., Albright, N., Ling, C.C., and Dewey, W.C. (1999). Using computerized video time lapse for quantifying cell death of X-irradiated rat embryo cells transfected with c-myc or c-Ha-ras. *Cancer Res.* **59**, 931–939.

Fry, R.C., Begley, T.J., and Leona D. Samson. (2005). Genome-wide responses to DNA-damaging agents. *Annu. Rev. Microbiol.* **59**, 357–377

Fung, T.K. and Poon, R.Y. (2005). A roller coaster ride with the mitotic cyclins. *Semin. Cell Dev. Biol.* **16**, 335–342.

Furet P. (2003). X-ray crystallographic studies of CDK2, a basis for cyclin-dependent kinase inhibitor design in anti-cancer drug research. *Curr. Med. Chem. — Anti-Cancer Agents.* **3**, 15–23.

Futreal, P.A., Coin, L., Marshall, M., Down, T., Hubbard, T., Wooster, R., Rahman, N., and Stratton, M.R. (2004). A census of human cancer genes. *Nat. Rev. Cancer*, **4**, 177–183.

Gao, M. and Karin, M. (2005). Regulating the regulators: control of protein ubiquitination and ubiquitin-like modifications by extracellular stimuli. *Mol. Cell.* **19**, 581–593.

Georgi, A.B., Stukenberg, P. T., and Kirschner, M.W. (2002). Timing of events in mitosis. *Current Biology* **12**, 105–114.

Goldie, J.H., Coldman, A.J., and Gudauskas, G.A. (1982). Rationale for the use of alternating non-cross-resistant chemotherapy. *Cancer Treat. Rep.* **66**, 439–449.

Goldie, J.H. and Coldman, A.J. (1979). A mathematic model for relating the drug sensitivity of tumors to their spontaneous mutation rate. *Cancer Treat. Rep.* **63**, 1727–1733.

Grünert, S., Jechlinger, M., and Beug, H. (2003). Diverse cellular and molecular mechanisms contribute to epithelial plasticity and metastasis. *Nat. Rev. Mol. Cell Biol.* **4**, 657–665.

Guo, W. and Giancotti, F.G. (2004). Integrin signalling during tumour progression. *Nat. Rev. Mol. Cell Biol.* **5**, 816–826.

Hagting, A., Jackman, M., Simpson, K., and Pines, J., (1999). Translocation of cyclin B1 to the nucleus at prophase requires a phosphorylation-dependent nuclear import signal. *Curr. Biol.* **9**, 680–689.

Hagting, A., Karlsson, C., Clute, P., Jackman, M., and Pines, J., (1998). MPF localization is controlled by nuclear export, *EMBO J.,* **17**, 4127–4138.

Hahntow, I.N., Schneller, F., Oelsner, M., Weick, K., Ringshausen, I., Fend, F., Peschel, C., and Decker, T. (2004). Cyclin-dependent kinase inhibitor Roscovitine induces apoptosis in chronic lymphocytic leukemia cells. *Leukemia* **18**, 747–755.

Hardcastle, I.R., Golding, B.T., and Griffin, R.J. (2002). Designing inhibitors of cyclin-dependent kinases. *Annu. Rev. Pharmacol. Toxicol.* **42**, 325–348.

Harris, S.L. and Levine, A.J. (2005). The p53 pathway: positive and negative feedback loops. *Oncogene* **24**, 2899–2908.

Hassan, K.A., Ang, K.K., El-Naggar, A.K., Story, M.D., Lee, J.I., Liu, D., Hong, W.K., and Mao, L. (2002). Cyclin B1 overexpression and resistance to radiotherapy in head and neck squamous cell carcinoma. *Cancer Res.* **62**, 6414–6417.

Hedley, B.D., Winquist, E., and Chambers, A.F. (2004). Therapeutic targets for antimetastatic therapy. *Expert Opin. Ther. Targets* **8**, 527–536.

Heinrich, T., Prowald, C., Friedl, R., Gottwald, B., Kalb, R., Neveling, K., Herterich, S., Hoehn, H., and Schindler, D. (2006). Exclusion/confirmation of Ataxia-telangiectasia via cell-cycle testing. *Eur. J. Pediatr* [Epub ahead of print].

Hershko, A. (2005). The ubiquitin system for protein degradation and some of its roles in the control of the cell division cycle. *Cell Death Differ.* **12**, 1191–1197.

Hirai, H., Kawanishi, N., and Iwasawa, Y. (2005). Recent advances in the development of selective small molecule inhibitors for cyclin-dependent kinases. *Curr. Top. Med. Chem.* **5**, 167–179.

Hodkinson, P.S., Elliott, T., Wong, W.S., Rintoul, R.C., Mackinnon, A.C., Haslett, C., and Sethi, T. (2006). ECM overrides DNA damage-induced cell cycle arrest and apoptosis in small-cell lung cancer cells through beta1 integrin-dependent activation of PI3-kinase. *Cell Death Differ* [Epub ahead of print].

Hopwood, L.E. and Tolmach, L.J. (1971). Deficient DNA synthesis and mitotic death in x-irradiated Hela cells. *Radiat. Res.* **46**, 70–84.

Hurwitz, C. and Tolmach, L.J. (1969a). Time lapse cinemicrographic studies of X-irradiated HeLa S3 cells I. Cell progression and cell disintegration. *Biophys. J.* **9**, 607–633.

Hurwitz, C. and Tolmach L.J. (1969b). Time-lapse cinemicrograhic studies of X-irradiated HeLa S3 cells. II. Cell fusion. *Biophys. J.* **9**, 1131–1143.

Huwe, A., Mazitschek, R., and Giannis, A. (2003). Small molecules as inhibitors of cyclin-dependent kinases. *Angew. Chem. Int. Ed. English* **42**, 2122–2138,

Incles, C.M., Schultes, C.M., Kelland, L.R., and Neidle, S. (2003). Acquired cellular resistance to flavopiridol in a human colon carcinoma cell line involves up-regulation of the telomerase catalytic subunit and telomere elongation: sensitivity of resistant cells to combination treatment with a telomerase inhibitor. *Mol. Pharmacol.* **64**, 1101–1108.

Jechlinger, M., Grunert, S., Tamir, I.H., Janda, E., Ludemann, S., Waerner, T., Seither, P., Weith, A., Beug, H., and Kraut, N. (2003). Expression profiling of epithelial plasticity in tumor progression. *Oncogene* **22**, 7155–7169.

Jenkins, D.E., Hornig, Y.S., Oei, Y., Dusich, J., and Purchio, T. (2005). Bioluminescent human breast cancer cell lines that permit rapid and sensitive in vivo detection of mammary tumors and multiple metastases in immune deficient mice. *Breast Cancer Res.* **7**, R444–R454.

Ji, P., Agrawal, S., Diederichs, S., Baumer, N., Becker, A., Cauvet, T., Kowski, S., Beger, C., Welte, K., Berdel, W.E., Serve, H., and Muller-Tidow, C. (2005). Cyclin A1, the alternative A-type cyclin, contributes to G1/S cell cycle progression in somatic cells. *Oncogene* **24**, 2739–2744.

Kelland, L.R. (2000). Flavopiridol, the first cyclin-dependent kinase inhibitor to enter the clinic: current status. *Expert Opin. Invest. Drugs* **9**, 2903–2911.

Kinyamu, H.K., Chen, J., and Archer, T.K. (2005). Linking the ubiquitin-proteasome pathway to chromatin remodeling/modification by nuclear receptors. *J. Mol. Endocrinol.* **34**, 281–297.

Knockaert, M., Greengard, P., and Meijer, L. (2002). Pharmacological inhibitors of cyclin-dependent kinases. *Trends Pharmacol. Sci.* **23**, 417–425,

Koscielny, S., Tubiana, M., and Valleron, A.J. (1985). A simulation model of the natural history of human breast cancer. *Br. J. Cancer* **52**, 515–524.

Kucia, M., Reca, R., Miekus, K., Wanzeck, J., Wojakowski, W., Janowska-Wieczorek, A., Ratajczak, J., and Ratajczak, M.Z. (2005). Trafficking of normal stem cells and metastasis of cancer stem cells involve similar mechanisms: pivotal role of the SDF-1-CXCR4 axis. *Stem Cells* **23**, 879–894.

Landberg, G. (2002). Multiparameter analyses of cell cycle regulatory proteins in human breast cancer: a key to definition of separate pathways in tumorigenesis. *Adv. Cancer Res.* **84**, 35–56.

Landers, J.E., Cassel, S.L., and George, D.L. (1997). Translational enhancement of mdm2 oncogene expression in human tumor cells containing a stabilized wild-type p53 protein. *Cancer Res.* **57**, 3562–3568.

Lau, L.F., Seymour, P.A., Sanner, M.A., and Schachter, J.B. (2002). Cdk5 as a drug target for the treatment of Alzheimer's disease. *J. Mol. Neurosci.* **19**, 267–273.

Lee, M.H. and Yang, H.Y. (2001). Negative regulators of cyclin-dependent kinases and their roles in cancers. *Cell. Mol. Life Sci.* **58**, 1907–1922.

Lim, Y.P., Diong, L.S., Qi, R., Druker, B.J., and Epstein, R.J. (2003). Phosphoproteomic fingerprinting of EGF signaling and anticancer drug action in human tumor cells. *Mol. Cancer Ther.* **2**, 1369–1377.

Litz, J., Carlson, P., Warshamana-Greene, G.S., Grant, S., and Krystal, G.W. (2003). Flavopiridol potently induces small cell lung cancer apoptosis during S phase in a manner that involves early mitochondrial dysfunction. *Clin. Cancer Res.* **9**, 4586–4594.

Mani, A. and Gelmann, E.P. (2005). The ubiquitin-proteasome pathway and its role in cancer. *J Clin Oncol.* **23**, 4776–4789.

Marchesi, F., Monti, P., Leone, B.E., Zerbi, A., Vecchi, A., Piemonti, L., Mantovani, A. and Allavena, P. (2004). Increased survival, proliferation, and migration in metastatic human pancreatic tumor cells expressing functional CXCR4. *Cancer Res.* **64**, 8420–8427.

Marin, G. and Bender, M.A. (1966). Radiation-induced mammalian cell death: lapse-time cinemicrographic observations. *Exp. Cell Res.* **43**, 413–423.

Marquez, N., Chappell, S.C., Sansom, O.J., Clarke, A.R., Court, J., Errington, R.J., and Smith, P.J. (2003). Single cell tracking reveals that Msh2 is a key component of an early-acting DNA damage-activated G2 checkpoint. *Oncogene* **22**, 7642–7648.

Mayer, F., Mueller, S., Malenke, E., Kuczyk, M., Hartmann, J.T., and Bokemeyer, C. (2005). Induction of apoptosis by flavopiridol unrelated to cell cycle arrest in germ cell tumour derived cell lines. *Invest. New Drugs.* **23**, 205–211.

McClue, S.J., Blake, D., Clarke, R., Cowan, A., Cummings, L., Fischer, P.M., MacKenzie, M., Melville, J., Stewart, K., Wang, S., Zhelev, N., Zheleva, D., Lane, D.P. (2002). *In vitro* and *in vivo* antitumor properties of the cyclin dependent kinase inhibitor CYC202 (R-roscovitine). *Int. J. Cancer.* **102**, 463–468.

Meehan, W.J., Samant, R.S., Hopper, J.E., Carrozza, M.J., Shevde, L.A., Workman, J.L., Eckert, K.A., Verderame, M.F., Welch, D,R. (2004). Breast cancer metastasis suppressor 1 (BRMS1) forms complexes with retinoblastoma-binding protein 1 (RBP1) and the mSin3 histone deacetylase complex and represses transcription. *J. Biol Chem.* **279**, 1562–1569.

Meuwissen, R. and Berns, A. (2005). Mouse models for human lung cancer. *Genes Dev.* **19**, 643–664.

Meuwissen, R., Linn, S.C., Linnoila, R.I., Zevenhoven, J., Mooi, W.J., and Berns, A. (2003). Induction of small cell lung cancer by somatic inactivation of both Trp53 and Rb1 in a conditional mouse model. *Cancer Cell* **4**, 181–189.

Mitsiades, C.S., Mitsiades, N., Hideshima, T., Richardson, P.G., and Anderson, K.C. (2005). Proteasome inhibitors as therapeutics. *Essays Biochem.* **41**, 205–218.

Mohapatra, S., Chu, B., Zhao, X., and Pledger, W.J. (2005). Accumulation of p53 and reductions in XIAP abundance promote the apoptosis of prostate cancer cells. *Cancer Res.* **65**, 7717–7723.

Mookherjee, P. and Johnson, G.V. (2001). Tau phosphorylation during apoptosis of human SH-SY5Y neuroblastoma cells. *Brain Res.* **921**, 31–43.

Nabel, E.G. (2002). CDKs and CKIs: molecular targets for tissue remodelling. *Nat. Rev. Drug Discovery.* **1**, 587–598.

Nakanishi, H., Ito, S., Mochizuki, Y., and Tatematsu, M. (2005). Evaluation of chemosensitivity of micrometastases with green fluorescent protein gene-tagged tumor models in mice. *Methods Mol. Med.* **111**, 351–362.

Nakayama, K.I. and Nakayama, K. (2005). Regulation of the cell cycle by SCF-type ubiquitin ligases. *Semin. Cell. Dev. Biol.* **16**, 323–333.

Napolitano, G., Majello, B., and Lania, L. (2003). Catalytic activity of Cdk9 is required for nuclear co-localization of the Cdk9/cyclin T1 (P-TEFb) complex. *J. Cell. Physiol.* **197**, 1–7.

Nigg, E.A. (1995). Cyclin-dependent protein kinases: key regulators of the eukaryotic cell cycle. *Bioessays* **17**, 471–480.

Norton, L. and Simon, R. (1986). The Norton-Simon hypothesis revisited. *Cancer Treat. Rep.* **70**, 163–169.

Norton, L. and Simon, R. (1977). Tumor size, sensitivity to therapy, and design of treatment schedules. *Cancer Treat. Rep.* **61**, 1307–1317.

Norton, L.A. (1988). Gompertzian model of human breast cancer growth. *Cancer Res.* **48**, 7067–7071.

Onn, A., Isobe, T., Itasaka, S., Wu, W., O'Reilly, M.S., Ki Hong, W., Fidler, I.J., and Herbst, R.S. (2003). Development of an orthotopic model to study the biology and therapy of primary human lung cancer in nude mice. *Clin Cancer Res.* **9**, 5532–5539.

Panetta, J.C. (1998). A mathematical model of drug resisitance: heterogeneous tumors. *Math. Biosci.* **147**, 41–61.

Pang, L., Reddy, P.V., McAuliffe, C.I., Colvin, G., and Quesenberry, P.J. (2003). Studies on BrdU labeling of hematopoietic cells: stem cells and cell lines. *J. Cell. Physiol.* **197**, 251–260.

Pantel, K. and Brakenhoff, R.H. (2004). Dissecting the metastatic cascade. *Nat. Rev. Cancer* **4**, 448–456.

Pei, X.H. and Xiong, Y. (2005). Biochemical and cellular mechanisms of mammalian CDK inhibitors: a few unresolved issues. *Oncogene* **24**, 2787–2795.

Peng, J., Marshall, N.F., and Price, D.H. (1998). Identification of a cyclin subunit required for the function of Drosophila P-TEFb. *J Biol Chem.* **273**, 13855–13860.

Perkins, N.D. (2002). Not just a CDK inhibitor: regulation of transcription by p21(WAF1/CIP1/SDI1). *Cell Cycle* **1**, 39–41.

Polymenis, M. and Schmidt, E.V. (1999). Coordination of cell growth with cell division. *Curr. Opin. Genet. Dev.* **9**, 76–80.

Prindull, G. and Zipori, D. (2004). Environmental guidance of normal and tumor cell plasticity: epithelial mesenchymal transitions as a paradigm. *Blood* **103**, 2892–2899.

Prindull, G. (2005). Hypothesis: cell plasticity, linking embryonal stem cells to adult stem cell reservoirs and metastatic cancer cells? *Exp. Hematol.* **33**, 738–746.

Raj, K., Ogston, P., and Beard, P. (2001). Virus-mediated killing of cells that lack p53 activity. *Nature* **412**, 914–917.

Raju, U., Nakata, E., Mason, KA., Ang, KK., and Milas, L. (2003). Flavopiridol, a cyclin-dependent kinase inhibitor, enhances radiosensitivity of ovarian carcinoma cells. *Cancer Res.* **63**, 3263–3267.

Ramaswamy, S., Ross, K.N., Lander, E.S. and Golub, T.R. (2003). A molecular signature of metastasis in primary solid tumors. *Nat Genet.* **33**, 49–54.

Ravi, R.K., Thiagalingam, A., Weber, E., McMahon, M., Nelkin, B.D., and Mabry, M. (1999). Raf-1 causes growth suppression and alteration of neuroendocrine markers in DMS53 human small-cell lung cancer cells. *Am J Respir Cell Mol Biol.* **20**, 543–549.

Retsky, M., Bonadonna, G., Demicheli, R., Folkman, J., Hrushesky, W., and Valagussa, P. (2004) Hypothesis: induced angiogenesis after surgery in premenopausal node-positive breast cancer patients is a major underlying reason why adjuvant chemotherapy works particularly well for those patients. *Breast Cancer Res.* **6**, R372–R374.

Retsky, M.W., Demicheli, R., Swartzendruber, D.E., Bame, P.D., Wardwell, R.H., Bonadonna, G., Speer, J.F., and Valagussa, P. (1997). Computer simulation of a breast cancer metastasis model. *Breast Cancer Res. Treat.* **45**, 193–202.

Sausville, E.A. (2002). Complexities in the development of cyclin-dependent kinase inhibitor drugs. *Trends Mol. Med.* **8**, S32–S37.

Schang, L.M. (2002). Cyclin-dependent kinases as cellular targets for antiviral drugs. *J. Antimicrob. Chemother.* **50**, 779–792.

Schrump, D.S., Matthews, W., Chen, G.A., Mixon, A., and Altorki, N.K. (1998). Flavopiridol mediates cell cycle arrest and apoptosis in esophageal cancer cells. *Clin. Cancer Res.* **4**, 2885–2890.

Schutte, B., Nieland, L., van Engeland, M., Henfling, M.E., Meijer, L., and Ramaekers, F.C. (1997). The effect of the cyclin-dependent kinase inhibitor olomoucine on cell cycle kinetics. *Exp. Cell Res.* **236**, 4–15.

Schwartz, G.K. (2002). CDK inhibitors: cell cycle arrest vs. apoptosis. *Cell Cycle.* **1**, 122–123.

Senderowicz, A.M. (2002). The cell cycle as a target for cancer therapy: basic and clinical findings with the small molecule inhibitors flavopiridol and UCN-01. *Oncologist* **7**, Suppl 3, 12–9.

Senderowicz, A.M. (2003). Novel small molecule cyclin-dependent kinases modulators in human clinical trials. *Cancer Biol. Ther.* **2**, S084–S095.

Senderowicz, A.M. (2004). Assays for cyclin-dependent kinase inhibitors. *Methods Mol. Biol.* **285**, 69–78.

Senderowicz, A.M. (2005). Inhibitors of cyclin-dependent kinase modulators for cancer therapy, *Prog Drug Res.* **63**, 183–206.

Seong, Y.S., Kamijo, K., Lee, J.S., Fernandez, E., Kuriyama, R., Miki, T., and Lee, K.S. (2002). A spindle checkpoint arrest and a cytokinesis failure by the dominant-negative polo-box domain of Plk1 in U-2 OS cells. *J. Biol. Chem.* **277**, 32282–32293.

Sgambato, A., Camerini, A., Pani, G., Cangiano, R., Faraglia, B., Bianchino, G., De Bari, B., Galeotti, T., and Cittadini, A. (2003). Increased expression of cyclin E is associated with an increased resistance to doxorubicin in rat fibroblasts. *Br. J. Cancer* **88**, 1956–1962.

Shah, M.A. and Schwartz, G.K. (2005). Cyclin-dependent kinases as targets for cancer therapy. *Cancer Chemother. Biol. Response Modifiers* **22**, 135–162.

Shapiro, G.I., Koestner, D.A., Matranga, C.B., and Rollins, B.J. (1999) Flavopiridol induces cell cycle arrest and p53-independent apoptosis in non-small cell lung cancer cell lines. *Clin. Cancer Res.* **5**, 2925–2938.

Shen, M., Feng, Y., Gao, C., Tao, D., Hu, J., Reed, E., Li, Q.Q., and Gong, J. (2004). Detection of cyclin B1 expression in G(1)-phase cancer cell lines and cancer tissues by post-sorting Western blot analysis. *Cancer Res.* **64**, 1607–1610.

Shimizu, E., Zhao, M., Shinohara, A., Namikawa, O., Ogura, T., Masuda, N., Takada, M., Fukuoka, M., and Sone, S. (1997). Differential expressions of cyclin A and the retinoblastoma gene product in histological subtypes of lung cancer cell lines. *J. Cancer Res. Clin. Oncol.* **123**, 533–538.

Shnyder, S.D., Hasan, J., Cooper, P.A., Pilarinou, E., Jubb, E., Jayson, G.C., and Bibby, M.C. (2005). Development of a modified hollow fibre assay for studying agents targeting the tumour neovasculature. *Anticancer Res.* **25**, 1889–1894.

Sierocki, J.S., Hilaris, B.S., Hopfan, S., Martini, N., Barton, D., Golbey, R.B., and Wittes, R.E. (1979). *cis*-Dichlorodiammineplatinum and VP-16-213: an active induction regimen for small cell carcinoma of the lung. *Cancer Treat. Rep.* **63**, 1593–1597.

Skipper, H.E., Schabel, F.M. Jr., and Wilcox, W.S. (1964). Experimental evaluation of potential anticancer agents: XIII. On the criteria and kinetics associated with 'curability' of experimental leukemia. *Cancer Chemother. Rep.* **35**, 1–111.

Smith, M.C., Luker, K.E., Garbow, J.R., Prior, J.L., Jackson, E., Piwnica-Worms, D., and Luker, G.D. (2004). CXCR4 regulates growth of both primary and metastatic breast cancer. *Cancer Res.* **64**, 8604–8612.

Smith, P.J., Blunt, N., Wiltshire, M., Hoy, T., Teesdale-Spittle, P., Craven, M.R., Watson, J.V. Amos, W.B., Errington, R.J., and Patterson, L.H. (2000). Characteristics of a novel deep red/infrared fluorescent cell-permeant DNA probe, DRAQ5TM, in intact human cells analyzed by flow cytometry, confocal and multiphoton microscopy. *Cytometry* **40**, 280–291.

Speer, J.F., Petrosky, V.E., Retsky, M.W., and Wardwell, R.H. (1984). A stochastic numerical model of breast cancer growth that simulates clinical data. *Cancer Res.* **44**, 4124–4130.

Spratt, J.A., von Fournier, D., Spratt, J.S., and Weber, E.E. (1993). Decelerating growth and human breast cancer. *Cancer* **71**, 2013–2019.

Steeg, P.S., Bevilacqua, G., Kopper, L., Thorgeirsson, U.P., Talmadge, J.E., Liotta, L.A., and Sobel, M.E. (1988). Evidence for a novel gene associated with low tumor metastatic potential. *J. Natl. Cancer Inst.* **80**, 200–204.

Steeg, P.S., Ouatas, T., Halverson, D., Palmieri, D., and Salerno, M. (2003). Metastasis suppressor genes: basic biology and potential clinical use. *Clin. Breast Cancer* **4**, 51–62.

Steeg, P.S. (2004). Perspectives on classic article: metastasis suppressor genes. *J. Natl. Cancer Inst.* **96**, E4.

Steen, H., Jebanathirajah, J.A., Springer, M., and Kirschner, M.W. (2005). Stable isotope-free relative and absolute quantitation of protein phosphorylation stoichiometry by MS. *Proc. Natl. Acad. Sci. U. S. A.* **102**, 3948–3953.

Steff, A.M., Fortin, M., Arguin, C., and Hugo, P. (2001). Detection of a decrease in green fluorescent protein fluorescence for the monitoring of cell death: an assay amenable to high-throughput screening technologies. *Cytometry* **45**, 237–243.

Takizawa, C.G. and Morgan, D.O. (2000). Control of mitosis by changes in the sub-cellular location of cyclin-B1-Cdk1 and Cdc25c. *Curr. Opin. Cell Biol.* **12**, 658–665.

Thomas, N. and Goodyer, I.D. (2003). Stealth sensors: real-time monitoring of the cell cycle. *Targets* **2**, 26–33.

Thomas, N., Kenrick, M., Giesler, T., Kiser, G., Tinkler, H., and Stubbs, S. (2005). Characterization and gene expression profiling of a stable cell line expressing a cell cycle GFP sensor. *Cell Cycle* **4**, 191–195.

Thomas, N. (2003). Lighting the circle of life: fluorescent sensors for covert surveillance of the cell cycle. *Cell Cycle* **2**, 545–549.

Thompson, L.H. and Suit H.D. (1969). Proliferation kinetics of X-irradiated mouse L cells studied with time-lapse photography II. *Int. J. Radiat. Biol.* **15**, 347–362.

Thompson, L.H. and Suit, H.D. (1967). Proliferation kinetics of X-irradiated mouse L cells studied with time-lapse photography: I. Experimental methods and data analysis. *Int. J. Radiat. Biol.* **13**, 391–397.

Tsai, L.H.. Lee, M.S., and Cruz, J. (2004). Cdk5, a therapeutic target for Alzheimer's disease? *Biochim. Biophys. Acta.* **1697**, 137–142.

Tsien, R.Y. (1998). The green fluorescent protein. *Annu Rev Biochem.* **67**, 509–544.

Uhr, J.W., Scheuermann, R.H., Street, N.E., and Vitetta ES. (1997). Cancer dormancy: opportunities for new therapeutic approaches. *Nat. Med.* **3**, 505–509.

Vanoosthuyse, V. and Hardwick, K.G. (2005). Bub1 and the multilayered inhibition of Cdc20-APC/C in mitosis. *Trends Cell Biol.* **15**, 231–233.

Vogelstein, B., Lane, D., and Levine, A.J. (2000). Surfing the p53 network. *Nature* **408**, 307–310.

Wang, W., Goswami, S., Sahai, E., Wyckoff, J.B., Segall, J.E., and Condeelis, J.S. (2005). Tumor cells caught in the act of invading: their strategy for enhanced cell motility. *Trends Cell Biol.* **15**, 138–145.

Wei, W., Ayad, N.G., Wan, Y., Zhang, G.J., Kirschner, M.W., and Kaelin, W.G. Jr. (2004). Degradation of the SCF component Skp2 in cell-cycle phase G1 by the anaphase-promoting complex. *Nature* **428**, 194–198.

Weinberg, W.C. and Denning, M.F. (2002). P21Waf1 control of epithelial cell cycle and cell fate. *Crit. Rev. Oral Biol. Med.* **13**, 453–464.

Widrow, R.J., Rabinovitch, P.S., Cho, K., and Laird, C.D. (1997). Separation of cells at different times within G2 and mitosis by cyclin B1 flow cytometry. *Cytometry* **27**, 250–254.

Yamano, H., Tsurumi, C., Gannon, J., and Hunt, T., (1998). The role of the destruction box and its neighbouring lysine residues in cyclin B for anaphase ubiquitin-dependent proteolysis in fission yeast: defining the D-box receptor. *EMBO J.* **17**, 5670–5678.

Zackrisson, B., Flygare, P., Gustafsson, H., Sjostrom, B., and Wilson, G.D. (2002). Cell kinetic changes in human squamous cell carcinomas during radiotherapy studied using the in vivo administration of two halogenated pyrimidines. *Eur. J. Cancer.* **38**, 1100–1106.

Zariwala, M., Liu, J., and Xiong, Y. (1998). Cyclin E2, a novel human G1 cyclin and activating partner of CDK2 and CDK3, is induced by viral oncoproteins. *Oncogene* **17**, 2787–2798.

Zavrski, I., Jakob, C., Schmid, P., Krebbel, H., Kaiser, M., Fleissner, C., Rosche, M., and Possinger, K. (2005). Proteasome: an emerging target for cancer therapy. *Anticancer Drugs* **16**, 475–481.

Zhang, G.-J. and Kaelin, W.G., Jr. (2005) Bioluminescent imaging of ubiquitin ligase activity: measuring Cdk2 activity in vivo through changes in p27 turnover. *Methods Enzymol.* **399**, 530–549.

Zhang, G.-J., Safran, M., Wei, W., Sorensen, E., Lassota, P., Zhelev, N., Neuberg, D.S., Shapiro, G., and Kaelin, W.G., Jr. (2004). Bioluminescent imaging of Cdk2 inhibition in vivo. *Nat. Med.* **10**, 643–648.

Zhu, Y.M., Webster, S.J., Flower, D., and Woll, P.J. (2004). Interleukin-8/CXCL8 is a growth factor for human lung cancer cells. *Br. J. Cancer* **91**, 1970–1976.

2 Functional Regulation of CIP/KIP CDK Inhibitors

Mong-Hong Lee and Ruiying Zhao

CONTENTS

2.1 OVERVIEW

Cell cycle deregulation is one of the most frequent sources of tumor development and stems from either the overexpression of positive regulators or underexpression of negative regulators. Positive regulators consist of cyclins and cyclin-dependent kinases (CDKs), and negative regulators consist of CDK inhibitors (CDKIs). There are two families of CDKIs: the INK4 family (consisting of p16 INK4a, p15 INK4b, p18INK4c, and p19INK4d) and the CIP/KIP family (consisting of p21^{CIP1}, p27^{KIP1}, and p57^{KIP2}). CDKIs are a particular focus of cancer research because they can control cell proliferation, and hence serve as rational cancer therapies. In this chapter we discuss the recently discovered molecular regulation of the CIP/KIP family members. We hope that these insights will be useful for developing strategies to eliminate cancer.

2.2 INTRODUCTION

It is clear that the cell cycle machinery is brought into play during cell proliferation, differentiation, and apoptosis. Its deregulation is also what drives cancer formation. Many researchers have determined how different signals regulate the important CIP/KIP proteins that control the cell cycle and hence control important biological functions. Given that the CIP/KIP proteins are negative regulators of cell cycle progression, an understanding of their regulation is particularly important to the development of treatments for cancer. In this chapter, we discuss the many molecular mediators that control the expression or activity of CIP/KIP proteins and describe their specific roles in regulating the cell cycle or important cell functions.

2.3 p21$^{\text{CIP1/WAF1}}$ REGULATORS

p21$^{\text{CIP1/WAF1/SDI}}$, a transcriptional target of p53, is a critical determinant of the G_1 arrest that occurs in response to DNA damage (el-Deiry et al., 1993) and also plays an important role in the G_2/M phase transition. To further understand p21's role in cancer, it is important to know something about several different signals that regulate p21 activity (Table 2.1), especially the recently identified signal mediators or transcriptional regulators that regulate p21 activity.

2.3.1 SIGNAL MEDIATORS

Many signal mediators involved in cell growth regulate the activity of p21. These include Akt, Rho, SCF, SET, Pim-1, and PP5. Akt is an important kinase involved in regulating cell growth and apoptosis (Burgering and Kops, 2002). Akt can phosphorylate p21 at threonine 145, resulting in cytoplasmic localization of p21Cip1 and subsequent antagonizing of the function of p21 in the nucleus (Zhou et al., 2001). Importantly, in the presence of activated HER2/neu and/or Akt, cytoplasmic p21 and overexpression of phospho-p21 (T145) are observed (Zhou et al., 2001), suggesting that one of the important functions of HER2/neu or Akt oncogene is to target and block p21 activity for growth advantage. Furthermore, the Akt can mediate phosphorylation of Mdm2 and promote the nuclear localization of Mdm2 to antagonize p53's function (Zhou et al., 2001), thereby decreasing p53 stability in inducing p21. This finding suggests that Akt can also regulate the Mdm2-p53 pathway and subsequently decrease the expression of p21.

Interestingly, another kinase, Pim-1, which is upregulated in prostate cancer, can also phosphorylate p21 on Thr(145), which causes the cytoplasmic localization of p21 (Wang et al., 2002). This points up the importance of the phosphorylation on Thr(145) in determining the cytoplasmic location of p21. In addition, Pim-1 can phosphorylate and increase the levels of the HDM2 protein (Ionov et al., 2003), which may in turn prevent p53-mediated p21 activation. Thus, Pim1 may deregulate p21 to exert its oncogenic signaling. Cytoplasmic location of p21 seems to inhibit apoptosis (Schepers et al., 2003). It was shown that cytoplasmic p21 can interact with the apoptosis signal-regulating kinase 1 (ASK1) and inhibits stress-activated

TABLE 2.1
Selected Signal Mediators/Transcriptional Activators That Regulate p21

Protein	Function/Characteristics	References
Akt	Akt phosphorylates p21Cip1/WAF1 at Thr(145), resulting in cytoplasmic localization	Zhou et al., 2001
Pim-1	Pim-1, protein kinase upregulated in prostate cancer, can phosphorylate p21 on Thr(145) and cause the cytoplasmic localization of p21	Wang et al., 2002
Rho/Rac1/ Cdc42	p21 interacts with Rho and inhibits its activity. Conversely, integrin-induced Cdc42/Rac1 signaling activates proteasomal degradation of p21 and may lead to anchorage-dependent cell cycle control	Bao et al., 2002; Tanaka et al., 2002
Skp2	Skp2 containing Skp1-Cullin1-F-box protein ubiquitin ligase facilitates the proteasome-mediated degradation of p21	Bornstein et al., 2003; Yu et al., 1998
WISp39	A tetratricopeptide repeat protein that binds to p21 and prevents its proteasomal degradation	Jascur et al., 2005
SET	SET, a positive regulator of CDK2, is an oncoprotein and binds to p21 at the carboxyl-terminal region to reduce the inhibitory effect of p21 on cyclin E-CDK2	Estanyol et al., 1999
TOK-1	TOK-1 (twenty-one and CDK-associated protein-1), also a carboxyl-terminal p21-associated protein, binds to CDK2 via p21 TOK-1 enhances the inhibitory effect of p21 on the kinase activity of CDK2	Ono et al., 2000
PP5	PP5 regulates the phosphorylation of p53, which in turn affects the ability of p53 to induce p21	Zuo et al., 1998
p53/14-3-3σ	14-3-3σ, a p53 gene induced in response to DNA damage, interacts with p53 and potentiates p53-dependent transcription of the p21 gene	Hermeking et al., 1997; Yang et al., 2003
Hdaxx	Hdaxx, a cytoplasmic mediator of Fas signaling and a transcriptional repressor, interacts with p53 at the oligomerization domain and represses p53-dependent transcription of the p21 gene	Gostissa et al., 2004
MUC1	Carboxyl-terminal MUC1 interacts with p53 in response to DNA damage. The cytoplasmic domain of MUC1 binds directly to p53 and coactivates p21 gene transcription	Wei et al., 2005
TGF-β, Smad FOXO	TGF-β signaling pathway leads to formation of a Smad3–Smad4 complex in the nucleus for target gene expression. FOXO, Forkhead transcription factor, can associate with the Smad complex to activate the expression of the p21Cip1 gene	Seoane et al., 2002
Mitf	Microphthalmia-associated transcription factor (Mitf)-mediates activation of p21Cip1 expression and consequent hypophosphorylation of Rb1	Carreira et al., 2005

(*continued*)

TABLE 2.1 (CONTINUED)
Selected Signal Mediators/Transcriptional Activators That Regulate p21

Protein	Function/Characteristics	References
INF-γ/ P202	p202, an IFN-inducible phosphoprotein (52-kDa), can upregulate p21 and subsequently inhibit transformation. p202 regulates the expression of p21 protein independently of p53 protein	Gutterman and Choubey, 1999
HNF4	HNF-4α is a transcription factor of the nuclear hormone receptor family that upregulates expression of p21 in a p53-independent manner	Chiba et al., 2005
c-Myc	Overexpression of c-Myc reduces the endogenous p21 mRNA levels and represses a p21-promoter of a luciferase-reporter gene	Claassen and Hann, 2000

MAP kinase cascade and apoptosis (Asada et al., 1999). Thus, Akt-mediated cytoplasmic location of p21 may contribute to Akt's antiapoptotic signal.

p21 also interacts with Rho, which inhibits Rho's activity (Tanaka et al., 2002). Members of the Rho family of small GTP-binding proteins, including Rho, Rac, and Cdc42, regulate a wide variety of cell responses, such as cell proliferation, apoptosis, the formation of the cytoarchitecture, cell polarity, cell adhesion, and membrane trafficking. When activated, Cdc42 and Rac1 can downregulate p21 (Bao et al., 2002). In particular, integrin-induced Cdc42/Rac1 signaling activates the proteasomal degradation of p21 (Bao et al., 2002), which may lead to anchorage-dependent cell cycle control. It is not clear how Cdc42 and Rac1 cause the proteasomal degradation pathway to degrade p21. It is important to determine this, however, given that Rho overexpression is involved in tumor initiation, progression, and metastasis.

p21 is posttranslationally controlled by the proteasome, and the Skp2-containing Skp1-Cullin1-F-box protein (SCF) ubiquitin ligase facilitates the proteasome-mediated degradation of p21 (Bornstein et al., 2003; Yu et al., 1998). A recent study indicated that WAF1/CIP1 stabilizing protein (WISp39), a tetratricopeptide repeat protein, can bind to p21 and block its proteasomal degradation (Jascur et al., 2005). WISp39 recruits Hsp90, a molecular chaperone, to form a trimeric complex with p21. This interaction is important in regulating p21 stability, as preventing the interaction between WISp39 and Hsp90 leads to WISp39's failure in stabilizing p21. The downregulation of WISp39 has also been observed to prevent the accumulation of p21 and cell cycle arrest in response to ionizing radiation. Importantly, hsp90 is also involved in positively modulating p53 DNA binding (Walerych et al., 2004). It was further shown that geldanamycin, a specific inhibitor of hsp90, abolished p53 binding to the p21 promoter sequence. Consequently, Hsp90 regulates p21 transcriptionally and posttranscriptionally. Interestingly, MDM2, a p53 ubiquitin ligase, binds to p21, which promotes its binding to the proteasomal C8 subunit (Zhang et al., 2004b). This physical interaction facilitates p21 degradation independent of the E3 ubiquitin ligase function of MDM2.

SET, a positive regulator of CDK2, is an oncoprotein that binds to p21 at the carboxyl-terminal region, which reduces p21's ability to inhibit cyclin E/CDK2 (Estanyol et al., 1999). Interestingly, SET does not bind to p27 to block its activity, suggesting that p21 is a specific target. Whether SET is involved in tumorigenesis warrants further investigation. Interestingly, the p21- and CDK-associated protein-1 (TOK-1), which is also a carboxyl-terminal p21-associated protein, binds to CDK2 via p21 (Ono et al., 2000). This enhances the inhibitory effect of p21 on the kinase activity of CDK2. The signals that regulate TOK-1 activity are not clear. Because, as described earlier, SET blocks the inhibitory effect of p21 on cyclin E/CDK2, it will be interesting to see whether these two p21-associated molecules functionally interact and whether this has an antagonistic effect. Finally, serine/threonine protein phosphatase type 5 (PP5), which regulates the phosphorylation of p53, in turn affects the ability of p53 to induce p21. The inhibition of PP5 gene expression induces p21 (WAF1/Cip1) in a p53-dependent manner (Zuo et al., 1998), which is followed by the arrest of cell growth.

2.3.2 TRANSCRIPTIONAL REGULATORS

p53 is an important transcriptional activator that was first identified as a regulator of p21 expression (el-Deiry et al., 1993). However, p21 also responds to p53-independent signals. Several such transcriptional regulators of p21 expression that have recently been identified include Hdaxx, MUC1, p202, HNF4α, STAT, FOXO, 14-3-3σ, and Myc.

14-3-3σ, a gene induced by p53 in response to DNA damage, works by interacting with p53, which potentiates p53-dependent transcription of the p21 gene (Yang et al., 2003; Hermeking et al., 1997). In particular, 14-3-3σ interacts with p53 in response to the DNA-damaging agent Adriamycin (doxorubicin). Importantly, 14-3-3σ expression leads to stabilized expression of p53 (Yang et al., 2003). 14-3-3σ can also antagonize the biological functions of Mdm2 by blocking Mdm2-mediated p53 ubiquitination and nuclear export (Yang et al., 2003). In addition, 14-3-3σ facilitates the oligomerization of p53 and enhances p53's transcriptional activity (Yang et al., 2003). As a target gene of p53, 14-3-3σ has a positive feedback effect on p53 activity. On the other hand, Hdaxx, which is a cytoplasmic mediator of Fas signaling and a transcriptional repressor, interacts with p53 at the oligomerization domain and represses p53-dependent transcription of the p21 gene (Gostissa et al., 2004). In fact, Hdaxx can repress the transcriptional activity of various p53 family members, including p53, p63, and p73 (Kim et al., 2003b).

The MUC1 oncoprotein, a transmembrane mucin glycoprotein, is overexpressed in most human carcinomas, and this overexpression of MUC1 leads to cell transformation (Ren et al., 2004). It has also been demonstrated that p53 binds to the carboxyl-terminal of MUC1 in response to DNA damage (Wei et al., 2005). This binding in turn leads to binding of the p53 to the p53-responsive elements of the p21 gene promoter, which activates p21 gene transcription (Wei et al., 2005). It is possible that this MUC1-mediated induction of p21 is involved in preventing apoptosis. Indeed, this is indicated by the fact that MUC1 attenuates p53-mediated Bax transcriptional activation (Wei et al., 2005). Thus, human tumors overexpressing MUC1 may have a survival advantage when exposed to genotoxic stress.

The transforming growth factor-β (TGF-β) signaling pathway leads to the formation of a Smad3–Smad4 complex in the nucleus for target gene expression. For example, the Forkhead transcription factor FOXO can associate with this Smad complex to activate the expression of the p21[CIP1] gene (Pardali et al., 2000; Seoane et al., 2002). Importantly, FOXO is antagonized by Akt kinase (Burgering and Kops, 2002). Specifically, Akt phosphorylates FOXO, which inhibits FOXO's localization to the nucleus and thereby blocks FOXO transcriptional activity on the p21 promoter (Seoane et al., 2002). In this way, Akt signaling antagonizes TGF-β-mediated p21 transcriptional activation.

The microphthalmia-associated transcription factor (Mitf) plays a fundamental role in melanocyte development. Mitf can induce the expression of p21 and cause cell cycle arrest (Carreira et al., 2005). Interestingly, however, a high percentage of melanomas show an increased Mitf1 turnover rate. It has also been shown that p21 expression is inversely correlated with melanoma progression (Du et al., 2004), suggesting that Mitf's impact on p21 expression is important during melanoma tumorigenesis.

p202, an interferon-inducible phosphoprotein (52 kDa) that regulates the expression of the p21 protein independent of the p53 protein, can upregulate p21 and subsequently inhibit transformation (Gutterman and Choubey, 1999). STATs, a family of latent cytoplasmic proteins that are activated to enter the nucleus, work with other transcription factors to transcriptionally activate downstream genes in cells treated with various cytokines and growth factors (Chen et al., 2000). STAT1 can bind to the promoter region of p21, which is an essential step in mediating cell growth suppression of the interferon- signal transduction pathway. Hepatocyte nuclear factor (HNF)-4α, a transcription factor of the nuclear hormone receptor family, is expressed in the hepatic diverticulum and involved in liver development. HNF-4α can upregulate the expression of p21 in a p53-independent manner (Chiba et al., 2005), which may play an important role in liver cell differentiation.

Finally, p21 is an important target of c-Myc (Claassen and Hann, 2000). In particular, the overexpression of c-Myc reduces the levels of endogenous p21 mRNA and represses p21-promoter luciferase-reporter gene expression (Claassen and Hann, 2000). In addition, c-Myc, which is induced by estrogen and suppresses the expression of p21, is critical for estrogen-stimulated breast cancer cell proliferation (Mukherjee and Conrad, 2005). Collectively these studies have shown p21[WAF1/CIP1] to be an important target of c-Myc in breast cancer cells, providing a link between estrogen, c-Myc, and the cell cycle machinery. It is further possible that aberrant c-Myc expression, which is frequently observed in human breast cancers, contributes to resistance to antiestrogen therapy by altering p21[WAF1/CIP1] regulation. It should also be noted that c-Myc negatively regulates the activity of CDKIs, including p15 (INK4b) (Seoane et al., 2001; Staller et al., 2001) and p27 (Yang et al., 2001). These findings indicate that c-Myc has profound negative effects on CDKIs, which promotes tumorigenesis. Future research will need to determine how these events are coordinated.

2.4 p27 KIP1 REGULATORS

The p27 protein is a CDKI that causes G_1 phase arrest by inhibiting G_1 cyclin/CDK activities (Polyak et al., 1994; Toyoshima and Hunter, 1994). As a negative regulator of the cell cycle, p27 is a member of new class of haplo-insufficient tumor suppressors (Fero et al., 1998). Because of its important role in inhibiting cancer cell growth, p27 is regulated by many oncogenic signal mediators (Table 2.2).

2.4.1 SIGNAL MEDIATORS

Akt also has an impact on p27 expression or its subcellular localization. For example, in breast cancer cells, Akt-mediated p27 phosphorylation at Thr(157) (within the nuclear localization signal sequence of p27, amino acids 153–166) impairs the nuclear import of p27 (Liang et al., 2002; Shin et al., 2002). In addition, Akt phosphorylates p27 at Thr(198) (Fujita et al., 2002), which results in the nuclear export and degradation of p27. It has been shown that the 14-3-3θ, η, and ε proteins bind to p27 through Thr198 to facilitate the nuclear export of p27 when this residue is phosphorylated by Akt. Interestingly, ribosomal protein S6 kinase, a downstream target of Akt/mTOR signaling, was shown to bind to and phosphorylate p27KIP1 at Thr198 (Fujita et al., 2003), which in turn promotes the binding of p27 and 14-3-3 and subsequent cytoplasmic localization of p27. Together, Akt and the ribosomal protein S6 kinases can phosphorylate p27 at Thr198, which causes p27 to be localized to the cytoplasm, suggesting that Akt signaling can affect p27 activity directly or indirectly through its downstream mediator.

The human kinase interacting stathmin (hKIS) has been found to be capable of Ser10 phosphorylation of p27, and this appears to regulate cell cycle progression (Boehm et al., 2002). hKIS phosphorylates p27 following mitogen stimulation and induces the export of p27 from the nucleus (Boehm et al., 2002). It is noteworthy that p27 can bind to and impair the function of stathmin, a microtubule-destabilizing protein, thereby increasing tubulin polymerization (Baldassarre et al., 2005). Accordingly, there appears to be evidence that the upregulation of p27 can affect cell migration. In keeping with this concept, high stathmin expression and low p27 expression correlate with the metastatic potential of human sarcomas (Baldassarre et al., 2005). Furthermore, p27 can bind to and inhibit RhoA activation by preventing the interaction of RhoA and the guanine-nucleotide exchange factors (Besson et al., 2004), which are RhoA activators. This observation also suggests that, by inhibiting the Rho pathway, p27 has an extra role in regulating cell migration. Extending this link between p27 and cell migration, focal adhesion kinase (FAK), a regulator of cell motility, can posttranscriptionally affect the expression of Skp2, an F-box protein involved in p27 degradation (Bond et al., 2004). It has also been shown that a dominant-negative FAK mutant (Y397F) can inhibit the expression of Skp2. Conversely, the proteasome inhibitor MG-132 can reverse the downregulation of Skp2 mediated by the dominant-negative FAK (Bond et al., 2004). It remains to be determined how FAK affects the expression of Skp2 by affecting the proteasome-mediated pathway. It is possible that FAK downregulates p27 expression, which promotes cell growth.

TABLE 2.2
Selected Signal Mediators/Transcriptional Activators That Regulate p27

Protein	Function/Characteristics	References
Akt/S6 kinase	Akt-mediated p27 phosphorylation at Thr(157) (within the NLS sequence of p27, aa153–166) impairs the nuclear import of p27. Akt and ribosomal protein S6 kinases can phosphorylate p27 at Thr(198) and mediate p27 cytoplasmic localization	Liang et al., 2002; Shin et al., 2002; Fujita et al., 2002; Fujita et al., 2003
KIS	KIS, the kinase interacting with stathmin, can phosphorylate p27 on Ser(10) following mitogen stimulation and induce p27 export from the nucleus	Boehm et al., 2002
RhoA	p27 can bind to and inhibit RhoA activation by interfering with the interaction between RhoA and the guanine-nucleotide exchange factors, which are RhoA activators	Besson et al., 2004
FAK	Focal adhesion kinase (FAK) promotes cell proliferation by reducing the expression of p27. FAK can posttranscriptionally affect the expression of Skp2, a protein involved in p27 degradation	Bond et al., 2004
Mirk/Dyrk1B	Mirk/dyrk1B, arginine-directed kinase, can phosphorylate p27 at Ser(10), which in turn leads to p27 stabilization	Deng et al., 2004
Skp2, Cdc34	Skp2 binds to phosphorylated p27, which is phosphorylated on Thr(187) by Cdk2, which facilitates the ubiquitination of p27. Downregulation of the ubiquitin-conjugating enzyme Cdc34 increases p27 expression	Carrano et al., 1999; Butz et al., 2005
KPC	Kip1 ubiquitination-promoting complex (KPC) can mediate the degradation of p27 in G_1, which depends on the export of p27 to the cytoplasm	Kamura et al., 2004; Kotoshiba et al., 2005
CSN	COP9 signalsome (CSN) regulates the ubiquitination and degradation of p27. However, the detailed mechanism is not defined. CSN5/JAB1, the fifth subunit of CSN complex and a c-Jun coactivator protein, is a p27 nuclear exporter and involved in p27 ubiquitination	Tomoda et al., 1999; Yang et al., 2002
Bcr-abl	Bcr-Abl, a tyrosine kinase, can facilitate the degradation of p27 by regulating the complex formation of CSN5/JAB1 through the MAP kinase and PI3 kinase signaling pathways	Gesbert, 2000; Tomoda et al., 2005
FRS2	Fibroblast growth-factor-dependent activation of FGFR tyrosine kinases leads to FRS2 phosphorylation, which in turn causes release of Cks1 from FRS2 and facilitates the degradation of p27	Zhang et al., 2004a

TABLE 2.2 (CONTINUED)
Selected Signal Mediators/Transcriptional Activators That Regulate p27

Protein	Function/Characteristics	References
Grb2	p27 binds to and inhibits Grb2 function by blocking Grb2's association with the guanine-nucleotide exchange factor SOS. Reciprocally, overexpression of Grb2 accelerates CSN5/JAB1-mediated degradation of p27	Moeller et al., 2003; Sugiyama et al., 2001
GR	Glucocorticoids, acting through the glucocorticoid receptor, can activate the expression of p27	Wang and Garabedian, 2003
Spy1	P27 and Spy1, and CDK form a complex. Also Spy1 antagonizes p27-mediated inhibition of CDK2 activity	Porter et al., 2003
Cx32	Connexin 32 (Cx32), a gap junction protein, is involved in regulating p27. Tumors from Cx32-deficient mice show increased activation of MAP kinase, which may accelerate the turnover of p27	King and Lampe, 2004b
PTEN	PTEN pathway upregulates p27 protein stability through the reduction of Skp2	Mamillapalli et al., 2001
RET/PTC	RET/PTC, an oncogene rearranged in transformation/papillary thyroid carcinomas, mediates MAP kinase activation and subsequent p27Kip1 deregulation	Vitagliano et al., 2004
BRCA1	BRCA1, which functions as a tumor suppressor in human breast cancer cells, is able to transactivate the expression of p27	Williamson et al., 2002
CBP	The transcriptional coactivator CREB-binding protein (CBP), a tumor suppressor, is important for p27 expression. Levels of p27Kip1 are reduced in CBP$^{-/-}$ tumors.	Kang-Decker et al., 2004
E2F1	E2F1 can bind to the promoter of p27 gene and activate p27 gene expression	Wang et al., 2005
MEN1	Menin, a tumor suppressor of multiple endocrine neoplasia type-1, can transactivate the expression of p27	Milne et al., 2005
PTB	Polypyrimidine tract-binding protein (PTB) can enhance the internal ribosomal entry site (IRES) activity of p27 (KIP1) mRNA	Cho et al., 2005
FOXO	The Forkhead FOXO transcription factor family is involved in regulating cell cycle progression. FOXO blocks G_1 cell cycle progression by transcriptionally activating p27 by binding to multiple FOXO4 binding sites present in the p27 promoter	Medema et al., 2000
c-Myc	c-Myc can transcriptionally repress expression of the p27 gene. The c-Myc repression region is located at the bp 20 to +20 region of the p27 gene	Yang et al., 2001

Mirk/dyrk1B, an arginine-directed kinase, can also phosphorylate p27 at Ser10 (Deng et al., 2004), which stabilizes p27. However, a recent p27 S10A knock-in study showed that Ser10 is a dispensable component of the nuclear export process because the nuclear export of p27 occurred normally in the embryonic fibroblasts of p27S10A knock-in mice (Kotake et al., 2005). It remains to be shown in detail how these different kinases regulate the Ser10 of p27 and subsequent protein stability.

It is clear that p27 is regulated posttranscriptionally. Recently, more factors involved in regulating p27 protein stability have been recognized. For example, Skp2, a subunit of SCF protein ubiquitin ligase complexes involved in p27 degradation, was found to be required for the ubiquitination and consequent degradation of p27 (Carrano et al., 1999). It was further noted that Skp2 binds to phosphorylated p27, which is phosphorylated on Thr(187) by CDK2, and this facilitates the ubiquitination of p27. In animal studies, Skp2 knockout mice were found to grow more slowly and have smaller organs than control (Nakayama et al., 2000). Also, high levels of p27 were found in Skp2-deficient cells, which demonstrated polyploidy and centrosome overduplication. Thus, loss of Skp2 in the mouse leads to changes in cell proliferation. It was further shown that the loss of p27 almost completely rescued the defects observed in Skp2$^{-/-}$ mice, suggesting that p27 is the principal target of the SCF (Skp2) ubiquitin ligase (Nakayama et al., 2000). In addition, downregulation of the ubiquitin-conjugating enzyme Cdc34 causes an increase in p27 but not other targets, suggesting that p27 is the major target of Cdc34 (Butz et al., 2005).

Recently, the KIP1 ubiquitination-promoting complex (KPC) was found to mediate the degradation of p27 as well (Kamura et al., 2004; Kotoshiba et al., 2005). KPC is composed of of KPC1 and KPC2. KPC1 is a RING-finger protein, whereas KPC2 contains a ubiquitin-like domain and two ubiquitin-associated domains. KPC is located in the cytoplasm, where it interacts with and ubiquitinates p27. Such KPC-mediated p27 degradation requires the export of p27 to the cytoplasm in the G_1 phase, but this process is phosphorylation independent. In contrast, SCF (Skp2) mediates p27 degradation in the nucleus during the S phase, a process that depends on the phosphorylation of p27 on Thr187.

Although it is known that the COP9 signalsome (CSN) is involved in the ubiquitination process of p27 (Tomoda et al., 1999; Yang et al., 2002), the detailed mechanism has not been defined. CSN5/JAB1, the fifth subunit of the CSN complex and a c-Jun co-activator protein, is a p27 nuclear exporter and involved in p27 ubiquitination. CSN5 is overexpressed in a high percentage of invasive breast cancers (Esteva et al., 2003; Kouvaraki et al., 2003). CSN5 overexpression is also associated with a poor prognosis for ovarian cancer patients (Sui et al., 2001). These data therefore indicate a link between CSN5 overexpression and p27 downregulation in cancers. Interestingly, the downregulation of p27 also correlates with HER2 over-expression in primary breast carcinomas (Newman et al., 2001). There is evidence that these reduced p27 levels are caused by enhanced ubiquitin-mediated degradation and that the HER-2/Grb2/MAPK pathway is involved in the decrease in p27 stability (Yang et al., 2000). In such a process, HER2 activity causes p27 and CSN5/JAB1 to be mislocated into the cytoplasm, thereby facilitating p27 degradation (Yang et al., 2000).

Bcr-Abl, a tyrosine kinase, can facilitate the degradation of p27 by regulating the formation of the CSN5/JAB1 complex through the MAP kinase and phosphatidylinositol 3 (PI3) kinase signaling pathways (Gesbert et al., 2000; Tomoda et al., 2005). The expression of Bcr-Abl leads to the formation of the small Jab1 complex, which is responsible for p27 degradation. Conversely, the inhibition of Bcr-Abl kinase by STI571 results in reduced formation of the small Jab1 complex and subsequent recovery of the p27 level (Tomoda et al., 2005), which are also caused by blockade of the MAP kinase and PI3 kinase with specific inhibitors. It remains to be determined how the MAP kinase and PI3 kinase pathways regulate the formation of the small Jab1 complex.

The unphosphorylated form of the fibroblast growth factor receptor (FGFR) substrate 2 (FRS2), an adaptor protein phosphorylated by FGFR kinases, has been shown to interact with Cks1, which is involved in the ubiquitination and degradation of p27 (Zhang et al., 2004a). Further, the fibroblast growth-factor-dependent activation of FGFR tyrosine kinases leads to FRS2 phosphorylation, which in turn causes the release of Cks1 from FRS2 and facilitates the degradation of p27. p27 also binds to and inhibits Grb2 function by blocking Grb2's association with the guanine-nucleotide exchange factor SOS (Moeller et al., 2003; Sugiyama et al., 2001). Reciprocally, the overexpression of Grb2 accelerates the CSN5/JAB1-mediated degradation of p27 (Moeller et al., 2003). Importantly, Grb2 is required for the downregulation of p27, as the downregulation of Grb2 leads to the upregulation of p27. Glucocorticoids, acting through the glucocorticoid receptor, can activate the expression of p27 (Wang and Garabedian, 2003). Because glucocorticoids have a growth inhibitory effect, they are used to treat certain malignancies. Because p27$^{-/-}$ MEFs are resistant to the growth inhibitory effects of glucocorticoids, p27 is likely critical for the growth inhibitory effect of glucocorticoids.

Speedy (Spy1), a cell cycle regulator involved in activating CDK (Porter et al., 2002), associates with p27 and blocks p27-mediated cell cycle arrest (Porter et al., 2003), which promotes cell cycle progression through the G_1/S phase transition. p27, Spy1, and CDK form a complex. Also, Spy1 antagonizes the p27-mediated inhibition of CDK2. It remains to be determined, however, whether Spy1 is involved in cancer formation.

There is wider evidence of p27 involvment in tumor suppressor and oncogene action. Connexin 32, a gap junction protein, is a lung tumor suppressor (King and Lampe, 2004a). This was validated by the finding that connexin 32 knockout mice (Cx32-KO) exposed to chemical carcinogens and radiation showed a greater tendency to form tumors (King and Lampe, 2004b). Importantly, these tumors exhibit decreased levels of p27, suggesting that connexin 32 is involved in regulating the p27 (King and Lampe, 2004b). In addition, tumors from connexin-32-deficient mice show increased activation of MAP kinase, which may accelerate the turnover of p27 (King and Lampe, 2004b). The tumor suppressor gene phosphatase and tensin homologue (PTEN) (Simpson and Parsons, 2001), one of the most commonly inactivated genes in human cancer, is associated with the increased expression of p27 (Gottschalk et al., 2001; Li and Sun, 1998). PTEN can reduce CDK2 activity, causing cell cycle arrest. Actually, the PTEN pathway upregulates p27 protein stability by reducing the expression of Skp2, an important component of p27 ubiquitinase

(Mamillapalli et al., 2001). Further, antisense p27 oligonucleotides can abrogate the growth arrest mediated by PTEN, suggesting that p27 is a critical mediator of PTEN-induced G_1 arrest and that p27 plays a role in the PTEN regulatory cascade. Mutations in the PTEN gene are seen in brain, breast, prostate, endometrial, and skin cancers (Simpson and Parsons, 2001). It will be interesting to determine whether p27 is deregulated in these tumors. RET/PTC, an oncogene that is rearranged in transformation/papillary thyroid carcinomas—indeed, it shows the most frequent genetic alterations in papillary thyroid carcinoma—mediates MAP kinase activation and subsequent p27[KIP1] deregulation (Vitagliano et al., 2004). RET/PTC can also induce the phosphorylation of PDK1 at the Y9 residue and thereby increase the activity of PDK1 (Kim et al., 2003a), a kinase involved in activating Akt. It is possible that RET/PTC helps destabilize p27 through either the PDK1-Akt or MAP kinase pathways.

2.4.2 TRANSCRIPTIONAL REGULATORS

BRCA1, which functions as a tumor suppressor in human breast cancer cells, is able to transactivate the expression of p27 (Williamson et al., 2002), but this depends on BRCA1 having a functional carboxyl-terminal transactivation domain. Promoter analysis has shown that position -615 to -511 of the p27 promoter contains a putative BRCA1-responsive element (Williamson et al., 2002). Consistent with this, p27 is downregulated in cells expressing mutated BRCA1. It is noteworthy that BRCA1 is also involved in regulating p21 expression (Somasundaram et al., 1997). The transcriptional coactivator CREB-binding protein (CBP), a tumor suppressor, can acetylate p53 and thereby enhance p53 DNA-binding activity. CBP is also important for p27 expression, as indicated by the finding that the levels of p27 are reduced in CBP-null tumors (Kang-Decker et al., 2004). It appears that Skp2 is involved in degrading p27 because the Skp2 level is upregulated in CBP$^{-/-}$ tumors. It is not clear how CBP regulates the expression of Skp2.

The E2F1 transcription factor is a critical cell cycle regulator promoting S phase entry. E2F1 activates p27 gene expression by binding to the promoter of p27 gene (Wang et al., 2005). On the other hand, the inhibition of p27 expression by small interfering RNA leads to E2F1 transcription and subsequent E2F1-regulated cell cycle progression. This observation indicates that p27 acts as a negative feedback regulator of E2F1 activity.

Menin, a tumor suppressor of multiple endocrine neoplasia type-1, can recruit the mixed-lineage leukemia (MLL) protein, which suppresses the growth of a variety of cell types, to transactivate the expression of p27 and p18 (Ink4c) (Milne et al., 2005). Indeed, menin is a component of the 1-MDa MLL complex. The induction of p27 and p18 could account for menin's tumor-suppressive effect. A recent study has indicated that polypyrimidine tract binding protein (PTB) can enhance the activity of the internal ribosomal entry site (IRES) of p27KIP1 mRNA (Cho et al., 2005). The IRES, a specialized RNA structure, can recruit ribosomes to an mRNA in a cap-independent manner. In particular, the IRES site allows for efficient p27 translation when cap-dependent translation is reduced. IRES-dependent translation of the p27 mRNA is inhibited when PTB expression is reduced. On the other hand, the neuronal ELAV protein HuD is a specific binding factor of the p27

5'UTR IRES site. The increased expression of HuD inhibits p27 translation and p27 IRES activity (Kullmann et al., 2002).

The Forkhead FOXO transcription factors, which are involved in regulating longevity in *Caenorhabditis elegans*, are regulated by the PI3K-Akt pathway. These transcription factors play important roles in cell cycle progression (Alvarez et al., 2001; Kops et al., 2002; Medema et al., 2000), apoptosis (Brunet et al., 1999), oxidative stress (Kops et al., 2002), and DNA repair (Tran et al., 2003). Mammalian cells contain three FOXOs, FOXO1 (FKHR), FOXO3a (FKHRL1), and FOXO4 (AFX). Importantly, when Akt is activated by particular extracellular signals, it can directly phosphorylate all three FOXOs at three conserved serine/threonine residues (Burgering and Kops, 2002). Phosphorylation of a FOXO leads to its release from its DNA and subsequent retention in the cytoplasm, where it binds to the 14-3-3 protein, which silences its transcriptional activity and suppresses the expression of cell cycle and apoptotic genes, such as the gene that encodes the p27 or the Fas ligand (reviewed in Burgering and Kops, 2002; Tran et al., 2003). FOXO4 blocks G_1 cell cycle progression by binding to multiple FOXO4 binding sites present in the p27 promoter, which transcriptionally activates p27 (Medema et al., 2000). p27 then inhibits CDK, blocking cell cycle progression. Because of the positive impact of FOXOs on p27 transcription, they are potential anticancer agents. Indeed, a constitutively active FOXO4 (which has three Akt phosphorylation sites — Thr32, Ser187, and Ser252 — mutated to alanine, FOXO4A3) can increase p27 expression and suppress HER2 oncogene-mediated cell growth, transformation, and tumorigenicity (Yang et al., 2005). In the study showing this, FOXO4A3's impact on p27 upregulation was attributed, at least in part, to posttranscriptional regulation, as demonstrated by the blocking of the HER2-mediated cytoplasmic mislocation of p27 and by the inhibition of CSN5-mediated p27 ubiquitination (Yang et al., 2005). Interestingly, a recent observation that FOXO1aΔ 256, a dominant-negative version of FOXO1a, can inhibit the nuclear localization of p27 constitutes additional evidence that the modified FOXO protein can affect the location of p27 (Cunningham et al., 2004). On the other hand, PAX3-FKHR, which results from a t(2,13) chromosomal translocation and is a genetic marker of alveolar rhabdomyosarcoma, can reduce the expression of p27. The PAX3-FKHR fusion protein, a truncated form of FKHR, has been found to destabilize the p27 protein but has no effect in transactivating the expression of p27 total mRNA (Zhang and Wang, 2003). PAX3-FKHR causes the destabilization of p27(Kip1) through an uncharacterized mechanism.

Finally, Myc can antagonize the expression of p27. For example, c-Myc can transcriptionally repress the gene expression of p27 (Yang et al., 2001). The c-Myc repression region is located at bp 20 to +20 region of the p27 gene. Taken together, these findings show that p27 is a tumor suppressor protein and an important target of different signals.

2.5 p57 KIP2 REGULATORS

The p57 protein is a maternally expressed, paternally imprinted CDKI located on chromosome 11p15.5 (Lee et al., 1995; Matsuoka et al., 1995). p57 is a candidate tumor suppressor gene because of its location, biochemical activities, and imprinting expression. This has been pointed up in animal studies demonstrating that p57 knockout mice have altered cell proliferation and differentiation and a variety of other abnormalities (Yan et al., 1997; Zhang et al., 1997). Many of these defective phenotypes are also seen in patients with Beckwith–Wiedemann syndrome, a child-hood overgrowth syndrome, suggesting that the loss of p57 plays a role in it (Yan et al., 1997; Zhang et al., 1997). Recently, more regulatory signals of p57 have been characterized, which has furthered our understanding of its functional roles in cancers (Table 2.3).

2.5.1 Signal Mediators

The myogenic factor MyoD can induce the expression of p57, which produces cell cycle withdrawal in differentiating myoblasts. Interestingly, MyoD is also able to induce p21. However, MyoD's ability to induce p57 is restricted to cells lacking p21 (Figliola and Maione, 2004), indicating that the two CDKIs function differently in muscle cells of different lineages. Reciprocally, p57 can stabilize MyoD to transac-tivate muscle-specific genes, including myosin heavy chain, creatine kinase, and myosin light chain 1. It has been shown that p57 stabilizes the MyoD protein by inhibiting cyclin E/CDK2 kinase activity in myoblasts (Reynaud et al., 1999). Such CDK2-dependent phosphorylation of MyoD is involved in regulating the turnover of the MyoD protein. Interestingly, the phosphorylation of mutant MyoD at Ser200 can still be stabilized by p57, suggesting that p57 can stabilize MyoD without inhibiting CDK's phosphorylation on MyoD (Reynaud et al., 2000). It has been shown that p57 physically interacts with MyoD through the N-terminal domain. However, major gaps exist in our understanding of how p57 stabilizes MyoD by not inhibiting CDK.

An accumulation of p57 in Skp2$^{-/-}$ cells indicates that p57 is ubiquitinated and regulated by Skp2. The SCF-type E3 ubiquitin ligase complex SCFSkp2 is involved in ubiquitinating p57 and targeting it for proteasome-mediated degrada-tion (Kamura et al., 2003). The phosphorylation of p57 at Thr310 is required for SCFSkp2-mediated ubiquitination to occur (Kamura et al., 2003), as shown by the fact that mutation of this site prevents the effect of Skp2 on p57 ubiquitination. Skp2 is overexpressed in many cancers, but it is not known whether these tumors have low levels of p57 and whether p57 expression is correlated with clinical outcome.

p57 interacts with c-Jun NH2-terminal kinase/stress-activated protein kinase (JNK/SAPK) through the carboxyl-terminal QT domain. Importantly, p57 can inhibit JNK in order to regulate stress-activated signaling (Chang et al., 2003). In addition, p57 can regulate the activity of LIM-kinase 1 (LIMK-1), a downstream effector of the Rho family of GTPases (Yokoo et al., 2003). In this setting, p57 binds to and translocates LIMK-1 from the cytoplasm into the nucleus (Yokoo et al., 2003). Because LIMK1 can phosphorylate and inactivate cofilin, an actin depolymerization factor, p57 in effect reorganizes the actin fiber by affecting the location of LIMK1.

The pituitary adenylate cyclase activating polypeptide (PACAP) can increase p57Kip2 activity in the embryonic cortex, thereby negatively regulating the cell cycle (Carey et al., 2002). Interestingly, a PACAP-treated precursor cell culture

TABLE 2.3
Selected Signal Mediators/Transcriptional Activators That Regulate p57

Protein	Function/Characteristics	References
MyoD	The myogenic factor MyoD can induce the expression of p57 to promote cell cycle withdrawal in differentiating myoblasts. On the other hand, p57 can stabilize MyoD without inhibiting CDK's phosphorylation on MyoD	Figliola and Maione, 2004; Reynaud et al., 2000
Skp2	57 is ubiquitinated and regulated by Skp2 because p57 is accumulated in Skp2$^{-/-}$ cells. Phosphorylation of p57 at Thr310 is required for SCFSkp2-mediated ubiquitination	Kamura et al., 2003
JNK	p57 interacts with JNK through the carboxyl-terminal QT domain. Importantly, p57 can inhibit JNK to regulate stress-activated signaling	Chang et al., 2003
LIMK1	LIMK1 can phosphorylate and inactivate cofilin, an actin depolymerization factor. p57 binds to and translocates LIMK-1 from the cytoplasm into the nucleus	Yokoo et al., 2003
PACAP	Pituitary adenylate cyclase activating polypeptide (PACAP) can increase p57^{KIP2} activity in embryonic cortex. p57 is involved in PACAP's antimitogenic effects	Carey et al., 2002
PThrP	Parathyroid hormone-related peptide (PTHrP), a positive regulator of chondrocyte proliferation, can have an opposing effect on p57 expression. p57 mRNA and protein are upregulated in proliferative chondrocytes in the absence of PTHrP	MacLean et al., 2004
Nurr1	Nurr1, an orphan nuclear receptor expressed in the embryonic ventral midbrain, can induce p57 expression and is important for the development of dopamine neurons	Joseph et al., 2003
GR	p57^{KIP2} protein is regulated in the glucocorticoid-induced antiproliferative effect p57 promoter contains a known glucocorticoid response element for glucocorticoid receptor (GR) binding	Alheim et al., 2003
TGF-β	p57 promoter extending from 165 to 77 is important for p57 transactivation through TGF β-Smad3 signaling	Scandura et al., 2004
Miz1	Miz1 is a member of the POZ domain/zinc finger transcription factor family. Importantly, p57KIP2 is not expressed in Miz1$^{-/-}$ embryos, suggesting that p57 is a target gene of Miz1 and is required for early embryonic development	Adhikary et al., 2003
B-Myb	In addition to its function as a transcription factor, B-Myb can prevent p57 from binding to cyclin A and promoting cell proliferation by a nontranscriptional mechanism	Joaquin and Watson, 2003
p73	p73 can induce cell cycle arrest and apoptosis. Importantly, p73β, but not p53 or 73-α, can activate the expression of p57. p57 is involved in p73β-mediated apoptosis	Blint et al., 2002

showed a dramatic decrease in DNA synthesis and CDK2 kinase activity. The decreased CDK2 kinase activity was caused by a twofold increase in levels of p57Kip2 protein. The levels of p21 and p27 were not affected. These data suggest that p57 is involved in the antimitogenic effects of PACAP. Future research will need to determine which signals regulate PACAP activity and how PACAP enhances the expression of p57 but not p21 and p27.

p57 is genomic imprinted, and it is silenced in a variety of human malignancies (Lee and Yang, 2001). In most cases, p57-associated mutations are not found, but p57 mRNA levels are significantly diminished in cancer cells compared to levels in normal cells. This transcriptional repression is a result of DNA methylation of the promoter because p57 gene expression is activated after treatment with a methylation inhibitor. Recently, the p57[KIP2] promoter was found to be methylated in a high percentage of patients with acute lymphocytic leukemia and was associated with a poor prognosis in these patients (Kikuchi et al., 2002).

Parathyroid-hormone-related peptide (PTHrP), a positive regulator of chondrocyte proliferation, may have an opposing effect on p57 expression (MacLean et al., 2004). This is indicated by the fact that p57 mRNA and protein are upregulated in proliferative chondrocytes in the absence of PTHrP. Importantly, p57$^{-/-}$ mice have bone defects resulting from a defect in chondrocyte proliferation (Yan et al., 1997). It remains to be determined how PTHrP causes the downregulation of p57.

2.5.2 TRANSCRIPTIONAL REGULATORS

Nurr1, an orphan nuclear receptor expressed in the embryonic ventral midbrain, can induce p57 expression and is important for the development of dopamine neurons (Joseph et al., 2003). Consistent with this observation, p57[KIP2] has been found to be critical for the maturation of midbrain dopamine neuronal cells in p57KIP2 knockout mice. Interestingly, p57 can interact with the amino-terminal domain of Nurr1, which enhances dopamine neuron differentiation. Glucocorticoids exert an antiproliferative effect, and the p57[KIP2] protein is regulated as part of this effect (Samuelsson et al., 1999). Because no *de novo* protein synthesis is required, this indicates that the glucocorticoid-mediated induction of p57 is regulated at the transcriptional level. In addition, glucocorticoids must stimulate the expression of p57 because the p57 promoter contains a known glucocorticoid response element for glucocorticoid receptor binding (Alheim et al., 2003).

In human hematopoietic cells, p57 is induced by TGF-β, which is essential for TGF-β-induced cell cycle arrest in these cells. The p57 promoter region, which extends from -165 to -77, is important for p57 transactivation through TGF-β-Smad3 signaling (Scandura et al., 2004). A reduction in the basal expression of p57 produced by RNA interference allows hematopoietic cells to proliferate more readily in the absence of TGF-β. TGF-β can also destabilize Skp2 (Wang et al., 2004), which otherwise destabilizes p57. Therefore, TGF-β's effect on Skp2 stability may stabilize p57. It is thus possible that TGF-β regulates p57 both transcriptionally and post-transcriptionally.

Miz1, a member of the POZ domain/zinc finger transcription factor family, can form a complex with the Myc oncoprotein, which represses Miz1-mediated transcriptional activation. A recent study showed that the Miz1$^{-/-}$ genotype in mice is embryonically lethal. Importantly, p57KIP2 is not expressed in Miz1$^{-/-}$ mice embryos (Adhikary et al., 2003), suggesting that p57 is a target gene of Miz1 and is required for early embryonic development. B-Myb, a transcription factor thought to be involved in regulating cell growth and differentiation, is a member of the Myb family of oncogenes that either activate or repress gene transcription. The cyclinA-mediated phosphorylation of B-Myb enhances Myb's transactivation ability. Because B-Myb can interact with p57 at the cyclin-binding domain, B-Myb competes with cyclin A for binding to p57 (Joaquin and Watson, 2003), which releases active cyclinA/CDK2 kinase from p57. Thus, in addition to its function as a transcription factor, B-Myb can prevent p57 from binding to cyclin A and promoting cell proliferation by a nontranscriptional mechanism. It is noteworthy that cyclin A release may result in the phosphorylation and subsequent activation of Myb.

p73, a member of the p53 family of proteins that can induce cell cycle arrest and apoptosis, functions as a transcriptional factor and is upregulated in response to DNA damage. Importantly, p73-β, but not p53 or 73-α, can activate the expression of p57 (Blint et al., 2002). It stands to reason that p57 is involved in p73-β-mediated apoptosis because apoptosis is dramatically reduced when p57 is downregulated by small interfering RNA. Consistent with this, p73-β-mediated apoptosis is considerably reduced in p57$^{-/-}$ embryo fibroblasts (Gonzalez et al., 2005). It remains unclear how p57 is involved in apoptosis, but it may relate to p57 increasing the activity of the proapoptotic enzyme caspase-3 (Samuelsson et al., 2002).

2.6 CONCLUSIONS

In this chapter, we have discussed many modes of regulating CDKIs. We expect greater characterization of the functional regulation of CDKIs in the future. For example, we know that p57 can inhibit kinases other than CDK, such as JNK and LIMK1, but it is important to determine whether these kinases have a feedback effect on p57. Also, as more transcription factors, such as FOXO and p73, have been found to be involved in the expression of CDKI, it is becoming apparent that transcriptional regulation also plays an important role in CDKI expression, in addition to the posttranscriptional regulation mainly regulated by ubiquitination. Because CDKIs have been shown to be capable of acting as tumor suppressors, it is important to learn how these transcriptional regulators affect the expression of CDKI during tumorigenesis. Most importantly, the multiple layers of controls and different modes of regulation discussed in this chapter may help identify new targets for agents that prevent deregulated CIP/KIP activities.

ACKNOWLEDGMENTS

We would like to thank the William McGowan Charitable Foundation, the Susan Komen Breast Cancer Foundation, and NIH grant (RO1CA 089266) for research support. M.-H. Lee is a recipient of the Flemin and Davenport research award. We regret that the excellent work of many of our colleagues in elucidating the functional regulation of CKI could not be summarized here owing to space constraints.

REFERENCES

Adhikary, S., Peukert, K., Karsunky, H., Beuger, V., Lutz, W., Elsasser, H. P., Moroy, T., and Eilers, M. (2003). Miz1 is required for early embryonic development during gastrulation. *Mol Cell Biol 23*, 7648–7657.

Alheim, K., Corness, J., Samuelsson, M.K., Bladh, L.G., Murata, T., Nilsson, T., and Okret, S. (2003). Identification of a functional glucocorticoid response element in the promoter of the cyclin-dependent kinase inhibitor p57Kip2. *J Mol Endocrinol 30*, 359–368.

Alvarez, B., Martinez, A.C., Burgering, B.M., and Carrera, A.C. (2001). Forkhead transcription factors contribute to execution of the mitotic programme in mammals. *Nature 413*, 744–747.

Asada, M., Yamada, T., Ichijo, H., Delia, D., Miyazono, K., Fukumuro, K., and Mizutani, S. (1999). Apoptosis inhibitory activity of cytoplasmic p21(Cip1/WAF1) in monocytic differentiation. *EMBO J 18*, 1223–1234.

Baldassarre, G., Belletti, B., Nicoloso, M. S., Schiappacassi, M., Vecchione, A., Spessotto, P., Morrione, A., Canzonieri, V., and Colombatti, A. (2005). p27(Kip1)-stathmin interaction influences sarcoma cell migration and invasion. *Cancer Cell 7*, 51–63.

Bao, W., Thullberg, M., Zhang, H., Onischenko, A., and Stromblad, S. (2002). Cell attachment to the extracellular matrix induces proteasomal degradation of p21(CIP1) via Cdc42/Rac1 signaling. *Mol Cell Biol 22*, 4587–4597.

Besson, A., Gurian-West, M., Schmidt, A., Hall, A., and Roberts, J.M. (2004). p27Kip1 modulates cell migration through the regulation of RhoA activation. *Genes Dev 18*, 862–876. Epub 2004 Apr 2012.

Blint, E., Phillips, A.C., Kozlov, S., Stewart, C.L., and Vousden, K.H. (2002). Induction of p57(KIP2) expression by p73beta. *Proc Natl Acad Sci U S A 99*, 3529–3534. Epub 2002 Mar 3512.

Boehm, M., Yoshimoto, T., Crook, M.F., Nallamshetty, S., True, A., Nabel, G.J., and Nabel, E.G. (2002). A growth factor-dependent nuclear kinase phosphorylates p27(Kip1) and regulates cell cycle progression. *EMBO J 21*, 3390–3401.

Bond, M., Sala-Newby, G.B., and Newby, A.C. (2004). Focal adhesion kinase (FAK)-dependent regulation of S-phase kinase-associated protein-2 (Skp-2) stability: a novel mechanism regulating smooth muscle cell proliferation. *J Biol Chem 279*, 37304–37310. Epub 32004 Jun 37318.

Bornstein, G., Bloom, J., Sitry-Shevah, D., Nakayama, K., Pagano, M., and Hershko, A. (2003). Role of the SCFSkp2 ubiquitin ligase in the degradation of p21Cip1 in S phase. *J Biol Chem 278*, 25752–25757. Epub 22003 May 25752.

Brunet, A., Bonni, A., Zigmond, M.J., Lin, M.Z., Juo, P., Hu, L.S., Anderson, M.J., Arden, K.C., Blenis, J., and Greenberg, M.E. (1999). Akt promotes cell survival by phosphorylating and inhibiting a Forkhead transcription factor. *Cell 96*, 857–868.

Burgering, B. M., and Kops, G. J. (2002). Cell cycle and death control: long live Forkheads. *Trends Biochem Sci 27*, 352–360.

Butz, N., Ruetz, S., Natt, F., Hall, J., Weiler, J., Mestan, J., Ducarre, M., Grossenbacher, R., Hauser, P., Kempf, D., and Hofmann, F. (2005). The human ubiquitin-conjugating enzyme Cdc34 controls cellular proliferation through regulation of p27Kip1 protein levels. *Exp Cell Res 303*, 482–493.

Carey, R.G., Li, B., and DiCicco-Bloom, E. (2002). Pituitary adenylate cyclase activating polypeptide anti-mitogenic signaling in cerebral cortical progenitors is regulated by p57Kip2-dependent CDK2 activity. *J Neurosci 22*, 1583–1591.

Carrano, A.C., Eytan, E., Hershko, A., and Pagano, M. (1999). SKP2 is required for ubiquitin-mediated degradation of the CDK inhibitor p27. *Nat Cell Biol 1*, 193–199.

Carreira, S., Goodall, J., Aksan, I., La Rocca, S.A., Galibert, M.D., Denat, L., Larue, L., and Goding, C.R. (2005). Mitf cooperates with Rb1 and activates p21Cip1 expression to regulate cell cycle progression. *Nature 433*, 764–769.

Chang, T.S., Kim, M.J., Ryoo, K., Park, J., Eom, S.J., Shim, J., Nakayama, K.I., Nakayama, K., Tomita, M., Takahashi, K. et al. (2003). p57KIP2 Modulates stress-activated signaling by inhibiting c-Jun NH2-terminal kinase/stress-activated protein kinase. *J Biol Chem 278*, 48092–48098. Epub 42003 Sep 48098.

Chen, B., He, L., Savell, V.H., Jenkins, J.J., and Parham, D.M. (2000). Inhibition of the interferon-gamma/signal transducers and activators of transcription (STAT) pathway by hypermethylation at a STAT-binding site in the p21WAF1 promoter region. *Cancer Res 60*, 3290–3298.

Chiba, H., Itoh, T., Satohisa, S., Sakai, N., Noguchi, H., Osanai, M., Kojima, T., and Sawada, N. (2005). Activation of p21CIP1/WAF1 gene expression and inhibition of cell proliferation by overexpression of hepatocyte nuclear factor-4alpha. *Exp Cell Res 302*, 11–21.

Cho, S., Kim, J.H., Back, S.H., and Jang, S.K. (2005). Polypyrimidine tract-binding protein enhances the internal ribosomal entry site-dependent translation of p27Kip1 mRNA and modulates transition from G1 to S phase. *Mol Cell Biol 25*, 1283–1297.

Claassen, G.F. and Hann, S.R. (2000). A role for transcriptional repression of p21CIP1 by c-Myc in overcoming transforming growth factor beta-induced cell-cycle arrest. *Proc Natl Acad Sci U S A 97*, 9498–9503.

Cunningham, M.A., Zhu, Q., and Hammond, J.M. (2004). FoxO1a can alter cell cycle progression by regulating the nuclear localization of p27kip in granulosa cells. *Mol Endocrinol 18*, 1756–1767. Epub 2004 Apr 1715.

Deng, X., Mercer, S.E., Shah, S., Ewton, D.Z., and Friedman, E. (2004). The cyclin-dependent kinase inhibitor p27Kip1 is stabilized in G(0) by Mirk/dyrk1B kinase. *J Biol Chem 279*, 22498–22504. Epub 22004 Mar 22499.

Du, J., Widlund, H.R., Horstmann, M.A., Ramaswamy, S., Ross, K., Huber, W.E., Nishimura, E.K., Golub, T.R., and Fisher, D.E. (2004). Critical role of CDK2 for melanoma growth linked to its melanocyte-specific transcriptional regulation by MITF. *Cancer Cell 6*, 565–576.

el-Deiry, W.S., Tokino, T., Velculescu, V.E., Levy, D.B., Parsons, R., Trent, J.M., Lin, D., Mercer, W.E., Kinzler, K.W., and Vogelstein, B. (1993). WAF1, a potential mediator of p53 tumor suppression. *Cell 75*, 817–825.

Estanyol, J.M., Jaumot, M., Casanovas, O., Rodriguez-Vilarrupla, A., Agell, N., and Bachs, O. (1999). The protein SET regulates the inhibitory effect of p21(Cip1) on cyclin E-cyclin-dependent kinase 2 activity. *J Biol Chem 274*, 33161–33165.

Esteva, F.J., Sahin, A.A., Rassidakis, G.Z., Yuan, L.X., Smith, T.L., Yang, Y., Gilcrease, M.Z., Cristofanilli, M., Nahta, R., Pusztai, L., and Claret, F.X. (2003). Jun activation domain binding protein 1 expression is associated with low p27(Kip1)levels in node-negative breast cancer. *Clin Cancer Res 9*, 5652–5659.

Fero, M.L., Randel, E., Gurley, K.E., Roberts, J.M., and Kemp, C.J. (1998). The murine gene p27Kip1 is haplo-insufficient for tumour suppression. *Nature 396*, 177–180.

Figliola, R. and Maione, R. (2004). MyoD induces the expression of p57Kip2 in cells lacking p21Cip1/Waf1: overlapping and distinct functions of the two cdk inhibitors. *J Cell Physiol 200*, 468–475.

Fujita, N., Sato, S., Katayama, K., and Tsuruo, T. (2002). Akt-dependent phosphorylation of p27Kip1 promotes binding to 14-3-3 and cytoplasmic localization. *J Biol Chem 277*, 28706–28713.

Fujita, N., Sato, S., and Tsuruo, T. (2003). Phosphorylation of p27Kip1 at threonine 198 by p90 ribosomal protein S6 kinases promotes its binding to 14-3-3 and cytoplasmic localization. *J Biol Chem 278*, 49254–49260. Epub 42003 Sep 49222.

Gesbert, F., Sellers, W.R., Signoretti, S., Loda, M., and Griffin, J.D. (2000). BCR/ABL regulates expression of the cyclin-dependent kinase inhibitor p27Kip1 through the phosphatidylinositol 3-Kinase/AKT pathway. *J Biol Chem 275*, 39223–39230.

Gonzalez, S., Perez-Perez, M. M., Hernando, E., Serrano, M., and Cordon-Cardo, C. (2005). p73beta-Mediated apoptosis requires p57kip2 induction and IEX-1 inhibition. *Cancer Res 65*, 2186–2192.

Gostissa, M., Morelli, M., Mantovani, F., Guida, E., Piazza, S., Collavin, L., Brancolini, C., Schneider, C., and Del Sal, G. (2004). The transcriptional repressor hDaxx potentiates p53-dependent apoptosis. *J Biol Chem 279*, 48013–48023. Epub 42004 Aug 48031.

Gottschalk, A.R., Basila, D., Wong, M., Dean, N.M., Brandts, C.H., Stokoe, D., and Haas-Kogan, D.A. (2001). p27Kip1 is required for PTEN-induced G1 growth arrest. *Cancer Res 61*, 2105–2111.

Gutterman, J.U. and Choubey, D. (1999). Retardation of cell proliferation after expression of p202 accompanies an increase in p21(WAF1/CIP1). *Cell Growth Differ 10*, 93–100.

Hermeking, H., Lengauer, C., Polyak, K., He, T.C., Zhang, L., Thiagalingam, S., Kinzler, K.W., and Vogelstein, B. (1997). 14-3-3 sigma is a p53-regulated inhibitor of G2/M progression. *Mol Cell 1*, 3–11.

Ionov, Y., Le, X., Tunquist, B.J., Sweetenham, J., Sachs, T., Ryder, J., Johnson, T., Lilly, M.B., and Kraft, A.S. (2003). Pim-1 protein kinase is nuclear in Burkitt's lymphoma: nuclear localization is necessary for its biologic effects. *Anticancer Res 23*, 167–178.

Jascur, T., Brickner, H., Salles-Passador, I., Barbier, V., El Khissiin, A., Smith, B., Fotedar, R., and Fotedar, A. (2005). Regulation of p21(WAF1/CIP1) stability by WISp39, a Hsp90 binding TPR protein. *Mol Cell 17*, 237–249.

Joaquin, M. and Watson, R.J. (2003). The cell cycle-regulated B-Myb transcription factor overcomes cyclin-dependent kinase inhibitory activity of p57(KIP2) by interacting with its cyclin-binding domain. *J Biol Chem 278*, 44255–44264. Epub 42003 Aug 44228.

Joseph, B., Wallen-Mackenzie, A., Benoit, G., Murata, T., Joodmardi, E., Okret, S., and Perlmann, T. (2003). p57(Kip2) cooperates with Nurr1 in developing dopamine cells. *Proc Natl Acad Sci U S A 100*, 15619–15624. Epub 12003 Dec 15611.

Kamura, T., Hara, T., Kotoshiba, S., Yada, M., Ishida, N., Imaki, H., Hatakeyama, S., Nakayama, K., and Nakayama, K.I. (2003). Degradation of p57Kip2 mediated by SCFSkp2-dependent ubiquitylation. *Proc Natl Acad Sci U S A 100*, 10231–10236. Epub 12003 Aug 10218.

Kamura, T., Hara, T., Matsumoto, M., Ishida, N., Okumura, F., Hatakeyama, S., Yoshida, M., Nakayama, K., and Nakayama, K.I. (2004). Cytoplasmic ubiquitin ligase KPC regulates proteolysis of p27(Kip1) at G1 phase. *Nat Cell Biol 6*, 1229–1235. Epub 2004 Nov 1227.

Kang-Decker, N., Tong, C., Boussouar, F., Baker, D.J., Xu, W., Leontovich, A.A., Taylor, W.R., Brindle, P.K., and van Deursen, J.M. (2004). Loss of CBP causes T cell lymphomagenesis in synergy with p27Kip1 insufficiency. *Cancer Cell 5*, 177–189.

Kikuchi, T., Toyota, M., Itoh, F., Suzuki, H., Obata, T., Yamamoto, H., Kakiuchi, H., Kusano, M., Issa, J. P., Tokino, T., and Imai, K. (2002). Inactivation of p57KIP2 by regional promoter hypermethylation and histone deacetylation in human tumors. *Oncogene 21*, 2741–2749.

Kim, D.W., Hwang, J.H., Suh, J.M., Kim, H., Song, J.H., Hwang, E.S., Hwang, I.Y., Park, K.C., Chung, H.K., Kim, J.M. et al. (2003a). RET/PTC (rearranged in transformation/papillary thyroid carcinomas) tyrosine kinase phosphorylates and activates phospho-inositide-dependent kinase 1 (PDK1): an alternative phosphatidylinositol 3-kinase-independent pathway to activate PDK1. *Mol Endocrinol 17*, 1382–1394. Epub 2003 May 1388.

Kim, E.J., Park, J.S., and Um, S.J. (2003b). Identification of Daxx interacting with p73, one of the p53 family, and its regulation of p53 activity by competitive interaction with PML. *Nucl Acids Res 31*, 5356–5367.

King, T.J. and Lampe, P.D. (2004a). The gap junction protein connexin32 is a mouse lung tumor suppressor. *Cancer Res 64*, 7191–7196.

King, T.J. and Lampe, P.D. (2004b). Mice deficient for the gap junction protein Connexin32 exhibit increased radiation-induced tumorigenesis associated with elevated mitogen-activated protein kinase (p44/Erk1, p42/Erk2) activation. *Carcinogenesis 25*, 669–680. Epub 2004 Jan 2023.

Kops, G.J., Medema, R.H., Glassford, J., Essers, M.A., Dijkers, P.F., Coffer, P.J., Lam, E.W., and Burgering, B.M. (2002). Control of cell cycle exit and entry by protein kinase B-regulated forkhead transcription factors. *Mol Cell Biol 22*, 2025–2036.

Kotake, Y., Nakayama, K., Ishida, N., and Nakayama, K.I. (2005). Role of serine 10 phos-phorylation in p27 stabilization revealed by analysis of p27 knock-in mice harboring a serine 10 mutation. *J Biol Chem 280*, 1095–1102. Epub 2004 Nov 1093.

Kotoshiba, S., Kamura, T., Hara, T., Ishida, N., and Nakayama, K. I. (2005). Molecular dissection of the interaction between p27 and kip1 ubiquitylation-promoting complex, the ubiquitin ligase that regulates proteolysis of p27 in G1 phase. *J Biol Chem 280*, 17694–17700. Epub 12005 Mar 17693.

Kouvaraki, M.A., Rassidakis, G.Z., Tian, L., Kumar, R., Kittas, C., and Claret, F.X. (2003). Jun activation domain-binding protein 1 expression in breast cancer inversely corre-lates with the cell cycle inhibitor p27(Kip1). *Cancer Res 63*, 2977–2981.

Kullmann, M., Gopfert, U., Siewe, B., and Hengst, L. (2002). ELAV/Hu proteins inhibit p27 translation via an IRES element in the p27 5'UTR. *Genes Dev 16*, 3087–3099.

Lee, M.H., Reynisdottir, I., and Massague, J. (1995). Cloning of p57KIP2, a cyclin-dependent kinase inhibitor with unique domain structure and tissue distribution. *Genes Dev 9*, 639–649.

Lee, M.H. and Yang, H.Y. (2001). Negative regulators of cyclin-dependent kinases and their roles in cancers. *Cell Mol Life Sci 58*, 1907–1922.

Li, D.M. and Sun, H. (1998). PTEN/MMAC1/TEP1 suppresses the tumorigenicity and induces G1 cell cycle arrest in human glioblastoma cells. *Proc Natl Acad Sci U S A 95*, 15406–15411.

Liang, J., Zubovitz, J., Petrocelli, T., Kotchetkov, R., Connor, M.K., Han, K., Lee, J.H., Ciarallo, S., Catzavelos, C., Beniston, R. et al. (2002). PKB/Akt phosphorylates p27, impairs nuclear import of p27 and opposes p27-mediated G1 arrest. *Nat Med 8*, 1153–1160.

MacLean, H.E., Guo, J., Knight, M.C., Zhang, P., Cobrinik, D., and Kronenberg, H.M. (2004). The cyclin-dependent kinase inhibitor p57(Kip2) mediates proliferative actions of PTHrP in chondrocytes. *J Clin Invest 113*, 1334–1343.

Mamillapalli, R., Gavrilova, N., Mihaylova, V.T., Tsvetkov, L.M., Wu, H., Zhang, H., and Sun, H. (2001). PTEN regulates the ubiquitin-dependent degradation of the CDK inhibitor p27(KIP1) through the ubiquitin E3 ligase SCF(SKP2). *Curr Biol 11*, 263–267.

Matsuoka, S., Edwards, M.C., Bai, C., Parker, S., Zhang, P., Baldini, A., Harper, J.W., and Elledge, S.J. (1995). p57KIP2, a structurally distinct member of the p21CIP1 Cdk inhibitor family, is a candidate tumor suppressor gene. *Genes Dev 9*, 650–662.

Medema, R.H., Kops, G.J., Bos, J.L., and Burgering, B.M. (2000). AFX-like Forkhead transcription factors mediate cell-cycle regulation by Ras and PKB through p27kip1. *Nature 404*, 782–787.

Milne, T.A., Hughes, C.M., Lloyd, R., Yang, Z., Rozenblatt-Rosen, O., Dou, Y., Schnepp, R.W., Krankel, C., Livolsi, V.A., Gibbs, D. et al. (2005). Menin and MLL cooperatively regulate expression of cyclin-dependent kinase inhibitors. *Proc Natl Acad Sci U S A 102*, 749–754. Epub 2005 Jan 2007.

Moeller, S.J., Head, E.D., and Sheaff, R.J. (2003). p27Kip1 inhibition of GRB2-SOS formation can regulate Ras activation. *Mol Cell Biol 23*, 3735–3752.

Mukherjee, S. and Conrad, S.E. (2005). c-Myc Suppresses p21WAF1/CIP1 expression during estrogen signaling and antiestrogen resistance in human breast cancer cells. *J Biol Chem 280*, 17617–17625. Epub 12005 Mar 17618.

Nakayama, K., Nagahama, H., Minamishima, Y.A., Matsumoto, M., Nakamichi, I., Kitagawa, K., Shirane, M., Tsunematsu, R., Tsukiyama, T., Ishida, N. et al. (2000). Targeted disruption of Skp2 results in accumulation of cyclin E and p27(Kip1), polyploidy and centrosome overduplication. *EMBO J 19*, 2069–2081.

Newman, L., Xia, W., Yang, H.Y., Sahin, A., Bondy, M., Lukmanji, F., Hung, M.C., and Lee, M.H. (2001). Correlation of p27 protein expression with HER-2/neu expression in breast cancer. *Mol Carcinog 30*, 169–175.

Ono, T., Kitaura, H., Ugai, H., Murata, T., Yokoyama, K.K., Iguchi-Ariga, S.M., and Ariga, H. (2000). TOK-1, a novel p21Cip1-binding protein that cooperatively enhances p21-dependent inhibitory activity toward CDK2 kinase. *J Biol Chem 275*, 31145–31154.

Pardali, K., Kurisaki, A., Moren, A., ten Dijke, P., Kardassis, D., and Moustakas, A. (2000). Role of Smad proteins and transcription factor Sp1 in p21(Waf1/Cip1) regulation by transforming growth factor-beta. *J Biol Chem 275*, 29244–29256.

Polyak, K., Lee, M.H., Erdjument-Bromage, H., Koff, A., Roberts, J.M., Tempst, P., and Massague, J. (1994). Cloning of p27Kip1, a cyclin-dependent kinase inhibitor and a potential mediator of extracellular antimitogenic signals. *Cell 78*, 59–66.

Porter, L.A., Dellinger, R.W., Tynan, J.A., Barnes, E.A., Kong, M., Lenormand, J.L., and Donoghue, D.J. (2002). Human Speedy: a novel cell cycle regulator that enhances proliferation through activation of Cdk2. *J Cell Biol 157*, 357–366. Epub 2002 Apr 2029.

Porter, L.A., Kong-Beltran, M., and Donoghue, D.J. (2003). Spy1 interacts with p27Kip1 to allow G1/S progression. *Mol Biol Cell 14*, 3664–3674. Epub 2003 Jul 3611.

Ren, J., Agata, N., Chen, D., Li, Y., Yu, W. H., Huang, L., Raina, D., Chen, W., Kharbanda, S., and Kufe, D. (2004). Human MUC1 carcinoma-associated protein confers resistance to genotoxic anticancer agents. *Cancer Cell 5*, 163–175.

Reynaud, E.G., Leibovitch, M.P., Tintignac, L.A., Pelpel, K., Guillier, M., and Leibovitch, S.A. (2000). Stabilization of MyoD by direct binding to p57(Kip2). *J Biol Chem 275*, 18767–18776.

Reynaud, E.G., Pelpel, K., Guillier, M., Leibovitch, M.P., and Leibovitch, S.A. (1999). p57(Kip2) stabilizes the MyoD protein by inhibiting cyclin E-Cdk2 kinase activity in growing myoblasts. *Mol Cell Biol 19*, 7621–7629.

Samuelsson, M.K., Pazirandeh, A., Davani, B., and Okret, S. (1999). p57Kip2, a glucocorticoid-induced inhibitor of cell cycle progression in HeLa cells. *Mol Endocrinol 13*, 1811–1822.

Samuelsson, M.K., Pazirandeh, A., and Okret, S. (2002). A pro-apoptotic effect of the CDK inhibitor p57(Kip2) on staurosporine-induced apoptosis in HeLa cells. *Biochem Biophys Res Commun 296*, 702–709.

Scandura, J.M., Boccuni, P., Massague, J., and Nimer, S.D. (2004). Transforming growth factor beta-induced cell cycle arrest of human hematopoietic cells requires p57KIP2 up-regulation. *Proc Natl Acad Sci U S A 101*, 15231–15236. Epub 12004 Oct 15211.

Schepers, H., Geugien, M., Eggen, B.J., and Vellenga, E. (2003). Constitutive cytoplasmic localization of p21(Waf1/Cip1) affects the apoptotic process in monocytic leukaemia. *Leukemia 17*, 2113–2121.

Seoane, J., Le, H.V., and Massague, J. (2002). Myc suppression of the p21(Cip1) Cdk inhibitor influences the outcome of the p53 response to DNA damage. *Nature 419*, 729–734.

Seoane, J., Pouponnot, C., Staller, P., Schader, M., Eilers, M., and Massague, J. (2001). TGFbeta influences Myc, Miz-1 and Smad to control the CDK inhibitor p15INK4b. *Nat Cell Biol 3*, 400–408.

Shin, I., Yakes, F.M., Rojo, F., Shin, N.Y., Bakin, A.V., Baselga, J., and Arteaga, C.L. (2002). PKB/Akt mediates cell-cycle progression by phosphorylation of p27Kip1 at threonine 157 and modulation of its cellular localization. *Nat Med 8*, 1145–1152.

Simpson, L. and Parsons, R. (2001). PTEN: life as a tumor suppressor. *Exp Cell Res 264*, 29–41.

Somasundaram, K., Zhang, H., Zeng, Y.X., Houvras, Y., Peng, Y., Zhang, H., Wu, G.S., Licht, J.D., Weber, B.L., and El-Deiry, W.S. (1997). Arrest of the cell cycle by the tumour-suppressor BRCA1 requires the CDK-inhibitor p21WAF1/CiP1. *Nature 389*, 187–190.

Staller, P., Peukert, K., Kiermaier, A., Seoane, J., Lukas, J., Karsunky, H., Moroy, T., Bartek, J., Massague, J., Hanel, F., and Eilers, M. (2001). Repression of p15INK4b expression by Myc through association with Miz-1. *Nat Cell Biol 3*, 392–399.

Sugiyama, Y., Tomoda, K., Tanaka, T., Arata, Y., Yoneda-Kato, N., and Kato, J. (2001). Direct binding of the signal-transducing adaptor Grb2 facilitates down-regulation of the cyclin-dependent kinase inhibitor p27Kip1. *J Biol Chem 276*, 12084–12090. Epub 12001 Jan 12022.

Sui, L., Dong, Y., Ohno, M., Watanabe, Y., Sugimoto, K., Tai, Y., and Tokuda, M. (2001). Jab1 expression is associated with inverse expression of p27(kip1) and poor prognosis in epithelial ovarian tumors. *Clin Cancer Res 7*, 4130–4135.

Tanaka, H., Yamashita, T., Asada, M., Mizutani, S., Yoshikawa, H., and Tohyama, M. (2002). Cytoplasmic p21(Cip1/WAF1) regulates neurite remodeling by inhibiting Rho-kinase activity. *J Cell Biol 158*, 321–329. Epub 2002 Jul 2015.

Tomoda, K., Kato, J.Y., Tatsumi, E., Takahashi, T., Matsuo, Y., and Yoneda-Kato, N. (2005). The Jab1/COP9 signalosome subcomplex is a downstream mediator of Bcr-Abl kinase activity and facilitates cell-cycle progression. *Blood 105*, 775–783. Epub 2004 Sep 2007.

Tomoda, K., Kubota, Y., and Kato, J. (1999). Degradation of the cyclin-dependent-kinase inhibitor p27Kip1 is instigated by Jab1. *Nature 398*, 160–165.

Toyoshima, H. and Hunter, T. (1994). p27, A novel inhibitor of G1 cyclin-Cdk protein kinase activity, is related to p21. *Cell 78*, 67–74.

Tran, H., Brunet, A., Griffith, E.C., and Greenberg, M.E. (2003). The many forks in FOXO's road. *Sci STKE 2003*, RE5.

Vitagliano, D., Carlomagno, F., Motti, M.L., Viglietto, G., Nikiforov, Y.E., Nikiforova, M.N., Hershman, J.M., Ryan, A.J., Fusco, A., Melillo, R.M., and Santoro, M. (2004). Regulation of p27Kip1 protein levels contributes to mitogenic effects of the RET/PTC kinase in thyroid carcinoma cells. *Cancer Res 64*, 3823–3829.

Walerych, D., Kudla, G., Gutkowska, M., Wawrzynow, B., Muller, L., King, F.W., Helwak, A., Boros, J., Zylicz, A., and Zylicz, M. (2004). Hsp90 chaperones wild-type p53 tumor suppressor protein. *J Biol Chem 279*, 48836–48845. Epub 42004 Sep 48839.

Wang, C., Hou, X., Mohapatra, S., Ma, Y., Cress, W.D., Pledger, W.J., and Chen, J. (2005). Activation of p27Kip1 Expression by E2F1. A negative feedback mechanism. *J Biol Chem 280*, 12339–12343. Epub 12005 Feb 12314.

Wang, W., Ungermannova, D., Jin, J., Harper, J.W., and Liu, X. (2004). Negative regulation of SCFSkp2 ubiquitin ligase by TGF-beta signaling. *Oncogene 23*, 1064–1075.

Wang, Z., Bhattacharya, N., Mixter, P.F., Wei, W., Sedivy, J., and Magnuson, N.S. (2002). Phosphorylation of the cell cycle inhibitor p21Cip1/WAF1 by Pim-1 kinase. *Biochim Biophys Acta 1593*, 45–55.

Wang, Z. and Garabedian, M. J. (2003). Modulation of glucocorticoid receptor transcriptional activation, phosphorylation, and growth inhibition by p27Kip1. *J Biol Chem 278*, 50897–50901. Epub 52003 Oct 50897.

Wei, X., Xu, H., and Kufe, D. (2005). Human MUC1 oncoprotein regulates p53-responsive gene transcription in the genotoxic stress response. *Cancer Cell 7*, 167–178.

Williamson, E.A., Dadmanesh, F., and Koeffler, H.P. (2002). BRCA1 transactivates the cyclin-dependent kinase inhibitor p27(Kip1). *Oncogene 21*, 3199–3206.

Yan, Y., Frisen, J., Lee, M.H., Massague, J., and Barbacid, M. (1997). Ablation of the CDK inhibitor p57Kip2 results in increased apoptosis and delayed differentiation during mouse development. *Genes Dev 11*, 973–983.

Yang, H., Zhao, R., Yang, H.Y., and Lee, M.H. (2005). Constitutively active FOXO4 inhibits Akt activity, regulates p27 Kip1 stability, and suppresses HER2-mediated tumorigenicity. *Oncogene 24*, 1924–1935.

Yang, H.Y., Wen, Y.Y., Chen, C.H., Lozano, G., and Lee, M.H. (2003). 14-3-3sigma Positively regulates p53 and suppresses tumor growth. *Mol Cell Biol 23*, 7096–7107.

Yang, H.Y., Zhou, B.P., Hung, M.C., and Lee, M.H. (2000). Oncogenic signals of HER-2/neu in regulating the stability of the cyclin-dependent kinase inhibitor p27. *J Biol Chem 275*, 24735–24739.

Yang, W., Shen, J., Wu, M., Arsura, M., FitzGerald, M., Suldan, Z., Kim, D.W., Hofmann, C.S., Pianetti, S., Romieu-Mourez, R. et al. (2001). Repression of transcription of the p27(Kip1) cyclin-dependent kinase inhibitor gene by c-Myc. *Oncogene 20*, 1688–1702.

Yang, X., Menon, S., Lykke-Andersen, K., Tsuge, T., Di, X., Wang, X., Rodriguez-Suarez, R.J., Zhang, H., and Wei, N. (2002). The COP9 signalosome inhibits p27(kip1) degradation and impedes G1-S phase progression via deneddylation of SCF Cul1. *Curr Biol 12*, 667–672.

Yokoo, T., Toyoshima, H., Miura, M., Wang, Y., Iida, K.T., Suzuki, H., Sone, H., Shimano, H., Gotoda, T., Nishimori, S. et al. (2003). p57Kip2 regulates actin dynamics by binding and translocating LIM-kinase 1 to the nucleus. *J Biol Chem 278*, 52919–52923. Epub 52003 Oct 52916.

Yu, Z.K., Gervais, J.L., and Zhang, H. (1998). Human CUL-1 associates with the SKP1/SKP2 complex and regulates p21(CIP1/WAF1) and cyclin D proteins. *Proc Natl Acad Sci U S A 95*, 11324–11329.

Zhang, L. and Wang, C. (2003). PAX3-FKHR transformation increases 26 S proteasome-dependent degradation of p27Kip1, a potential role for elevated Skp2 expression. *J Biol Chem 278*, 27–36.

Zhang, P., Liegeois, N. J., Wong, C., Finegold, M., Hou, H., Thompson, J.C., Silverman, A., Harper, J.W., DePinho, R.A., and Elledge, S.J. (1997). Altered cell differentiation and proliferation in mice lacking p57KIP2 indicates a role in Beckwith-Wiedemann syndrome. *Nature 387*, 151–158.

Zhang, Y., Lin, Y., Bowles, C., and Wang, F. (2004a). Direct cell cycle regulation by the fibroblast growth factor receptor (FGFR) kinase through phosphorylation-dependent release of Cks1 from FGFR substrate 2. *J Biol Chem 279*, 55348–55354. Epub 52004 Oct 55327.

Zhang, Z., Wang, H., Li, M., Agrawal, S., Chen, X., and Zhang, R. (2004b). MDM2 is a negative regulator of p21WAF1/CIP1, independent of p53. J Biol Chem *279*, 16000–16006. Epub 12004 Feb 16003.

Zhou, B.P., Liao, Y., Spohn, B., Lee, M.-H., and Hung, M.-C. (2001). HER-2/neu induces cytoplasmic retention of p21 Cip1/WAF1 via akt. *Nat Cell Biol 3*, 245–252.

Zuo, Z., Dean, N.M., and Honkanen, R.E. (1998). Serine/threonine protein phosphatase type 5 acts upstream of p53 to regulate the induction of p21(WAF1/Cip1) and mediate growth arrest. *J Biol Chem 273*, 12250–12258.

<h1>3   Mouse Models to Study the *In Vivo* Function of Cyclin-Dependent Kinases in Normal Homeostasis and Tumor Development</h1>

Marcos Malumbres, Pierre Dubus,
and Sagrario Ortega

CONTENTS

3.1 INTRODUCTION

The control of cell proliferation by the extracellular environment takes place mostly at the G$_1$/S transition of the cell cycle. The importance of this control is underscored by the finding that this process is frequently altered in cancer cells. From yeast to humans, cell cycle entrance from quiescence (G$_0$), as well as progression through the different stages of the cell cycle (G$_1$-S-G$_2$ and M phases), is controlled by the sequential activation of a family of serine-threonine kinases called cyclin-dependent kinases (CDKs).[1,2] These kinases are heterodimeric enzymes, formed by a catalytic subunit, the CDK, and a regulatory subunit, the cyclin. Oscillatory levels of the cyclin subunits lead to the assembly of different CDK–cyclin complexes along the cell cycle. In mammals, the CDK protein family includes about 20 members, 11 of which are structurally related to the first discovered CDK — Cdc2 in *S. pombe* or Cdc28 in *S. cerevisiae*.[3,4] The functions of many of these CDKs have not yet been studied in detail.[5] At least three CDKs (CDK4, CDK6, and CDK2), and their corresponding regulatory subunits, the D-type- and E-type-cyclins, are involved in the control of the G_1/S transition in mammalian cells and have long been considered essential regulators of cell proliferation as well as potential targets for cancer therapy. Recently, it has been possible to inactivate specifically each one of these genes in the mouse, by gene targeting in embryonic stem cells. Conclusions from the phenotypic characterization of genetically modified mice lacking cyclins and CDKs have challenged many previous assumptions related to cell cycle control in mammals. Most of the G_1 cyclins and CDKs previously thought to be essential for the control of cell proliferation have turned out to be dispensable in most cell types. This is probably due to functional redundancy, activation of compensatory mechanisms, and/or promiscuous protein–protein interactions. These mouse models have also revealed unexpected new functions for the cyclins and CDKs, as well as a high level of cell-type-dependent specificity in cell cycle regulation and the control of proliferation. The lessons learned from these models and their implications for therapeutic intervention in cancer will be discussed in this chapter.

3.2 G_1 CDKs AND THE CONTROL OF CELL PROLIFERATION

During the G_1 phase of the cell cycle, cells can either stay quiescent or enter S phase and replicate their genomes. Progression throughout the G_1 phase is regulated by a complex mechanism that involves at least three CDKs, CDK4, CDK6, and CDK2[6,7] (Figure 3.1). An additional kinase, CDK3, can also function at this stage by binding cyclin C and phosphorylating pRb at the entrance into the G_1 phase from the previous mitosis (M).[8] However, its physiological role is not clear, because most laboratory mice are deficient in this kinase owing to a spontaneous mutation that introduces a premature stop codon in the CDK3 open reading frame.[9]

Whether cells proliferate or not depends on the balance of mitogenic and anti-mitogenic signals that they receive from their environment and evaluate during the G_1 phase. Proliferation signals, such as those mediated by growth factors, frequently result in the induction of several signal transduction pathways such as the Ras- or PI3K-dependent cascades. Activation of these mitogenic pathways and the corresponding mitogen-activated protein kinases (MAPKs) result in increased levels and nuclear localization of D-type cyclins (cyclin D1, D2, or D3), which bind and activate CDK4 and CDK6, thereby acting as sensors of extracellular mitogenic factors. In fact, these D-type cyclins display some mitogen-response elements in their promoters that control their induction under the appropriate stimulation.[10] Turnover of the D-cyclins

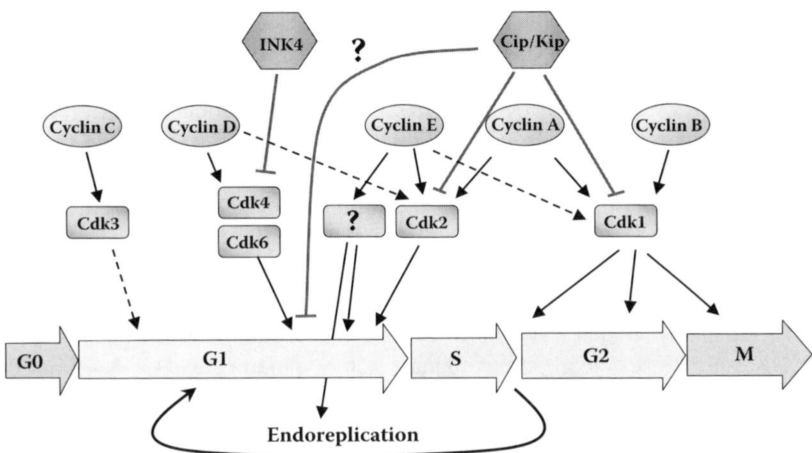

FIGURE 3.1 Mammalian CDKs and their role in cell cycle progression. In mammals, at least four CDKs (CDK4, CDK6, CDK2, and CDK1) are involved in promoting progression of the cell cycle through its different stages. At least four cyclin families (cyclins D, E, A, and B) activate CDK4/6, CDK2 and CDK1, respectively. A-cyclins bind to and activate both CDK2, and CDK1, at different points in the cell cycle. CDKs are negatively controlled by the INK4 and CIP/KIP families of cell cycle inhibitors. Discontinuous arrows indicate recently found CDK activities whose physiological role is still unknown, such as the promiscuous cyclin D/CDK2 or cyclin E/CDK1 (Cdc2) complex formation and the function of the CDK3/cyclin C complex in G_1 progression. Cyclin E is also involved in promoting endoreplication (consecutive rounds of DNA synthesis without intervening mitosis) independently of its catalytic partner CDK2. Whether the role of cyclin E in endoreplication involves activation of a kinase other than CDK2 remains to be determined.

is also inhibited by growth factors, as proteasome-mediated degradation of the D-cyclins depends on the phosphorylation of specific residues by the glycogen synthase kinase 3 beta, a process that is repressed by the PI3K/Akt signaling pathway.[11] Cyclin-D-bound CDK4 and CDK6, once fully activated, are then able to partially phosphorylate the retinoblastoma family members pRb, p107, and p130 (called the *pocket proteins*), the major substrates of these kinases. The pRb proteins function to repress transcription through several mechanisms. First, through the binding and inactivation of transcription factors, such as E2F family members (E2F1–E2F5), but also through the binding to histone deacetylases (HDACs) and chromatin remodeling complexes.[12–14] The importance of the pRb pathway in the control of G_1 phase has been underscored by the recent identification, after a long search, of the pRb ortholog in *S. cerevisiae* — the Whi5 protein.[15,16]

Phosphorylation of pRb by CDK4 and CDK6 is thought to partially alleviate pRb-mediated transcriptional repression, allowing the expression of some target genes. Among these genes are the E-type cyclins (E1 and E2), which in turn bind and activate CDK2 (Figure 3.1). CDK2/cyclin E complexes are able to further phosphorylate the pRb protein, canceling the pRb-mediated repression of many genes whose activities are necessary for S phase entry. Further activity of CDK2 bound to cyclin A is required to progress through S phase and to prepare cells for mitosis. Another member of the CDK family, CDK1 (Cdc2), is mostly responsible for the control of the G_2 and M phases, mainly through binding to cyclin A2 and B1 (Figure 3.1).

The activity of these cell cycle CDKs is tightly controlled at different levels, including interaction with their activating partners (the cyclins), binding to negative regulators (CDK inhibitors), phosphorylation of specific residues, and subcellular localization.[6,17] Two families of CDK inhibitors have been described: the INK4 family (p16[INK4a], p15[INK4b], p18[INK4c], and p19[INK4d]) and the CIP/KIP family (p21[CIP1], p27[KIP1], and p57[KIP2]). INK4 proteins specifically bind to and inhibit CDK4 and CDK6 by allosteric competition with D-cyclins for binding to the CDK subunit. CIP/KIP proteins, on the other hand, are able to bind all CDKs. Whereas their interaction with CDK2 complexes clearly blocks kinase activity, their role in CDK4/6 inhibition is unclear. Indeed, CDK4/6 and cyclin D heterodimers can bind CIP/KIP inhibitors at stoichiometric concentrations without losing their kinase activity.[6,18] It has been proposed that this interaction titrates these inhibitors away from CDK2/cyclin E complexes, facilitating the activation of the CDK2 kinase. The physiological relevance of this function, however, remains controversial.[6,19]

To understand how cyclins and CDKs control cell cycle progression, it is instructive to analyze their substrates. Immunoprecipitation and *in vitro* kinase studies in the early 1990s soon identified the pRb protein as a substrate of CDK1 and CDK2.[20,21] Later it was found that pRb, and the other members of the pocket family, p107 and p130, are substrates of most CDKs, including CDK4 and CDK6. In fact, pRb proteins are thought to be the major substrates of these D-type CDKs. In addition to these pocket proteins, only Smad3 has been very recently reported to be phosphorylated and inactivated by CDK4.[22] Although the physiological relevance of this interaction in cell cycle progression still remains unclear, it seems to be a method of antagonizing the antiproliferative activity of TGF-ß. Unlike for CDK4 and CDK6, many targets for CDK2 kinase activity have been identified by using different approaches. Apart from pRb and E2F5 (both involved in gene transcription), CDK2 as CDK4 is able to phosphorylate Smad3, another transcriptional repressor. Moreover, NPAT/p220 and CBP/p300 (both involved in histone function), Cdt1 (involved in DNA replication), centrosome proteins (CP110, Mps1, nucleophosmin/B23), Rad6/Ubc2 and Cdh1 (involved in ubiquitin-mediated proteolysis), and p27[KIP1], a CDK inhibitor, among others, have also been shown to be phosphorylated by CDK2.[23,24] Recently, proteins involved in double-strand break repair such as Ku70 and BRCA2 have been shown to be potential targets of CDK2 as well.[25,26] CDK1 activity appears to be more ubiquitous, at least in yeast, because around 11% of the yeast proteome might serve as a substrate for this kinase.[27]

Most data described earlier have been obtained from experiments in which exogenous proteins, either wild type or some mutated forms such as kinase-dead (dominant-negative) proteins, have been overexpressed in cells in culture, or their activities have been interrupted by antibodies, interfering RNAs, or antisense oligonucleotides. However, these experimental approaches have intrinsic limitations. Given the high level of structural and functional redundancy among cyclins and CDKs, it is practically impossible to interfere specifically with the expression or activity of one of these proteins without disturbing the normal function or regulation of other family members. In the last 10 years we have experienced a tremendous breakthrough in technology for the controlled manipulation of the mouse genome. Gene-targeting techniques in mouse embryonic stem cells make it possible to modify specifically any given gene in the mouse genome to create loss-of-function mouse models (knockouts), or introduce any subtle modification of a particular gene (knockin models). These genetic tools have been applied by us and others to study the function of mammalian cyclins and CDKs by generating knockout or knockin mouse strains in which the genes encoding each one of these proteins have been specifically ablated or modified. In many cases, the phenotypic characterization of these gene-targeted mice have challenged the currently accepted canonical model of cell cycle regulation, providing new insights into how cell proliferation and cell cycle are regulated in mammals.

3.3 LOSS-OF-FUNCTION MOUSE MODELS OF G$_1$ CYCLINS AND CDKS

3.3.1 MICE LACKING D-TYPE CYCLINS: HOW MANY ARE TOO FEW?

D-cyclins bind to and activate CDK4 and CDK6, acting as sensors that couple the mitogenic signals from the extracellular environment to the progression of the cell cycle. Throughout most embryonic development and in the adult, the three D-type cyclins, D1, D2, and D3, are expressed in a tissue-specific but highly overlapping pattern.[28] Therefore, it is not surprising that ablation of each one of the D-type cyclins in the mouse does not result in major developmental alterations. The three single-knockout strains are viable and show only defects in specific cell types (Table 3.1). Thus, cyclin-D1-null mice are viable, and although smaller than wild-type siblings, exhibit only focal developmental abnormalities in the retina and breast tissue as well as mild neurological disorders.[29,30] Similarly, cyclin-D2-deficient mice exhibit compromised fertility and postnatal cerebellar and B-cell development.[31] Cyclin-D3-null mice, on the other hand, display specific deficiencies in the maturation of T-cell lymphocytes.[32]

Functional redundancy among the different D-type cyclins *in vivo* has been further explored using mouse models. A knockin strain in which the *cyclin D1* gene is replaced by the gene encoding cyclin D2 has been generated by gene targeting.[33] In this strain, the defective proliferation of the breast epithelium of pregnant cyclin-D1-null females is fully rescued, whereas the other phenotypes are rescued only partially. This result suggests that subtle functional differences exist among the different D-type cyclins *in vivo* that are consistent with the complex control of cell proliferation in so many diverse cell types that coexist temporally and spatially in a mammalian organism.

Another level of complexity in cell cycle regulation that has been highlighted by the cyclin D knockout models is *embryonic plasticity or developmental compensation*, a term used to describe the ability to activate potentially compensatory pathways when an essential function is missing during embryonic development. This phenomenon has been unambiguously demonstrated by the generation of double-knockout mice lacking two of the three D-type cyclins (single-cyclin mice; Table 3.1).[34] In these strains, the tissue-specific pattern of expression of the remaining cyclin gene is

TABLE 3.1
Major Phenotypes of the Different Knockout and Knockin Models of Cell Cycle Cyclins, CDKs, and Their Inhibitors

Genotype	Life Span	Main Phenotype	Reference
cyclin D1[−/−]	Reduced viability	Reduced size. Hypoplastic retina. Pregnancy-insensitive mammary gland. Neurological disorders. Malformation of the jaw. Impaired proliferation of Schwann cells after injury.	**29,30,**92,93
cyclin D1 replaced by *cyclin E* (knockin)	Normal	Rescues most of the *cyclin D1*[−/−] phenotype except for some defects in the breast epithelium.	92,**94**
cyclin D1 replaced by *cyclin D2* (knockin)	Normal	Rescues the *cyclin D1*[−/−] phenotypes to a different extent. Only the phenotype in the mammary gland is fully rescued.	33
cyclin D2[−/−]	Normal	Female sterility, defective proliferation of ovarian granulosa and Sertoli cells in response to hormones. Small testis. Impaired proliferation of B-lymphocytes. Impaired proliferation of pancreatic beta-cells. Developmental abnormalities in the cerebellum.	**31,**41,42,95–98
cyclin D3[−/−]	Normal	Thymic atrophy with reduced expansion of immature T lymphoid cells.	32
cyclin D1[−/−]; *cyclin D2*[−/−]	Reduced viability (die within 3 weeks)	Retarded growth. Hypoplasia of the cerebellum.	34
cyclin D1[−/−]; *cyclin D3*[−/−]	Neonatal lethality A fraction survive up to 2 months	Reduced size. Hypoplastic retina. Neurological disorders.	34
cyclin D2[−/−]; *cyclin D3*[−/−]	Late embryonic lethality (E17,5–18,5)	Reduced embryo size. Decreased erythropoiesis and megaloblastic anemia.	34
cyclin D1[−/−]; *cyclin D2*[−/−]; *cyclin D3*[−/−]	Embryonic lethality (E15,5–E16,5)	Decreased size of the embryo. Severe megaloblastic anemia and multilineage hematopoietic failure. Developmental heart defects (in a fraction of the embryos).	35

TABLE 3.1 (CONTINUED)
Major Phenotypes of the Different Knockout and Knockin Models of Cell Cycle Cyclins, CDKs, and Their Inhibitors

Genotype	Life Span	Main Phenotype	Reference
$CDK4^{-/-}$	Decreased viability	Reduced body size. Pituitary atrophy. Female sterility and decreased fertility in males. Insulin-dependant diabetes due to a reduced number of beta cells.	**36,37,**38–40,114
$CDK6^{-/-}$	Normal	Decreased female fertility. Decrease in the splenic hematopoiesis with megaloblastic erythrocytes. Partial thymic atrophy and delayed T-cell response to stimulation.	45
$CDK4^{-/-}$ $CDK6^{-/-}$	Late embryonic lethality (E14.5–E18.5)	Severe megaloblastic anemia. Multilineage hematopoietic failure in the liver.	45
cyclin $E1^{-/-}$	Normal	No defect.	**47,48**
cyclin $E2^{-/-}$	Normal	Testicular atrophy, reduced male fertility with abnormal meiotic features.	**47,48**
cyclin $E1^{-/-}$; cyclin $E2^{-/-}$	Embryonic lethality (E10.5–E11.5)	Placental defect, failure of the trophoblast giant cells to undergo normal endoreplication.	**47,48**
cyclin $E1^{-/-}$; cyclin $E2^{-/-}$, after tetrapoid rescue	Perinatal lethality	Frequent cardiovascular abnormalities. Reduced endoreplication of megakaryocytes.	47
$CDK2^{-/-}$	Normal	Sterility with atrophy of the gonads. Defective spermatogenesis and oogenesis due to a block in the first meiotic division.	**52,53**
$CDK2^{-/-}$; $CDK6^{-/-}$	Normal	Sum of the single-knockout phenotypes.	45
cyclin $A1^{-/-}$	Normal	Testicular atrophy and male sterility due to a block in spermatogenesis.	54
cyclin $A2^{-/-}$	Embryonic lethality (E5.5)	Death shortly after implantation	99
cyclin $B1^{-/-}$	Embryonic lethality (earlier than E10.5)	Death shortly after implantation	100
cyclin $B2^{-/-}$	Normal	No abnormalities	100

(continued)

TABLE 3.1 (CONTINUED)
Major Phenotypes of the Different Knockout and Knockin Models of Cell Cycle Cyclins, CDKs, and Their Inhibitors

Genotype	Life Span	Main Phenotype	Reference
cyclin F$^{-/-}$	Embryonic lethality (E10.5)	Abnormal development of extraembryonic tissues. Delayed embryonic development.	101
p16$^{INK4a-/-}$	Viable. Some spontaneous tumors with ageing	Thymic hyperplasia. Tumors: soft tissue sarcoma, lymphoma, and melanoma. Increased sensitivity to carcinogen-induced cancers.	**69,70**
p15$^{INK4b-/-}$	Viable. Some spontaneous tumors with aging	Increased extramedullary hematopoiesis and lymphoproliferative disorders. Tumors: angiosarcoma and lymphoma.	71
p18$^{INK4c-/-}$	Viable. Some spontaneous tumors with aging	Increased body size, multiple cysts in the kidney and mammary glands. Leydig cell hyperplasia. Tumors: mainly pituitary adenoma and some other malignancies (adrenal medulla tumors, thyroid tumors). Haplo-insufficiency for carcinogen-induced tumor suppression.	**71,72**,102,104
p19$^{INK4d-/-}$	Normal	Testicular atrophy but conserved fertility.	103
p15$^{INK4b-/-}$; *p18$^{INK4c-/-}$*	Viable. Some spontaneous tumors with aging	Similar to the addition of single-knockout phenotypes. Multiple cysts in pancreas and testis.	71
p18$^{INK4c-/-}$; *p19$^{INK4d-/-}$*	Viable. Some spontaneous tumors with aging	Added phenotypes of the individual knockout strains.	102
p16$^{INK4a-/-}$; *p19$^{ARF-/-}$*	Viable. High level of spontaneous tumors	Tumors: lymphoma and sarcoma.	105

TABLE 3.1 (CONTINUED)
Major Phenotypes of the Different Knockout and Knockin Models of Cell Cycle Cyclins, CDKs, and Their Inhibitors

Genotype	Life Span	Main Phenotype	Reference
$CDK4^{R24C/R24C}$ (knockin)	Viable. Spontaneous tumors with aging	Beta islet cell hyperplasia in the pancreas. Hyperplasia of the Leydig cells and pituitary gland. Variety of tumors with long latency (angiosarcomas, other sarcomas, pancreatic endocrine tumors, testicular Leydig cell tumors, pituitary tumors, adenomas, and carcinomas). Increased sensitivity to carcinogen-induced tumors.	**36**,43,73,106
$p21^{CIP1-/-}$	Viable. Some spontaneous tumors with aging	Normal development. Tumors: histiocytic sarcoma, hemangioma, B-cell lymphoma, lung carcinoma.	81,**107,108**
$p27^{KIP1-/-}$	Viable. Some spontaneous tumors with aging	Increased body size and organomegaly. Female sterility. Retinal dysplasia. Pituitary hyperplasia and adenomas of the intermediate lobe. Intestinal adenocarcinomas. Haplo-insufficiency for tumor suppression.	77,78,**79,80**,109
$p57^{KIP2-/-}$ or $p57^{KIP2m-/+}$ $^{m-}$(imprinted)	Neonatal lethality	Several developmental defects in the gastrointestinal tract and cleft palate. Abnormal cell proliferation in placenta, cartilage, and lens.	**110,111**,112,113
$p21^{CIP1-/-}$; $p27^{KIP1-/-}$	Viable	Similar phenotype to that of $p27^{KIP1-/-}$. More pronounced hyperplasia of the ovaries (granulosa cells).	115
$p21^{CIP1-/-}$; $p57^{KIP2m-/+}$	Late embryonic lethality (E16.5–E18.5)	Abnormal skeletal musculature (failure to form myotubes). Abnormal development of the lung alveoli. Abnormal skeletal development.	113
$p27^{KIP1-/-}$; $p57^{KIP2m-/+}$	Embryonic lethality (E12–E16.5)	Abnormal placental and lens development due to increased cell proliferation.	112

(continued)

TABLE 3.1 (CONTINUED)
Major Phenotypes of the Different Knockout and Knockin Models of Cell Cycle Cyclins, CDKs, and Their Inhibitors

Genotype	Life Span	Main Phenotype	Reference
CDK4[R24C/R24C]; *p27*[KIP1]-/-	Reduced viability (up to 3 months)	Decreased size. Curved spinal cords, lordokyphosis. Undifferentiated pituitary tumors.	116
CDK4 [R24C/R24C]; *p18*[INK4c]-/-	Viable. Spontaneous tumors with aging	Added phenotypes of each single-mutant strain.	116
p18[INK4c-/-]; *p21*[CIP1-/-]	Viable	Accelerated development of pituitary tumors. Multifocal gastric neuroendocrine hyperplasia. Bronchioloalveolar adenomas.	117
p18[INK4c-/-]; *p27*[KIP1-/-]	Viable	Accelerated development of pituitary tumors. Other hyperplasias or tumors in the thyroid, parathyroid, adrenal gland, endocrine pancreas, testis, and duodenum.	**72**,117
p18[INK4c-/-]; *CDK4*[-/-]	Viable	Similar to the *CDK4*[-/-] phenotype.	118
p19[INK4d-/-]; *p27*[KIP1-/-]	Postnatal lethality (3 weeks)	Neurological disorders. Abnormal proliferation of neuronal populations in the central nervous system.	119
p27[KIP1-/-]; *CDK4*[-/-]	Viable	Increased body weight (but smaller than *p27*[KIP1-/-] alone).	118
p27[KIP1-/-]; *cyclin D1*[-/-]	Viable	Rescue of the *cyclin D1*[-/-] phenotypes but not of those of the *p27*[KIP1-/-] mice.	**120,121**

Note: When several references are cited for one strain, the original description of that particular strain appears in bold type.

altered, and it becomes almost ubiquitously expressed in most tissues in the embryo and in the adult. The molecular mechanisms underlying the ectopic expression of the single-cyclin gene varies in different tissues but relies mostly on transcriptional or posttranslational processes. As a consequence of this developmental compensation, cyclin D1-only mice die at late gestation, whereas cyclin-D2-only or cyclin-D3-only mice are viable and survive for several weeks, demonstrating that in most cells the D-type cyclins are functionally interchangeable. The inability to efficiently upregulate the remaining cyclin in a particular tissue or cell type appears to be the main cause of the tissue-specific phenotypes observed in the various single-cyclin mice. However, it is also possible that subtle functional differences, as described earlier may contribute to the incomplete rescue of the phenotype in certain tissues. Nevertheless, the major conclusion from the single-cyclin D models is that proliferation of most cells relies on the net D-type cyclin activity present at a given time. The question is, to what extent can cells still proliferate in the total absence of D-type cyclins? To address this issue, Sicinski and coworkers intercrossed the three individual D-cyclin knockout strains to generate embryos lacking the three D-type cyclins. As D-cyclins had long been considered essential mediators of extracellular mitogenic signaling, the observation that embryos could develop extensively without D-cyclin activity was a surprise.[35] In fact, embryos lacking D-type cyclins develop to mid gestation, and start to die at embryonic day E14.5 when most organs are well developed. Some of them progress to E17.5. Death is caused in the embryos by multilineage hematopoietic abnormalities. The proliferation of hematopoietic stem cells in the fetal liver is drastically diminished, resulting in severe anemia in the embryo. Fetal hematopoietic cells are therefore dependent on cyclin-D-mediated signaling for proliferation, whereas most other cell types in the embryo are not. Additionally, heart malformations are observed, but only in a fraction of these embryos.

3.3.2 MICE LACKING CDK4 AND CDK6: FETAL HEMATOPOIETIC FAILURE

In parallel to the generation of D-cyclin knockout mice, their catalytic partners CDK4 and CDK6 have been also ablated in mice by gene targeting. As in the case of the D-cyclins, neither CDK4 nor CDK6 is essential for cell proliferation, but the lack of either causes tissue-specific phenotypes (Figure 3.2 and Table 3.1). Mice lacking CDK4 are viable, although they show reduced numbers of certain endocrine cell types such as prolactin-producing cells in the anterior lobe of the adenohypophysis and Leydig cells in the testis.[36,37] Low numbers of lactotroph cells and prolactin deficiency lead to female infertility in the absence of CDK4.[38–40] Male fertility is also reduced, probably due to the decrease in the number of testosterone-producing Leydig cell and hormonal defects, although this has not yet been unequivocally proved in males. However, the most striking phenotype of the CDK4-null mice is a severe defect in another endocrine cell type, the insulin-producing beta cells of the pancreas. CDK4 is required for the postnatal proliferation of beta cells, although it is dispensable for their neogenesis from precursor cells during embryonic development.[40] As a consequence, CDK4-null mice show a dramatic reduction in the number of beta cells (Figure 3.2) that results in early onset of insulin-dependent diabetes and reduced lifespan. This strong dependency of beta cells on CDK4 for proliferation may be a consequence of the absence of CDK6 in this cell type.[40] A similar, although less severe, phenotype in beta cells is produced by the absence of cyclin D2,[41] indicating that cyclin D2/CDK4 is the main complex involved in the postnatal control of beta cell proliferation. However, the phenotype of beta cells is more severe in double-knockout mice lacking both cyclin D1 and D2,[42] showing

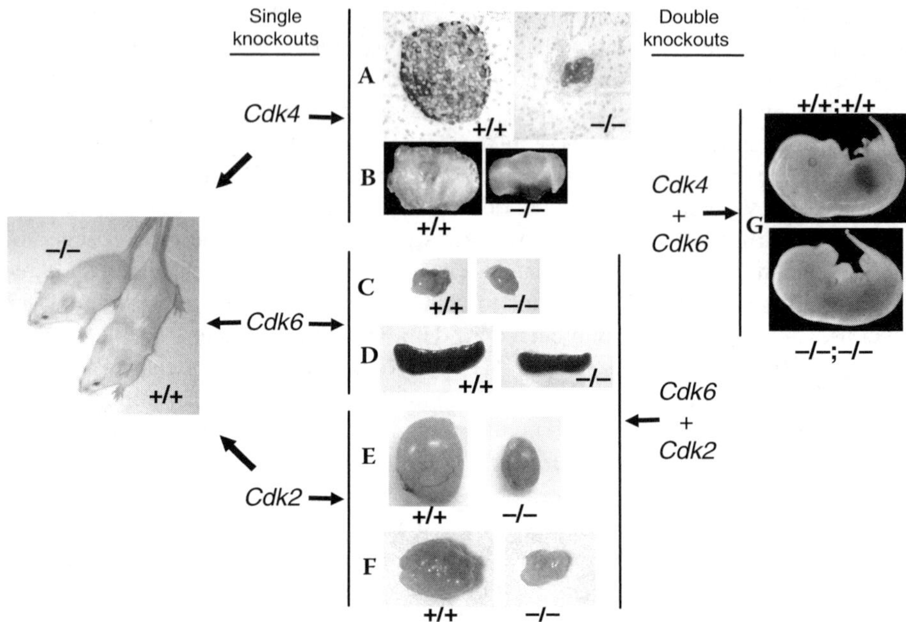

FIGURE 3.2 Tissue-specific phenotypes of G_1 CDK knockout mice. The appearance of tissues and organs where a phenotype is evident in each individual G_1 CDK knockout mouse strain is shown. Lack of CDK4 results in a dramatic reduction of the number of beta cells in the pancreatic islets, as it is shown by insulin immunostaining of pancreas sections (A), as well as in the size of the hypophysis, particularly in the anterior lobe (B). Lack of CDK6 results in thymus (C) and spleen (D) hypoplasia, whereas absence of CDK2 leads to a testis (E)- and ovary (F)-specific phenotype consisting in the absence of mature germ cells. The three individual knockout mice are viable but show a reduction in size with different penetrance, being more severe in the case of CDK4 knockout mice. Double-knockout mice lacking both CDK4 and CDK6 develop up to mid gestation, at which stage they die of hematopoietic failure and severe anemia (G). Double-knockoutsfor CDK2 and CDK6 are viable and show the sum of phenotypes of the single-knockout mice.

that cyclin D1, also expressed in beta cells albeit at lower levels than cyclin D2, may partially compensate for the lack of cyclin D2.

Why endocrine cells are more dependent than other cell types on CDK4 for proliferation remains to be established, but this observation is further supported by the phenotype of a knockin mouse strain expressing endogenous levels of a form of CDK4 that is insensitive to INK4 inhibitors (CDK4R24C).[36] These mice develop hyperplasias and multiple tumors that are mostly of endocrine cell origin, such as pituitary and Leydig cell tumors and insulinomas with almost complete penetrance.[43] Moreover, hyperproliferation of beta cells is detected from the early postnatal period with complete penetrance in these mice.[40] The phenotype of this strain is discussed later in this article.

Similar to cyclin D1 knockout mice, mice lacking CDK4 are smaller than wild-type mice (Figure 3.2). A similar phenotype is observed in *Drosophila*, where ablation of CDK4 (there is no CDK6 in flies) leads to normal but smaller individuals at birth.[44] This phenotype in flies and mice is not the consequence of cell size reduction but of reduced cell numbers. Moreover, in mice the smaller size is not caused by endocrine dysfunction, as reexpression of CDK4 in pituitary and beta cells rescues the diabetes and the female sterility but not the reduced size.[40] Therefore, it seems that mammalian homeostatic cell number and body size are regulated by CDK4/cyclin D1 complexes in a cell-autonomous fashion.

Mice lacking CDK6 also show cell-type-specific phenotypes. They exhibit hypoplastic thymus and spleen (Figure 3.2), which indicates a specific dependence of lymphoid cells on CDK6 for proliferation and/or development. In general, CDK6 knockout mice are overtly normal, fertile, and have normal lifespan.[45] A small reduction in body size is observed in females, but the cause of this phenotype remains to be established. As both CDKs are coexpressed in most normal tissues, the mild phenotypes observed in the single-knockout mice could in principle be explained by functional redundancy. However, the generation of mice lacking both CDK4 and CDK6 has shown that, rather unexpectedly, most cells are independent of CDK4 and CDK6 for proliferation, at least during embryonic development.[45] In fact, double-null mice for CDK4 and CDK6 exhibit a phenotype that mirrors that of mice lacking the three D-type cyclins. They die at mid/late gestation from defects in the hematopoietic tissues and severe anemia. However, most of the fetal organs exhibit normal morphological development at the time of death (Figure 3.2). As in mice lacking D-cyclins, the liver hematopoietic stem cells of the double-knockout mice show impaired proliferation both *in vivo* and *in vitro*. The fact that the absence of D-cylins or of CDK4 and CDK6 leads to virtually identical phenotypes provides genetic evidence that the only function of the D-cyclins during embryonic development is to activate CDK4 and CDK6. The strong dependence of the fetal hematopoietic cells on CDK4 or CDK6 activation for proliferation suggests that highly proliferative tissues with a rapid cellular turnover, such as the hematopoietic system, require cyclin-D-mediated kinase activation for proliferation, probably to reach a threshold of kinase activity higher than the one needed by other, less active proliferating cells. Yet, it remains to be established whether other cell types in the embryo have compensatory mechanisms that either do not exist or cannot be activated in hematopoietic cells, to overcome the lack of cyclin-D-mediated signaling. The generation of double-knockout mice lacking CDK4 and CDK2 as well as the triple knockouts lacking the three G_1 CDKs has not been reported yet, although work is in progress (M. Barbacid, personal communication).

3.3.3 MICE LACKING E-CYCLINS: ENDOREPLICATION DEFECTS

Cyclin E1 and E2 are coexpressed in virtually all proliferating cells in embryos and adult.[46] The only known function of these cyclins up to now is to bind and activate their catalytic partner, CDK2. No other CDK has so far been described to be activated by binding cyclin E, at least under physiological conditions. Ablation of each individual E-type cyclin in the mouse has no phenotypic consequences except for an approximately 50% reduction in fertility in males lacking cyclin E2.[47,48] A similar, although more severe, phenotype is observed in mice carrying a single cyclin E1 allele and null for cyclin E2, suggesting that either cyclin E1 is haplosufficient for normal cell proliferation or most cells can proliferate in the absence of E-type cyclins. Ablation of both cyclin E1 and E2 in the mouse showed that, unexpectedly, this seems to be the case. Embryos lacking both E-type cyclins die *in utero* at mid gestation (E11.5). At this stage mutant embryos look normal, albeit smaller than wild-type littermates, display normal organogenesis, and show normal cell proliferation rates.[47,48] However the mutant placentas showed severe defects, caused by the absence of trophoblast giant cells, that resulted in reduced vascularization in the yolk sac and in the embryos. This phenotype can be rescued by a tetraploid embryo complementation assay.[47] In this assay embryonic stem cells obtained from cyclin E1 and E2 double-knockout mice are microinjected into wild-type blastocysts in which cells have been made tetraploid by electrofusion of the two-cells at the two cell stage of the embryo.[49] After tetraploid rescue, embryos develop to term, demonstrating in a very elegant way that in fact E-type cyclins are dispensable for development of the embryo proper, but are essential for placental function.[47] The rescued double-knockout mice die early after birth owing to multiple cardiovascular defects. However, this cannot be attributed unequivocally to the lack of E-cyclins in the newborns, as mice born from tetraploid rescue very rarely survive to adulthood. Whether the E-type cyclins indeed play a role in cardiovascular development during embryogenesis remains to be established. Nevertheless, the most important conclusion of these mouse models is that most cells can proliferate and develop properly in the absence of E-type cyclins.

The phenotype observed in the placentas of double-null mice deserves further attention. The lack of trophoblast giant cells in the placenta indicates a failure in the capacity of the cells to go through endoreplication cycles that lead to an increase in DNA content of up to 1000N. This increase in DNA content is not observed in the cells of the mutant placentas, suggesting that cyclin E is essential for endoreplication.[47,48] Moreover, in the tetraploid rescued embryos, megakaryocytes, another cell type known to undergo endoreplication cyles, do not exhibit this increase in DNA content and are reduced in numbers, which further supports a function of E-type cyclins in endoreplication in mammals (Figure 3.1). Interestingly, this function of cyclin E has been previously described in *Drosophila*, where it has been implicated in the control of the endoreplication cycles of the salivary gland cells.[50]

3.3.4 Mice Lacking CDK2: CDK2 Essential No More

Early experiments had shown that ablation of CDK2 in normal and tumor cell lines by different approaches interfered with cell proliferation. These experiments, together with the identification of CDK2 kinase activity targets that are involved in critical processes at the G_1/S transition of the cell cycle, DNA synthesis, and activation of histone gene transcription, have led to the conclusion that CDK2 is essential for cell cycle progression from G_1 to S phase. More recently, unlike what had been previously described, it was shown that CDK2 was not required for the proliferation of colon cancer cell lines.[51] However, these experiments could not be taken as evidence of the dispensability of CDK2 for cell cycle progression, as other pathways may be deregulated in tumor cells that may compensate for the lack of CDK2. The issue was resolved when the viability of mice in which CDK2 had been specifically inactivated by gene targeting was reported.[52,53] Rather unexpectedly, mice lacking

CDK2 are fully viable and overtly normal, with the exception of a dramatic failure in the meiotic cycle that results in the absence of mature germ cells in both males and females, gonad hypoplasia (Figure 3.2), and therefore complete sterility in both genders.[52] Although ablation of cell cycle regulatory genes often leads to alterations in germ cell development and partial or complete sterility, this is the first case in which a cell cycle gene is shown to be essential for meiosis in both sexes. It is important to mention that cyclin A1 knockout males are also 100% sterile,[54] although females are fully fertile. Because cyclin A1 is a catalytic partner of CDK2 and specifically expressed in the male germ line, it could be assumed that the phenotype of mice lacking either cyclin A1 or CDK2 during spermatogenesis would be the same. However, this is not the case. In the absence of cyclin A1, spermatocytes progress to the transition from prophase to metaphase I, where they stop owing to the inability to activate CDK1–cyclin B complexes in the absence of cyclin A1,[55] whereas spermatocytes lacking CDK2 stop earlier, at the pachytene stage of prophase I where they show aberrant synaptonemal complex formation and unpaired fibers. Therefore, CDK2 must have another cyclin-A1-independent function earlier in prophase I of the meiotic cycle in spermatocytes.

The lack of phenotypes other than the meiotic failure in CDK2-null mice can be interpreted in two ways: either CDK2 is dispensable for somatic cell proliferation, or compensatory mechanisms and embryonic plasticity account for the lack of phenotype in the majority of the cells lacking CDK2. However, although the latter explanation cannot be completely ruled out, overexpression of other CDKs that could assume the function of CDK2, such as CDK4, CDK6, or CDK1, has not been detected in CDK2 knockout mice. Up to now the regulatory subunits, the E- or A- type cyclins, have not been found to be upregulated. Moreover, Ortega et al. designed a conditional knockout allele of CDK2 (CDK2^{fl}) with the purpose of inactivating CDK2 selectively in adult tissues and in a cell-type-specific manner.[52] Cre-mediated inactivation of the CDK2^{fl} allele in primary cultures of mouse embryonic fibroblasts (MEFs) from CDK2$^{fl/fl}$ mice had no effect on cell proliferation. Because these cells are established from embryos with normal levels of active CDK2 and CDK2 is ablated once the cells are in culture, the lack of phenotype cannot be attributed in this case to compensatory mechanisms activated during embryonic development. Rather, they provide evidence that CDK2 is in fact dispensable for normal cell proliferation.

No endoreplication defects have been observed in CDK2 knockout mice, in contrast with what has been described in mice lacking E-type cyclins. This is a very important and intriguing finding as no other functions have so far been described for E-cyclins other than CDK2 activation. The contrast between cyclin E and CDK2 knockout mouse phenotypes unambiguously indicates that cyclin E has CDK2-independent functions in endoreplication. Whether cyclin E acts in this process by activating a kinase other than CDK2 or has CDK-independent functions remains to be established. A recent report shows that cyclin E can form complexes with CDK1 *in vivo*.[56] The physiological relevance of these complexes and whether or not they can phosphorylate the same targets as CDK2–cyclin E complexes is not known, but this finding suggests that CDK1 may act as a catalytic partner of cyclin E in the absence of CDK2.

On the other hand, it is tempting to speculate that other G_1 CDK could also play CDK2 functions. For instance, CDK4 can phosphorylate pRb in residues that are considered "CDK2 specific" in cells lacking CDK2 activity.[51] The opposite is also true because pRb is phosphorylated, although to a lesser extent in the absence of cyclin-D-dependent signaling.[35,45] Moreover, in the absence of CDK4 and CDK6, proliferation becomes dependent on CDK2, as expression of CDK2-interfering RNA in mouse embryonic fibroblasts (MEFs) lacking CDK4 and CDK6 reduces the proliferation capacity of these cells.[45] Interestingly, in these cells CDK2 has been found forming complexes with cyclin D,[45] indicating that, as in the case of CDK1–cyclin E complexes, promiscuity in terms of partner choice may actually be a way by which cell cycle cyclins and CDKs may compensate for the lack of each other, adding one more level of complexity to the established model of cell cycle regulation. Finally, CDKs other than the "canonical" cell cycle CDKs may come into play when these are absent. For instance, CDK9 is able to bind to and phosphorylate pRb,[57] and CDK3 has been shown recently to play a role in promoting G_0 exit.[8] The possibility that other kinases that show structural homology with the CDKs could be responsible for such compensation needs to be further explored.[5,58]

3.4 CDKs AND TUMOR DEVELOPMENT

3.4.1 ALTERATION OF G_1/S REGULATION IN HUMAN CANCER

Most of the proteins involved in G_1-CDKs regulation and some of their substrates have been linked with human tumor development, underscoring the importance of these cell cycle regulators in maintaining appropriate proliferation rates.[17] Thus, D- and E-type cyclins are frequently overexpressed in human tumors, whereas CDK inhibitors (such as p16[INK4a], p15[INK4b], p21[CIP1], or p27[KIP1]) or CDK substrates (such as pRb) are frequently inactivated by different mechanisms.[59,60] Genetic alterations of CDKs themselves are far more rare in tumors, although the CDK4 gene is amplified and overexpressed in a wide variety of tumors and tumor cell lines. A point mutation in CDK4 has also been described in spontaneous and familiar melanomas.[61,62] This mutation, substitution of Arg24 by Cys (R24C), leads to misregulation of the kinase activity by preventing its binding to the INK4 family of cell cycle inhibitors without affecting its affinity for cyclin D1. Similarly, CDK6 is frequently overexpressed in hematopoietic malignancies, in some cases as a consequence of a translocation that places the CDK6 gene under the control of strong promoters in these cells.[63,64] Although CDK2 is not frequently altered in human cancer, two of their regulators, cyclin E and p27[KIP1], have specific prognostic value in many different tumor types.[24,65]

3.4.2 LESSONS FROM CDKs GAIN-OF-FUNCTION MOUSE MODELS

In addition to the molecular analysis of human tumors, animal models have been used to demonstrate the causal involvement of CDK dysregulation in tumor development (Table 3.1 and Table 3.2). Classical transgenic mouse models have been generated to overexpress cyclins and CDKs in specific tissues and study the consequences of this overexpression in tumor development. Table 3.2 summarizes the more relevant phenotypes of some of these transgenic models of cyclins and CDKs.

TABLE 3.2
Phenotype of Transgenic Mouse Models of Cyclins and CDKs

Transgene	Promoter/Tissue	Phenotype	Reference
K5-cyclin D1	Keratin 5/basal layer of the skin	Basal cell hyperplasia, increased epidermal proliferation. Severe thymic hyperplasia	122,123
K5-cyclin D2	Keratin 5/basal layer of the skin	Basal cell hyperplasia, increased epidermal proliferation. Mild thymic hyperplasia	88
K5-cyclin D3	Keratin 5/basal layer of the skin	Basal cell hyperplasia, increased epidermal proliferation	88
K5-CDK4	Keratin 5/basal layer of the skin	Epidermal hyperplasia and dermal fibrosis. Increased sensitivity to carcinogen-induced skin tumors	68,124
K5-cyclin D1 and *K5-CDK4* double transgenic	Keratin 5/basal layer of the skin	Similar phenotype to that of the *keratin 5-CDK4* transgenic line	124
MMTV-cyclin D1	MMTV LTR/breast	Constitutive mammary hyperplasia. Multifocal carcinoma and metastasis with age albeit with low incidence	66
Eμ-cyclin D1	Eμ/lymphocytes	Impaired lymphocyte maturation. No lymphocyte hyperproliferation	67
Eμ-cyclin D1 and *Eμ-myc* double transgenic	Eμ/lymphocytes	Accelerated development of lymphoma induced by overexpression of *c-myc*	67
MHC-cyclin D1	Alpha-cardiac myosin heavy chain/ differentiated cardyomyocytes	Polyploidy and multinucleation, but not proliferation of cardyomyocites. The expression of myocardial markers is not affected	125
AlphaA-crystallin-cyclin D1	AlphaA-crystallin/eye lens fiber cells	Does not promote S phase entrance	126
AlphaA-crystallin-cyclin D1 and *alphaA-crystallin-CDK4*	AlphaA-crystallin/eye lens fiber cells	Promotes entrance in S phase	126
MMTV-cyclin D2	MMTV LTR/breast	Mammary hyperplasia and carcinomas with age. Inhibition of normal alveologenesis and nursing in pregnant females	127

(continued)

TABLE 3.2 (CONTINUED)
Phenotype of Transgenic Mouse Models of Cyclins and CDKs

Transgene	Promoter/Tissue	Phenotype	Reference
AlphaA-crystallin -cyclin D2	AlphaA-crystallin/eye lens fiber cells	Does not promote S phase entrance	126
AlphaA-crystallin -cyclin D2 and *alphaA-crystallin-CDK4*	AlphaA-crystallin/eye lens fiber cells	Promotes entrance in S phase	126
AlphaA-crystallin -cyclin D3	AlphaA-crystallin/eye lens fiber cells	Promotes entrance in S phase	126
Beta-lactoglobulin- -cyclin E	Beta-lactoglobulin/mammary epithelium after pregnancy	Epithelial cell hyperplasia after the first lactation. Mammary gland carcinomas with low penetrance	74
CD2-cyclin E	CD2/Tcells	Does not develop lymphoid neoplasia. More susceptible to tumors induced by carcinogenic treatment	75
CD2-cyclin E; $p27^{KIP1-/-}$	CD2/Tcells	Monoclonal T-cell lymphoma	76
Beta-lactoglobulin-cyclin A1	Beta-lactoglobulin/mammary epithelium after pregnancy	Multinucleation and karyomegaly	128
Beta-lactoglobulin-cyclin A1 and *Beta-lactoglobulin-CDK2* double transgenic	Beta-lactoglobulin/mammary epithelium after pregnancy	More severe phenotype than single *cyclin A1* transgenic	128

For instance, transgenic mice overexpressing cyclin D1 in breast tissue, under the control of the MMTV promoter, display mammary hyperplasia and succumb to breast cancer,[66] demonstrating that overexpression of cyclin D1 in breast predisposes to tumor development, although other genetic alterations are also required. However, not all tissues are equally sensitive to the overexpression of cyclin D1. Transgenic mice that express cyclin D1 in lymphocytes under the control of the Eμ promoter do not develop lymphocyte hyperplasia. However, overexpression of cyclin D1 in Eμ-c-myc transgenic mice accelerates lymphoma development.[67]

Similarly, overexpression of CDK4 in the skin of transgenic mice under the control of the keratin 5 promoter does not lead to tumor development, but makes mice more sensitive to tumors induced by classical skin-carcinogenic protocols that result in oncogenic *H-ras* activation,[68] demonstrating that overexpression of CDK4 cooperates with oncogenic ras to induce cell transformation in the skin.

Inactivating mutations in single INK4 inhibitors in the mouse, including p16[INK4a],[69,70] p15[INK4b],[71] and p18[INK4c],[71,72] also result in different levels of tumor susceptibility (Table 3.1). Indeed, knockin mice expressing the CDK4R24C mutant develop a wide spectrum of spontaneous tumors, confirming the central role of CDK4 in the entry into the cell cycle.[43,73]

Two animal models of cyclin-E-associated cancer have been reported. Mice carrying a lactoglobulin-cyclin E transgene develop mammary gland hyperplasia and carcinomas.[74] Similarly, mice engineered to express cyclin E under the T-cell-specific CD2 promoter are susceptible to developing clonal lymphomas after specific carcinogenic or genetic stress.[75,76] Finally, p21[CIP1]- and p27[KIP1]-deficient mice display diverse susceptibility to tumor development. Targeted disruption of the murine p27[KIP1] gene in mice caused a gene dose-dependent increase in animal size and sterility in females without other gross morphological abnormalities.[77–80] In addition, p27[KIP1] deletion caused neoplastic growth of the pituitary pars intermedia. Similarly, p21[CIP1]-deficient mice are more susceptible to spontaneous tumor development.[81]

3.4.3 Lessons from Mice and Cells Lacking Cyclins and CDKs

The dispensability of the G_1 cyclins and CDKs for normal proliferation of most cells during embryogenesis *in vivo* is also confirmed by cell culture assays in MEFs isolated from the knockout mouse strains (Table 3.3). Thus, individual loss of any of the D- or E-type cyclins or of CDK4, CDK6, or CDK2 results in either no phenotype or minor delays in cell proliferation in tissue-cultured MEFs.[29,30,35,45,47,48,52] Furthermore, D-type cyclin-null or CDK4/6-null cells not only do proliferate but are also able to reenter S phase from quiescence, albeit with decreased kinetics, demonstrating that CDK4/6 signaling is not required to restart the cell cycle in response to mitogenic stimulation.[35,45] Similarly, E-type cyclin-deficient or CDK2-deficient MEFs proliferate normally *in vitro*.[47,48,52] Intriguingly, double cyclin E1-E2-deficient MEFs, unlike CDK2-null MEFs, are not able to respond to serum stimulation and reenter S phase from G_0. This is not due to defective pRb phosphorylation, as pRb is phosphorylated normally in these cells, but to impaired loading of MCM complexes to the chromatin during the G_0/G_1 transition.[47] Loading of the MCM complex is required for the proper assembly of the prereplication complexes at the origins of replication. Therefore, these results demonstrate that E-cyclins play an essential role in promoting DNA synthesis that is independent of CDK2. Whether cyclin E acts in complex with another CDK or has another noncatalytic function in this process remains to be determined.

Wild-type MEFs can be transformed by the combination of *ras* and other oncogenes such as *myc, E1A,* or dominant negative (DN) *p53.* A widely used and relatively simple assay to study the implication of cell cycle regulatory genes in cellular transformation is to test the sensitivity of MEFs obtained from these cell cycle gene knockout mouse strains to transformation mediated by any of these oncogene combinations. Thus, MEFs lacking all three D-cyclins are resistant to transformation by any oncogene combination whereas MEFs lacking only cyclin D1 are not,[35] probably because of functional redundancy with cyclin D2 and D3. Similarly, MEFs lacking CDK4, but expressing CDK6, become resistant to transformation by *ras* and DN *p53.*[82] This result indicates that cyclin-D-dependent signaling through

TABLE 3.3
Properties of the Mouse Embryonic Fibroblasts (MEFs) Derived from Knockout and Knockin Mouse Strains of Cyclins, CDKs, and CDK Inhibitors

Genotype	Properties of MEFs	Reference
cyclin D1$^{-/-}$	Similar to wild-type MEFs.	29
cyclin D1$^{-/-}$	Reduced fraction of cells in S phase.	35
cyclin D2$^{-/-}$	Slight decrease in the overall proliferating rate.	
cyclin D3$^{-/-}$	Delay in S phase reentry from G_0 when stimulated with low serum concentration.	
	Resistance to oncogenic transformation by H-*ras* + c-*myc*, H-*ras*+DN*p53*, or H-*ras*+*E1A*.	
CDK4$^{-/-}$	Normal proliferation rate.	36,37
	Delay in S phase reentry from G_0.	
CDK4$^{-/-}$; *p27*$^{KIP1-/-}$	Rescued the delay in S phase reentry of the CDK4$^{-/-}$ phenotype.	37,118
CDK4$^{-/-}$; *p18*$^{INK4c-/-}$	Accelerated S phase reentry from G_0 similar to the *p18*$^{INK4c-/-}$ phenotype.	118
CDK6$^{-/-}$	Similar to wild-type control.	45
CDK4$^{-/-}$; *CDK6*$^{-/-}$	Normal proliferation at early passages. Delayed proliferation at later passages and premature replicative senescence.	45
	Delayed proliferation in low serum conditions.	
	Delay in S phase reentry from G_0.	
cyclin E1$^{-/-}$	Similar to wild-type control.	47,48
cyclin E2$^{-/-}$	Similar to wild-type control.	47,48
cyclin E1$^{-/-}$; *cyclin E2*$^{-/-}$ (tetrapoid rescued)	Slight decrease in the overall proliferation rate.	47
	Premature replicative senescence.	
	Impaired S phase reentry from G_0.	
	Resistance to oncogenic transformation by c-*myc*, H-*ras* + c-*myc*, H-*ras*+DN*p53*, H-*ras*+*E1A*.	
CDK2$^{-/-}$	Normal proliferation at early passages but premature replicative senescence.	52,53
	Slight delay in S phase reentry from G_0.	
	Mild resistance to oncogenic transformation by H-*ras* + *E1A* and to genotoxic insults.	
p16$^{INK4a-/-}$	Normal proliferation and senescence.	69,70
	Increased number of immortalized clones.	
p15$^{INK4b-/-}$	Higher proliferation rate and plating efficiency.	71
	More efficient S phase reentry from G_0.	
	Increased susceptibility to transformation by H-*ras* or H-*ras* + c-*myc*.	
p18$^{INK4c-/-}$	Normal proliferation, spontaneous senescence.	71,118
	More efficient S phase entry after serum deprivation.	

TABLE 3.3 (CONTINUED)
**Properties of the Mouse Embryonic Fibroblasts (MEFs) Derived
from Knockout and Knockin Mouse Strains of Cyclins, CDKs,
and CDK Inhibitors**

Genotype	Properties of MEFs	Reference
$p15^{INK4b-/-}$; $p18^{INK4c-/-}$	Similar to $p15^{INK4b-/-}$ MEFs.	71
$CDK4^{R24C/R24C}$	Higher proliferation rate and plating efficiency. Higher cell density at confluency. More efficient S phase reentry from G_0. Do not enter replicative senescence. Increase susceptibility to transformation by H-*ras*, H-*ras* + c-*myc*, or H-*ras* + *E1A*.	43,106
$p21^{CIP1-/-}$	Higher cell density at confluency. Increased plating efficiency when plated at low density. More efficient S phase reentry from G_0. Impaired arrest after *gamma*-irradiation or etoposide treatment.	89,107,108
$p27^{KIP1-/-}$	Proliferate slightly faster than control and reach higher cell density at confluency. Increased plating efficiency when plated at low density. More efficient S phase reentry from G_0.	80,89
$p27^{KIP1-/-}$; $p21^{CIP1-/-}$	Normal S phase reentry from G_0. Impaired assembly of CDK4–cyclin D complexes. Impaired nuclear localization of cyclin D1.	129
$p21^{CIP1-/-}$; $CDK2^{-/-}$	Similar to $p21^{CIP1-/-}$.	89
$p27^{IPp1-/-}$; $CDK2^{-/-}$	Similar to $p27^{KIP1-/-}$.	89

CDK4, but not CDK6, is important for cells to respond efficiently to strong constitutive mitogenic stimulation at least in MEFs.

MEFs lacking E-cyclins are also resistant to transformation induced by combinations of ras and another oncogene.[47] However, this result is in sharp contrast to the mild resistance of CDK2 knockout MEFs to the same transforming protocols,[52] indicating that cyclin E plays additional roles mediating oncogenic transformation other than activation of CDK2. In fact, previous results already pointed in this direction, as a mutant form of cyclin E unable to form active complexes with CDK2 is still able to transform rat embryo fibroblast in the presence of activated *H-ras*.[83]

In vivo, ablation of each of the D-type cyclins makes mice resistant to tumors driven by certain oncogenes but not by others. For instance, mice lacking cyclin D1 are resistant to breast tumors driven by *neu* or *ras* activation but not to breast tumors driven by *c-myc* or *Wnt-1*,[84] indicating that in the mammary epithelium cyclin D1 is the downstream target of the *neu-ras* pathway for promotion of cell cycle entrance and cell proliferation. Cyclin D1 knockout mice are also resistant to intestinal tumors driven by *apc* loss or beta catenin activation[85] or to skin papillomas induced by *H-ras* activation.[84,86] Similarly, cyclin D2 ablation confers resistance to gonadal tumors,[87] and mice lacking cyclin D3 are more resistant to leukemias driven by *notch* and thymomas induced by p56LCK.[32] In general, the tissue specificity of the phenotypes of each of the individual cyclin D knockout strains is mirrored by the tissue specificity of resistance to tumors conferred by the loss of each of them. CDK4 ablation also confers resistance to tumors, for instance, those induced by *ras* or *c-myc* activation in the skin.[88]

The tumor susceptibility of CDK2-null mice has been assayed by crossing these animals with mice deficient in the CIP/KIP inhibitors. CDK2 deficiency is not able to rescue the tumor phenotype induced by p27^{KIP1} deficiency.[56,89] Because it was thought that these inhibitors act by inhibiting CDK2 activity, these data suggest that CIP/KIP inhibitors might have CDK2-independent ways to efficiently inhibit the cell cycle and suppress tumor growth. CDK4/6 complexes seem not to be the target of CIP/KIP inhibitors, because CDK4 activity is not dramatically inhibited after p21^{CIP1}/p27^{KIP1} overexpression and these inhibitors are also active in CDK4/6 double-knockout cells.[45] CDK1, on the other hand, is efficiently inhibited by p21^{CIP1} and p27^{KIP1} in both wild-type and CDK2-null cells, and its inhibition might be sufficient to arrest the cell cycle. This is also supported by the fact that CDK1, at least under specific circumstances, may form active complexes with E-type cyclins.[56] However, the unavailability of CDK1-null cells and the fact that CIP/KIP inhibitors retain the ability to arrest cells in G$_1$[89] raise some questions regarding the mechanism that these inhibitors use for cell cycle inhibition (Figure 3.1).

As most of the knockout strains generated are constitutive, where the gene is ablated from developmental day 1, embryonic lethality in some strains such as double cyclin E1 and E2 or CDK4 and CDK6 or triple cyclin D1, D2, and D3 knockout strains prevents the study of the effect of these genotypes in tumor development. These important studies require the future generation of conditional knockout mouse strains for these genes.

3.5   IMPLICATIONS FOR CANCER THERAPY

The phenotypes resulting from genetic manipulation of CDKs and their regulators in the mouse have important implications for cancer therapy. As described earlier, in gain-of-function mouse models, activation of the CDK4/6 pathway dramatically decreases the requirements for cells to enter the cell cycle and form tumors in response to oncogenic stimuli. Similarly, activation of CDK2, by overexpression of cyclin E or by inactivation of p27^{KIP1}/p21^{CIP1}, might force the cells to enter into S phase and commit to progress through the mitotic cell cycle. These data, together with the finding of CDK activation in human tumors, have stimulated the design and development of small molecules that inhibit CDKs as new drugs for cancer therapy, and in the last few years, a plethora of CDK inhibitors have been analyzed *in vitro*, in mouse models or even in clinical trials.[17,90] However, the fact that the absence of each of the G$_1$ CDKs (CDK4, CDK6, or CDK2) in loss-of-function mouse models has no major effects on cell cycle progression *in vivo* (see earlier text) has underscored the need for caution with respect to the efficacy of specific CDK inhibition as an anticancer therapy. For instance, specific inhibitors might have little effect on G_1 progression given the fact that all the G_1 CDKs can phosphorylate the retinoblastoma protein and individual defects might be compensated by other CDKs. Although it has been long accepted that CDK4/6 and CDK2 phosphorylate different specific residues in pRb, the absence of one of these kinases could alter the affinities of other CDKs for certain sites in the protein. In fact, CDK4 is quite efficient at phosphorylating CDK2-specific sites in CDK2-depleted cells.[51] Moreover, not only the canonical G_1 CDKs, but also CDK1,[91] CDK3,[8] and CDK9[57] are also able to phosphorylate pRb.

Alternatively, the control of pRb function might be modulated by the overall phosphorylation state rather than by phosphorylation of specific sites, and a single CDK could achieve the appropriate levels of phosphorylation in most cell types. However, it is more difficult to explain these types of "compensatory roles" among CDKs for those substrates that seem to be specific for each of the G_1 kinases, particularly CDK2. Although the finding that cells can proliferate without these kinases, whatever the compensatory molecular mechanisms, could in principle make G_1 CDK inhibitors less attractive for cancer therapy, several considerations have to be taken into account. First, the molecular effects resulting from the enzymatic inhibition of one protein might differ from those caused by its absence. Thus, lack of CDK4 and CDK6 favor the presence of complexes between CDK2 and D-type cyclins, and these complexes could function to promote G_1 progression.[45] Small ATP analogs should elicit different results because they would allow the formation of inactive CDK4/6 and cyclin D complexes without inducing redistribution of the cyclins to other kinase-active complexes. Moreover, even when normal cells do not require these proteins for proliferation, tumor cells may need a higher threshold of kinase activation to keep up with constitutive mitogenic stimulation. In fact, in tissue culture assays, some tumor cell types are insensitive to the lack of CDK2 but require CDK4,[51] and lack of D-cyclins impairs oncogenic transformation of MEFs.[34]

The results from *in vivo* loss-of-function mouse models also suggest that future design of small molecule CDK inhibitors could benefit if specificity toward a particular CDK is reduced. In general, one could assume that drugs inhibiting all cell cycle CDKs could be more effective than specific ones. In fact, most of the CDK inhibitors that have shown some activity in clinical trials have reduced specificity.[90] Obviously, concomitant inhibition of multiple CDKs might produce undesired results that are difficult to predict until a detailed characterization of the function of these proteins *in vivo* becomes available. The combination of biochemical approaches, analysis of human tumors, and genetic analysis in mouse models will continue to improve our knowledge about CDK function and hopefully benefit cancer patients in the near future.

3.6 SUMMARY AND PERSPECTIVES

Since the early molecular characterization of yeast cell division mutants, our understanding of the cell cycle has been overloaded by the existence of multiple and complex levels of regulation. In mammals, this complexity is increased by the presence of multiple members of the different protein families involved in cell cycle regulation. Although some of these members have similar functions *in vitro*, it has become clear from genetic analysis in the mouse that we need to characterize their possible compensatory or cell-type-specific roles to understand how the cell cycle is regulated and how we can use that information to design appropriate therapeutic strategies for cancer therapy. The use of mouse models has allowed us to demonstrate that deregulation of cell cycle proteins does indeed promote tumor development and that inhibition of some of these proteins might be beneficial for therapy of specific tumor types. Yet, these studies have challenged some important aspects of the current model of cell cycle regulation. The design of more accurate mouse models in which, for instance, one or more cell cycle regulators could be specifically ablated or induced in tumor cells at a given stage of tumor development, along with further biochemical and tissue culture analysis, will improve our understanding of cell cycle regulation in mammalian cells and how to modulate it for cancer therapy.

ACKNOWLEDGMENTS

The authors thank Mariano Barbacid for his continuous support. This work has been funded by grants from C.A.M (GR/SAL/0223/2004), MCYT (BMC2003-06098), Foundation Ramón Areces, Foundation La Caixa, and AECC (to M.M.); A.R.C., INSERM (CreS/4CR03G), and the Ligue contre la Cancer (to P.D.); and C.A.M. (GR/SAL/0206/2004), MCYT (SAF2003-06280), and BFU2005-05668-C03-02/BMC (to S.O.).

REFERENCES

1. Norbury, C. and Nurse, P., Animal cell cycles and their control, *Annu. Rev. Biochem.*, 61, 441, 1992.
2. Morgan, D.O., Cyclin-dependent kinases: engines, clocks and microprocessors, *Annu. Rev. Cell. Dev. Biol.*, 13, 261, 1997.
3. Manning, G. et al., Evolution of protein kinase signalling from yeast to man, *Trends. Biochem. Sci.*, 27, 514, 2002.
4. Caenepeel S. et al., The mouse kinome: discovery and comparative genomics of all mouse protein kinases, *Proc Natl Acad Sci USA*, 101, 11707, 2004.
5. Malumbres, M., Revisiting the "CDK-centric" view of the mammalian cell cycle, *Cell Cycle* 4, e19, 2005.
6. Sherr, C.J. and Roberts, J.M., CDK inhibitors: positive and negative regulators of G_1-phase progression, *Genes Dev.*, 13, 1501, 1999.
7. Sherr, C.J., Cancer cell cycles revisited, *Cancer. Res.*, 60, 3689, 2000.
8. Ren, S. and Rollins B.J., Cyclin C/CDK3 promotes Rb-dependent G_0 exit, *Cell*, 117, 239, 2004.
9. Ye, X., Zhu, C., and Harper, J.W., A premature-termination mutation in the *Mus musculus* cyclin-dependent kinase 3 gene, *Proc. Natl. Acad. Sci. USA*, 98, 1682, 2001.
10. Amanatullah, D.F. et al., Ras regulation of cyclin D1 promoter. *Methods Enzymol.* 333, 116, 2001.
11. Jones, S.M. and Kazlauskas, A., Growth-factor dependent mitogenesis requires two distinct phases of signaling, *Nat. Cell. Biol.*, 3, 165, 2001.
12. Dyson, N., The regulation of E2F by pRb family proteins, *Genes Dev.*, 12, 2245, 1998.

13. Brehm, A. and Kouzarides, T., Retinoblastoma protein meets chromatin, *Trends Biochem. Sci.*, 24, 142, 1999.

14. Harbour, J.W. and Dean, D.C., The Rb/E2F pathway: expanding roles and emerging paradigms. *Genes Dev.*, 14, 2393, 2000.

15. Constanzo, M. et al. CDK activity antagonizes Whi5, an inhibitor of G_1/S transition in yeast, *Cell*, 117, 2173, 2004.

16. de Bruin, R.A. et al., Cln3 activates G_1-specific transcription via phosphorylation of the SBF bound repressor Whi5, *Cell,* 117, 887, 2004.

17. Malumbres, M. and Barbacid, M., To cycle or not to cycle: a critical decision in cancer, *Nat. Rev. Cancer,* 1, 222, 2001.

18. Pavletich, N.P., Mechanisms of cyclin-dependent kinase regulation: structures of cdks, their cyclin activators and Cip and INK inhibitors, *J. Mol. Biol.*, 287, 821, 1999.

19. Olashaw, N., Bagui, T.K., and Pledger, W.J., Cell cycle control: a complex issue, *Cell Cycle*, 3, 263, 2004.

20. Akiyama, T. et al., Phosphorylation of the retinoblastoma protein by cdk2, *Proc. Natl. Acad. Sci USA*, 89, 7900, 1992.

21. Hinds, P.W. et al., Regulation of the retinoblastoma protein functions by ectopic expression of human cyclins, *Cell,* 70, 993, 1992.

22. Matsuura, I. et al., Cyclin-dependent kinases regulate the antiproliferative function of Smads, *Nature*, 430, 226, 2004.

23. Möroy, T. and Geisen, C., Cyclin E, *Int. J. Biochem. Cell. Biol.*, 36, 1424, 2004.

24. Hwang, H.C. and Clurman, B.E., Cyclin E in normal and neoplastic cell cycles, *Oncogene*, 24, 2776, 2005.

25. Müller-Tidow, C. et al., The Cyclin A1-CDK2 complex regulates DNA double-strand break repair, *Mol. Cell. Biol.*, 24, 8917, 2004.

26. Esashi, F. et al., CDK-dependent phosphorylation of BRCA2 as a regulatory mechanism for recombination and repair, *Nature,* 434, 598, 2005.

27. Ubersax, J.A. et al., Targets of the cyclin-dependent kinase CDK1, *Nature*, 425, 859, 2003.

28. Wianny, F. et al., G_1-phase regulators, cyclin D1, cyclin D2 and cyclin D3: up-regulation at gastrulation and dynamic expression during neurulation. *Dev. Dyn.*, 212, 49, 1998.

29. Fantl, V. et al., Mice lacking cyclin D1 are small and show defects in eye and mammary gland development, *Genes Dev.*, 9, 2364, 1995.

30. Sicinski P. et al., Cyclin D1 provides a link between development and oncogenesis in the retina and breast, *Cell*, 82, 621, 1995.

31. Sicinski, P. et al., Cyclin D2 is an FSH-responsive gene involved in gonadal cell proliferation and oncogenesis, *Nature*, 384, 470, 1996.

32. Sicinska, E. et al., Requirement for Cyclin D3 in lymphocyte development and T cell leukemias, *Cancer Cell*, 4, 451, 2003.

33. Carthon, B.C. et al., Genetic replacement of Cyclin D1 function in mouse development by cyclin D2, *Mol Cell Biol*, 25, 1081, 2005.

34. Ciemerych, M.A. et al., Development of mice expressing a single D-type cyclin, *Genes Dev.* 16, 3277, 2002.

35. Kozar, K. et al., Mouse development and cell proliferation in the absence of D-cyclins, *Cell,* 118, 477, 2004.

36. Rane, S.G. et al., Loss of CDK4 expression causes insulin-deficient diabetes and CDK4 activation results in ß-cell hyperplasia, *Nat. Genet.*, 22, 44, 1999.

37. Tsutsui, T. et al., Targeted disruption of CDK4 delays cell cycle entry with enhanced p27Kip1 activity, *Mol. Cell. Biol.*, 19, 7011, 1999.

38. Moons, D.S. et al., Intact follicular maturation and defective luteal function in mice deficient for cyclin-dependent kinase-4, *Endocrinology*, 143, 647, 2002.

39. Moons, D.S. et al., Pituitary hypoplasia and lactotroph dysfunction in mice deficient for cyclin-dependent kinase-4, *Endocrinology*, 143, 3001, 2002.

40. Martín, J. et al., Genetic rescue of CDK4 null mice restores pancreatic beta-cell proliferation but not homeostatic cell number, *Oncogene*, 22, 5261, 2003.

41. Georgia S. and Bhushan, A., Beta cell replication is the primary mechanism for maintaining postnatal beta cell mass, *J. Clin. Invest.*, 114, 963, 2004.

42. Kushner, J.A. et al., Cyclins D2 and D1 are essential for postnatal pancreatic beta-cell growth. *Mol. Cell. Biol.*, 25, 3752, 2005.

43. Sotillo, R. et al., Wide spectrum of tumors in knock in mice carrying a CDK4 protein insensitive to INK4 inhibitors, *EMBO J.*, 20, 6637, 2001.

44. Meyer, C.A. et al., Cyclin D-cdk4 is not a master regulator of cell multiplication in Drosophila embryos, *Curr. Biol.*, 12, 661, 2002.

45. Malumbres, M. et al., Mammalian cells cycle without the D-type cyclin-dependent kinases CDK4 and CDK6, *Cell*, 118, 493, 2004.

46. Geng, Y. et al., Expression of cyclins E1 and E2 during mouse development and in neoplasia. *Proc. Natl. Acad. Sci. USA*, 98, 13138, 2001.

47. Geng, Y. et al., Cyclin E ablation in the mouse, *Cell*, 114, 431, 2003.

48. Parisi, T. et al., Cyclins E1 and E2 are required for endoreplication in placental trophoblast giant cells, *EMBO J.*, 22, 4794, 2003.

49. Tanaka, M. et al., Mash2 acts cell autonomously in mouse spongiotrophoblast development. *Dev. Biol.*, 190, 55, 1997.

50. Su, T.T. et al., Chromosome association of minichromosome maintenance proteins in Drosophila endoreplicative cycles, *J. Cell. Biol.*, 140, 451, 1998.

51. Tetsu, O. and McCormick, F., Proliferation of cancer cells despite CDK2 inhibition, *Cancer Cell*, 3, 233, 2003.

52. Ortega, S. et al., Cyclin-dependent kinase 2 is essential for meiosis but not for mitotic cell division in mice, *Nat. Genet.*, 35, 25, 2003.

53. Berthet, C. et al., CDK2 knockout mice are viable, *Curr. Biol.*, 13, 1775, 2003.

54. Liu, D. et al., Cyclin A1 is required for meiosis in the male mouse, *Nat. Genet.*, 20, 377, 1998.

55. Liu, D., Liao, C. and Wolgemuth, D.J., A role for cyclin A1 in the activation of MPF and G_2-M transition during meiosis of male germ cells in mice, *Dev. Biol.*, 224, 388, 2000.

56. Aleem, E. et al., Cdc2-cyclin E complexes regulate the G_1/S phase transition, *Nat. Cell Biol.*, 7, 831, 2005.

57. Simone, C. et al., Physical interaction between pRb and cdk9/cyclinT2 complex, *Oncogene*, 21, 4158, 2002.

58. Manning, G. et al., The protein kinase complement of the human genome, *Science*, 298, 1912, 2002.

59. Ruas, M. and Peters, G., The p16INK4a/CDKN2A tumor suppressor and its relatives. *Biochim Biophys Acta.* 1378, F115, 1998.

60. Ortega, S., Malumbres, M., and Barbacid, M., Cyclin D-dependent kinases, INK4 inhibitors and cancer, *Biochim. Biophys. Acta,* 1602, 73, 2002.

61. Wolfel, T. et al., A p16Ink4a — insensitive CDK4 mutant targeted by cytolytic T lymphocytes in a human melanoma, *Science*, 269, 1281, 1995.

62. Zuo, L. et al., Germline mutations in the p16INK4a binding domain of CDK4 in familial melanoma. *Nat. Genet.* 12, 97, 1996.

63. Corcoran, M.M. et al., Dysregulation of cyclin dependent kinase 6 expression in splenic marginal zone lymphoma through chromosome 7q translocations. *Oncogene* 18, 6271, 1999.

64. Brito-Babapulle, V. et al., Translocation t(2;7)(p12;q21-22) with dysregulation of the CDK6 gene mapping to 7q21-22 in a non-Hodgkin's lymphoma with leukemia. *Haematologica* 87, 357, 2002.

65. Bloom, J. and Pagano, M., Deregulated degradation of the cdk inhibitor p27 and malignant transformation. *Semin. Cancer Biol.* 13, 41, 2003.

66. Wang, T.C. et al., Mammary hyperplasia and carcinoma in MMTV-Cyclin D1 transgenic mice, *Nature,* 369, 669, 1994.

67. Bodrug, S.E. et al., Cyclin D1 transgene impedes lymphocyte maturation and collaborates in lymphomagenesis with the myc gene, *EMBO. J.* 13, 2124, 1994.

68. Miliani de Marval, P.L. et al., Transgenic expression of cyclin-dependent kinase 4 results in epidermal hyperplasia, hypertrophy, and severe dermal fibrosis, *Am. J. Pathol.,* 159, 369, 2001.

69. Krimpenfort, P. et al., Loss of p16^{Ink4a} confers susceptibility to metastatic melanoma in mice, *Nature,* 413, 83, 2001.

70. Sharpless, N.E. et al., Loss of p16^{INK4a} with retention of p19^{ARF} predisposes to tumorigenesis, *Nature,* 413, 86, 2001.

71. Latres, E. et al., Limited overlapping roles of p15^{INK4b} and p18^{INK4c} cell cycle inhibitors in proliferation and tumorigenesis, *EMBO J.,* 19, 3496, 2000.

72. Franklin, D.S. et al., CDK inhibitors p18INK4c and p27Kip1 mediate two separate pathways to collaborative suppress pituitary tumorigenesis, *Genes. Dev.* 12, 2899, 1998.

73. Sotillo, R. et al., Invasive melanoma in CDK4 targeted mice, *Proc. Natl. Acad. Sci. USA,* 98, 13312, 2001.

74. Bortner, D.M. and Rosenberg, M.P., Induction of mammary gland hyperplasia and carcinomas in transgenic mice expressing human cyclin E, *Mol. Cell. Biol.,* 17, 453, 1997.

75. Karsunky, H.C. et al., Oncogenic potential of Cyclin E in T-cell lymphomagenesis in transgenic mice: evidence for the cooperation between Cyclin E and Ras but not Myc, *Oncogene,* 18, 7816, 1999.

76. Geisen, C. et al., Loss of p27(kip1) cooperates with Cyclin E in T-cell lymphomagenesis, *Oncogene,* 22, 1724, 2003.

77. Fero, M.L. et al., A syndrome of multiorgan hyperplasia with features of gigantism, tumorigenesis, and female sterility in p27(Kip1)-deficient mice, *Cell,* 85, 733, 1996.

78. Fero, M.L. et al., The murine gene p27Kip1 is haplo-insufficient for tumor suppression, *Nature,* 396, 177, 1998.

79. Kiyokawa, H. et al., Enhanced growth of mice lacking the cyclin-dependent kinase inhibitor function of p27(Kip1), *Cell,* 85, 721, 1996.

80. Nakayama, K. et al., Mice lacking p27(Kip1) display increased body size, multiple organ hyperplasia, retinal dysplasia, and pituitary tumors, *Cell,* 85, 707, 1996.

81. Martin-Caballero, J. et al., Tumor susceptibility of p21(Waf1/Cip1)-deficient mice, *Cancer Res,* 61, 6234, 2001.

82. Zou, X. et al., CDK4 disruption renders primary mouse cells resistant to oncogenic transformation, leading to Arf/p53-independent senescence, *Genes. Dev.,* 16, 2923, 2002.

83. Geisen, G. and Möröy, T., The oncogenic activity of cyclin E is not confined to CDK2 activation alone but relies on several other, distinct functions of the protein, *J. Biol. Chem.,* 277, 39909, 2002.

84. Yu, Q., Geng, Y., and Sicinski, P., Specific protection against breast cancers by cyclin D1 ablation, *Nature*, 411, 1017, 2001.

85. Hulit, J.C. et al., Cyclin D1 genetic heterozygosity regulates colonic epithelial cell differentiation and tumor number in ApcMin mice, *Mol. Cell. Biol.* 24, 7598, 2004.

86. Robles, A.I. et al., Reduced skin tumor development in Cyclin-D1 deficient mice highlights the oncogenic ras pathway in vivo, *Genes. Dev.,* 12, 2469, 1998.

87. Burns, K.H. et al., Cyclin D2 and p27 are tissue-specific regulators of tumorigenesis in inhibin alpha knockout mice, *Mol. Endocrinol.,* 10, 2053, 2003.

88. Rodríguez-Puebla, M.L. et al., CDK4 deficiency inhibits skin tumor development but does not affect keratinocyte proliferation, *Am. J. Pathol.,* 161, 405, 2002.

89. Martín, A. et al., CDK2 is dispensable for cell cycle inhibition and tumor suppression mediated by p27(Kip1) and p21(Cip1). *Cancer Cell* 7, 591, 2005.

90. Senderowicz, A.M., Small-molecule cyclin-dependent kinase modulators, *Oncogene*, 22, 6609, 2003.

91. Lin, B.T. et al., Retinoblastoma cancer suppressor gene product is a substrate of the cell cycle regulator cdc2 kinase. *EMBO J.* 10, 857, 1991.

92. Kim, H.A. et al., A developmentally regulated switch directs regenerative growth of Schwann cells through cyclin D1, *Neuron,* 26, 405, 2000.

93. Atanasoski, S. et al., Differential cyclin D1 requirements of proliferating Schwann cells during development and after injury, *Mol Cell Neurosci,* 18, 581, 2001.

94. Geng, Y. et al., Rescue of cyclin D1 deficiency by knockin cyclin E, *Cell,* 97, 767, 1999.

95. Huard, J.M. et al., Cerebellar histogenesis is disturbed in mice lacking cyclin D2, *Development,* 126, 1927, 1999.

96. Lam, E.W. et al., Cyclin D3 compensates for loss of cyclin D2 in mouse B-lymphocytes activated via the antigen receptor and CD40, *J. Biol. Chem.,* 275, 3479, 2000.

97. Solvason, N. et al., Cyclin D2 is essential for BCR-mediated proliferation and CD5 B cell development, *Int. Immunol.,* 12, 631, 2000.

98. Kowalczyk, A. et al., The critical role of cyclin D2 in adult neurogenesis, *J. Cell. Biol.,* 167, 209, 2004.

99. Murphy, M. et al., Delayed early embryonic lethality following disruption of the murine cyclin A2 gene, *Nat. Genet.,* 15, 83, 1997.

100. Brandeis, M. et al., Cyclin B2-null mice develop normally and are fertile whereas cyclin B1-null mice die in utero, *Proc. Natl. Acad. Sci. USA,* 95, 4344, 1998.

101. Tetzlaff, M.T. et al., Cyclin F disruption compromises placental development and affects normal cell cycle execution, *Mol. Cell. Biol.,* 24, 2487, 2004.

102. Zindy, F. et al., Control of spermatogenesis in mice by the cyclin D-dependent kinase inhibitors p18(Ink4c) and p19(Ink4d), *Mol. Cell. Biol.,* 21, 3244, 2001.

103. Zindy, F. et al., INK4d-deficient mice are fertile despite testicular atrophy, *Mol. Cell. Biol.,* 20, 372, 2000.

104. Bai, F. et al., Haploinsufficiency of p18(INK4c) sensitizes mice to carcinogen-induced tumorigenesis, *Mol. Cel. Biol.,* 23, 1269, 2003.

105. Serrano, M. et al., Role of the INK4a locus in tumor suppression, *Cell*, 85, 27, 1996.

106. Rane, S.G. et al., Germ line transmission of the CDK4(R24C) mutation facilitates tumorigenesis and escape from cellular senescence, *Mol. Cell. Biol.,* 22, 644, 2002.

107. Deng, C. et al., Mice lacking p21CIP1/WAF1 undergo normal development, but are defective in G_1 checkpoint control, *Cell,* 82, 675, 1995.

108. Brugarolas, J. et al., Radiation-induced cell cycle arrest compromised by p21 deficiency. *Nature* 377, 1995.

109. Yang, W. et al., Targeted inactivation of p27kip1 is sufficient for large and small intestinal tumorigenesis in the mouse, which can be augmented by a Western-style high-risk diet, *Cancer Res.*, 63, 4990, 2003.

110. Yan, Y. et al., Ablation of the CDK inhibitor p57Kip2 results in increased apoptosis and delayed differentiation during mouse development, *Genes Dev.*, 11, 973, 1997.

111. Zhang, P. et al., Altered cell differentiation and proliferation in mice lacking p57KIP2 indicates a role in Beckwith-Wiedemann syndrome, *Nature*, 387, 151, 1997.

112. Zhang, P. et al., Cooperation between the CDK inhibitors p27(KIP1) and p57(KIP2) in the control of tissue growth and development, *Genes Dev.*, 12, 3162, 1998.

113. Zhang, P. et al., p21(CIP1) and p57(KIP2) control muscle differentiation at the myogenin step, *Genes Dev.*, 13, 213, 1999.

114. Jirawatnotai, S. et al., CDK4 is indispensable for postnatal proliferation of the anterior pituitary, *J. Biol. Chem.*, 279, 51100, 2004.

115. Jirawatnotai, S. et al., The cyclin-dependent kinase inhibitors p27Kip1 and p21Cip1 cooperate to restrict proliferative life span in differentiating ovarian cells, *J. Biol. Chem.*, 278, 17021, 2003.

116. Sotillo, R. et al., Cooperation between CDK4 and p27Kip1 in tumor development: a preclinical model to evaluate cell cycle inhibitors with therapeutic activity, *Cancer Res.*, 65, 3846, 2005.

117. Franklin, D.S. et al., Functional collaboration between different cyclin-dependent kinase inhibitors suppresses tumor growth with distinct tissue specificity, *Mol. Cell. Biol.*, 20, 6147, 2000.

118. Pei, X.H. et al., Genetic evidence for functional dependency of p18Ink4c on CDK4, *Mol. Cell. Biol.*, 24, 6653, 2004.

119. Zindy, F. et al., Postnatal neuronal proliferation in mice lacking Ink4d and Kip1 inhibitors of cyclin-dependent kinases, *Proc. Natl. Acad. Sci. U S A*, 96, 13462, 1999.

120. Geng, Y. et al., Deletion of the p27Kip1 gene restores normal development in cyclin D1-deficient mice, *Proc. Natl. Acad. Sci. U S A*, 98, 194, 2001.

121. Tong, W. and Pollard, J.W., Genetic evidence for the interactions of Cyclin D1 and p27 (Kip1) in mice, *Mol. Cell. Biol.* 21, 1319, 2001.

122. Robles, A.I. et al., Expression of cyclin D1 in epithelial tissues of transgenic mice results in epidermal hyperproliferation and severe thymic hyperplasia, *Proc. Natl. Acad. Sci. USA*, 93, 7634, 1996.

123. Rodriguez-Puebla, M.L., LaCava, M., and Conti, C.J., Cyclin D1 overexpression in mouse epidermis increases cyclin-dependent kinase activity and cell proliferation in vivo but does not affect skin tumor development, *Cell Growth Differ.*, 10, 467, 1999.

124. Miliani de Marval, P.L. et al., Enhanced malignant tumorigenesis in CDK4 transgenic mice, *Oncogene*, 23, 1863, 2004.

125. Soonpaa, M.H. et al., Cyclin D1 overexpression promotes cardiomyocyte DNA synthesis and multinucleation in transgenic mice, *J. Clin. Invest.*, 99, 2644, 1997.

126. Gómez LaHoz, E. et al., CyclinD- and E-dependent kinases and the p57Kip2 inhibitor: cooperative interactions in vivo, *Mol. Cell. Biol.*, 19, 353, 1999.

127. Kong, G. et al., Functional analysis of cyclin D2 and p27Kip1 in cyclin D2 transgenic mouse mammary gland during development, *Oncogene*, 21, 7214, 2002.

128. Bortner, D.M. and Rosenberg, M.P., Overexpression of cyclin A in the mammary glands of transgenic mice results in the induction of nuclear abnormalities and increased apoptosis, *Cell Growth Differ.*, 6, 1579, 1995.

129. Cheng, M. et al., The p21 (Cip1) and p27(Kip1)CDK "inhibitors" are essential activators of cyclin D-dependent kinases in murine fibroblasts, *EMBOJ.*, 15, 1571, 1999.

4 Cyclin-Dependent Kinase Inhibitors and Vascular Disease

Martin F. Crook and Manfred Boehm

CONTENTS

4.1 INTRODUCTION

As with other tissues and organs, cellular proliferation in the cardiovascular system is a tightly regulated process that is coordinated on the molecular level by the cyclins, cyclin-dependent kinases (CDKs), and the cyclin-dependent kinase inhibitors (CDKIs). In normal healthy arteries and veins, the vast majority of vascular smooth muscle cells (VSMCs) in the medial layer are quiescent and proliferate at very low indices (less than 2% of medial cells). Likewise, in the adult myocardium there are few proliferating cells as cardiomyocytes are usually in a postmitotic terminally differentiated state, and are unable to proliferate. Following chronic or acute injury to the vasculature, a reparative process ensues, which results in VSMC proliferation. In contrast, myocardial injury following myocardial infarction or infection does not result in cardiomyocyte proliferation. Instead a process called ventricular remodeling occurs that is characterized by progressive fibrosis, myocyte hypertrophy, and apoptosis. However, cell cycle proteins also regulate the myocardial repair process through mechanisms that are distinct from repair processes in the vasculature.

In this chapter, we present the major cardiovascular diseases in which manipulation of cell division can have beneficial or pathological consequences on cardiovascular function. We focus on two major topics in cardiovascular medicine, the excessive cellular growth in vascular proliferative diseases and the inability of adult cardiomyocytes to proliferate and regenerate. The causal role of the cyclins, CDKs and CKIs, in the pathophysiology of these diseases will be discussed. Finally, understanding the molecular mechanisms of growth control in cardiovascular cells has led to the development of exciting new molecular therapies for cardiovascular diseases.

4.2 CELL CYCLE REGULATION OF VASCULAR INJURY

The CIP/KIP CKIs are important regulators of the tissue-remodeling process in the vasculature. Normally, endothelial cells lining the lumen of the vessel wall and VSMCs in the medial layer of arteries are quiescent or in the G_0/G_1 phase of the cell cycle. The reparative process that ensues after vascular injury comprises several important events that replace damaged cells. After vascular injury, damage to both the endothelium and underlying VSMCs leads to increased expression and release of inflammatory cytokines and chemokines. The vessel wall is infiltrated by inflammatory cells that by autocrine and paracrine mechanisms lead to an increase in tissue levels of several additional cytokines, mitogens, and proteolytic enzymes. Active proteases in the vessel wall degrade the extracellular matrix, including the collagen basement membrane that surrounds VSMCs, thus relieving the cell of its inhibitory effects on proliferation. Subsequently, VSMCs dedifferentiate and migrate from the medial layer into the intimal layer, where they are exposed to several mitogens, which in combination with the lack of antiproliferative signals from the extracellular matrix, stimulate VSMCs to enter the cell cycle and divide. After several rounds of cell divisions, proliferation ceases and the inflammation resolves, leading to completion of arterial wound repair. The molecular basis underlying these events has been under intense scrutiny to help identify and develop novel drug targets for treating chronic diseases of vascular injury such as atherosclerosis or more acute vascular pathologies such as restenosis and in-stent restenosis. Interestingly, the CKI p27^{KIP1} — an endogenous inhibitor of cyclin E/CDK2 — plays an important role in the molecular mechanisms that regulate several of the processes just described.

4.2.1 CKI EXPRESSION AND VASCULAR DISEASE

The CKIs have distinct temporal and spatial patterns of expression in normal, injured, and diseased arteries *in vivo* (Figure 4.1). p27^{KIP1} is constitutively expressed in endothelial cells and VSMCs of normal arteries and contributes to their quiescent state by inhibiting cyclin E/CDK2. However, p27^{KIP1} levels are rapidly reduced after vascular injury with a concomitant increase in VSMC proliferation.[1] In contrast, p21^{CIP1} protein is not observed in VSMCs of normal arteries, but is upregulated along with p27^{KIP1} in the later phases of arterial wound repair. p16^{INK4} expression is low in normal and injured arteries. These patterns of CKI protein expression are observed in several animal models of vascular disease, including mice, rats, and pigs.[2,3]

In human coronary arteries p27^{KIP1} is expressed within medial and intimal VSMCs of normal and atherosclerotic arteries, including the VSMCs of new blood vessels within the atherosclerotic plaque.[4] In contrast, p21^{CIP1} is present only in advanced atherosclerotic lesions. These distinct temporal and spatial patterns of expression

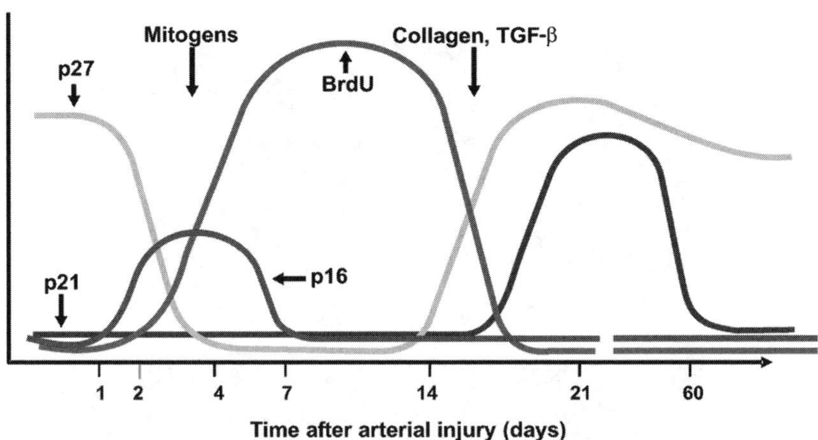

FIGURE 4.1 CKIs and vascular wound repair. Protein levels of various CKIs at different stages of vascular wound repair program in pig artery. p27[KIP1] levels are high in normal arteries, and are rapidly downregulated after vascular injury. p21[CIP1] is barely detectable in uninjured arteries. The protein levels of p27[KIP1] and p21[CIP1] are upregulated in the later phases of vascular wound repair, leading to successful vessel remodeling. Proliferation of intimal and medial cells is indicated by BrdU incorporation. Vascular smooth muscle cells secrete collagen, TGF-, and other molecules that signal back to the vascular smooth muscle cell to complete arterial wound repair. (Adapted from Tanner, F.C. et al. Expression of cyclin-dependent kinase inhibitors in vascular disease. *Circ Res* 82, 396–403, 1998. With permission.)

suggest that the CIP/KIP CKIs regulate G_1 to S phase progression in vascular cells and promote favorable vascular remodeling.

4.2.2 p27[KIP1] AND VASCULAR REPAIR

One of the initial phases of the vascular repair process includes the recruitment of inflammatory cells to the site of injury. This is mediated by the activation of the endothelium, the induction of adhesion molecules, and chemokine expression. Previous work in our laboratory has shown that arterial tissue levels of chemokines and cytokines that activate adhesion molecule expression are elevated in p27[KIP1]-deficient arteries that were subjected to vascular injury compared to wild-type controls.[5] We also observed a significant increase in the recruitment of bone-marrow-derived cells to vascular lesions of p27[KIP1]-deficient mice (Figure 4.2). Furthermore, in agreement with previously published reports, we observed an increased accumulation of macrophages in p27-deficient models of atherosclerosis.[6] It is not clear whether p27[KIP1] protects against the inflammatory response by inhibition of cyclin E/CDK2 or by some other molecular mechanism, although it is widely known p27[KIP1] is important in regulating the proliferation of inflammatory cells such as T lymphocytes and monocytes/macrophages.[7–10]

Many of the inflammatory genes that mediate the recruitment of leukocytes to the vascular wall are regulated by the transcriptional activity of NFkappaB. It is interesting

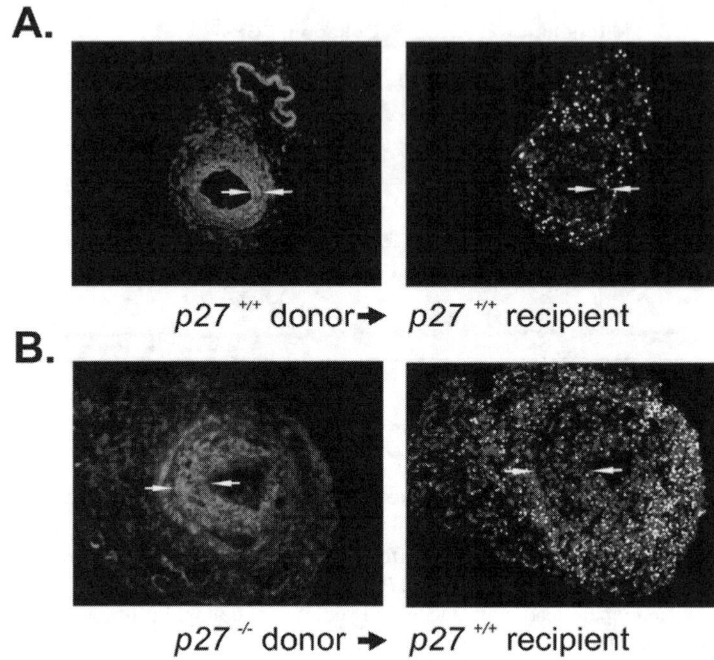

FIGURE 4.2 (See color insert following page 142.) p27[KIP1] modulates the contribution of bone-marrow-derived cells to vascular lesions. Cross sections of recipient p27[+/+]female arteries following transplantation of male p27[+/+] or p27[-/-] donor marrow: Y chromosome[+] (Yellow), alpha-actin[+] (smooth muscle specific) cells (red), and nuclei (DAPI stain, blue). Arrows indicate the margins of the intima as determined by the internal and external elastic lamina. (Adapted from Boehm, M. et al. *J Clin Invest* 114, 419–426, 2004. With permission.)

to note that the transcription factor E2F-1, which is regulated by cyclin E/CDK2 activity, also has diverse effects on NFkappaB signaling in vascular cells.[11,12] However, it is as yet unclear whether there are any changes in NFkappaB signaling in p27[KIP1]-deficient cells and if this contributes to the excessive inflammation seen after vascular injury in p27[KIP1]-deficient mice.

The CKI proteins also regulate VSMC migration through p27[KIP1]. VSMC migration *in vitro* is inhibited by treatment of cells with rapamycin.[13] Rapamycin, a macrolide antibiotic, prevents growth-factor-dependent downregulation of p27[KIP1] through inhibition of mTOR and the serine/threonine kinase p70[S6K].[14] Although rapamycin inhibits VSMC migration in wild-type mice, it also reduces VSMC migration to a certain degree in cells derived from p27[-/-] mice, suggesting both a p27[KIP1]-independent and -dependent mechanism.[15] More recently, a study suggests that p27[KIP1] directly modulates cell migration through regulation of RhoA activation.[16] Interestingly, a human organ culture model of arterial in-stent restenosis showed that neointima development was accompanied by a decrease in p27[KIP1] protein levels.[17] Furthermore, the reduction in p27[KIP1] protein levels paralleled an increase in RhoA activity, and this could be inhibited by treatment with rapamycin.

Although VSMCs of normal healthy arteries express high levels of p27[KIP1], these levels rapidly fall upon vascular injury.[1] Furthermore, the overexpression of p27[KIP1] in injured arteries significantly reduces both the degree of VSMC proliferation and lesion size.[1] Additionally, mice that are deficient in p27[KIP1] have larger lesions after vascular injury than their wild-type controls, and this is due in part to a greater degree of VSMC proliferation (Figure 4.3).[5] These studies highlight the important role p27[KIP1] plays in suppressing VSMC proliferation.

This inflammatory phase is followed by a synthetic phase in which stimuli such as TGF-beta activate VSMCs synthesis of extracellular matrix proteins, including collagen. Polymerized type 1 collagen fibrils, a mature form of collagen, increase p27[KIP1] protein levels and may serve as a mechanism that suppresses the VSMC proliferative response.

As mentioned earlier, VSMCs in normal healthy arteries are maintained in a quiescent state by their native extracellular matrix. There is some evidence that suggests that even in the presence of mitogens, VSMCs surrounded by their native basement membrane do not undergo proliferation.[18,19] It is believed that these anti-proliferative signals are mediated by fibrillar type I collagen and alpha2 integrin outside-in signaling pathways that result in rapamycin-sensitive changes in p70[S6K] activity.[20] The p70[S6K] protein regulates the translation of several cycle genes and has been suggested to regulate p27[KIP1] protein levels.[21,22]

Additional molecular mechanisms may also exist that contribute to the decrease in p27[KIP1] levels observed after extracellular matrix degradation. Interestingly, the extracellular matrix regulates Skp2 protein levels in VSMCs via a Focal Adhesion Kinase-dependent mechanism.[23] Furthermore, as discussed in more detail later, Skp2 directly regulates p27[KIP1] and p21[CIP1] protein levels by targeting them for ubiquitination and degradation via the proteasome.[24,25] Collectively, these signaling pathways regulate protein levels of the CKI's p27[KIP1] and p21[CIP1] in vascular cells.

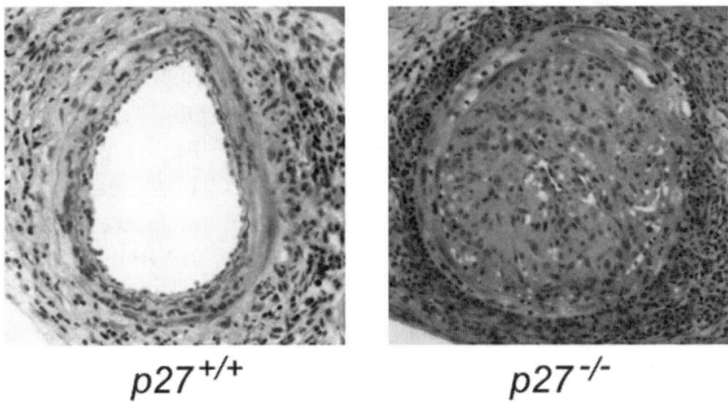

FIGURE 4.3 (See color insert.) p27[KIP1] modulates neointima formation during vascular wound repair. Increased vascular lesions in p27[-/-] arteries compared to p27[+/+] arteries in H&E cross sections of murine arteries 2 weeks after mechanical injury.

4.3 THE ROLE OF CKIs AND BONE-MARROWED STEM CELLS NICHE

Traditionally, postnatal neovascularization was believed to occur through the proliferation and migration of local endothelial cells that would rapidly recover an injured/denuded area of a vessel or sprout into new vessels.[26] There is now increasing evidence that stem and progenitor cells may participate in the repair processes of blood vessels. These stem/progenitor cells may be derived from local cells residing within the vessel itself or may be circulating in the blood. Circulating endothelial progenitor cells (EPCs) were first described by Asahara et al. in 1997 and are a fraction of circulating hematopoietic progenitor cells, which express the hematopoietic stem cell markers CD133/CD34 and various endothelial markers (Ve-Cadehrin,[28–29] CD31, VWF, VEGFR2, Tie2, and CD146).[28] These cells can incorporate themselves into sites of neoangiogenesis, where they adopt an endothelial-like phenotype. There is also evidence that endothelial-like cells may originate from circulating myelo-monocytic cells, too but these cells may have a reduced angiogenic potential and not be true EPCs.[28] Numerous studies have shown that these bone marrow/ blood-derived cells can participate in neoangiogensis at sites of vascular injury or ischemia, not only in animal models but also in clinical studies.[29] For example, patients implanted with left-ventricular assist devices were found to have CD34+ VEGFR2+ bone-marrow-derived cells lining the artificial surface of these devices.[30]

Although there have been fewer studies on smooth muscle progenitor cells, it appears that there are progenitor cells circulating in the blood and also locally in some blood vessels. Mouse hematopoietic stem cells can differentiate into SMCs and under certain circumstances can contribute to postangioplasty restenosis, graft vasculopathy, and atherosclerosis.[31] Although human data at present are limited, it appears that cells with a VSMC phenotype can be cultured from blood.[32] In patients who have received BM transplants, donor-derived neointimal cells were found within vascular lesions at necroscopy.[33]

The frequency of circulating EPCs within the blood is extremely low, although levels rise after mobilization from the bone marrow under appropriate stimuli. Mobilization of stem cells from their niche within the bone marrow is determined by the local environment, which includes fibroblasts, osteoblasts, endothelial cells, and local and systemic cytokines. Mobilization of EPCs has been found to be increased by a number of cytokines such as VEGF, EPO, G-CSF, SDF-1, and also by the use of statins as well as by exercise, trauma, and surgery.[29] Physiologically, ischemia is believed to be the predominant signal to induce EPC mobilization from the bone marrow. It is speculated that stem cells reside in a nonproliferative state. EPCs are mobilized by appropriate stimuli such as VEGF or PlGF. VEGF induces the production of MMP-9 within the BM, which causes an increase in soluble Kit ligand (stem cell factor). Soluble Kit ligand results in increased cycling of quiescent stem and progenitor cells and enhanced translocation to a vascular niche that is conducive to stem-cell proliferation, differentiation, and mobilization to the peripheral circulation.[34] Angiogenic cytokine-mediated MMP-9 activation is a crucial checkpoint in controlling the cell turnover and, hence, mobilization of EPCs.

The cell cycle inhibitors p21[CIP1] and p27[KIP1] are important regulators of the bone marrow stem cell compartment and are therefore important in controlling the proliferation and mobilization of EPC- and BM-derived smooth muscle cells. Interestingly, these two structurally similar CKIs have very distinct functions in modulating the bone marrow stem cell compartment. In the absence of p21[CIP1], hematopoietic stem cell number and proliferation activity are increased.[35] However, the self-renewal potential of hematopoietic stem cells is impaired, leading to hematopoietic failure in serially transplanted bone marrow from p21[−/−] mice. p21[CIP1] appears to be a molecular switch governing the entry of hematopoietic stem cells into the cell cycle. In the same model p21[CIP1] does not regulate the progenitor cell's pool size within the bone marrow as well as in the circulation. A different regulatory mechanism was postulated to control progenitor pool size and shortly afterward was identified by the same group.[36] They reported that p27[KIP1] does not affect stem cell number, cell cycling, or self-renewal but markedly modulates progenitor cell proliferation and pool size. Modulation of p27[KIP1] expression in stem cells may lead to increased production of transient amplifying cells and, as a consequence, vast numbers of maturing cells are released into the circulation. Modulation of p27[KIP1] and p21[CIP1] expression and function thereby provides a promising strategy in stem cell therapy.

4.4 MOLECULAR REGULATION OF p27[KIP1] FUNCTION

Understanding the signaling pathways that regulate CKI function has led to new candidate molecular targets for vascular and cardiac diseases. The CKI p27[KIP1] is regulated by both transcriptional and posttranscriptional mechanisms. TGF-beta, which is expressed in the vessel wall during the later stages of the wound-healing process, has been shown to lead to an increase in p27[KIP1] gene expression.[37] Posttranscriptional mechanisms of p27[KIP1] function include the translation of p27[KIP1] mRNA, serine/threonine phosphorylation of p27[KIP1] protein, and p27[KIP1] subcellular localization and its ubiquination and degradation by the proteasome.[38–40]

Growth-factor-dependent downregulation of p27[KIP1] levels results from proteasomal degradation (Figure 4.4).[38] There are at least two phosphorylation sites on p27[KIP1] that regulate p27[KIP1] turnover. Threonine187 is phosphorylated by cyclinE/CDK2 and mediates Skp2-dependent degradation of p27[KIP1].[40–42] Serine10 phosphorylation mediates the growth-factor-dependent translocation of p27[KIP1] from the nucleus to the cytoplasm. A Skp2-independent mechanism of p27[KIP1] proteasomal degradation has been identified, although it is not yet clear whether serine10 phosphorylation is required.[43] KIS (kinase-interacting stathmin) is a recently identified serine/threonine protein kinase that phosphorylates p27[KIP1] at serine10.[39] It is interesting to note that following growth factor stimulation, the Ets transcription factor GABPalpha upregulates both KIS and Skp2 gene expression (Crook, M.F. et al., unpublished data 2005).[44] KIS promotes cell cycle progression in VSMCs and other cell types through phosphorylation of serine10 on p27[KIP1], and nuclear export of the protein, where it can no longer inhibit nuclear cyclin/E:cdk2. This kinase is a promising molecular target for manipulating p27[KIP1] expression and function.[39] Additionally, studies have reported threonine157 phosphorylation of human p27[KIP1] by Akt and it's role in

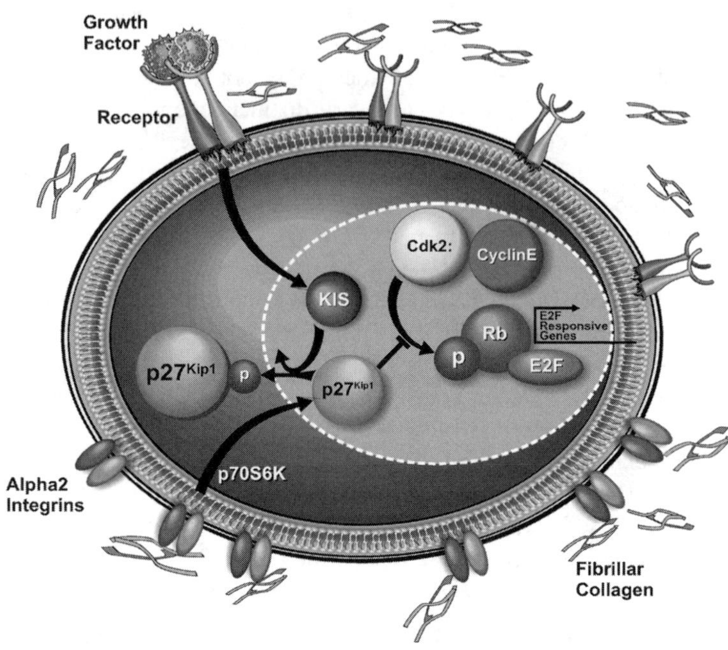

FIGURE 4.4 Growth factor stimulation and extracellular matrices regulate p27[KIP1] levels and cellular proliferation. Fibrillar collagen maintains p27[KIP1] levels and cellular quiescence through an outside-in signaling pathway involving alpha 2 integrin and p70[S6K] pathways. Growth factor stimulation leads to an increase in serine10 phosphorylation of p27[KIP1] on KIS, leading to its nuclear-to-cytoplasmic redistribution, thereby promoting CDK2/cyclinE activity and E2F activation.

nuclear import.[45] However, it should be mentioned that the threonine157 residue of p27[KIP1] is not conserved in other mammalian species and is therefore difficult to investigate experimentally *in vivo*.

There is some evidence that suggests nuclear export of p27[KIP1] is followed by proteasomal degradation, although it still remains possible that other cellular events may recycle p27[KIP1] and lead to its reimport into the nucleus. Understanding the precise spatial and temporal events that regulate p27[KIP1] function remains technically challenging. Sophisticated cellular imaging of wild-type and mutant p27[KIP1] will assist our understanding of the complex molecular and subcellular dynamics that regulate p27[KIP1] function.

4.5 MOLECULAR TARGETS FOR VASCULAR THERAPIES

Recent therapies for vascular proliferative diseases have focused on cell cycle targets. Restenosis and in-stent restenosis have been treated in animal models by the overexpression of the CKIs p27[KIP1] and p21[CIP1] using adenoviral gene transfer. These studies, which demonstrate proof-of-principle for the rationale of cell cycle inhibition, coupled with advances in stent technology have provided the platform for a new class of therapeutic devices and agents. Recent attention has focused on stents coated with drugs that inhibit VSMC proliferation and migration. Two drugs are worthy of discussion: rapamycin and paclitaxel.

4.6 PHARMACOLOGICAL AGENTS USED TO INHIBIT VSMC PROLIFERATION

Rapamycin is a natural fermentation product of *Streptomyces hygroscopicus*. Although it has antifungal properties, it was not developed as an antibiotic because of its potent immunosuppressive effects. Subsequent molecular and cellular studies have elucidated its properties. Rapamycin binds the cytosolic receptor FKBP12, leading to inhibition of the protein kinase TOR (target of rapamycin), elevation of $p27^{KIP1}$, and G_1 arrest.[46,47] Rapamycin blocks G_1-S phase transition in T-cell lymphocytes, and more recently these observations have been extended to VSMCs.[48] In addition to its antiproliferative properties, rapamycin inhibits VSMC migration.[13] The FDA approved rapamycin in 1999 for treatment of acute rejection in renal transplantation. Rapamycin-coated stents have been approved for use in Europe and are still under investigation in the U.S. Phase I/II studies of patients undergoing *de novo* angioplasty treated with a rapamycin-coated stent indicate that the formulation is safe and well tolerated.[49] Phase III studies in Europe report a reduction in in-stent restenosis in the rapamycin-coated stent group compared to the noncoated stent group.[50]

Paclitaxel is an equally promising approach that is also undergoing clinical testing. This agent is a derivatized diterpenoid that was isolated from the bark of a yew, *Taxus brevifolia*.[51] Paclitaxel induces tubulin polymerization, which results in unstable microtubules. Microtubules are an essential component of the mitotic spindle and are required for cell division and maintaining cell shape, along with other cellular functions including motility, anchorage, and intracellular signal transduction. Taxol and other antimicrotubule drugs have a principal role in the treatment of certain malignancies, and more recently this agent has been shown to inhibit VSMC migration and proliferation. The drug inhibits intimal hyperplasia in an animal model of restenosis,[52] and the results of a Phase II/III clinical trial in Europe indicate comparable efficacy in in-stent restenosis in patients to rapamycin.[53]

Synthetic inhibitors of cdk activity have been developed, including flavopiridol.[54,55] Treatment of human VSMCs with flavopiridol leads to potent inhibition of cellular proliferation and migration.[55] Furthermore, Flavopiridol-coated stents implanted in rat carotid arteries lead to a significant decrease in in-stent restenosis.

Another approach used in the treatment of vascular graft hyperplasia is oligonucleotides (ODNs) that act as a decoy and scavenge specific transcription factors, thereby selectively modulating gene transcription. For example, ODNs have been designed with sequence specificity to bind transcription factors, such as E2F, in order to prevent S phase entry. In contrast, ODNs that bind NFkappaB have been developed as an anti-inflammatory treatment. In animal models of vascular disease in which vein grafts are placed around arterial lesions, introduction of E2F decoy ODNs prevents the development of intimal hyperplasia within the vein graft.[56] This therapy has been evaluated in a Phase I human study of peripheral bypass surgery in which a vein graft is placed around a lesion in the femoral artery. The ODNs are safe and well tolerated in patients, and the results of efficacy studies are pending.[57]

Local radiation therapy with β- and δ-particle-emitting stents is being offered to patients with severe restenosis who have failed other forms of therapy. Radiotherapy induces DNA damage and p53-mediated G_1 arrest and induction of apoptosis.[58] Clinical studies have demonstrated initial success in reducing the hyperproliferative lesion, but lesion recurrence still occurs, often with a delayed onset compared to stenting in the absence of radiation.

Other RNA approaches are under development. The rapid advances in siRNA and shRNA technology might also increase the spectrum of effective useful agents delivered locally at the treated artery.[59] Identification and characterization of signaling pathways regulating VSMC proliferation using transcription profiling and proteomics will identify new targets for the treatment of vascular proliferative disease.

Acute myocardial infarction (AMI) results from occlusion of one of the major coronary arteries, leading to ischemia of the underlying myocardium. Cardiac ischemia that is transient, defined as lasting no longer than 30 min after the inciting event, is potentially reversible. Progressive loss of myocardial viability due to myocardial necrosis is complete by 6 to 12 h.[60] Depending on the location and extent of the infarction, myocardial necrosis results in a loss of functional myocardium. A long-held concept is that adult myocytes are terminally differentiated cells and do not undergo proliferation and repopulation of the infracted myocardium. Recent studies are challenging this concept. Restoration of myocytes or other cell types within the infracted tissue may favorably impact ventricular remodeling and improve the morbidity and mortality following AMI.

During cardiac differentiation and early heart development, cardiomyocytes are characterized by a high DNA synthesis and proliferative rate.[61,62] In the mouse embryo at stage E8, the precardiac mesoderm has a labeling index of up to 70%, as determined by thymidine incorporation. At later stages proliferation of cardiomyocytes gradually decreases.[63] Interestingly, proliferation activity varies greatly depending on the location of cells, providing insights into the interplay of cell cycle and morphogenic changes in the heart. The transcriptional programs regulating cell cycle progression in cardiomyocytes at different locations within the heart are not known. A dramatic change in cardiomyocyte cell cycle regulation occurs in the transition from prenatal to postnatal period. During this time, myocardial growth shifts from a hyperplastic to a hypertrophic phenotype.[63] The changes in terminally differentiated cardiomyocytes are characterized by an increase in myofibril density, the appearance of mature intercalated discs, and the formation of binucleated cells. The gradual decrease in DNA synthesis is inversely related to the appearance of binucleated cardiomyocytes.[64] This suggests that DNA replication and karyokinesis occur in the absence of cytokinesis. Terminally differentiated adult cardiomyocytes do not have significant proliferative activity, with >80% of cells arrested in G_0/G_1 phase and >15% of cells arrested in G_2/M phase. The inability of cardiomyocytes to proliferate and regenerate following injury results in an impairment of cardiac function associated with physical impediment and may lead to death. The genetic program in the cardiomyocytes that leads to their inability to proliferate and regenerate is not understood, but if identified could lead to therapies aimed at reinitiating cell cycle and proliferation in cardiomyocytes.

In addition, an understanding of myocyte growth regulation would also lead to an appreciation of the underlying mechanisms of myocardial hypertrophy. It appears that cellular stimuli mediating cardiomyocyte proliferation during fetal cardiac development induce hypertrophic growth in the adult heart.

Several theories speculate on the causes of cell cycle exit in cardiomyocytes shortly after birth. One hypothesis is that the presence and density of highly organized myofibers permit karyokinesis but not cytokinesis. Alternatively, an intrinsic timer may limit the number of cell divisions.[65] However, rounds of cell division in several genetically altered mice are not compatible with this hypothesis.[66] A very intriguing concept is that cardiomyocytes cease proliferation because one or more cell cycle checkpoints are triggered by the absence or presence of as yet unidentified molecules. Expression studies of cell cycle regulatory proteins during cardiac development focus on the transition from a hyperplastic to hypertrophic phenotype. Cell cycle regulatory proteins, including cyclins, CDKs, and transcription factors such as Rb, p107, and p130, are present during embryonic heart development but are absent in the adult heart.[67–69] Interestingly, p21^{CIP1} and p27^{KIP1}, as well as TSC2, a regulator of p27^{KIP1}, were increasingly expressed during the transition period and continuously expressed in the adult heart.[70,71] The CIP/KIP CKIs may be required to keep cardiomyocytes quiescent, but the role of CKIs in a cell type that does not express their inhibitory targets, the CDKs, is not known.

Primary cultures of cardiomyocytes from adult, neonatal, and fetal heart have been established to study cell cycle regulation. Growth factors such as FGF, IGF, and EGF, induce DNA synthesis and proliferation in embryonic and neonatal cardiomyocytes, but they only induce DNA synthesis and hypertrophy in adult cardiomyocytes.[72,73] Forced expression of cell cycle regulators, such as the D-cyclins, E2F1, or c-myc, drives DNA replication, genomic replication, and karyokinesis but not cell proliferation in adult cardiomyocytes.[74–76] It appears that DNA replication and karyokinesis are not sufficient to force adult cardiomyocytes through the cell cycle. Additional checkpoints at G_2/M may be triggered that block cellular propagation. Several transgenic mouse models have been established to test this theory. Overexpression of cyclin D, calmodulin, and myc ER results in developmental hyperplasia, but not in stimulation of myocyte proliferation in the adult heart.[77] However, it is possible to bypass cell cycle checkpoints in adult cardiomyocytes. Transgenic mice expressing the SV40 large T antigen under a heart-specific promoter develop tumors in the heart.[78] Mice with deletion of the p27^{KIP1} loci exhibit enhanced growth and enlargement of multiple organs, including the heart.[79] Interestingly, loss of p27^{KIP1} in the mouse heart results in two to three more rounds of myocyte proliferation during transition from the fetal to the adult heart.[69] These data suggest that p27^{KIP1} may regulate terminal differentiation in cardiomyocytes.

A loss of myocardium is associated with functional impairment of cardiac contractility. It has been a long-standing concept that the loss of functional cardiac tissue is irreversible because of the inability of cardiomyocytes to regenerate. Regeneration of cardiac tissue and recovery of cardiac contractility are desirable therapeutic goals. These goals have been elusive because the complex biological program that regulates the terminal differentiation of cardiomyocytes is poorly understood. Considerable research must be conducted on cell cycle regulation in the heart to understand the mechanisms leading to the transition from a proliferative hyperplasic phenotype to a nondividing hypertrophic cardiac phenotype. The recent identification of multipotent adult stem cells from the bone marrow raises the possibility of myogenesis in the adult heart.[80,81] Bone-marrow-derived precursor cells have been shown to repopulate infarcted mouse heart.[82–84] These early studies need to be duplicated in other animal species. Yet, they raise the possibility and promise of tissue regeneration in the heart, an organ once thought to undergo irreversible injury.

REFERENCES

1. Tanner, F.C. et al. Differential effects of the cyclin-dependent kinase inhibitors p27(Kip1), p21(Cip1), and p16(Ink4) on vascular smooth muscle cell proliferation. *Circulation* **101**, 2022–2025, 2000.

2. Yang, Z.Y. et al. Role of the p21 cyclin-dependent kinase inhibitor in limiting intimal cell proliferation in response to arterial injury. *Proc Natl Acad Sci U S A* **93**, 7905–7910, 1996.

3. Chen, D. et al. Downregulation of cyclin-dependent kinase 2 activity and cyclin A promoter activity in vascular smooth muscle cells by p27(KIP1), an inhibitor of neointima formation in the rat carotid artery. *J Clin Invest* **99**, 2334–2341, 1997.

4. Tanner, F.C. et al. Expression of cyclin-dependent kinase inhibitors in vascular disease. *Circ Res* **82**, 396–403, 1998.

5. Boehm, M. et al. Bone marrow-derived immune cells regulate vascular disease through a p27(Kip1)-dependent mechanism. *J Clin Invest* **114**, 419–426, 2004.

6. Diez-Juan, A. et al. Selective inactivation of p27(Kip1) in hematopoietic progenitor cells increases neointimal macrophage proliferation and accelerates atherosclerosis. *Blood* **103**, 158–161, 2004.

7. Tsukiyama, T. et al. Down-regulation of p27Kip1 expression is required for development and function of T cells. *J Immunol* **166**, 304–312, 2001.

8. Mohapatra, S., Agrawal, D., and Pledger, W.J. p27Kip1 regulates T cell proliferation. *J Biol Chem* **276**, 21976–21983, 2001.

9. Boussiotis, V.A. et al. p27kip1 functions as an energy factor inhibiting interleukin 2 transcription and clonal expansion of alloreactive human and mouse helper T lymphocytes. *Nat Med* **6**, 290–297, 2000.

10. Matsuoka, M., Nishimoto, I., and Asano, S. Interferon-gamma impairs physiologic downregulation of cyclin-dependent kinase inhibitor, p27Kip1, during G1 phase progression in macrophages. *Exp Hematol* **27**, 203–209, 1999.

11. Chen, M., Capps, C., Willerson, J.T., and Zoldhelyi, P. E2F-1 regulates nuclear factor-kappaB activity and cell adhesion: potential antiinflammatory activity of the transcription factor E2F-1. *Circulation* **106**, 2707–2713, 2002.

12. Goukassian, D.A. et al. Engineering the response to vascular injury: divergent effects of deregulated E2F1 expression on vascular smooth muscle cells and endothelial cells result in endothelial recovery and inhibition of neointimal growth. *Circ Res* **93**, 162–169, 2003.

13. Poon, M. et al. Rapamycin inhibits vascular smooth muscle cell migration. *J Clin Invest* **98**, 2277–2283, 1996.

14. Nourse, J. et al. Interleukin-2-mediated elimination of the p27Kip1 cyclin-dependent kinase inhibitor prevented by rapamycin. *Nature* **372**, 570–573, 1994.

15. Sun, J. et al. Role for p27(Kip1) in vascular smooth muscle cell migration. *Circulation* **103**, 2967–2972, 2001.

16. Besson, A., Gurian-West, M., Schmidt, A., Hall, A., and Roberts, J.M. p27Kip1 modulates cell migration through the regulation of RhoA activation. *Genes Dev* **18**, 862–876, 2004.

17. Guerin, P. et al. Stent implantation activates RhoA in human arteries: inhibitory effect of rapamycin. *J Vasc Res* **42**, 21–28, 2005.

18. Newby, A.C. and Zaltsman, A.B. Fibrous cap formation or destruction—the critical importance of vascular smooth muscle cell proliferation, migration and matrix formation. *Cardiovasc Res* **41**, 345–360, 1999.

19. Izzard, T.D., Taylor, C., Birkett, S.D., Jackson, C.L., and Newby, A.C. Mechanisms underlying maintenance of smooth muscle cell quiescence in rat aorta: role of the cyclin dependent kinases and their inhibitors. *Cardiovasc Res* **53**, 242–252, 2002.

20. Koyama, H., Raines, E.W., Bornfeldt, K.E., Roberts, J.M., and Ross, R. Fibrillar collagen inhibits arterial smooth muscle proliferation through regulation of Cdk2 inhibitors. *Cell* **87**, 1069–1078, 1996.

21. Gallo, R. et al. Inhibition of intimal thickening after balloon angioplasty in porcine coronary arteries by targeting regulators of the cell cycle. *Circulation* **99**, 2164–2170, 1999.

22. Braun-Dullaeus, R.C. et al. Cell cycle protein expression in vascular smooth muscle cells in vitro and in vivo is regulated through phosphatidylinositol 3-kinase and mammalian target of rapamycin. *Arterioscler Thromb Vasc Biol* **21**, 1152–1158, 2001.

23. Bond, M., Sala-Newby, G.B., and Newby, A.C. Focal adhesion kinase (FAK)-dependent regulation of S-phase kinase-associated protein-2 (Skp-2) stability. A novel mechanism regulating smooth muscle cell proliferation. *J Biol Chem* **279**, 37304–37310, 2004.

24. Sutterluty, H. et al. p45SKP2 promotes p27Kip1 degradation and induces S phase in quiescent cells. *Nat Cell Biol* **1**, 207–214, 1999.

25. Bornstein, G. et al. Role of the SCFSkp2 ubiquitin ligase in the degradation of p21Cip1 in S phase. *J Biol Chem* **278**, 25752–25757, 2003.

26. Folkman, J. Seminars in Medicine of the Beth Israel Hospital, Boston. Clinical applications of research on angiogenesis. *N Engl J Med* **333**, 1757–1763, 1995.

27. Asahara, T. et al. Isolation of Putative Progenitor Endothelial Cells for Angiogenesis. *Science* **275**, 964–966, 1997.

28. Rafii, S. and Lyden, D. Therapeutic stem and progenitor cell transplantation for organ vascularization and regeneration. *Nat Med* **9**, 702–712, 2003.

29. Urbich, C. and Dimmeler, S. Endothelial progenitor cells: characterization and role in vascular biology. *Circ Res* **95**, 343–353, 2004.

30. Peichev, M. et al. Expression of VEGFR-2 and AC133 by circulating human CD34(+) cells identifies a population of functional endothelial precursors. *Blood* **95**, 952–958, 2000.

31. Sata, M. et al. Hematopoietic stem cells differentiate into vascular cells that participate in the pathogenesis of atherosclerosis. *Nat Med* **8**, 403–409, 2002.

32. Simper, D., Stalboerger, P.G., Panetta, C.J., Wang, S., and Caplice, N.M. Smooth Muscle Progenitor Cells in Human Blood. *Circulation* **106**, 1199–1204, 2002.

33. Caplice, N.M. et al. Smooth muscle cells in human coronary atherosclerosis can originate from cells administered at marrow transplantation. *Proc Natl Acad Sci U S A* **100**, 4754–4759, 2003.

34. Rabbany, S.Y., Heissig, B., Hattori, K., and Rafii, S. Molecular pathways regulating mobilization of marrow-derived stem cells for tissue revascularization. *Trends Mol Med* **9**, 109–117, 2003.

35. Cheng, T. et al. Hematopoietic stem cell quiescence maintained by p21cip1/waf1. *Science* **287**, 1804-8 (2000).

36. Cheng, T., Rodrigues, N., Dombkowski, D., Stier, S., and Scadden, D.T. Stem cell repopulation efficiency but not pool size is governed by p27(kip1). *Nat Med* **6**, 1235–1240, 2000.

37. Polyak, K. et al. p27Kip1, a cyclin-Cdk inhibitor, links transforming growth factor-beta and contact inhibition to cell cycle arrest. *Genes Dev* **8**, 9–22, 1994.

38. Pagano, M. et al. Role of the ubiquitin-proteasome pathway in regulating abundance of the cyclin-dependent kinase inhibitor p27. *Science* **269**, 682–685, 1995.

39. Boehm, M. et al. A growth factor-dependent nuclear kinase phosphorylates p27(Kip1) and regulates cell cycle progression. *EMBO J* **21**, 3390–3401, 2002.

40. Ishida, N., Kitagawa, M., Hatakeyama, S., and Nakayama, K. Phosphorylation at serine10, a major phosphorylation site of p27(Kip1), increases its protein stability. *J Biol Chem* **275**, 25146–25154, 2000.

41. Carrano, A.C., Eytan, E., Hershko, A., and Pagano, M. SKP2 is required for ubiquitin-mediated degradation of the CDK inhibitor p27. *Nat Cell Biol* **1**, 193–199, 1999.

42. Malek, N.P. et al. A mouse knock-in model exposes sequential proteolytic pathways that regulate p27Kip1 in G1 and S phase. *Nature* **413**, 323–327, 2001.

43. Kamura, T. et al. Cytoplasmic ubiquitin ligase KPC regulates proteolysis of p27(Kip1) at G1 phase. *Nat Cell Biol* **6**, 1229–1235, 2004.

44. Imaki, H. et al. Cell cycle-dependent regulation of the Skp2 promoter by GA-binding protein. *Cancer Res* **63**, 4607–4613, 2003.

45. Viglietto, G. et al. Cytoplasmic relocalization and inhibition of the cyclin-dependent kinase inhibitor p27(Kip1) by PKB/Akt-mediated phosphorylation in breast cancer. *Nat Med* **8**, 1136–1144, 2002.

46. Hung, D.T. and Schreiber, S.L. cDNA cloning of a human 25 kDa FK506 and rapamycin binding protein. *Biochem Biophys Res Commun* **184**, 733–738, 1992.

47. Luo, Y. et al. Rapamycin resistance tied to defective regulation of p27Kip1. *Mol Cell Biol* **16**, 6744–6751, 1996.

48. Marx, S.O., Jayaraman, T., Go, L.O., and Marks, A.R. Rapamycin-FKBP inhibits cell cycle regulators of proliferation in vascular smooth muscle cells. *Circ Res* **76**, 412–417, 1995.

49. Sousa, J.E. et al. Lack of neointimal proliferation after implantation of sirolimus-coated stents in human coronary arteries: a quantitative coronary angiography and three-dimensional intravascular ultrasound study. *Circulation* **103**, 192–195, 2001.

50. Morice, M.C. et al. A randomized comparison of a sirolimus-eluting stent with a standard stent for coronary revascularization. *N Engl J Med* **346**, 1773–1780, 2002.

51. Wani, M.C., Taylor, H.L., Wall, M.E., Coggon, P., and McPhail, A.T. Plant antitumor agents. VI. The isolation and structure of taxol, a novel antileukemic and antitumor agent from Taxus brevifolia. *J Am Chem Soc* **93**, 2325–2327, 1971.

52. Sollott, S.J. et al. Taxol inhibits neointimal smooth muscle cell accumulation after angioplasty in the rat. *J Clin Invest* **95**, 1869–1876, 1995.

53. Heldman, A.W. et al. Paclitaxel stent coating inhibits neointimal hyperplasia at 4 weeks in a porcine model of coronary restenosis. *Circulation* **103**, 2289–2295, 2001.

54. Senderowicz, A.M. Small-molecule cyclin-dependent kinase modulators. *Oncogene* **22**, 6609–6620, 2003.

55. Jaschke, B. et al. Local cyclin-dependent kinase inhibition by flavopiridol inhibits coronary artery smooth muscle cell proliferation and migration: implications for the applicability on drug-eluting stents to prevent neointima formation following vascular injury. *FASEB J* **18**, 1285–1287, 2004.

56. Morishita, R. et al. A gene therapy strategy using a transcription factor decoy of the E2F binding site inhibits smooth muscle proliferation in vivo. *Proc Natl Acad Sci U S A* **92**, 5855–5859, 1995.

57. Mann, M.J. et al. Ex-vivo gene therapy of human vascular bypass grafts with E2F decoy: the PREVENT single-centre, randomized, controlled trial. *Lancet* **354**, 1493–1498, 1999.

58. el-Deiry, W.S. et al. WAF1, a potential mediator of p53 tumor suppression. *Cell* **75**, 817–825, 1993.

59. Hutvagner, G. and Zamore, P.D. RNAi: nature abhors a double-strand. *Curr Opin Genet Dev* **12**, 225–232, 2002.

60. Gertz, S.D., Kalan, J.M., Kragel, A.H., Roberts, W.C., and Braunwald, E. Cardiac morphologic findings in patients with acute myocardial infarction treated with recombinant tissue plasminogen activator. *Am J Cardiol* **65**, 953–961, 1990.

61. Anversa, P., Fitzpatrick, D., Argani, S., and Capasso, J.M. Myocyte mitotic division in the aging mammalian rat heart. *Circ Res* **69**, 1159–1164, 1991.

62. Marino, T.A. et al. Proliferating cell nuclear antigen in developing and adult rat cardiac muscle cells. *Circ Res* **69**, 1353–1360, 1991.

63. Wang, X., Li, F., Said, S., Capasso, J.M., and Gerdes, A.M. Measurement of regional myocardial blood flow in rats by unlabeled microspheres and Coulter channelyzer. *Am J Physiol* **271**, H1656–H1665, 1996.

64. Winick, M. and Noble, A. Quantitative changes in DNA, RNA, and protein during prenatal and postnatal growth in the rat. *Dev Biol* **12**, 451–466, 1965.

65. Poolman, R.A. and Brooks, G. Expressions and activities of cell cycle regulatory molecules during the transition from myocyte hyperplasia to hypertrophy. *J Mol Cell Cardiol* **30**, 2121–2135, 1998.

66. Soonpaa, M.H., Kim, K.K., Pajak, L., Franklin, M., and Field, L.J. Cardiomyocyte DNA synthesis and binucleation during murine development. *Am J Physiol* **271**, H2183–H2189, 1996.

67. Burton, P.B., Yacoub, M.H., and Barton, P.J. Cyclin-dependent kinase inhibitor expression in human heart failure: a comparison with fetal development. *Eur Heart J* **20**, 604–611, 1999.

68. Burton, P.B., Raff, M.C., Kerr, P., Yacoub, M.H., and Barton, P.J. An intrinsic timer that controls cell-cycle withdrawal in cultured cardiac myocytes. *Dev Biol* **216**, 659–670, 1999.

69. Poolman, R.A., Li, J.M., Durand, B., amd Brooks, G. Altered expression of cell cycle proteins and prolonged duration of cardiac myocyte hyperplasia in p27KIP1 knockout mice. *Circ Res* **85**, 117–127, 1999.

70. Koh, K.N. et al. Persistent and heterogenous expression of the cyclin-dependent kinase inhibitor, p27KIP1, in rat hearts during development. *J Mol Cell Cardiol* **30**, 463–474, 1998.

71. Poolman, R.A., Gilchrist, R., and Brooks, G. Cell cycle profiles and expressions of p21CIP1 AND P27KIP1 during myocyte development. *Int J Cardiol* **67**, 133–142, 1998.

72. Reiss, K. et al. Insulin-like growth factor-1 receptor and its ligand regulate the reentry of adult ventricular myocytes into the cell cycle. *Exp Cell Res* **235**, 198–209, 1997.

73. Goldman, B., Mach, A., and Wurzel, J. Epidermal growth factor promotes a cardiomyoblastic phenotype in human fetal cardiac myocytes. *Exp Cell Res* **228**, 237–245, 1996.

74. Soonpaa, M.H. et al. Cyclin D1 overexpression promotes cardiomyocyte DNA synthesis and multinucleation in transgenic mice. *J Clin Invest* **99**, 2644–2654, 1997.

75. von Harsdorf, R. et al. E2F-1 overexpression in cardiomyocytes induces downregulation of p21CIP1 and p27KIP1 and release of active cyclin-dependent kinases in the presence of insulin-like growth factor I. *Circ Res* **85**, 128–136, 1999.

76. Machida, N., Brissie, N., Sreenan, C., and Bishop, S.P. Inhibition of cardiac myocyte division in c-myc transgenic mice. *J Mol Cell Cardiol* **29**, 1895–1902, 1997.

77. Xiao, G. et al. Inducible activation of c-Myc in adult myocardium in vivo provokes cardiac myocyte hypertrophy and reactivation of DNA synthesis. *Circ Res* **89**, 1122–1129, 2001.

78. Claycomb, W.C. et al. HL-1 cells: a cardiac muscle cell line that contracts and retains phenotypic characteristics of the adult cardiomyocyte. *Proc Natl Acad Sci U S A* **95**, 2979–2984, 1998.

79. Kiyokawa, H. et al. Enhanced growth of mice lacking the cyclin-dependent kinase inhibitor function of p27(Kip1). *Cell* **85**, 721–732, 1996.

80. Clarke, D.L. et al. Generalized potential of adult neural stem cells. *Science* **288**, 1660–1663, 2000.

81. Toma, C., Pittenger, M.F., Cahill, K.S., Byrne, B.J., and Kessler, P.D. Human mesenchymal stem cells differentiate to a cardiomyocyte phenotype in the adult murine heart. *Circulation* **105**, 93–98, 2002.

82. Condorelli, G. et al. Cardiomyocytes induce endothelial cells to trans-differentiate into cardiac muscle: implications for myocardium regeneration. *Proc Natl Acad Sci U S A* **98**, 10733–10738, 2001.

83. Orlic, D. et al. Mobilized bone marrow cells repair the infarcted heart, improving function and survival. *Proc Natl Acad Sci U S A* **98**, 10344–10349, 2001.

84. Orlic, D. et al. Bone marrow cells regenerate infarcted myocardium. *Nature* **410**, 701–705, 2001.

Section II

CDK Inhibitors: Targets and Their Evaluation

5 Evaluation of CDK Inhibitor Selectivity: From Affinity Chromatography to Yeast Genetics

Stéphane Bach, Marc Blondel,
and Laurent Meijer

CONTENTS

5.1   OVERVIEW

Abnormal phosphorylation of cellular proteins is frequently observed in human disease. For this reason, there has been a growing interest in the discovery and optimization of protein kinase inhibitors (review in Cohen, 2002). Indeed, after G-protein-coupled receptors, protein kinases constitute the second class of drug discovery screening targets. Fifty-five kinase inhibitors are currently undergoing clinical evaluation against various human diseases. Some, such as imatinib (Gleevec) or gefitinib (Iressa), have actually achieved real market success and have provided proof-of-principle that small molecule kinase inhibitors can be effective drugs. Among the human 518+ kinases, cyclin-dependent kinases (CDKs) regulate the cell division cycle, apoptosis, transcription, differentiation, and many functions in the nervous system. CDKs are low-molecular-weight serine or threonine kinases (34–40 kDa) that are inactive in a monomeric state. To be active, they require an association with one of the cyclins, their regulatory subunits. Frequent deregulation of CDKs in cancers, neurodegenerative diseases, and other pathologies justify the active search for chemical inhibitors able to reversibly and selectively inhibit this class of enzymes. Intensive screening of collections of natural and synthetic compounds has led to the identification of several families of CDK inhibitors, most of which act by competition with ATP for binding at the catalytic site. Although the therapeutic potential of the most promising compounds is currently being evaluated in preclinical and clinical trials, their exact mechanism of action and the real spectrum of their intracellular targets remain largely unknown. Determination of the *in vivo* selectivity of the compounds and identification of their intracellular targets constitute a prerequisite to understand their cellular effects and to improve their efficiency on a rational basis.

5.2　INTRODUCTION

Efforts to identify chemical CDK inhibitors of pharmacological interest tend to identify molecules with high efficiency and selectivity to meet the two criteria required for clinical development of a drug: efficiency and harmlessness. Drug candidates are expected to inhibit efficiently the target they have been optimized against (CDKs), but also to not interfere (or interfere as little as possible) with the activity of other cellular enzymes and proteins, to limit undesirable side effects. Determination of the selectivity of the compounds and identification of their intracellular targets thus constitute a crucial step in understanding their cellular effects.

The classical method for determining a compound's selectivity involves testing the compound against a panel of kinases (Davies et al., 2000; Bain et al., 2003). Notably, using various academic or commercially available kinase panels (ProQinase selectivity panel [85 kinases], Invitrogen SelectScreen™ Kinase Profiling panel [70 kinases], the Cerep kinase panel [50 kinases], (R)-roscovitine (CYC202, Seliciclib, Cyclacel Ltd., Dundee, U.K.)), a member of the 2,6,9-trisubstituted purines family (Meijer et al., 1997) has been tested against 151 different purified kinases (Bach et al., 2005). IC_{50} values were below 1 μM only for CDK1, CDK2, CDK3, CDK5, CDK7, and CDK9. Only a few kinases were sensitive to roscovitine in the 1 to 40 μM range (CaM Kinase 2, CSNK1A1, CSNK1D, DYRK1A, EPHB2, ERK1, ERK2, FAK, and IRAK4). Most other kinases were insensitive to roscovitine. Based on this analysis, (R)-roscovitine appears to be relatively selective for CDKs (Bach et al., 2005). However, this panel only reflects 29.2% of the reported 518+ kinases that constitute the human kinome. In addition, the chemical compound might interact with some of the other ~1500 ATP-utilizing enzymes and numerous other nucleotide-binding proteins present in the human proteome. Consequently, many potential targets are not evaluated using this method which, in addition, requires time- and budget-consuming expression, purification, and assay set up for each individual kinase (Davies et al., 2000).

Alternative methods to tackle the selectivity problem have been described in the literature. They take advantage of the recent progress in genomic and mass spectrometry detection and allow mapping of drug targets on a genomewide or proteomewide scale.

Here, we review several biochemical and yeast genetics-based approaches developed or applicable to investigate the range of true targets of CDK inhibitors. In this chapter, we will not present other methods such as computational approaches that have already been discussed elsewhere (Rockey and Elcock, 2005).

5.3 BIOCHEMICAL METHODS

5.3.1 Affinity Chromatography

This first alternative method to investigate the selectivity of an inhibitor is based on the affinity chromatography purification of targets on immobilized CDK inhibitors (review in Knockaert and Meijer, 2002). Compared to testing compounds against a panel of purified kinases, this approach should, in principle, allow a more comprehensive analysis of the targets of a given inhibitor. In particular, this method has been applied successfully to purvalanol (Knockaert et al., 2002a), paullones (Knockaert et al., 2002b), indirubins (Meijer et al., 2003), hymenialdisine (Wan et al., 2004), Iressa (Brehmer et al., 2005), and more recently to roscovitine (Bach et al., 2005).

The affinity purification method is summarized in Figure 5.1 using different purine compounds ((R)-roscovitine and purvalanol) as an example. Briefly, a poly-ethylene glycol linker is covalently attached to an appropriate position on the inhibitor, and the other extremity of the linker is then covalently bound to an agarose matrix. The choice of the appropriate position on the compound for attachment of the linker can largely benefit from 3-D structure determination of CDK cocrystallized with the inhibitor. For example, the crystal structures of roscovitine in complex with CDK2 (De Azevedo et al., 1997) and CDK5 (Mapelli et al., 2005) show that the benzyl ring substitution faces the outside of the ATP-binding pocket of the kinases. This is where a polyethylene glycol extension is attached (Figure 5.1). Indeed, addition of a linker on this site does not significantly modify the protein kinase inhibition properties, as is expected (Bach et al., 2005). The linker allows the covalent binding of roscovitine to agarose (Figure 5.1). Extracts from various cell types and tissues are batch-loaded on the inhibitor matrix. After extensive washing of the resin, the bound proteins are eluted with an electrophoresis sample buffer, separated by polyacrylamide gel electrophoresis, and can be identified by various approaches (Western blotting, peptide microsequencing, or MALDI-TOF peptide mapping). An example of the porcine brain proteins that bind to roscovitine and purvalanol matrices is shown in Figure 5.1. As expected, CDKs, but also p42/p44 MAPKs, are found to interact to the same extent with both matrices in most models tested (starfish oocytes, sea urchin eggs, porcine brain, and mammalian cell lines) (Knockaert et al., 2000; Knockaert et al., 2002a; Bach et al., 2005). The binding of p42/p44 MAPKs seems quite intriguing at first, given the significant difference between the IC_{50} values obtained for purvalanol on CDKs and MAPKs *in vitro*: 6 nM for CDK1/cyclin B and 1000 nM for p44 MAPK, respectively. This result probably partly reflects the relative abundance of the two types of kinases. In other words, purvalanol interacts in the cell with targets for which it has a strong affinity but that are not very abundant (CDKs), and also with targets for which it has a moderate affinity but that are much

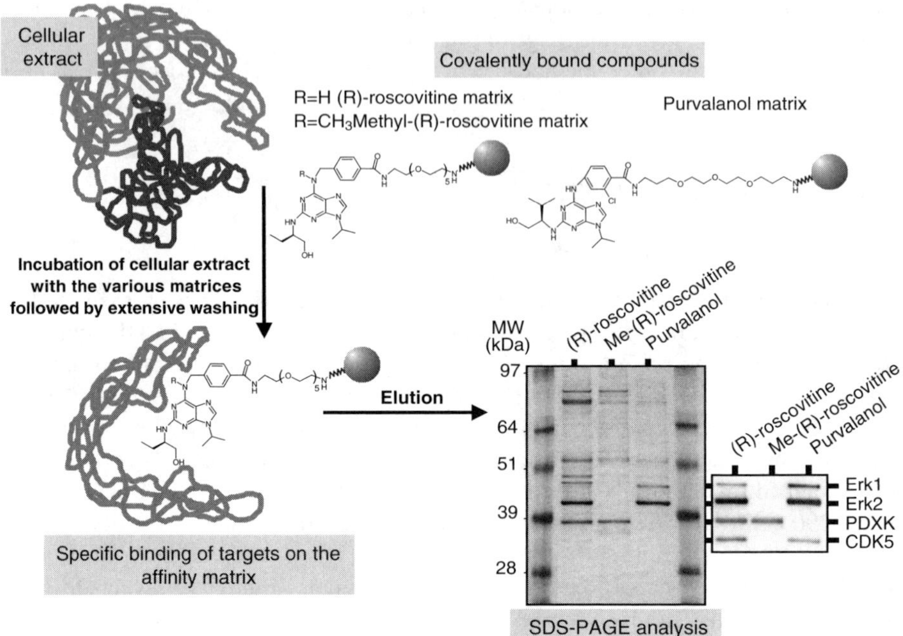

FIGURE 5.1 Affinity chromatography purification of the targets of CDK inhibitors (Adapted from Knockaert, M. and Meijer, L. (2002). *Biochem Pharmacol.* **64**, 819–825.) The inhibitors, (R)-roscovitine and purvalanol in this example, were covalently linked to agarose beads through a polyethylene glycol linker. A control resin for roscovitine was also prepared using N^6-methyl-(R)-roscovitine, a protein-kinase-inactive compound designed on the basis of PDXK/(R)-roscovitine and CDK/(R)-roscovitine crystal structures. Cell and tissue extracts were then incubated with these matrices. After extensive washing, the affinity matrix-bound proteins were resolved by SDS-PAGE. In this experiment, the identification of the proteins detected after silver staining was realized by Western blotting with specific antibodies. Alternatively, the proteins excised from the gel can also be identified by microsequencing of internal tryptic peptides or by MALDI-TOF peptide mapping. In this example, using a porcine brain extract, the targets of (R)-roscovitine were identified as p44MAPK/Erk1, p42MAPK/Erk2, pyridoxal kinase (PDXK), and CDK5. All the different protein kinases were absent from control beads (N^6-methyl-(R)-roscovitine). Note that, with the exception of PDXK, the targets of purvalanol are the same as those of roscovitine.

more abundant (MAPKs). Taken together, these results suggest that the cellular effects of purvalanol are probably more complex than were initially anticipated. Indeed, it has been shown that inhibition of p42/p44/MAPKs by purvalanol contributes to its observed cellular effects (Knockaert et al., 2002a).

Roscovitine has been found to interact with an unexpected nonprotein kinase, off-target, pyridoxal kinase (PDXK) (Figure 5.1) (Bach et al., 2005; Tang et al., 2005). PDXK is responsible for the phosphorylation and activation of vitamin B_6. In addition, structural analysis reveals that there are notable differences between the interactions of roscovitine with PDXK, CDK2, and CDK5. Comparison of these structures is very helpful in the design of different roscovitine derivatives (Tang et al., 2005). As shown in Figure 5.1, N^6-methyl-(R)-roscovitine (control beads) specifically targets PDXK while not interacting with protein kinases. Compared to (R)-roscovitine, N^6-methyl-(R)-roscovitine displays only a weak antiproliferative activity, but additional experiments are needed to completely understand the contribution of the PDXK inhibition to the antiproliferative effects of (R)-roscovitine (Bach et al., 2005). Nevertheless, one can imagine that PDXK may be able to titrate out roscovitine, thereby reducing its effects on proliferation-relevant targets such as CDKs. To eliminate the interaction with PDXK and thus increase roscovitine's intracellular availability for inhibition of CDKs, different roscovitine analogs specifically targeting the protein kinases are currently being synthesized. After analysis of their cellular effects, the same affinity chromatography approach will be performed to confirm their specificity toward CDKs and possibly toward other kinases.

Using the same strategy, the selectivity of different protein kinase inhibitors has been analyzed by the group of Henrik Daub at Axxima Pharmaceuticals AG. These include the following: SB203580, a p38 kinase inhibitor (Godl et al., 2003); PP58, a member of the pyrido[2,3-d]pyrimidine class of compounds used as alternative inhibitor of cellular Bcr-Abl tyrosine kinase activity (Wissing et al., 2004) and gefitinib (Iressa, ZD1839), an inhibitor of the EGFR tyrosine kinase (Brehmer et al., 2005). The ATP-competitive kinase inhibitor gefitinib was the first EGFR-directed small molecule that received approval for the treatment of non-small-cell lung cancer. The affinity chromatography approach has led to the identification of more than 20 other kinase targets, notably RICK and GAK serine and threonine kinases with IC_{50} values of about 50 and 90 nM, respectively. These results provide new insight into the potential cellular mode of action of gefitinib and will be used for the optimization of gefitinib-derived drugs (Brehmer et al., 2005).

5.3.2 Phage-Based Methods

Phage display, a well-described technique used to elucidate protein–protein interactions (Scott and Smith, 1990), is an *in vitro* technique that allows the expression, selection, and subsequent amplification of proteins on the surface of bacteriophage viral particles (Crameri et al., 1994). New phage-based techniques have been described recently that allow small molecule or target protein analysis: (i) phage display and (ii) competition-binding assay. These methods are summarized in Figure 5.2. All these assays are based on the use of small chemical compound ("bait") used in combination with (i) cDNA phage display library (Sche et al., 1999) or with (ii) particular protein kinases expressed as fusions to the T7 bacteriophage capsid protein (Fabian et al., 2005).

Display cloning is a method for the concomitant isolation and identification of protein cellular targets using a T7 cDNA phage display library and a biotinylated chemical compound as an affinity probe (Figure 5.2, panel a). Sche et al. (1999) have used an immunosuppressive drug, FK506, to validate this screening method. After several rounds of selection, they were able, in particular, to isolate a cDNA clone that encodes for the FKBP12 protein, the known and expected cellular target

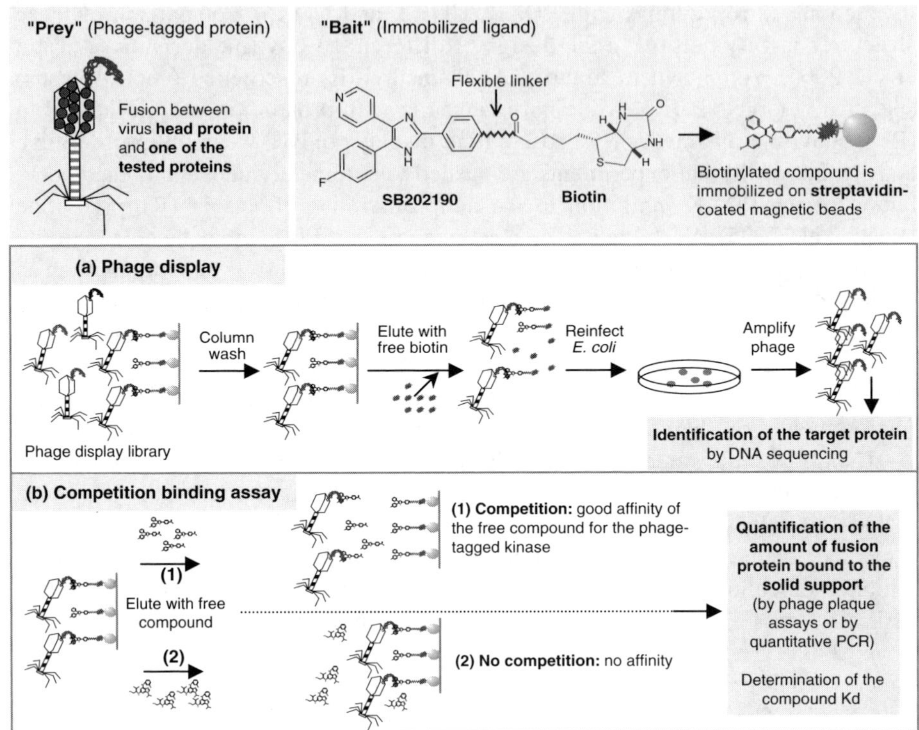

FIGURE 5.2 Schematic overview of phage-based methods used for the analysis of small chemical compounds' cellular targets. As shown in the upper panel, these two methods used a "prey" (a phage-tagged protein) and a "bait" (a biotinylated chemical compound immobilized on avidin- or streptavidin-coated beads). The SB202190 compound, known to interact through the ATP-binding site of the p38 MAP kinase, is used as an example. (a) Phages that display proteins with specific affinity for the "prey" compound remain on the column after extensive washing and are then selectively eluted with excess biotin. After phage amplification into *E. coli*, the target protein is identified by sequencing the phage DNA. (b) In the competition-binding assay, the affinity of different chemical compounds (or "free test compounds") for one protein kinase is tested. In this case, the amount of phage bound on the affinity column after competition indicates the affinity of the free test compound for the ATP site of the prey kinase.

of FK506. The display-cloning method can thus be used to select proteins directly from cDNA phage display libraries using small molecule probes (Sche et al., 1999). The second phage-based approach, developed by Ambit Biosciences, is a quantitative biochemical assay. It is an ATP site-dependent competition-binding assay and is designed for the study of small molecule–kinase interactions (Figure 5.2, panel b). After validation of this method using SB202190 as chemical bait, Fabian et al. (2005) have tested the affinity of 20 well-characterized clinical ATP-competitive inhibitors (e.g., staurosporine and roscovitine) in a panel of 113 different kinases. The results obtained show that molecular specificity varies widely among these inhibitors. Staurosporine is known to be a highly promiscuous inhibitor of many different kinases, it efficiently binds 104 of the 113 kinases. Other inhibitors (such as roscovitine or vatalanib) bind very few kinases, in addition to their known primary targets (Fabian et al., 2005). This assay is performed without any addition of ATP or substrate and measures binding rather than activity. The results therefore, depend neither on ATP concentration nor on the specific choice of substrate. In addition, there is a good correlation between the measured binding constants and published IC_{50} values for the inhibitor tested. This phage-based binding assay is thus a good alternative method to the traditional *in vitro* kinase selectivity panel assays to investigate selectivity. In conclusion, for genome screening of compound targets, phage display has a lot of advantages but is limited by the quality of the cDNA library.

5.4 YEAST-BASED METHODS

5.4.1 CHEMICAL-GENOMIC

The budding yeast *Saccharomyces cerevisiae* is a unicellular organism with a compact genome of approximately 6000 genes and is widely used as a model organism for investigating many aspects of eukaryotic cell biology. Indeed, a high level of conservation between its cellular processes (e.g., cell division cycle and basic cellular metabolism) and those of higher organisms can be observed. In addition, yeast displays very simple growth requirements, rapid cell division (90 min in rich medium), ease of genetic manipulation, and a wealth of experimental tools for genomewide analysis of biological functions (Simon and Bedalov, 2004). Recently, different yeast-based systems have been described for the discovery of new therapeutic drugs (Mager and Winderickx, 2005). For instance, 6-aminophenanthridine has been isolated using a yeast prion-based screening method and has then been shown to be active against mammalian prions (Bach et al., 2003). Concerning drug target identification, a number of genetic approaches based on the use of *S. cerevisiae* have been developed (Parsons et al., 2003).

5.4.1.1 Cell-Based High-Throughput Screening Methods

These cell-based high-throughput screening methods are: (1) haploinsufficiency profiling (HIP), (2) synthetic lethal screen (SLS), and (3) genomewide overexpression screen (OES). All these different approaches are presented in Figure 5.3.

In the HIP approach (Figure 5.3, part 1), a library of diploid yeast strains with heterozygous deletions of each single gene is screened for drug sensitivity either in a single culture with a competitive growth assay or by screening the 6000+ strains in parallel. This assay is based on the observation that reducing the copy number of a gene encoding a drug target from two copies (in diploid yeast strain) to one copy (in a heterozygous deletion strain) results in a strain that could be sensitized to the drug of interest (Giaever et al., 2004). To obtain *bona fide* proof of concept for this approach, two different groups, Lum et al. (2004) and Giaever et al. (2004), have

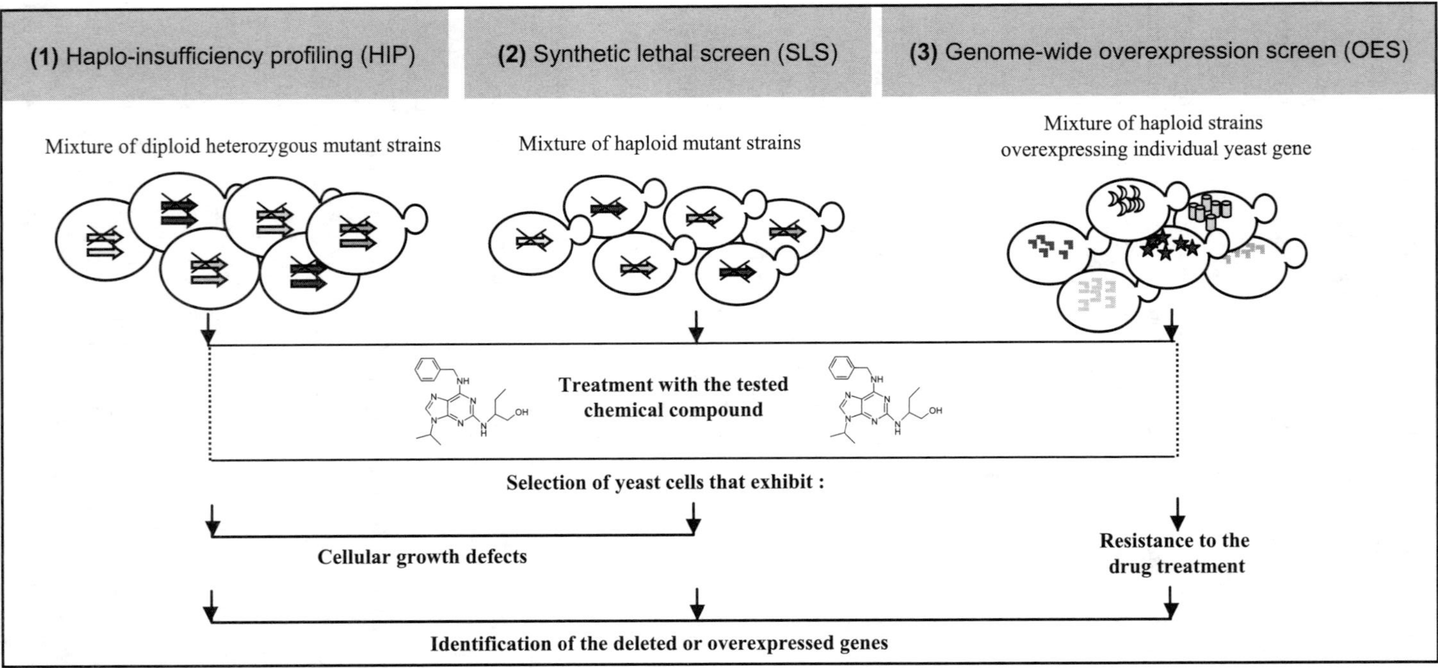

FIGURE 5.3 Cell-based high-throughput screening methods. Note that in (1) HIP and (2) SLS, cells that are sensitive to the drug of interest are selected, whereas in the case of (3) OES, resistant clones are analyzed.

assessed the cellular effect of 78 and 10 different chemical compounds (e.g., staurosporine), respectively. They used a mixture of isogenic yeast mutants generated by pooling 3503 (representing over half the genome) and ~6000 heterozygous deletion strains, respectively. Taking advantage of "molecular bar" codes, they were able to quantify the relative abundance of each strain. As reported in the two studies, some of the most sensitive heterozygous strains often carry a deletion in the gene whose product is known to interact directly with the test molecule (e.g., methotrexate with dihydrofolate reductase, *DHFR*) (Giaever et al., 2004). A likely explanation for this observation is that the compound inhibits cellular proliferation by reducing the activity of the remaining gene product of the heterozygous locus, thereby mimicking a complete deletion. In addition, they have discovered many potentially new drug targets: for example, the implication of different tricyclic antidepressants (e.g., chlorpromazine) in integral membrane ATPase activity (Lum et al., 2004). In the case of staurosporine, the failure to detect its precise protein target (protein kinase C1, *Pkc1*) can be attributed to the fact that the diploid strain deleted for one of the two copies of the *PKC1* gene was not present in the pool of screened strains (Lum et al., 2004). Indeed, in this study, screened heterozygotes represented only about half the yeast genome. It is important to note that a putative limitation for this method is the existence of genes that, although homologues of genes clearly important for cell viability in higher eukaryotes, could be neither essential, nor their inactivation detrimental in budding yeast, thus preventing their isolation despite the fact they could be actual targets of the studied drugs.

In the SLS approach (Figure 5.3, part 2), a drug is screened at a concentration that is normally sublethal against a library of haploid yeast strains with individual gene deletions. Genes whose deletion results in increased drug sensitivity might be potential direct drug targets or genes that are involved in the same cellular pathways as the drug target. Charles Boone's team has integrated data obtained using this chemical-genetic approach and genetic data from synthetic lethal interaction screens. Using this large-scale analysis, they were able to link 12 diverse and well-characterized growth-inhibitory compounds to their target pathways and proteins (e.g., calcineurin for FK506 compound) (Parsons et al., 2004).

The OES (Figure 5.3, part 3) is a complementary approach to the screens described earlier. It corresponds to the identification of genes, the overexpression of which (because of expression from a high copy number plasmid) confers resistance to an inhibitor. This screen is based on the principle that cells that express increased levels of a target should tolerate higher levels of the drug that interacts with it. Using this approach, Gray and coworkers have analyzed the yeast cellular targets of a kinase inhibitor of the phenylaminopyrimidine (PAP) structural family. They showed that specific resistance to PAP compounds is associated with the overexpression of Pkc1 and a subset of downstream protein kinases. The binding to Pkc1 was then checked using a biochemical approach based on affinity chromatography (described in Subsection 5.3.1), and the utility of this approach in the analysis of small molecule targets was demonstrated (Luesch et al., 2005).

The main limitations of using these yeast-based screening methods to identify drug targets is that only a subset of potentially interesting compounds has a growth-inhibitory phenotype in yeast (Luesch et al., 2005). Failure of a compound to affect the growth rate of a cell may indicate that (1) the drug target is not encoded by the yeast genome, (2) the drug's effects are masked by other proteins with redundant activities but that are not sensitive to its action, (3) the drug is metabolized, or (4) the drug is unable to enter the cells (Lum et al., 2004). Concerning this last point, yeast cell permeability (which is naturally relatively poor, probably because in addition to a plasma membrane, yeast cells have a cell wall) can be artificially increased by inactivating various genes involved either in the plasma membrane, cell wall metabolism, or in multidrug resistance (MDR). For instance, the *erg6* mutation is known to increase permeability across the yeast cell wall. In several cases, drugs that are inactive in WT cells exhibit clear activity in *erg6* mutant cells. For example, this mutation was successfully used in a yeast-based antiprion drugs screening assay (Bach et al., 2003). Despite these various limitations, approaches based on the use of budding yeast represent a powerful tool for drug target identification.

5.4.1.2 Transcriptional "Signature" Analysis

The cell-based high-throughput screening technologies described earlier (HIP, SLS, and OES) provide information about cellular pathways responsible for drug sensitivity; but because not all kinases are essential, they do not necessarily identify all targets in a cell. Recently, Kevan Shokat and colleagues have described a chemical genomic strategy that uses direct comparison between microarray transcriptional signatures elicited by highly specific pharmacological inhibition-engineered candidate kinase targets to identify the kinase targeted by protein kinase inhibitors (Kung et al., 2005). To summarize this screening method, a space-creating mutation was inserted in the ATP-binding pocket of each protein kinase. This mutation renders protein kinases susceptible to ATP-competitive inhibitors, such as 1-NA-PP1 and 1-NM-PP1, which have been designed to be poor inhibitors for nonengineered kinases. Importantly, it appears that most protein kinases are amenable to inhibitor sensitization at this site and still allow for sufficient kinase catalytic activity to retain normal cellular function (Kung and Shokat, 2005). Shokat and coworkers used a library of 126 different yeast strains in which the genes for all individual kinases were replaced with their inhibitor analog-sensitive counterparts. Profiling the transcriptional effect arising from inhibition of each kinase would identify specific sets of transcripts that could serve as a diagnostic transcriptional "signature." The study of the selectivity of a new kinase inhibitor of incompletely characterized specificity is then realized by comparison of the signature obtained by treatment of the yeast cells with this new inhibitor to "the reference profiles" elicited by specific inhibition of candidate analog-sensitive kinases (Kung and Shokat, 2005).

Using this strategy, the selectivity of GW400426, a CDK inhibitor, was analyzed in yeast. Cdk1 (yeast CDK2 homologue) and Pho85 were characterized as the targets of this inhibitor in yeast. Interestingly, this study also shows that the simultaneous inhibition by GW400426 of both CDK1 and Pho85 together, and not the mere inhibition of only one of these two kinases, controls the expression of specific transcripts involved in polarized cell growth, thus revealing a cellular process that is uniquely sensitive to the multiplex inhibition of these two kinases (Kung et al., 2005).

An exhaustive application of this technology would thus result in the identification of phenotypic changes elicited by inhibition of both individual or multiple kinases (Kung and Shokat, 2005). A drawback of this method is that it is probably efficient for highly selective compounds as poorly selective molecules would lead to very complicated overlapping signatures.

5.4.2 THREE-HYBRID SYSTEM

The yeast three-hybrid system (Y3H) is based on the yeast two-hybrid system (Y2H), which has proved to be a powerful tool for detecting protein–protein interactions (Fields and Song, 1989). In the Y2H system, protein–protein interactions lead to reconstitution of a transcriptional activator. Indeed, due to a close positioning in space of its DNA binding (DBD) and activation domain (AD) expressed separately as fusion with the two potentially interacting proteins, an active transcription factor can be reconstituted (Fields and Song, 1989). In 1996, Licitra and Liu extended this method to small molecule target identification by dimerization of two receptor proteins via a heterodimeric ligand (Licitra and Liu, 1996). As described in Figure 5.4, one ligand–receptor pair (e.g., DHFR-*methotrexate*) serves as an anchor, but the other ligand–receptor pair (e.g., (R)-roscovitine-target) is the small molecule–protein interaction of interest. In 2004, GPC Biotech reported the first application of this promising assay in a selectivity analysis of different ATP-competitive CDK inhibitors (Becker et al., 2004). As a new proof-of-principle, (R)-roscovitine scored positively with its known target CDK2. Interestingly, it also scored positively with other protein kinases: the casein kinase 1 CSNK1D and the CDK-like protein PCTK1; and comparatively weakened interactions with CSNK1E, CLK1, PAK4, PCTK3, PKWA, and GSK3α (Becker et al., 2004). However, given the variable parameters associated with expression, folding, and translocation of fusion proteins in yeast, it is too early to draw conclusions regarding potential differences in relative binding affinities of a given compound for different targets. Nevertheless, Becker et al. (2004) have confirmed many of the interactions involving putative novel kinase targets by secondary *in vitro* binding or enzyme inhibition assays; thus, Y3H can be considered another efficient way of scanning the proteome to identify the targets of kinase inhibitors (Becker et al., 2004).

5.5 CONCLUSIONS

Drug discovery progresses not through the use of a single technique, but by bringing together a host of tools. Efforts to identify CDK inhibitors first start with *in vitro* studies using purified native or recombinant enzymes. Following these *in vitro* studies, the properties of the most promising compounds are then generally evaluated in a cellular context. From a therapeutic perspective, performing thorough evaluation of CDK inhibitors' cellular mode of action is crucial as several factors are likely to interfere with the effects of a given compound: cell permeability, intracellular metabolism of the compound, intracellular distribution, competition with high intracellular concentration of ATP, and interaction with other kinase and nonkinase targets.

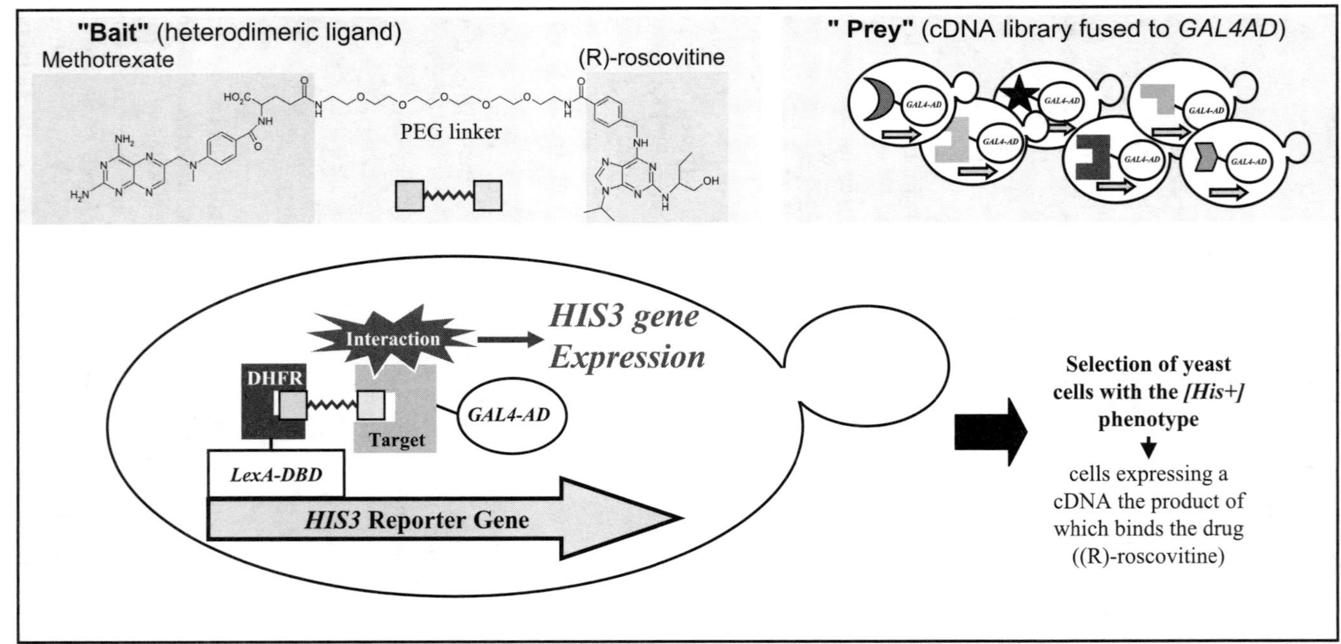

FIGURE 5.4 Three-hybrid assay (adapted from Lefurgy, S. and Cornish, V. (2004). *Chem Biol.* **11**, 151–153.) This method is based on the use of a heterodimeric ligand constituted by: (a) Methotrexate the ligand of the dihydrofolate reductase (DHFR) and; (b) the compound of interest, ((R)-roscovitine in this example). The heterodimeric ligand (Mtx/(R)-roscovitine) can be screened against a library of cDNA in fusion with the activation domain (AD) of a transcriptional activator (here, *GAL4*). The *HIS3* reporter gene will only be activated in cells expressing a cDNA encoding a protein to which the drug ((R)-roscovitine) binds, and thus only these cells will be able to grow in the absence of histidine in the medium.

TABLE 5.1
Selectivity Analysis of (R)-Roscovitine Using Different Methods

Method Used (Range of Putative Targets)	Targets Identified	References
Enzyme inhibition assay (151 protein kinases)	$IC_{50}<1\mu M$: CDK1, CDK2, CDK5, CDK7, and CDK9 $1<IC_{50}<40\ \mu M$: CaMK2, CSNK1A1, CSNK1D, DYRK1A, EPHB2, p42/p44 MAPKs, FAK, and IRAK4	Bach et al., 2005 Bain et al., 2003 Meijer et al., 1997
Phage-based competition binding assay (113 protein kinases)	CSNK1E, CLK2, PCTK1, CSNK1G2, CDK5, CLK1, TTK, CDK2, CSNK1G1, RPS6KA2 (Kin. Dom. 1), CLK4	Fabian et al., 2005
Affinity chromatography (Proteome)	CDKs (notably, CDK2 and CDK5), p42/p44 MAPKs, CaMK2 (isoforms α, β, δ, and γ), CSNK1A1, PDXK[a]	Bach et al., 2005
Yeast three-hybrid (Y3H) (Proteome)	CDK2, CSNK1D, PCTK1, and comparatively weaker with CSNK1E, CLK1, PAK4, PCTK3, PKWA, and GSK3α	Becker et al., 2004

[a] In this particular case (pyridoxal kinase, -PDXK-), roscovitine uses the substrate-binding site to inhibit PDXK activity.

After this validation, the use of different selectivity analysis methods presented here is crucial for a comprehensive exploration of the mode of action of the drug.

An example is given in Table 5.1 with the well-researched CDK inhibitor that is currently in phase 2 clinical trials, (R)-roscovitine (CYC-202, Seliciclib) (this book, Chapter 9). By using more techniques, a better understanding of the drug's cellular effects can be achieved. Indeed, the characterization of the 29 potential protein targets of (R)-roscovitine, shown in Table 5.1, was the result of four different approaches. By quantification of these potential interactions and inhibition, the main protein kinase targets of a drug can be defined. Here, in the case of (R)-roscovitine, the four studies have confirmed interaction with the anticipated targets: CDKs. In addition, several off-target proteins (protein kinases or nonprotein kinases, e.g., PDXK) have been identified. These other targets cannot be neglected as they may significantly contribute to the cellular response to (R)-roscovitine. For instance, high levels of PDXK in some tissues may reduce the availability of (R)-roscovitine for CDKs. Inhibition of these unexpected targets may also either contribute toward the desired pharmacological effect or cause undesirable side effects. All the data obtained should contribute to the development of second-generation compounds with improved pharmacological properties. Large-scale analysis, similar to the RNAi approach recently used to identify human kinases that regulate cell survival (MacKeigan et al., 2005), will probably help to determine positive or negative effects of multiplex inhibition on cellular growth inhibition.

Regarding the different available affinity-based methods (affinity chromatography, phage display-based methods, and Y3H), one limitation is that it is necessary to find ways to modify the structure of the tested compound to introduce the PEG-spacer or linker without interfering with the biological activity of the original "free" compound. *Saccharomyces cerevisiae*-based genetics approaches can be used to circumvent the chemical linker problem. Indeed, yeast research performed in the recent past has demonstrated the tremendous potential of this organism as a model system for medical and medicinal purposes. For instance, in 2000, Snyder and coworkers developed a protein chip technology that allows high-throughput biochemical analysis and used this approach to analyze nearly all protein kinases from *Saccharomyces cerevisiae* (119 out of the 122 known or predicted yeast protein kinases) (Zhu et al., 2000). With regard to drug selectivity analysis, at least four different elegant genomewide screening strategies have been designed using the yeast *Saccharomyces cerevisiae*. We anticipate that recent applications of this model organism in the characterization of the cellular role of key proteins involved, for instance, in complex neurodegenerative diseases (e.g., Huntingtin in Huntington's disease) (Giorgini et al., 2005) should lead to the elaboration of new, powerful yeast-derived screening methods in the near future.

ACKNOWLEDGMENTS

This article is dedicated to the memory of Dr. Jean-Claude Jauniaux. We acknowledge associates in our laboratories for helpful discussion and Damien Guiffant for critical reading of the manuscript. S. Bach is supported by a CNRS "Sciences de la Vie" grant; M. Blondel is supported by "Association pour la Recherche contre le Cancer" (contract ARC MB5812), by a PRIR from the "Conseil Regional de Bretagne" and by an ACI "Jeunes Chercheurs" from the French "Ministère de la Recherche"; L. Meijer is supported by the Ministère de la Recherche/INSERM/CNRS "Molécules et Cibles Thérapeutiques" program, a grant from the EEC (FP6-2002-Life Sciences and Health, PRO-KINASE Research Project) and the Canceropole Grand-Ouest.

REFERENCES

Bach, S., Knockaert, M., Reinhardt, J., Lozach, O., Schmitt, S., Baratte, B., Koken, M., Coburn, S.P., Tang, L., Jiang, T., Liang, D.C., Galons, H., Dierick, J.F., Pinna, L.A., Meggio, F., Totzke, F., Schachtele, C., Lerman, A.S., Carnero, A., Wan, Y., Gray, N., and Meijer, L. (2005). Roscovitine targets: protein kinases and pyridoxal kinase. *J Biol Chem.* June 23 **280**, 31208–31213

Bach, S., Talarek, N., Andrieu, T., Vierfond, J.-M., Mettey, Y., Galons, H., Dormont, D., Meijer, L., Cullin, C., and Blondel, M. (2003) Isolation of drugs active against mammalian prions using a yeast-based screening assay. *Nat. Biotechnol.* **21**, 1075–1081.

Bain, J., McLauchlan, H., Elliott, M., and Cohen, P. (2003) The specificities of protein kinase inhibitors: an update. *Biochem. J.* **371**, 199–204.

Becker, F., Murthi, K., Smith, C., Come, J., Costa-Roldan, N., Kaufmann, C., Hanke, U., Degenhart, C., Baumann, S., Wallner, W., Huber, A., Dedier, S., Dill, S., Kinsman, D., Hediger, M., Bockovich, N., Meier-Ewert, S., Kluge, AF., and Kley, N. (2004) A three-hybrid approach to scanning the proteome for targets of small molecule kinase inhibitors. *Chem Biol.* **11**, 211–223.

Brehmer, D., Greff, Z., Godl, K., Blencke, S., Kurtenbach, A., Weber, M., Muller, S., Klebl, B., Cotten, M., Keri, G., Wissing, J., and Daub, H. (2005) Cellular targets of gefitinib. *Cancer Res.* **65**, 379–382.

Cohen P. (2002) Protein kinases — the major drug targets of the twenty century? *Nat. Rev. Drug Discovery* **1**, 309–315.

Crameri, R., Jaussi, R., Menz, G., and Blaser, K. (1994) Display of expression products of cDNA libraries on phage surfaces: a versatile screening system for selective isolation of genes by specific gene-product/ligand interaction. *Eur J Biochem.* **226**, 53–58.

Davies, S.P., Reddy, H., Caivano, M., and Cohen, P. (2000) Specificity and mechanism of action of some commonly used protein kinase inhibitors. *Biochem. J.* **351**, 95–105.

De Azevedo, W.F., Leclerc, S., Meijer, L., Havlicek, L., Strnad, M., and Kim, S.H. (1997) Inhibition of cyclin-dependent kinases by purine analogues: crystal structure of human cdk2 complexed with roscovitine. *Eur J Biochem.* **243**, 518–526.

Fabian, M.A., Biggs, W.H., Treiber, D.K., Atteridge, C.E., Azimioara, M.D., Benedetti, M.G., Carter, T.A., Ciceri, P., Edeen, P.T., Floyd, M., Ford, J.M., Galvin, M., Gerlach, J.L., Grotzfeld, R.M., Herrgard, S., Insko, D.E., Insko, M.A., Lai, A.G., Lelias, J.M., Mehta, S.A., Milanov, Z.V., Velasco, A.M., Wodicka, L.M., Patel, H.K., Zarrinkar, P.P., and Lockhart, D.J. (2005) A small molecule-kinase interaction map for clinical kinase inhibitors. *Nat. Biotechnol.* **23**, 329–336.

Fields, S. and Song, O. (1989) A novel genetic system to detect protein-protein interactions. *Nature* **340**, 245–246.

Giaever, G., Flaherty, P., Kumm, J., Proctor, M., Nislow, C., Jaramillo, D.F., Chu, A.M., Jordan, M.I., Arkin, A.P., and Davis, R.W. (2004) Chemogenomic profiling: identifying the functional interactions of small molecules in yeast. *Proc Natl Acad Sci U S A* **101**, 793–798.

Giorgini, F., Guidetti, P., Nguyen, Q., Bennett, S.C., and Muchowski, P.J. (2005) A genomic screen in yeast implicates kynurenine 3-monooxygenase as a therapeutic target for Huntington disease. *Nat. Genet.* **37**, 526–531.

Godl, K., Wissing, J., Kurtenbach, A., Habenberger, P., Blencke, S., Gutbrod, H., Salassidis, K., Stein-Gerlach, M., Missio, A., Cotten, M., and Daub, H. (2003) An efficient proteomics method to identify the cellular targets of protein kinase inhibitors. *Proc Natl Acad Sci U S A* **100**, 15434–15439.

Knockaert, M., Gray, N., Damiens, E., Chang, YT., Grellier, P., Grant, K., Fergusson, D., Mottram, J., Soete, M., Dubremetz, JF., Le Roch, K., Doerig, C., Schultz, P., and Meijer, L. (2000) Intracellular targets of cyclin-dependent kinase inhibitors: identification by affinity chromatography using immobilised inhibitors. *Chem Biol* **7**, 411–422.

Knockaert, M., Lenormand, P., Baffet, G., Gray, N., Schultz, P., Pouyssegur, J., and Meijer, L. (2002a) p42/p44 MAPKs are intracellular targets of the CDK inhibitor purvalanol. *Oncogene* **21**, 6413–6424.

Knockaert, M. and Meijer, L. (2002) Identifying *in vivo* targets of cyclin-dependent kinase inhibitors by affinity chromatography. *Biochem. Pharmacol.* **64**, 819–825.

Knockaert, M., Viking, K., Schmitt, S., Leost, M., Mottram, J., Kunick, C., and Meijer, L. (2002b) Intracellular targets of paullones: identification by affinity chromatography using immobilized inhibitor. *J. Biol. Chem.* **277**, 25493–25501.

Kung, C., Kenski, D.M., Dickerson, S.H., Howson, R.W., Kuyper, L.F., Madhani, H.D., and Shokat, K.M. (2005) Chemical genomic profiling to identify intracellular targets of a multiplex kinase inhibitor. *Proc Natl Acad Sci U S A* **102**, 3587–3592.

Kung, C. and Shokat, K.M. (2005) Small-molecule kinase-inhibitor target assessment. *Chembiochemistry* **6**, 523–526.

Lefurgy, S. and Cornish, V. (2004) Finding Cinderella after the ball: a three-hybrid approach to drug target identification. *Chem Biol.* **11**, 151–153.

Licitra, E.J. and Liu, J.O. (1996) A three-hybrid system for detecting small ligand-protein receptor interactions. *Proc Natl Acad Sci U S A* **93**, 12817–12821.

Luesch, H., Wu, T.Y., Ren, P., Gray, N.S., Schultz, P.G., and Supek, F. (2005) A genome-wide overexpression screen in yeast for small-molecule target identification. *Chem Biol.* **12**, 55–63.

Lum, P.Y., Armour, C.D., Stepaniants, S.B., Cavet, G., Wolf, M.K., Butler, J.S., Hinshaw, J.C., Garnier, P., Prestwich, G.D., Leonardson, A., Garrett-Engele, P., Rush, C.M., Bard, M., Schimmack, G., Phillips, J.W., Roberts, C.J., and Shoemaker, D.D. (2004) Discovering modes of action for therapeutic compounds using a genome-wide screen of yeast heterozygotes. *Cell* **116**, 121–137.

Mackeigan, J.P., Murphy, L.O., and Blenis, J. (2005) Sensitized RNAi screen of human kinases and phosphatases identifies new regulators of apoptosis and chemoresistance. *Nat Cell Biol.* **7**, 591–600.

Mager, W.H. and Winderickx, J. (2005) Yeast as a model for medical and medicinal research. *Trends Pharmacol Sci.* **26**, 265–273.

Mapelli, M., Massimiliano, L., Crovace, C., Seeliger, M.A., Tsai, L.H., Meijer, L., and Musacchio, A. (2005) Mechanism of CDK5/p25 binding by CDK inhibitors. *J Med Chem.* **48**, 671–679.

Meijer, L., Borgne, A., Mulner, O., Chong, J.P.J., Blow, J.J., Inagaki, N., Inagaki, M., Delcros, J.G., and Moulinoux, J.P. (1997) Inhibition of cyclin-dependent kinases by purine analogues: crystal structure of human cdk2 complexed with roscovitine. *Eur. J. Biochem.* **243**, 527–536.

Meijer, L., Skaltsounis, A.L., Magiatis, P., Polychronopoulos, P., Knockaert, M., Leost, M., Ryan, X.P., Vonica, C.D., Brivanlou, A., Dajani, R., Tarricone, A., Musacchio, A., Roe, S.M., Pearl, L., and Greengard, P. (2003) GSK-3-Selective Inhibitors Derived from Tyrian Purple Indirubins. *Chem. Biol.* **10**, 1255–1266.

Parsons, A.B., Brost, R.L., Ding, H., Li, Z., Zhang, C., Sheikh, B., Brown, G.W., Kane, P.M., Hughes, T.R., and Boone, C. (2004) Integration of chemical-genetic and genetic interaction data links bioactive compounds to cellular target pathways. *Nat Biotechnol.* **22**, 62–69.

Parsons, A.B., Geyer, R., Hughes, T.R., and Boone, C. (2003) Yeast genomics and proteomics in drug discovery and target validation. *Prog Cell Cycle Res.* **5**, 159–166.

Rockey, W.M. and Elcock, A.H. (2005) Rapid computational identification of the targets of protein kinase inhibitors. *J. Med. Chem.*, published online DOI: 10.1021/jm049461b.

Sche, P.P., McKenzie, K.M., White, J.D., and Austin, D.J. (1999) Display cloning: functional identification of natural product receptors using cDNA-phage display. *Chem Biol.* **6**, 707–716. Erratum in *Chem Biol* (2001) **8**, 399–400.

Scott, J.K. and Smith, G.P. (1990) Searching for peptide ligands with an epitope library. *Science* **249**, 386–390.

Simon, J.A. and Bedalov, A. (2004) Yeast as a model system for anticancer drug discovery. *Nat Rev Cancer* **4**, 481–492.

Tang, L., Li, M.H., Cao, P., Wang, F., Chang, W.R., Bach, S., Reinhardt, J., Ferandin, Y., Galons, H., Wan, Y., Gray, N., Meijer, L., Jiang, T., and Liang, D.C. (2005) Crystal structure of pyridoxal kinase in complex with roscovitine and derivatives. *J Biol Chem.* June 28 **280**, 31220–31229.

Wan, Y., Hur, W., Cho, C.Y., Liu, Y., Adrian, F.J., Lozach, O., Bach, S., Mayer, T., Fabbro, D., Meijer, L., and Gray, N.S. (2004) Synthesis and target identification of hymenialdisine analogs. *Chem. Biol.* **11,** 247–259.

Wissing, J., Godl, K., Brehmer, D., Blencke, S., Weber, M., Habenberger, P., Stein-Gerlach, M., Missio, A., Cotten, M., Muller, S., and Daub, H. (2004) Chemical proteomic analysis reveals alternative modes of action for pyrido[2,3-d]pyrimidine kinase inhibitors. *Mol Cell Proteomics* **3**, 1181–1193.

Zhu, H., Klemic, J.F., Chang, S., Bertone, P., Casamayor, A., Klemic, K.G., Smith, D., Gerstein, M., Reed, M.A., and Snyder, M. (2000) Analysis of yeast protein kinases using protein chips. *Nat Genet.* **26**, 283–289.

6 Development of CDK Inhibitors as Anticancer Agents: From Rational Design to CDK4-Specific Inhibition

Sachin Mahale and Bhabatosh Chaudhuri

CONTENTS

6.1 INTRODUCTION

Pharmacological inhibition of cyclin-dependent kinases (CDKs) is possibly becoming one of the most productive strategies for the design and discovery of novel anticancer agents that specifically target the cell cycle. It is even more encouraging that the number of CDK inhibitors entering clinical trials has been increasing with time (Senderowicz and Sausville, 2000; Knockaert et al., 2002; Sausville, 2003; Senderowicz, 2003a; Fischer and Gianella-Borradori, 2005). Large numbers of potential CDK inhibitors have been identified, and currently there is an intense interest in developing selective inhibitors of the various CDK targets for prospective use in cancer therapeutics (Garrett and Fattaey, 1999; Sausville, 2003; Senderowicz, 2003b; Dai and Grant, 2004; Eastman, 2004; Liu et al., 2004; Swanton, 2004; Hirai et al., 2005; Benson et al., 2005; Senderowicz, 2005; Schwartz, 2005; Welburn and Endicott, 2005). Extensive biological studies on CDK inhibitors in the past decade have demonstrated their role in controlling the cell cycle, inhibiting cancer cell growth, inducing apoptosis or terminal differentiation (senescence), and other hyperproliferative disorders. CDK inhibitors may also have a profound role in abrogating apoptosis that is aberrantly induced in postmitotic neuronal cells and cardiomyocytes (Monaco and Vallano, 2003; O'Hare et al., 2002; Boehm and Nabel, 2003; Greene et al., 2004; Regula et al., 2004; Herrup et al., 2004; Sarkar et al., 2004; Johnson et al., 2005; Neve and McPhie, 2005; Verdaguer et al., 2005). It has also been reported that pharmacological inhibition of CDKs has a profound effect on viral replication and transcription (Schang, 2004).

6.1.1 INTRODUCTION TO THE CDKs AND THE CELL CYCLE

The normal mitotic cell division cycle, which is a clocklike mechanism, consists of four distinct phases: the gap phases G_1 and G_2, the DNA synthesis phase (S), and mitosis (M). A dormant, quiescent cell at G_0 must be reawakened with the help of growth-promoting factors and ushered into the G_1 phase of the cell cycle. During the G_1 phase, the cell first commits itself to completing a cell division cycle and then prepares for the synthesis of genomic DNA. The synthesis of genomic DNA takes place during the S phase, and once this is complete the cell enters the second gap phase, which is known as G_2. During the early G_2 phase, the cell repairs errors that may occur during DNA duplication and also prepares for cell division (i.e., mitosis, designated as M). At the M phase, the cell physically divides into two daughter cells, each containing one complete copy of genetic information that embodies the DNA of the parent cell (Figure 6.1; reviewed in Oshima and Campisi, 1991; Stewart et al., 2003; McGowan, 2003; Kastan and Bartek, 2004; Massague, 2004; Murray, 2004). Immediately after completion of mitosis, the daughter cell can reenter the G_1 phase or permanently exit the cell cycle to undergo either senescence (a G_0 state from which the cell cannot be normally reawakened) or a form of programmed cell death (often referred to as *apoptosis*). Aberrations at different stages of this normal cell cycle progression are responsible for most human malignancies.

All the major transitions in the cell division cycle, such as commitment to enter the cell cycle, initiation of DNA synthesis phase, and progression from G_2 into mitosis, are tightly controlled by the activity of specific CDKs along with their regulatory subunits, cyclins (Morgan, 1997; Grana and Reddy, 1995; Kastan and

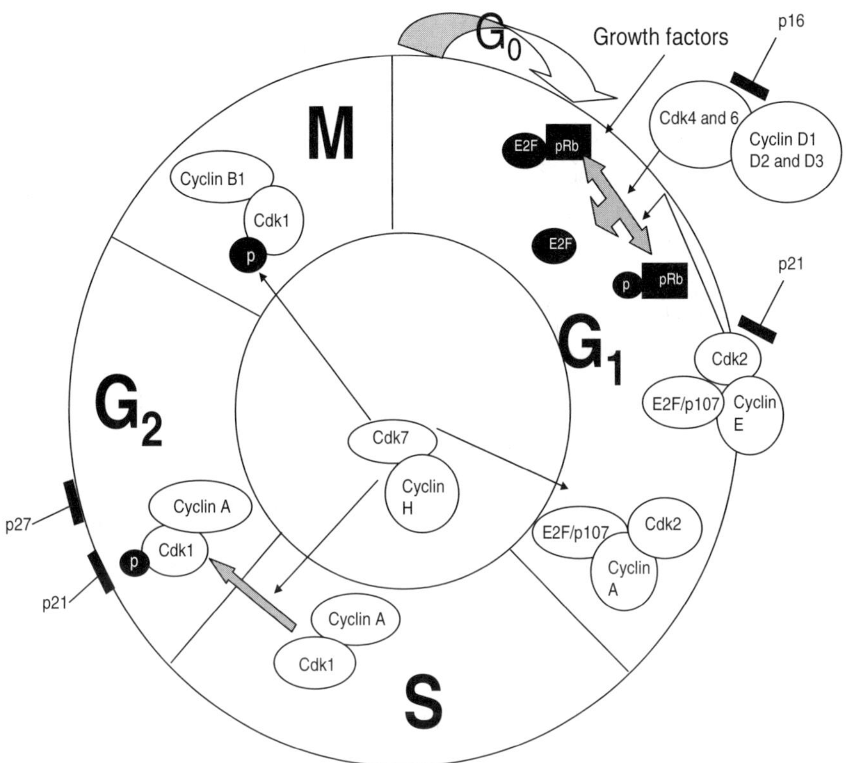

FIGURE 6.1 Systematic representation of the cell division cycle. It shows various cyclin-dependent kinases and their respective cyclin partners, their position at different phases of the cell division cycle, where they function, and where their activity is required for the controlled progression of the cell cycle.

Bartek, 2004; Massague, 2004; Murray, 2004). CDKs belong to a family of serine/threonine kinases, and the name is derived from the required association with cyclins for their enzymatic activity. At least 11 CDKs (CDK1–CDK11) and 15 cyclins (cyclin A–cyclin T) are known. Most of them are clearly involved in cell cycle regulation (Pines, 1999; Peng et al., 1998; Sherr, 1996; Trembley et al., 2003; Guo and Stiller, 2004; Sanchez and Dynlacht, 2005; Figure 6.1).

The transition of G_1 cells into the S phase is regulated by CDK4/CDK6-cyclin D activity. CDK6 is thought to be a functional homologue of CDK4, although recent reports indicate that CDK6 may also actively participate in differentiation (Grossel and Hinds, 2006). The most extensively studied substrate for CDK4/CDK6 is a product of the tumor suppressor gene retinoblastoma (pRb), which acts as repressor for a number of genes required for the G_1/S transition, initiation, and completion of DNA synthesis. The phosphorylation of pRb protein, which leads to its functional inactivation by cyclin-D-bound CDK4 (i.e., CDK4/cyclin D), is required to initiate a cascade of events that include further phosphorylation of pRb by CDK2/cyclin

E and CDK2/cyclin A. CDK4-mediated phosphorylation of pRb compels the cell to commit itself to completing one full cell division cycle (Bartek et al., 1996; Lundberg and Weinberg, 1998). Besides CDK4/CDK6 and CDK2, the other CDK that is involved directly in cell cycle control is CDK1, which is also known as the Cdc2 kinase. CDK1 binds to B-type cyclins, and its activity is required during mitosis (Sherr, 1996; Preisinger et al., 2005). Cyclin-bound CDK4/CDK6, CDK2, and CDK1 need activation by CDK7-cyclin H (also known as *CDK-activating kinase*, CAK) (Kastan and Bartek, 2004; Massague, 2004; Murray, 2004).

6.1.2 Efforts to Target CDK Activities: Promising Approach to Cancer Chemotherapeutics

The CDKs that play an integral part in controlling cell proliferation are of intrinsic interest as potential targets for drug design to fight diseases such as cancer, which are characterized by aberrant cell growth (Garrett and Fattaey, 1999; Cohen, 2002; Sausville, 2003; Senderowicz, 2003b; Dai and Grant, 2004; Eastman, 2004; Liu et al., 2004; Swanton, 2004; Hirai et al., 2005; Welburn and Endicott, 2005). There have been numerous scientific reports discussing the chemical syntheses and biological activities of CDK inhibitors. Most of the intense ongoing research is focused on finding ATP competitive compounds (using structure-based chemical design methods) that can selectively modulate ATP-binding activity of the CDKs (Hirai et al., 2005; Soni et al., 2000; Carini et al., 2001; Toogood, 2002; Beattie et al., 2003; Welburn and Endicott, 2005). The pharmacological inhibitors of CDKs exhibit great diversity in their chemical structures, and they also exhibit different specificity and potency toward different CDKs.

The initial attempts to obtain inhibitors that have structural similarities with adenosine 5-triphosphate (ATP) resulted in compounds, such as olomoucine, roscovitine, purvalanol A, purvalanol B, and CVT-313, that show remarkable inhibitory effects on the activity of CDKs 1, 2, 5, 7, and 9 but manifest very low potency toward CDK4 (Knockaert et al., 2002; De Azevedo et al., 1997; Schulze-Gahmen et al., 1995). Staurosporine and UCN-01 are alkaloids that were discovered as general protein kinase inhibitors that also inhibit CDKs 1, 2, and 5 with IC_{50} values in the nanomolar (nM) range (Lawrie et al., 1997; Wang et al., 1995).

The flavonoid class of compounds is also a relatively potent structural class of CDK inhibitors (Senderowicz, 2005). Flavopiridol is a semisynthetic flavonoid that has been extensively studied because of its inhibitory activity against CDKs and protein kinases in general. It potently blocks growth of a variety of cancer cell lines, and studies on its clinical efficacy are in progress (Carlson et al., 1996; Senderowicz and Sausville, 2000; Shapiro, 2004; Fischer and Gianella-Borradori, 2005). However, flavopiridol is not specific for any particular CDK and inhibits all CDKs tested so far, including CDK7-cyclin H (Schang, 2004). It even acts as a global transcription inhibitor by inhibiting CDK9-cyclin T1 (Dai and Grant, 2004).

Other classes of compounds have been reported to be selective for CDK4/CDK6, and these include fascaplysin (Soni et al., 2000), CINK4 (Soni et al., 2001), Compound 66 (Inayat-Hussain and Thomas, 2004), and Compound 26a (Knockaert et al., 2002). Fascaplysin, isolated from a marine sponge (Segraves et al., 2004), inhibits

CDK4 specifically, causing G_0/G_1 arrest of cancer cells. However, fascaplysin is highly toxic, and the potential for its planar structure to intercalate double-stranded DNA has been suggested as a possible explanation of its toxicity (Hormann et al., 2001). We have explored the possibility of separating the DNA-intercalating ability of fascaplysin from its potent CDK4-specific inhibitory activity (Aubry et al., 2004; Aubry et al., 2005; Aubry et al., 2006; Mahale et al., 2006).

Similar to fascaplysin, indirubin is also a natural product that is an active ingredient of a traditional Chinese herbal recipe used for the treatment of cancer. Unlike fascaplysin, it is not specific and is found to be a potent inhibitor of CDK1, CDK2, CDK4, and CDK5. Derivatives of indirubin such as indirubin-3-monoxime or indirubin-5-sulfonic acid have shown improved activity toward inhibition of a broad spectrum of CDKs with IC_{50} values for CDK inhibition in the nM range (Eisenbrand et al., 2004; Moon et al., 2006).

6.1.3 ENZYMATIC INHIBITION OF CDKs RESULTS IN THE BLOCK OF CELL GROWTH: GENOTYPIC AND PHENOTYPIC ANTIPROLIFERATIVE CONSEQUENCES

The biological action of CDK inhibitors on cultured cells (tumor cells as well as normal cells) is well documented and suggests their strong antiproliferative nature. Growth inhibition and cell cycle arrest are the main consequences of inhibition of CDKs at the cellular level. A number of small molecular CDK inhibitors arrest cycling cells at the G_0/G_1 and/or G_1/S and/or G_2/M phase boundaries, depending on their concentrations and the cell type used for the experiments, which clearly indicate their ability to inhibit more than one CDK.

Both the selective CDK4 inhibitors fascaplysin and CINK4 arrest cycling cells specifically at the G_0/G_1 boundary. The G_0/G_1 arrest correlates with the accumulation of hypophosphorylated pRb. CINK4 and fascaplysin prevent the initiation of pRb hyperphosphorylation that is specifically triggered by CDK4 activity (Paull et al., 1989; Meijer et al., 1997; Alessi et al., 1998; Carlson et al., 1996; Soni et al., 2000; Soni et al., 2001).

Amongst the phenotypic consequences, induction of apoptotic cell death, driving cells toward hypermitogenic arrest (which can lead to senescence), and induction of senescence are the reported results from CDK inhibition in cells (references in Soni, 2000; 2001). Interestingly, a number of compounds that are particularly potent CDK2 inhibitors have also been reported to induce apoptosis selectively in transformed cells (reviewed in Blagosklonny, 2003; Parker et al., 1998; Chen et al., 1999), although it has been recently suggested that CDK2 may not play any role in oncogenesis (Tetsu and McCormick, 2003; Martin et al., 2005).

6.1.4 CDK INHIBITORS AND *IN VIVO* TUMOR GROWTH

The use of CDK inhibitors in animal models in which human tumor cells were grown in nude or SCID mice and then administered with sublethal concentrations of compounds demonstrated significant anti-tumor activity and underlines the importance of these molecules for cancer chemotherapeutic applications (Table 6.1).

TABLE 6.1
A Summary of the Anti-Tumor Activity of Reported CDK Inhibitors in Animal Models to Underscore Their Potential Application in Clinical Studies

Inhibitor	*In Vitro* Activity Against	Tumor Model Used for the Investigation	Effect on the Tumor Growth	Reference
Flavopiridol	CDKs	Human tumor xenograft	Anti-tumor activity	Zhai et al., 2002; Drees et al., 1997
JNJ-7706621	CDKs	Human tumor xenograft	Anti-tumor activity	Emanuel et al., 2005
PHA-533533, 13	CDK2-cyclin A	Mouse xenograft (A2780)	Tumor growth inhibition by 70%	Pevarellao et al., 2005
Compound 20	CDKs	Mouse xenograft (A375)	Significant increase in survival	Kuo et al., 2005
Roscovitine	CDK1, CDK2, and CDK5	Mouse xenograft (ESFT)	Reduced tumor growth	Tirado et al., 2005
CINK4	CDK4/6-cyclin D1	Mouse xenograft (HCT116)	Reduced tumor volume	Soni et al., 2001
UCN-01	CDKs	Human tumor xenografts A431, HT1080 and HL-60	*In vivo* Anti-tumor activity	Akinaga et al., 1991
PD 0332991	CDK4 and CDK6	Human tumor xenograft Colo-205	Tumor regression and net reduction in some tumors	Fry et al., 2004

It has also been observed that CDK inhibitors, when used in combination with conventional cytotoxic drugs, potentiate cell death in a number of tumor models (Tenzer and Pruschy, 2003; Robson et al., 2005).

6.2 DEVELOPMENT OF NONTOXIC ANALOGS OF FASCAPLYSIN

Of all the CDKs, CDK4 is crucially important enzyme for a number of reasons. A cell's initial commitment to complete a cell division cycle depends on CDK4 activity. Therefore, activation of CDK4 at G_0/G_1 is also essential for the later G_1/S transition in the cell cycle, at least in cells that contain a functional copy of pRb. Moreover, many human cancers are characterized by either overexpression of its activating partner, cyclin D1, or loss of p16[INK4A] (often referred to as just p16), which is a natural CDK4 inhibitory protein (Bartek et al., 1996; Weinberg, 1995). Recent studies have demonstrated that mice lacking cyclin D1 are either refractory or resistant to tumor development induced by certain oncogenes, and it has also been confirmed that the tumorogenic activity of cyclin D1 during the development or maintenance of breast

  127

carcinoma is associated with the CDK4 kinase, confirming CDK4-cyclin D1 as a crucially important target for cancer therapy (Yu et al., 2006; Landis et al., 2006; Malumbres and Barbacid, 2006) Therefore, a small molecule that specifically inhibits the CDK4 enzyme activity *in vitro* and prevents cell growth and tumor volume *in vivo* could be of immense therapeutic value for the treatment of cancer.

Fascaplysin, a marine natural product, was identified in a large screening program as a small molecule that specifically inhibits the *in vitro* phosphorylation of the retinoblastoma protein pRb, a CDK4 substrate (Soni et al., 2000). Fascaplysin demonstrates pRb-dependent arrest of cancer and normal cells at the G_0/G_1 phase of the cell cycle. Recently, 1-deoxysecofascaplysin A, an analog of fascaplysin, has also been reported to inhibit cell growth of MCF-7 (human breast cancer) and OVCAR-3 (human ovarian cancer) cell lines (Charan et al., 2004).

However, fascaplysin cannot be therapeutically used as an anticancer agent because it is a highly toxic molecule. The potential for its planar structure to intercalate with double-stranded DNA has been suggested as a possible explanation for some of its unusual biological activity and toxicity. The DNA-binding property of fascaplysin is similar to structurally related DNA intercalating agents such as cryptolepine and ellipticine (Hormann et al., 2001).

The aim of our ongoing research is to develop nonplanar and less toxic compounds based on the structure of fascaplysin. We seek to identify compounds that inhibit specifically CDK4 enzyme activity *in vitro* (i.e., compounds that are at least tenfold more potent in inhibiting CDK4 than CDK2/CDK1), prevent cancer cell growth *in vitro,* and tumor growth *in vivo* and yet do not intercalate or damage DNA as fascaplysin does. Here, we review the biological activities of the newly synthesized structural analogs of fascaplysin, and show that fascaplysin's ability to inhibit CDK4 specifically can be separated from its deleterious DNA-intercalating characteristic. Some of these analogs manifest the expected CDK4-specific inhibition by blocking at the G_0/G_1 phase of the cell cycle, but surprisingly also inhibit the G_2/M phase in a CDK-independent manner (Aubry et al., 2004; Aubry et al., 2005; Aubry et al., 2006; Mahale et al., 2006).

6.2.1 MOLECULAR MODELING AND RATIONALE FOR CHEMICAL SYNTHESES

The CDK4 protein is highly prone to aggregation, and hence x-ray crystal coordinates, which are essential for rational design of inhibitors, are unavailable. Therefore, a homology model of CDK4 was produced using Modeller (Sali, 1993) with the 3-D structure of CDK2 (which shares 40% sequence identity with CDK4; PDB accession code 1HCK, containing ATP-Mg^{2+}: Berman et al., 2000; Schulze-Gahmen et al., 1996) and CDK6 (sharing 70% sequence identity with CDK4; 1BLX: Brotherton et al., 1998) as templates. Docking studies were performed on the basis of the homology-model-derived structure for CDK4 bound to fascaplysin (Figure 6.2). Solutions for each of the compounds **7a-7c**, **9a-9q**, **12a-12q,** and **15** (Figure 6.3 and Figure 6.4) were generated using GOLD v3.0 (Jones et al., 1997) with the Chemscore fitness function, to predict their putative ability to inhibit CDK4 (Aubry et al., 2006; Verdonk et al., 2003). The putative binding of fascaplysin analogs **9q** and **12q** to the CDK4 molecular homology model are depicted in Figure 6.2.

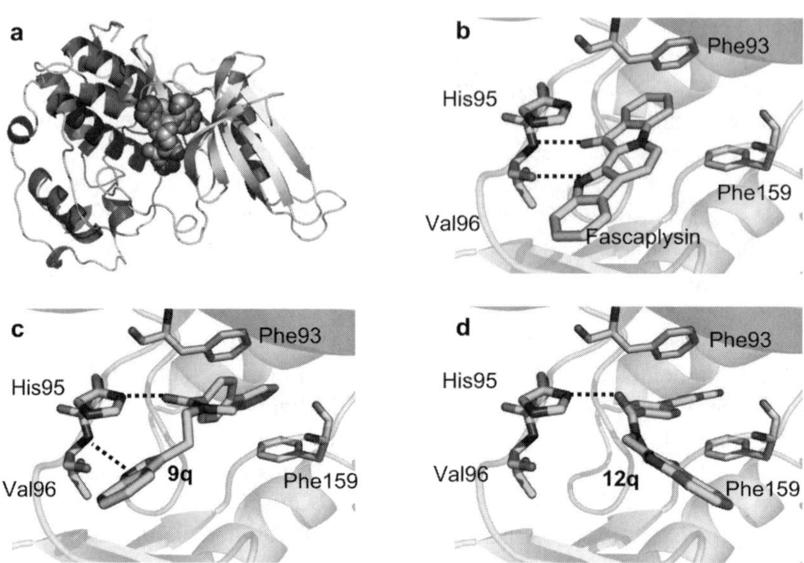

FIGURE 6.2 (See color insert following page 142.) The active site of the CDK4 homology model. (a) Cartoon representation of overall structure of the CDK4, with the positions of ATP/Mg^{2+} (C in grey, N in slate, O in salmon, P in yellow, Mg in black) and fascaplysin (C in magenta, N in cyan, O in red), is shown. (b) Predicted binding mode of fascaplysin. (c) Predicted binding mode of the most potent compound, **9q**, in the series **9a-q**; note the π-π interactions of the benzoid ring with Phe93 and Phe159 of CDK4. (d) Predicted binding mode of **12q** — this weaker inhibitor is structurally similar to, but more conformationally constrained, than **9q**. Hydrogen bonds are shown as dashed lines. Figures were produced using Pymol. (From Aubry, C. et al. (2006). *Org. Biomol. Chem.*, DOI: 10.1039/b518019h.)

The general chemical strategy that was adopted for removing toxicity (Aubry et al., 2004; Aubry et al., 2005; Aubry et al., 2006), consisted of releasing bonds *a, b* and changing double bond *c* into a single bond in fascaplysin **1** (Figure 6.3), leading to the derivatives **7a-7c, 9a-9q, 12a-12q,** and **15** (Figure 6.3 and Figure 6.4). The first-generation compounds that are nonplanar but still contain the basic structural components of fascaplysin are **7a-7c**. They have a great deal of flexibility owing to rotation around six bonds. The second-generation compounds **9a-9q** (Figure 6.3 and Figure 6.4) have a shorter chain between the indole and benzene ring, and an amide bond that results in less degrees of rotational freedom in the molecules. In the third-generation compounds, **12a-q** and **15** (Figure 6.3 and Figure 6.4), the rotational freedom of the structures was further reduced by incorporating the β-carboline structure.

Compounds **7a-7c, 9a-9q, 12a-12q,** and **15** are nonplanar, yet are predicted to be CDK4-specific inhibitors from the CDK4 homology model. Because of their nonplanar structures, it is very unlikely that they would intercalate with DNA.

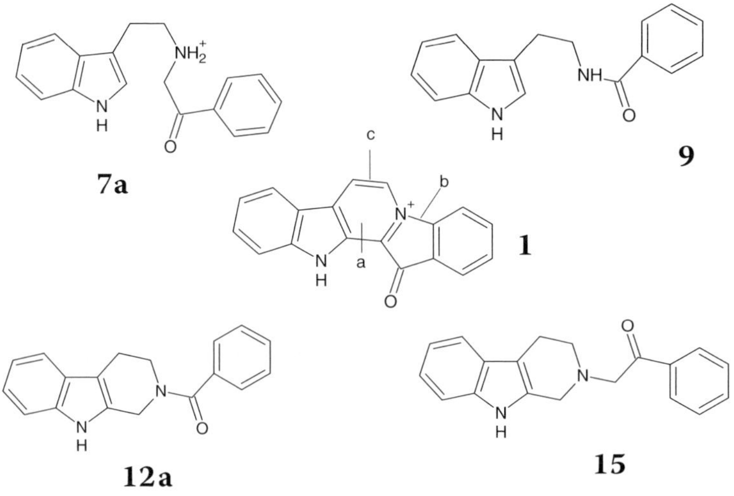

FIGURE 6.3 Strategy used to produce nonplanar fascaplysin derivatives. The basic structures **7a**, **9**, **12a,** and **15** derived from fascaplysin **1** are depicted. (From Aubry et al., 2004; Aubry et al., 2005; Aubry, C. et al. (2006). *Org. Biomol. Chem.*, DOI: 10.1039/b518019h.)

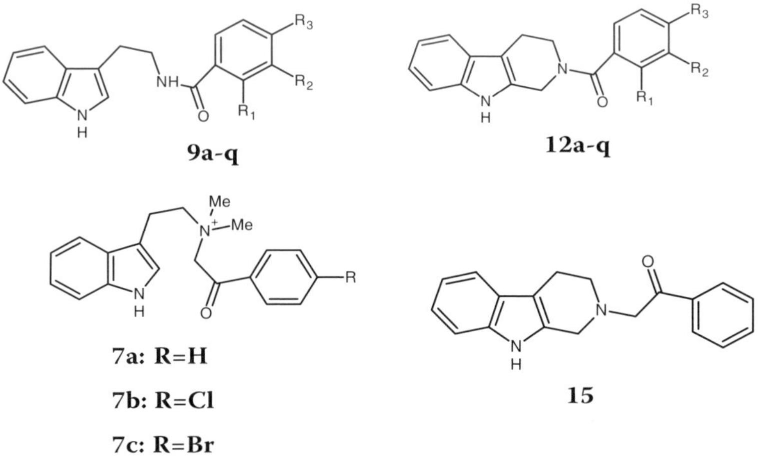

FIGURE 6.4 Chemical structures of the nonplanar derivatives of fascaplysin. Structures **7a-7c**, **9a-9q**, **12a-12q,** and **15** are represented with the help of a table (Aubry, C. et al. (2004). *Chem. Commun.* **15**, 1696–1697; Aubry, C. (2005). *Tetrahedron Lett.* **46**, 1423–1425; Aubry et al., 2006). Analogs of **15** are poorly active, and hence representative results are not shown.

Consequently, the biological activities of these nonplanar derivatives of fascaplysin, predicted by molecular modeling to be CDK4-specific inhibitors, were explored further.

6.2.2 Biological Methods

6.2.2.1 Expression and Purification of CDK4/GST–Cyclin D1, CDK2/GST–Cyclin A, CDK2/GST–Cyclin E, CDK1/GST–Cyclin B1, GST-pRb152, and GST-CTD

Fusion proteins of human cyclins A, B1, D1, and E, covalently linked to glutathione S-transferase (GST), were coexpressed with the catalytic subunits CDK2, CDK1, CDK4, and CDK2, respectively, in Sf-9 insect cells using protocols similar to those described previously (Phelps and Xiong, 1996; Sarcevic et al., 1997; Soni et al., 2000; Soni et al., 2001). Active enzyme complexes containing a catalytic subunit bound to GST-cyclin were loaded onto glutathione-agarose columns and eluted with reduced glutathione. The eluted proteins were further subjected to dialysis.

The plasmid encoding the GST-pRb152 fusion gene (that codes for the pRb amino acids 773-924 and contains the majority of CDK phosphorylation sites) was transformed into the *Escherichia coli* strain BL21 (DE3) pLysS. For expression of GST-pRb152, the cells were induced with isopropyl-β-thiogalactopyranoside, and further purification of the GST-pRb152 protein was carried out as described previously (Sarcevic et al., 1997). The active enzyme CDK9-cyclin T1, which participates in global transcription, was obtained by protocols similar to the ones used for the other CDKs. GST-CTD (i.e., a CDK9 substrate) was expressed in insect cells by single infection with a recombinant baculovirus that carries the *GST-CTD* gene.

6.2.2.2 Kinase Assays

The screening of fascaplysin analogs was carried out using kinase assays that do not use radioactive isotopes but are based on luminescence detection. The assays measure the depletion in ATP concentration as a result of phosphorylation by CDKs of different substrates, i.e., GST-pRb152 (substrate for CDK4/cyclin D1, CDK2/cyclin A, CDK2/cyclin E, and CDK1/cyclin B1), histone H1 (substrate for CDK2/cyclin A, CDK2/cyclin E, and CDK1/cyclin B1), and GST-CTD (substrate for CDK9-cyclin T1). The kinase assays were carried out in a 96-well format, and all steps using a particular enzyme were performed in a single white polystyrene plate. The compounds were dissolved in DMSO and preserved as 10 mM stock solutions. Compounds were further diluted in kinase buffer (40 mM Tris pH 7.5, 20 mM $MgCl_2$, and 0.1 mg/ml BSA) to obtain the desired serial dilutions (usually 10-fold) for determination of IC_{50} values. The kinase assay was performed in 50 μl of kinase buffer containing 2 μg of purified GST-pRb152 (for CDK4/cyclin D1 and CDK2/cyclin E), 3 μg of histone H1 (for CDK2/cyclin A), 10 μg of histone H1 (for CDK1/cyclin B1), 2 μg of GST-CTD (for CDK9/cyclin T1), and 6 μM ATP.

The phosphatase and protease inhibitor cocktail containing β-glycerophophate, sodium fluoride, and sodium orthovanadate in the presence of reducing agent dithiothreitol was added at a final concentration of 10 mM, 0.1 mM, 0.1 mM, and 1 mM, respectively. The assays were initiated by adding 200 ng of active enzyme complexes and the plates were incubated for 30 min at 30°C in a humidified incubator. The reaction was stopped by addition of an equal volume of Kinase Glo™ reagent (Promega). Luminescence was measured using a Packard Luminometer (Fusion 3.50). The rate of ATP depletion (i.e., the rate of reaction) in the control reactions with respect to the blank (without substrate or enzyme) reactions was calculated and used to determine the IC_{50} concentrations of compounds. Several CDK inhibitory compounds reported in the literature, i.e., fascaplysin, flavopiridol, roscovitine CINK4, and indirubin-5-sulfonic acid-sodium salt were used to validate the novel chemiluminescent assays.

6.2.2.3 Ethidium Bromide Displacement Assay

The DNA-binding affinities of fascaplysin and its nonplanar analogs were measured using an ethidium bromide fluorescence-quenching assay. This assay is based on the displacement of DNA-intercalating agent ethidium bromide from purified pBlue-Script plasmid DNA in the presence of compound (Geall et al., 1999; Brotz-Oesterhelt et al., 2003). The assays were performed in 96-well black plates and involve addition of 10 μl of 10X concentrated stock solution of compounds (dissolved in DMSO) to 90 μl of reaction mixture containing 6 μg of purified pBlueScript DNA and 1.3 μM ethidium bromide in a buffer (20 mM NaCl, 2 mM Hepes and 10 μM EDTA) with final pH 7.4. Equivalent amounts of DMSO were added to the vehicle controls. The reduction in relative fluorescence counts was monitored (λ_{excit} = 260 nM, λ_{emiss} = 600 nM) and recorded after 1 min equilibration time. Fascaplysin and actinomycin D, which are known to intercalate double-stranded DNA molecules, were used as standard compounds in the assay (Hormann et al., 2001).

6.2.2.4 Topoisomerase I DNA Unwinding Assay

The ability of fascaplysin and its analogs to intercalate into plasmid DNA was determined by a topoisomerase I unwinding assay (Fortune and Osheroff, 1998). Reactions contained 5 nM relaxed or negatively supercoiled pBlueScript plasmid DNA and 10 units of topoisomerase I (Invitrogen). DNA relaxation assays were performed in the presence or absence of compounds in 40 μl of DNA unwinding buffer (50 mM Tris-HCl pH 7.5, 50 mM KCl, 10 mM $MgCl_2$, 0.5 mM dithiothreitol, 0.1 mM EDTA, and 30 μg/ml bovine serum albumin). After 30 min incubation at 37°C, reaction mixtures were treated with 3 μl of 250 mM EDTA and extracted with phenol/chloroform. The DNA was disolved in Tris-EDTA buffer, pH 8. The samples (20 μl) were treated with 2 μl of 2.5% SDS, mixed with 2.5 μl agarose gel-loading buffer (10X) and subjected to electrophoresis on a 0.8% agarose gel. DNA bands were stained with 1 μg/ml ethidium bromide and visualized using a UV illuminator.

6.2.2.5 Cell Growth Inhibition *In Vitro*

All eight cell lines were maintained at 37°C in 5% CO_2 in RPMI-1640 medium, supplemented with 10% fetal calf serum and 100 µg/ml Normocin™. The eight cancer cell lines used for screening were the non-small-cell lung carcinoma (NSCLC; a form of cancer that is resistant to chemotherapy) lines NCI-H460 (pRb+, p53+), A549 (pRb+, p53+), Calu-1 (pRb+, p53-null), NCI-H1299 (pRb+, p53-null), NCI-H358 (pRb-null, p53-null); and the following chemotherapy-resistant lines: the colon carcinoma line LS174T (pRb+, p53+), the prostate carcinoma line PC3 (pRb+, p53-null), and the pancreatic cancer line MiaPaCa (pRb+, p53-mutant). The genotypes within brackets indicate the status of the tumor suppressor proteins pRb and p53.

Exponentially growing cells representing an asynchronous population were seeded in 96-well plates, at densities of 5000 to 10,000 cells per well (depending on the doubling time of the individual cell line) in 180 µl of complete growth medium. The wells at the extreme four corners were omitted to avoid the edge effect and variations in the assay. The plated cells were allowed to stabilize by incubating for 24 h. The compounds were dissolved in DMSO, and 10 mM stock solutions were prepared. After 24 h of stabilization, the stock solutions of compounds were diluted in medium without serum. 20 µl of 10X concentrated compounds were added into the wells in triplicates, while an equivalent amount of DMSO was added to the control wells. The plates were mixed gently and incubated further for 48 h. After drug exposure, 50 µl of 2 mg/ml MTT reagent was added and the plates were incubated for 2–3 h at 37°C in the dark. The medium containing MTT was removed. The blue-colored formazan formed was dissolved in 150 µl of DMSO per well. The absorbance was measured at 540 nm. The IC_{50} concentration of compound was calculated as the concentration at which 50% of cell growth was inhibited as compared to the control wells, which did not contain any drug.

6.2.2.6 FACS Analysis

The control and treated cultures were harvested by trypsinization, washed once with PBS, and then fixed in 70% chilled ethanol for a minimum time period of 1 h. After fixation, cells were centrifuged for 5 min at 3000X g at room temperature, and then the pellet was suspended in PBS containing 50 µg/ml propidium iodide and 0.5 mg/ml DNase free ribonuclease. The cells were stained for 1 h in the dark at 4°C. Cell cycle analyses were performed on a Beckman-Coulter (Epics® Altra™) fluorescence-activated cell sorter. To gate all the events that represent single cells, and not doublets or cell clumps, the following analyses were performed on the samples. Cytogram of propidium iodide fluorescence peak signal vs. integrated fluorescence or the linear signal were plotted. All data points on the straight line were isolated in a single gate, and the gated data further used for plotting a histogram that represents a complete cell cycle. The total number of events was not allowed to exceed 200 events/sec. Data acquisition was stopped after a minimum of 10,000 events had been collected.

6.2.2.7 Western Blot Analysis

A549 and Calu-1 cells were seeded in 25 cm² tissue culture flasks in complete medium. When the culture flasks reached 40–50% confluency, the cells were treated with the compound for 24 h. After treatment, the cells were harvested by trypsinization, washed in ice-cold PBS, and then lysed in a buffer containing a cocktail of protease inhibitors. The lysates were centrifuged at 14,000X g for 10 min at 4°C, and the amounts of proteins in the clear supernatant were estimated using the Bradford method. 50 µg of protein from each sample were subjected to SDS-PAGE separation. The proteins were transferred to a PVDF membrane and blocked with 5% milk. Membranes were probed with polyclonal antibodies raised against the full-length pRb protein, and the phospho-specific pRb epitopes, pRb (Ser780-P), pRb (Ser795-P), and pRb (Ser807/811-P). After overnight incubation at 4°C, membranes were exposed to appropriate HRP-conjugated secondary antibody at room temperature for 1 h. Immunoreactivity was visualized with the enhanced chemiluminescence Western Blot detection reagents.

6.2.2.8 Tubulin Polymerization *In Vitro*

The purified tubulin was obtained commercially (Cytoskeleton Inc., Denver, CO), and the polymerization assays were carried out according to the method previously described (Jordan et al., 2002). Tubulin polymerization assay is based on the adaptation of the original methods of Lee and Timasheff (1977), which demonstrated that light is scattered (which can be quantitated by measuring the absorbance) by tubulin protein to an extent that is proportional to the concentration of the polymer mass of the tubulin. The resulting polymerization curves (Figure 6.6H) is representative of three phases of tubulin polymerization, namely, nucleation, growth, and steady state equilibrium. Paclitaxel and nocodazole were used in the assay as a known enhancer and inhibitor of tubulin polymerization, respectively.

6.2.3 Biological Activities and Results

We have identified three distinct nonplanar, structurally unique classes of fascaplysin analogs, **7a-7c**, **9a-9q**, and **12a-12q**. Unlike fascaplysin, these analogs do not intercalate DNA (Figure 6.5B and Figure 6.5C), yet inhibit CDK4 relatively potently and quite specifically (Figure 6.5A). The best compound of each class has a CDK4 IC_{50} = 50, 6, 20 μM, respectively; IC_{50} for CDK2/CDK1/CDK9 is greater than 500 μM. Interestingly, the average IC_{50} for cell growth inhibition in a diverse panel of cancer lines averages 3.5 μM for **9q**, 7 μM for **12m,** and 50 μM for **7a** (Figure 6.5D).

The compound **7a** (of the **7a-7c** series) blocks growth of cancer cells at the G_0/G_1 phase of cell cycle (Figure 6.6E), and Western blot analysis indicates that the G_0/G_1 block is pRb dependent (Figure 6.6F) (manuscript in preparation).

The compounds **9q** (representative of the **9a-9q** series) and **12m** (representative of the **12a-12q** series) not only block growth at G_0/G_1 phase of the cell division

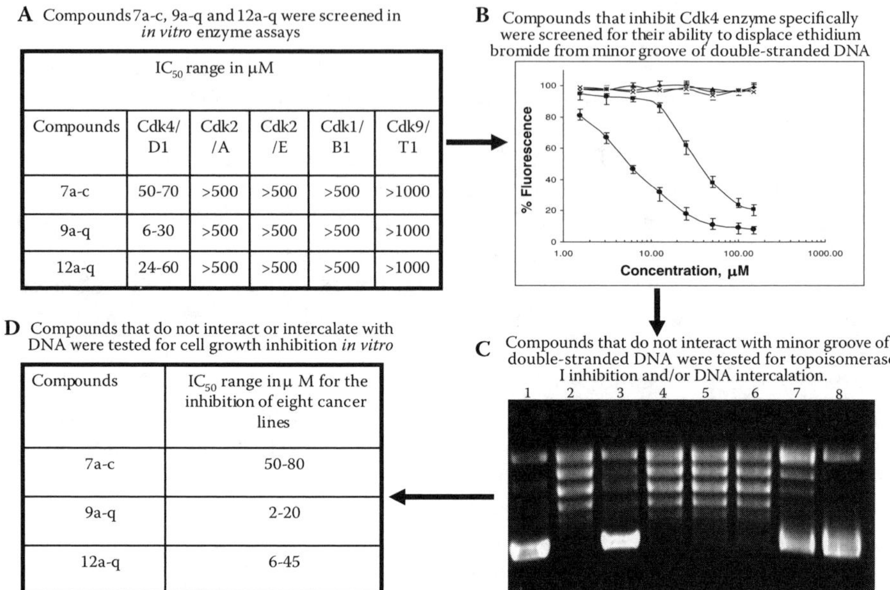

Panel A:

Compounds	Cdk4/D1	Cdk2/A	Cdk2/E	Cdk1/B1	Cdk9/T1
7a-c	50-70	>500	>500	>500	>1000
9a-q	6-30	>500	>500	>500	>1000
12a-q	24-60	>500	>500	>500	>1000

Panel D:

Compounds	IC50 range in μM for the inhibition of eight cancer lines
7a-c	50-80
9a-q	2-20
12a-q	6-45

FIGURE 6.5 The biological results obtained after screening of nonplanar fascaplysin analogs in various biochemical and cellular assays. (**A**) The *in vitro* potencies (IC$_{50}$ values) of compounds **7a-7c**, **9a-9q,** and **12a-12q** in different CDK assays, expressed in micromolar concentrations. It shows 10- to 100-fold specificity toward inhibition of CDK4/cyclinD1 compared to their inhibition of CDK2/cyclin A, CDK2/cyclin E, CDK1/cyclin B1, and CDK9/cyclin T1. (**B**) The ability of fascaplysin and its analogs to interact with the minor groove of double-stranded DNA was tested using the ethidium bromide displacement assay (Geall et al., 1999; Brotz-Oesterhelt et al., 2003). The graph shows that none of the analogs **7b** (cross), **9q** (triangle), and **12m** (unfilled squares) interact with the minor groove of DNA up to concentrations of 150 μM. This is in contrast to fascaplysin (filled circles) and actinomycin D (filled squares), which displace the bound ethidium bromide at relatively lower concentrations. (**C**) Representative picture shows that, unlike fascaplysin, the nonplanar analogs do not intercalate with double-stranded DNA. Lane 1: control plasmid DNA; lane 2: control plasmid DNA + toposiomerase I. The lanes 3–8 represent the products of the topoisomerase I reaction carried out in the presence of camtothecin (50 μM; lane 3), fascaplysin analogs 7a, 9q, and 12 m (all at 150 μM; lanes 4-6) and fascaplysin at 1 and 10 μM (lanes 7–8). (**D**) The table shows average IC$_{50}$ values of compounds **7a-7c**, **9a-9q**, and **10a-10q** for cancer cell growth inhibition *in vitro*.

cycle (Figure 6.6E) in a pRb-dependent manner (Figure 6.6G) but surprisingly also exhibit a profound block at G$_2$/M only in cancer cell lines in which the mitotic spindle checkpoint is unimpaired (Figure 6.6E), indicating that CDK4 inhibition may not be the only target for these compounds. Compounds **9q** and **12m** are found to inhibit tubulin polymerization *in vitro* and also in cancer cell lines. Their ability to affect tubulin polymerization seems to reflect the observed G$_2$/M arrest in the cell cycle analyses (Figure 6.6H) (manuscripts submitted for publication).

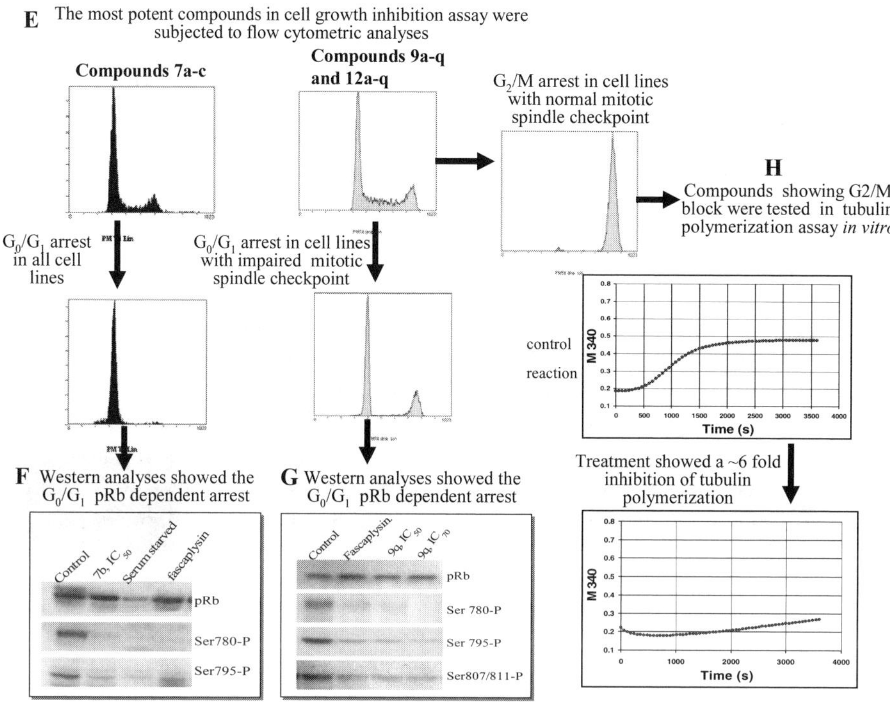

FIGURE 6.6 **(E)** Flow cytometric analyses show that the compounds **7a-c** arrest the growth of all cancer cell lines tested at the G_0/G_1 phase of cell cycle, whereas compounds **9a-9q** and **12a-12q** exhibit only G_0/G_1 block in the cell lines with impaired mitotic spindle checkpoint but show profound G_2/M block in cell lines containing normal mitotic spindle checkpoint. **(F)** Western blot analyses show that the G_0/G_1 block induced by compound **7a** is pRb dependent and that treatment with compound inhibit the CDK4-specific pRb phosphorylations at serine residues. **(G)** Western blotting confirms that the G_0/G_1 arrest induced by compounds **9q** and **12m** in the cell lines with impaired mitotic spindle checkpoint normal is pRb dependent, and treatment with compounds result again in the inhibition of CDK4-specific pRb phosphorylation at serine residues. **(H)** The compounds **9q** and **12m** inhibit the *in vitro* polymerization of purified tubulin at relatively low concentrations (i.e., at concentrations that are lower than the IC_{50} values obtained for CDK4/cyclin D1 inhibition in the *in vitro* enzyme assay). This is the likely explanation for the G_2/M arrest of those cancer cells in which the mitotic spindle checkpoint is intact.

6.3 CONCLUSIONS

9q, 12m, and **7a** are novel compounds that show anticancer effects. We have identified these molecules on the basis of an *in vitro* screen for CDK4 enzyme and found that they not only inhibit CDK4 specifically (i.e., do not inhibit CDK2, CDK1, and CDK9), but also the following characteristics:

1. None of the compounds intercalate with DNA.
2. Compound **7a** blocks the growth of cells at G_0/G_1 irrespective of the cells' mitotic spindle checkpoint status (manuscript in preparation).

3. Compounds **9q** and **12m** block cells at G_0/G_1 in cell lines in which the mitotic spindle checkpoint is impaired.
4. Compounds **9q** and **12m** profoundly block cells at G_2/M at comparatively low concentrations in cells with an intact mitotic spindle checkpoint.
5. Compounds **9q** and **12m** induce massive apoptosis in cancer cells (manuscripts submitted for publication).
6. Compounds **9q** and **12m** selectively induce apoptosis in SV40 large T-antigen-transformed cells and not perceptibly in untransformed cells (manuscripts submitted for publication).

We conclude that compounds **9q, 12m,** and **7a** are rather unusual molecules with profiles that are uniquely distinct from what has been reported in the literature. Therefore, the unique **9q, 12m,** and **7a** chemical scaffolds could be further exploited to rationally improve the therapeutic indices of these molecules for the prospective treatment of cancer. We have recently identified a close analog of **9q** that is much more potent in cellular assays, offering the hope that with our two-pronged chemical biological approach we shall be able to find molecules that are far superior in potency to **9q** and **12m.**

ACKNOWLEDGMENTS

The work in the authors' laboratory was supported by Cancer Research U.K. The authors also gratefully acknowledge the contributions of Paul Jenkins in chemical syntheses and Professor Michael Sutcliffe in molecular modeling.

REFERENCES

Akinaga, S., Gomi, K., Morimoto, M., Tamaoki, T., and Okabe, M. (1991). Antitumor activity of UCN-01 a selective inhibitor of protein kinase C, in murine and human tumor models. *Cancer Res.* **51**, 4888–4892.

Alessi, F., Quarta, S., Savio, M., Riva, F., Rossi, L., Stivala, L.A., Scovassi, A.I., Meijer, L., and Prosperi, E. (1998). The cyclin-dependent kinase inhibitors olomoucine and roscovitine arrest human fibroblasts in G1 phase by specific inhibition of CDK2 kinase activity. *Exp. Cell Res.* **245**, 8–18.

Aubry, C., Jenkins, P.R., Mahale, S., Chaudhuri, B., Marechal, J.-D., and Sutcliffe, M.J. (2004). New fascaplysin-based CDK4-specific inhibitors: design synthesis and biological activity. *Chem. Commun.* **15**, 1696–1697.

Aubry, C., Patel, A., Mahale, S., Chaudhuri, B., Maréchal, J.-D., Sutcliffe, M.J., and Jenkins, P.R. (2005). The design and synthesis of novel 3-[2-indol-1-yl-ethyl]-1*H*-indole derivatives as selective inhibitors of CDK4. *Tetrahedron Lett.* **46**, 1423–1425.

Aubry, C., Wilson,A.J., Jenkins, P.R., Mahale, S., Chaudhuri, B., Maréchal, J.-D., and Sutcliffe, M.J. (2006). Design, synthesis and biological activity of new CDK4-specific inhibitors, based on fascaplysin. *Org. Biomol. Chem.*, 4, 787–801.

Bartek, J., Bartkova, J., and Lukas, J. (1996). The retinoblastoma protein pathway and the restriction point. *Curr. Opin. Cell Biol.* **8**, 805–814.

Beattie, J.F., Breault, G.A., Ellston, R.P., Green, S., Jewsbury, P.J., Midgley, C.J., Naven, R.T., Minshull, C.A., Pauptit, R.A., Tucker, J.A., and Pease J.E. (2003). Cyclin dependent kinase 4 inhibitors as a treatment of cancer. Part 1: Identification and optimization of substituted 4, 6-Bis anilino pyrimidines. *Bioorg. Med. Chem. Lett.* **13**, 2955–2960.

Benson, C., Kaye, S., Workman, P., Garrett, M., Walton, M., and de Bono J. (2005). Clinical anticancer drug development: targeting the cyclin-dependent kinases. *Br. J. Cancer* **92**, 7–12.

Berman, H.M., Westbrook, J., Feng, Z., Gilliland, G., Bhat, T.N., Weissig, H., Shindyalov, I., and Bourne, P.E. (2000). The protein data bank. *Nucl. Acids Res.* **28**, 235–242.

Blagosklonny M.V. (2003). Cell senescence and hypermitogenic arrest. *EMBO Rep.* **4**, 358–362.

Boehm, M. and Nabel, E.G. (2003). The cell cycle and cardiovascular diseases. *Prog. Cell Cycle Res.* **5**, 19–30.

Brotherton, D.H., Dhanaraj, V., Wick, S., Brizuela, L., Domaille, P.J., Volyanik, E., Xu, X., Parisini, E., Smith, B.O., Archer, S.J., Serrano, M., Brenner, S.L., Blundell, T.L., and Laue, E.D. (1998). Crystal structure of the complex of the cyclin D-dependent kinase CDK6 bound to the cell-cycle inhibitor p19INK4d. *Nature* **395**, 244–250.

Brotz-Oesterhelt, H., Knezevic, I., Bartel, S., Lampe, T., Warnecke-Eberz, U., Ziegelbauer, K., Habich, D., Labischinski, H. (2003). Specific and potent inhibition of NAD+ dependent DNA ligase by pyridochromanones. J. Biol. Chem. 278, 39435–39442.

Carini, D.J., Kaltenbach, R.F., Liu, J., Benfield, P.A., Boylan, J., Boisclair, M., Brizuela, L., Burton, C.R., Cox, S., Grafstrom, R., Harrison, B.A., Harrison, K., Akamike, E., Markwalder, J.A., Nakano, Y., Seitz, S.P., Sharp, D.M., Trainor, G.L., and Sielecki T.M. (2001). Identification of selective inhibitors of cyclin dependent kinase 4. *Bioorg. Med. Chem. Lett.* **11**, 2209–2211.

Carlson, B.A., Dubay, M.M., Sausville, E.A., Brizuela, L., and Worland, P.J. (1996). Flavopiridol induces G1 arrest with inhibition of cyclin dependent kinase (CDK) 2 and CDK4 in human breast carcinoma cells. *Cancer Res.* **56**, 2973–2978.

Charan, R.D., McKee, T.C., and Boyd, M.R. (2004). Cytotoxic alkaloid from the marine sponge Thorectandra. *Nat. Prod. Res.* 18, 225–229.

Chen, Y.N., Sharma, S.K., Ramsey, T.M., Jiang, L., Martin, M.S., Baker, K., Adams, P.D., Bair, K.W., and Kaelin, W.G. (1999). Selective killing of transformed cells by cyclin/cyclin-dependent kinase 2 antagonists. *Proc. Natl. Acad. Sci. USA.* **96**, 4325–4329.

Cohen, P. (2002). Protein kinases–the major drug targets of the twenty first century? *Nat. Rev. Drug. Discovery* **1**, 309–315.

Dai, Y. and Grant, S. (2004). Small molecule inhibitors targeting cyclin dependent kinases as anticancer agents. *Curr. Oncol. Rep.* **6**, 123–130.

De Azevedo, W.F., Leclerc, S., Meijer, L., Havlicek, L., Strnad, M., and Kim, S.H. (1997). Inhibition of cyclin-dependent kinases by purine analogues: crystal structure of human CDK2 complexed with roscovitine. *Eur. J. Biochem.* **243**, 518–526.

Drees, M., Dengler, W.A., Roth, T., Labonte, H., Mayo, J., Malspeis, L., Grever, M., Sausville, E.A., and Fiebig, H.H. (1997). Flavopiridol (L86-8275): selective antitumor activity *in vitro* and activity *in vivo* for prostate carcinoma cells. *Clin. Cancer Res.* **3**, 273–279.

Eastman A. (2004). Cell cycle checkpoints and their impact on anticancer therapeutic strategies. *J. Cell Biochem.* **91**, 223–231.

Eisenbrand, G., Hippe, F., Jakobs, S., and Muehlbeyer, S. (2004). Molecular mechanisms of indirubin and its derivatives: novel anticancer molecules with their origin in traditional Chinese phytomedicine. *J. Cancer Res. Clin. Oncol.* **130**, 627–635.

Emanuel, E., Rugg, C.A., Gruninger, R.H., Lin, R., Fuentes-Pesquera, A., Connolly, P.J., Wetter, S.K., Hollister, B., Kruger, W.W., Napier, C., Jolliffe, L., and Middleton, S.A. (2005). The *in vitro* and *in vivo* effects of JNJ-7706621: a dual inhibitor of cyclin dependent kinases and aurora kinases. *Cancer Res.* **65**, 9038–9046.

Fischer, P.M. and Gianella-Borradori, A. (2005). Recent progress in the discovery and development of cyclin-dependent kinase inhibitors. *Expert. Opin. Invest. Drugs* **14**, 457–477.

Fortune, J.M. and Osheroff, N. (1998). Merbarone inhibits the catalytic activity of human topoisomerase II alpha by blocking DNA cleavage. *J. Biol. Chem.* 273, 17643–17650.

Fry, D.W., Harvey, P.J., Kellar, P.R., Elliott, W.L., Meade, M., Trachet, E., Albassam, M., Zheng, X., Leopold, W.R., Pryer, N.K., and Toogood, P.L. (2004). Specific inhibition of cyclin dependent kinase 4/6 by PD 0332991 and associated antitumor activity in human tumor xenografts. *Mol. Cancer Ther.* **3**, 1427–1438.

Garrett, M.D. and Fattaey, A. (1999). CDK inhibition and cancer therapy. *Curr. Opin. Genet. Dev.* **9**, 104–111.

Geall, A.J., Eaton, M.A., Baker, T., Catterall, C., and Blagbrough, I.S. (1999). The regiochemical distribution of positive charges along cholesterol polyamine carbamates plays significant roles in modulating DNA binding affinity and lipofection. *FEBS Lett.* 459, 337–342.

Grana, X. and Reddy, E.P. (1995). Cell cycle control in mammalian cells: role of cyclins, cyclin dependent kinases (CDKs), growth suppressor genes and cyclin dependent kinase inhibitors (CKIs). *Oncogene* **11**, 211–219.

Greene L.A, Biswas S.C., and Liu D.X. (2004). Cell cycle molecules and vertebrate neuron death: E2F at the hub. *Cell Death Differ.* **11**, 49–60.

Grossel, M.J. and Hinds, P.W. (2006). Beyond the cell cycle: a new role for CDK6 in differentiation. *J. Cell Biochem.* **97**, 485–493.

Guo, Z. and Stiller, J.W. (2004). Comparative genomics of cyclin-dependent kinases suggest co-evolution of the RNAP II C-terminal domain and CTD-directed CDKs. *BMC Genomics* **5**, 69–81.

Herrup, K., Neve, R., Ackerman, S.L., and Copani, A. (2004). Divide and die: cell cycle events as triggers of nerve cell death. *J. Neurosci.* **24**, 9232–9239.

Hirai, H., Kawanishi, N., and Iwasawa, Y. (2005). Recent advances in the development of selective small molecule inhibitors for cyclin dependent kinases. *Curr. Top. Med. Chem.* **5**, 167–179.

Hormann, A., Chaudhuri, B., and Fretz, H. (2001). DNA binding properties of the marine sponge pigment Fascaplysin. *Bioorg. Med. Chem.* **9**, 917–921.

Inayat-Hussain, S.H. and Thomas, N.F. (2004). Recent advances in the discovery and development of stilbenes and lactones in anticancer therapy. *Exp. Opin. Ther. Patents* **14**, 819–835.

Johnson, K., Liu, L., Majdzadeh, N., Chavez, C., Chin, P.C., Morrison, B., Wang, L., Park, J., Chugh, P., Chen, H.M., and D'Mello, S.R. (2005). Inhibition of neuronal apoptosis by the cyclin-dependent kinase inhibitor GW8510: identification of 3' substituted indolones as a scaffold for the development of neuroprotective drugs. *J. Neurochem.* **93**, 538–548.

Jones, G., Willett, P., Glen, R.C., Leach, A.R., and Taylor, R. (1997). Development and validation of a genetic algorithm for flexible docking. *J. Mol. Biol.* **267**, 727–748.

Jordan, M.A., Ojima, I., Rosas, F., Distefano, M., Wilson, L., Scambia, G., and Ferlini, C. (2002). Effects of novel taxanes SB-T-1213 and IDN5109 on tubulin polymerization and mitosis. *Chem. Biol.* **9**, 93–101.

Kastan, M.B. and Bartek J. (2004). Cell-cycle checkpoints and cancer. *Nature* **432**, 316–323.

Knockaert, M., Grrengard, P., and Meijer, L. (2002). Pharmacological inhibitors of cyclin dependent kinases. *Trends Pharmacol. Sci.* **23**, 417–425.

Kuo, G.H., Deangelis, A., Emanuel, S., Wang, A., Zhang, Y., Connolly, P.J., Chen, X., Gruninger, R.H., Rugg, C., Fuentes-Pesquera, A., Middleton, S.A., Jolliffe, L., and Murray, W.V. (2005). Synthesis and identification of (1, 3, 5) triazine pyridine biheteroaryl as a novel series of potent cyclin dependent kinase inhibitors. *J. Med. Chem.* **48**, 4535–4546.

Landis, M.W., Pawlyk, B.S., Li, T., Sicinski, P., and Hinds P.W. (2006). Cyclin D1-dependent kinase activity in murine development and mammary tumorigenesis. *Cancer Cell* **9**, 13–22.

Lawrie, A.M., Noble, M.E., Tunnah, P., Brown, N.R., Johnson, L.N., and Endicott, J.A. (1997). Protein kinase inhibition by staurosporine revealed in details of the molecular interactions. *Nat. Struct. Biol.* **4**, 796–801.

Lee, J.C. and Timasheff, S.N. (1977). *In vitro* reconstitution of calf brain microtubules: effects of solution variables. *Biochemistry* **16**, 1754–1764.

Liu, M.C., Marshall, J.L., and Pestell, R.G. (2004). Novel strategies in cancer therapeutics: targeting enzymes involved in cell cycle regulation and cellular proliferation. *Curr. Cancer Drug Targets* **4**, 403–424.

Lundberg, A.S. and Weinberg, R.A. (1998). Functional inactivation of retinoblastoma protein requires sequential modification by at least two distinct cyclin-cdk complexes. *Mol. Cell Biol.* **18**, 753–761.

Mahale, S., Aubry, C., Wilson, A.J., Jenkins, P.R., Maréchal, J.D., Sutcliffe, M.J., and Chaudhuri, B. (2006). CA224, a non-planar analogue of fascaplysin inhibits Cdk4 but not Cdk2 and arrests cells at G0/G1 inhibiting pRb phosphorylation. *Bioorg. Med. Chem. Lett.* [Epub ahead of print] PMID16750360.

Malumbres, M. and Barbacid M. (2006). Is Cyclin D1-CDK4 kinase a bona fide cancer target? *Cancer Cell* **9**, 2–4.

Martin, A., Odajima, J., Hunt, S.L., Dubus, P., Ortega, S., Malumbres, M., and Barbacid M. (2005). CDK2 is dispensable for cell cycle inhibition and tumor suppression mediated by p27(Kip1) and p21(Cip1). *Cancer Cell* **7**, 591–598.

Massague, J. (2004). G1 cell-cycle control and cancer. *Nature* **432**, 298–306.

McGowan, C.H. (2003). Regulation of the eukaryotic cell cycle. *Prog. Cell Cycle Res.* **5**, 1–4.

Meijer, L., Borgne, A., Mulner, O., Chong, J.P., Blow, J.J., Inagaki, N., Inagaki, M., Delcros, J.G., and Moulinoux, J.P. (1997). Biochemical and cellular effects of roscovitine, a potent and selective inhibitor of the cyclin-dependent kinases cdc2, cdk2 and cdk5. *Eur. J. Biochem.* **243**, 527–536.

Moon, M.J., Lee, S.K., Lee, J.W., Song, W.K., Kim, S.W., Kim, J.I., Cho, C., Choi, S.J., and Kim, Y.C. (2006). Synthesis and structure-activity relationship of novel indirubin derivatives as potent antiproliferative agents with CDK2 inhibitory activities. *Bioorg. Med. Chem.* **14**, 237–246.

Monaco, E.A. III and Vallano, M.L. (2003). Cyclin-dependent kinase inhibitors: cancer killers to neuronal guardians. *Curr. Med. Chem.* **10**, 367–379.

Morgan, D.O. (1997). Cyclin-dependent kinases: engines, clocks and microprocessors. *Annu. Rev. Cell Dev. Biol.* **13**, 261–291.

Murray, A.W. (2004). Recycling the cell cycle: cyclins revisited. *Cell* **116**, 221–234.

Neve, R.L. and McPhie, D.L. (2005). The cell cycle as a therapeutic target for Alzheimer's disease. *Pharmacol. Ther.* Epub ahead of print.

O'Hare, M., Wang, F., and Park, D.S. (2002). Cyclin-dependent kinases as potential targets to improve stroke outcome. *Pharmacol. Ther.* **93**, 135–143.

Oshima, J. and Campisi, J. (1991). Fundamentals of cell proliferation: control of the cell cycle. *J. Dairy Sci.* **74**, 2778–2787.

Parker, B.W., Kaur, G., Nieves-Neira, W., Taimi, M., Kohlhagen, G., Shimizu, T., Losiewicz, M.D., Pommier, Y., Sausville, E.A., and Senderowicz, A.M. (1998). Early induction of apoptosis in haematopoietic cell lines after exposure to flavopiridol. *Blood* **91**, 458–465.

Paull, K.D., Shoemaker, R.H., Hodes, L., Monks, A., Scudiero, D.A., Rubinstein, L., Plowman, J., and Boyd, M.R. (1989). Display and analysis of patterns of differential activity of drugs against human tumor cell lines: development of mean graph and COMPARE algorithm. *J. Natl. Cancer Inst.* **81**, 1088–1092.

Peng, J., Zhu, Y., Milton, J.T., and Price, D.H. (1998). Identification of multiple cyclin subunits of human P-TEFB. *Genes. Dev.* **12**, 755–762.

Pevarello, P., Brasca, M.G., Orsini, P., Traquandi, G., Longo, A., Nesi, M., Orzi, F., Piutti, C., Sansonna, P., Varasi, M., Cameron, A., Vulpetti, A., Roletto, F., Alzani, R., Ciomei, M., Albanese, C., Pastori, W., Marsiglio, O., Pesenti, E., Fiorentini, F., Bischoff, J.R., and Mercurio, C. (2005). 3-Aminopyrazole inhibitors of CDK2/Cyclin A as antitumor agents 2. Lead optimization. *J. Med. Chem.* **48**, 2944–2956.

Phelps, D.E. and Xiong, Y. (1996). Assay for activity of mammalian cyclin D dependent kinases CDK4 and CDK6. *Methods Enzymol.* **283**, 194–205.

Pines, J. (1999). Four-dimensional control of the cell cycle. *Nat. Cell. Biol.* **1**, E73–79.

Preisinger, C., Korner, R., Wind, M., Lehmann, W.D., Kopajtich, R., and Barr, F.A. (2005). Plk1 docking to GRASP65 phosphorylated by CDK1 suggests a mechanism for Golgi checkpoint signaling. *EMBO J.* **24**, 753–765.

Regula, K.M., Rzeszutek, M.J., Baetz, D., Seneviratne, C., and Kirshenbaum, L.A. (2004). Therapeutic opportunities for cell cycle re-entry and cardiac regeneration. *Cardiovasc. Res.* **64**, 395–401.

Robson, T., Worthington, J., McKeown, S.R., and Hirst, D.G. (2005). Radiogenic therapy: novel approaches for enhancing tumor radiosensitivity. *Technol. Cancer Res. Treat.* **4**, 343–361.

Sali, A. and Blundell, T.L. (1993). Comparative protein modelling by satisfaction of spatial restraints. *J. Mol. Biol.* **234**, 779–815.

Sanchez, I. and Dynlacht, B.D. (2005). New insights into cyclins, CDKs, and cell cycle control. *Semin. Cell Dev. Biol.* **16**, 311–321.

Sarcevic, B., Lilischkis, R., and Sutherland, R.L. (1997). Differential phosphorylation of T-47D human breast cancer cell substrates by D1-, D2-, D3-, E-, and A type cyclin CDK complexes. *J. Biol. Chem.* **273**, 33327–33337.

Sarkar, S., Chawla-Sarkar, M., Young, D., Nishiyama, K., Rayborn, M.E., Hollyfield, J.G., and Sen, S. (2004). Myocardial cell death and regeneration during progression of cardiac hypertrophy to heart failure. *J. Biol. Chem.* **279**, 52630–52642.

Sausville, E.A. (2003). Cyclin-dependent kinase modulators studied at the NCI: pre-clinical and clinical studies. *Curr. Med. Chem. Anticancer Agents* **3**, 47–56.

Schang, L.M. (2004). Effects of pharmacological cyclin-dependent kinase inhibitors on viral transcription and replication. *Biochim. Biophys. Acta.* **1697**, 197–209.

Schulze-Gahmen, U., Brandsen, J., Jones, H.D., Morgan, D.O., Meijer, L., Vesely, J., and Kim, S.H. (1995). Multiple modes of ligand recognition: crystal structures of cyclin-dependent protein kinase 2 in complex with ATP and two inhibitors, olomoucine and isopentenyladenine. *Proteins* **22**, 378–391.

Schulze-Gahmen, U., De Bondt, H.L., and Kim, S.H. (1996). High resolution crystal structures of human cyclin-dependent kinase 2 with and without ATP: bound waters and natural ligand as guides for inhibitor design. *J. Med. Chem.* **39**, 4540–4546.

Schwartz, G.K. (2005). Development of cell cycle active drugs for the treatment of gastrointestinal cancers: a new approach to cancer therapy. *J. Clin. Oncol.* **23**, 4499–4508.

Senderowicz, A.M. and Sausville, E.A. (2000). Preclinical and clinical development of cyclin dependent kinase modulators. *J. Natl. Cancer Inst.* **92**, 376–387.

Segraves, N.L., Robinson, S.J., Garcia, D., Said, S.A., Fu, X., Schmitz, F.J., Pietraszkiewicz, H., Valeriote, F.A., and Crews, P. (2004). Comparison of fascaplysin and related alkaloids: a study of structures, cytotoxicities, and sources. *J. Nat. Prod.* **67**, 783–792.

Senderowicz, A.M. (2003a). Novel direct and indirect cyclin-dependent kinase modulators for the prevention and treatment of human neoplasms. *Cancer Chemother. Pharmacol.* **52**(Suppl. 1), S61–73.

Senderowicz, A.M. (2003b). Small-molecule cyclin-dependent kinase modulators. *Oncogene* **22**, 6609–6620.

Senderowicz, A.M. (2005). Inhibitors of cyclin-dependent kinase modulators for cancer therapy. *Prog. Drug Res.* **63**, 183–206.

Shapiro, G.I. (2004). Preclinical and clinical development of the cyclin-dependent kinase inhibitor flavopiridol. *Clin. Cancer. Res.* **10**, 4270s–4275s.

Sherr, C.J. (1996). Cancer cell cycle. *Science* **274**, 1672–1677.

Soni, R., Muller, L., Furet, P., Schoepfer, J., Stephan, C., Zumstein-Mecker, S., Fretz, H., and Chaudhuri, B. (2000). Inhibition of cyclin dependent kinase 4 (CDK4) by fascaplysin, a marine natural product. *Biochem. Biophys. Res. Commun.* **275**, 877–884.

Soni, R., O'Reilly, T., Furet, P., Muller, L., Stephan, C., Zumstein-Mecker, S., Fretz, H., Fabbro, D., and Chaudhuri, B. (2001). Selective *in vivo* and *in vitro* effects of a small molecule inhibitor of cyclin dependent kinase 4. *J. Nat. Cancer Inst.* **93**, 436–446.

Stewart, Z.A., Westfall, M.D., and Pietenpol, J.A. (2003). Cell-cycle dysregulation and anti-cancer therapy. *Trends Pharmacol. Sci.* **24**, 139–145.

Swanton, C. (2004). Cell-cycle targeted therapies. *Lancet. Oncol.* **5**, 27–36.

Tenzer, A. and Pruschy, M. (2003). Potentiation of DNA-damage-induced cytotoxicity by G2 checkpoint abrogators. *Curr. Med. Chem. Anticancer Agents* **3**, 35–46.

Tetsu, O. and McCormick, F. (2003). Proliferation of cancer cells despite CDK2 inhibition. *Cancer Cell* **3**, 233–245.

Tirado, O.M., Mateo-Lozano, S., and Notario, V. (2005). Roscovitine is an effective inducer of apoptosis of Ewing's sarcoma family tumor cells *in vitro* and *in vivo*. *Cancer Res.* **65**, 9320–9327.

Toogood, P.L. (2002). Progress toward the development of agents to modulate the cell cycle. *Curr. Opin. Chem. Biol.* **6**, 472–478.

Trembley, J.H., Hu, D., Slaughter, C.A., Lahti, J.M., and Kidd, V.J. (2003). Casein kinase 2 interacts with cyclin-dependent kinase 11 (CDK11) in vivo and phosphorylates both the RNA polymerase II carboxyl-terminal domain and CDK11 in vitro. *J. Biol. Chem.* **278**, 2265–2270.

Verdaguer, E., Jorda, E.G., Alvira, D., Jimenez, A., Canudas, A.M., Folch, J., Rimbau, V., Pallas, M., and Camins, A. (2005). Inhibition of multiple pathways accounts for the antiapoptotic effects of flavopiridol on potassium withdrawal-induced apoptosis in neurons. *J. Mol. Neurosci.* **26**, 71–84.

Verdonk, M.L., Cole, J.C., Hartshorn, M.J., Murray, C.W., and Taylor R.D. (2003). Improved protein ligand docking using GOLD. *Proteins* **52**, 609–623.

Wang, Q., Worland, P.J., Clark, J.L., Carlson, B.A., and Sausville, E.A. (1995). Apoptosis in 7-hydroxystaurosporine treated T lymphoblasts correlates with activation of cyclin dependent kinases 1 and 2. *Cell Growth Differ.* **6**, 927–936.

Weinberg, R.A. (1995). The retinoblastoma protein and cell cycle control. *Cell* **81**, 323–330.

Welburn, J.P. and Endicott, J.A. (2005). Inhibition of the cell cycle with chemical inhibitors: a targeted approach. *Semin. Cell Dev. Biol.* **16**, 369–381.

Yu, Q., Sicinska, E., Geng, Y., Ahnstrom, M., Zagozdzon, A., Kong, Y., Gardner, H., Kiyokawa, H., Harris, L.N., Stal, O., and Sicinski., P. (2006). Requirement of CDK4 kinase function in breast cancer. *Cancer Cell* **9**, 23–32.

Zhai, S., Senderowicz, A.M., Sausville, E.A., and Figg, W.D. (2002). Flavopiridol, a novel cyclin-dependent kinase inhibitor, in clinical development. *Ann. Pharmacother.* **36**, 905–911.

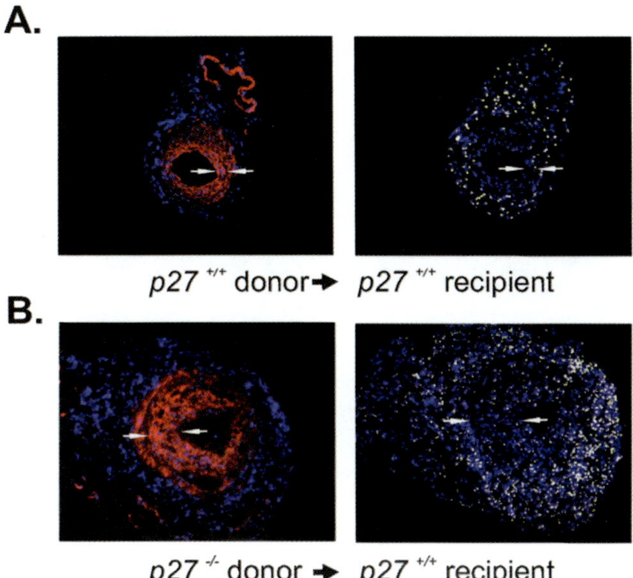

COLOR FIGURE 4.2 p27[KIP1] modulates the contribution of bone-marrow-derived cells to vascular lesions. Cross sections of recipient p27[+/+]female arteries following transplantation of male p27[+/+] or p27[-/-] donor marrow: Y chromosome[+] (Yellow), alpha-actin[+] (smooth muscle specific) cells (red), and nuclei (DAPI stain, blue). Arrows indicate the margins of the intima as determined by the internal and external elastic lamina. (Adapted from Boehm, M. et al. *J Clin Invest* 114, 419–426, 2004. With permission.)

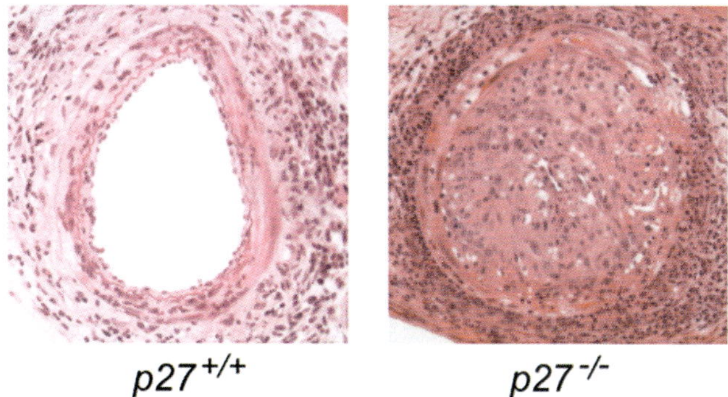

COLOR FIGURE 4.3 p27[KIP1] modulates neointima formation during vascular wound repair. Increased vascular lesions in p27[-/-] arteries compared to p27[+/+] arteries in H&E cross sections of murine arteries 2 weeks after mechanical injury.

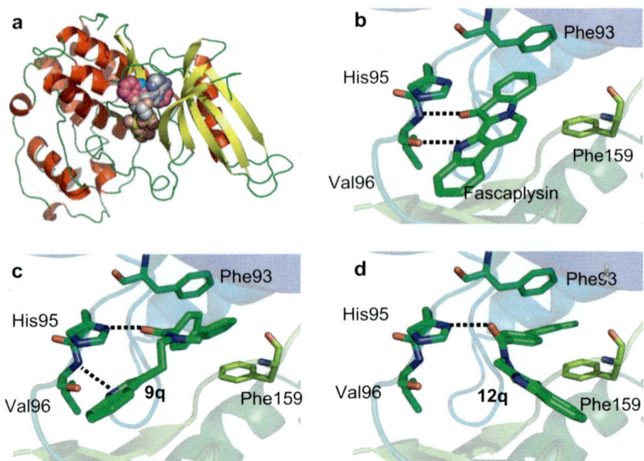

COLOR FIGURE 6.2 The active site of the CDK4 homology model. (a) Cartoon representation of overall structure of the CDK4, with the positions of ATP/Mg^{2+} (C in grey, N in slate, O in salmon, P in yellow, Mg in black) and fascaplysin (C in magenta, N in cyan, O in red), is shown. (b) Predicted binding mode of fascaplysin. (c) Predicted binding mode of the most potent compound, **9q**, in the series **9a-q**; note the π-π interactions of the benzoid ring with Phe93 and Phe159 of CDK4. (d) Predicted binding mode of **12q** — this weaker inhibitor is structurally similar to, but more conformationally constrained, than **9q**. Hydrogen bonds are shown as dashed lines. Figures were produced using Pymol. (From Aubry, C. et al. (2006). *Org. Biomol. Chem.*, DOI: 10.1039/b518019h.)

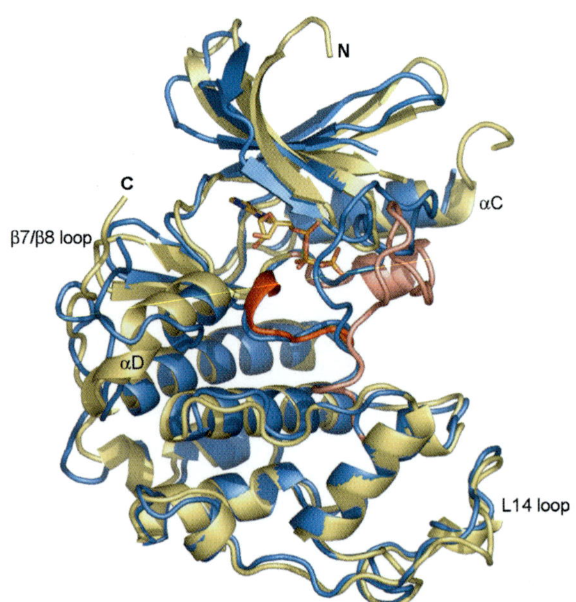

COLOR FIGURE 7.2 CDK2 and CDK7 apoenzymes superimposed on the C-terminal domain. CDK2 is shown in yellow, CDK7 in blue. The catalytic loop and activation loop in CDK2 are drawn in red and salmon, respectively.

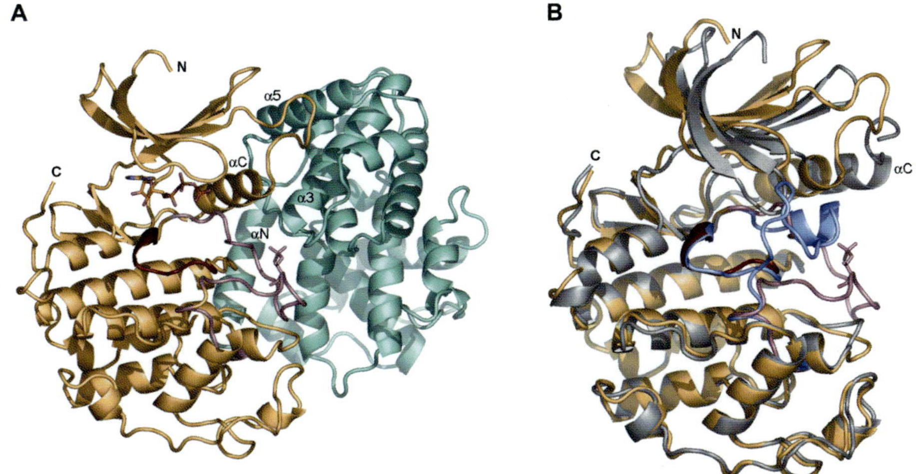

COLOR FIGURE 7.3 (A) Phosphorylated CDK2/cyclinA. (B) CDK2 apoenzyme in gray and activated CDK2 in yellow are superimposed. The catalytic loop and activation loop in the activated CDK2 are drawn in red and salmon (blue in the apoenzyme).

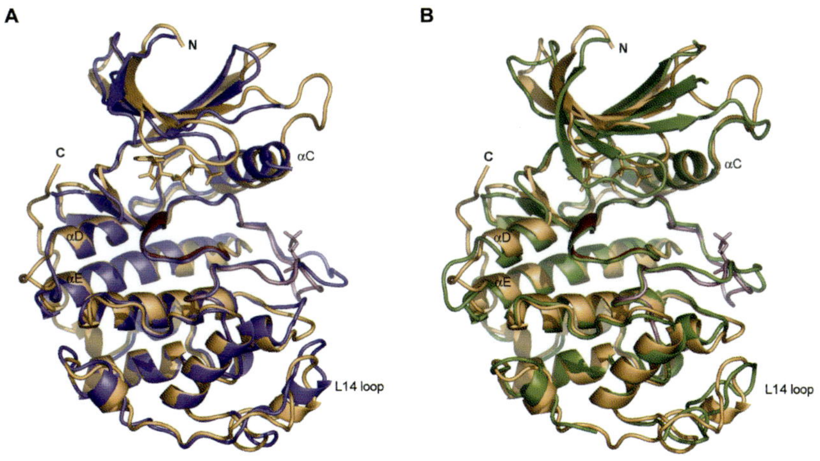

COLOR FIGURE 7.4 Superposition of activated CDK5 and CDK6 onto activated CDK2. CDK5 from the p25 complex (A) and CDK6 from the ternary CDK6/Vcyclin/fisetin complex (B) were superimposed onto the C-terminal domain of activated CDK2/cyclin A complex. CDK2 is shown in yellow, CDK5 in purple, and CDK6 in green. The catalytic loop of CDK2 is highlighted in red, and the activation loop in salmon. Atoms of the phosphorylated threonine residue in the activation loop are shown as stick model.

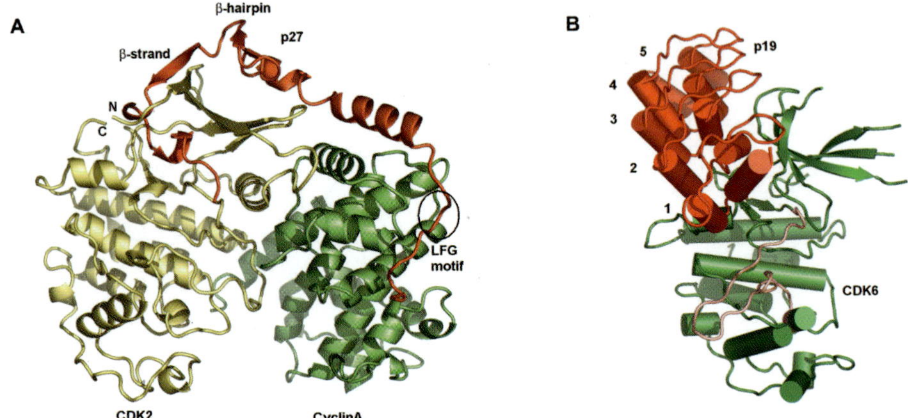

COLOR FIGURE 7.5 Structures of CDK/inhibitor complexes. (A) The p27^KIP1 inhibitor is bound in a ternary complex to CDK2 and cyclin A. The inhibitor is shown in red, CDK2 in yellow, and cyclin A in green. (B) The p19^INK4 inhibitor, shown in red, is bound to CDK6, shown in green.

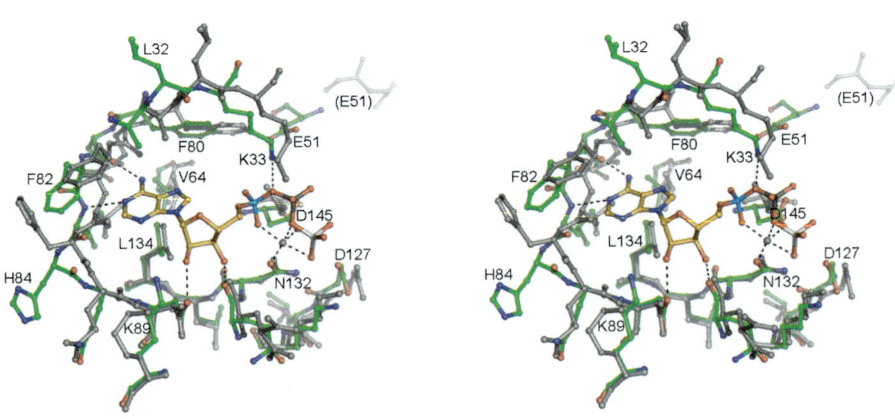

COLOR FIGURE 7.6 Stereo drawing of the ATP-binding pocket in CDK2. Residues 11 to 18 that form the glycine loop were omitted for clarity. The binding pocket in the activated CDK2 is shown in green with bound ATP in yellow and the Mg ion as a gray sphere. The structure of the binding pocket in the CDK2 apoenzyme is shown in gray. Hydrogen bonds are drawn as broken lines.

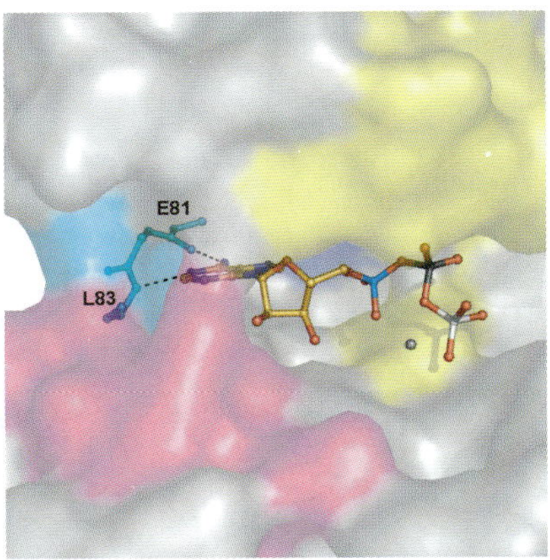

COLOR FIGURE 7.7 Surface representation of the ATP-binding pocket in CDK2 (Brown, N.R., Noble, M.E., Endicott, J.A., and Johnson, L.N. (1999a). The structural basis for specificity of substrate and recruitment peptides for cyclin-dependent kinases. *Nat Cell Biol 1*, 438–443). The surface is colored by pocket regions (Davies, T.G., Pratt, D.J., Endicott, J.A., Johnson, L.N., and Noble, M.E. (2002b). Structure-based design of cyclin-dependent kinase inhibitors. *Pharmacol Ther 93*, 125–133.) The hinge region is shown in cyan, with the backbone of residues 81 to 83 drawn as a stick model. Conserved hydrogen bonds with the hinge region are shown as broken lines. The F80 pocket is colored blue, the ribose and phosphate binding cleft is yellow, and the specificity region is purple. Bound ATP with one Mg ion is drawn as a stick model.

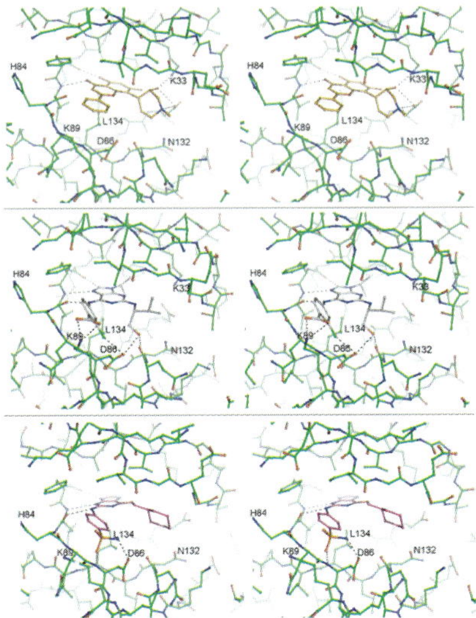

COLOR FIGURE 7.8 Structure of the binding pocket of three CDK2/inhibitor complexes. The top shows the complex with the nonspecific inhibitor des-chloro-flavopiridol. There are few contacts of the inhibitor with residues in the specificity region of the binding pocket. The middle panel shows the interactions of CDK2 with the highly specific inhibitor purvalanol B. There are many interactions with the specificity region. However, the inhibitor fills very little space in the area of the binding pocket that binds the ribose and phosphates of ATP. The bottom panel shows the interactions with NU6102, another CDK2-specific inhibitor. There are several interactions with the specificity region. The inhibitor also fills the ribose-binding pocket with the cyclo hexylmethyl group of the inhibitor. Hydrogen bonds between inhibitors and CDK2 are shown in black broken lines.

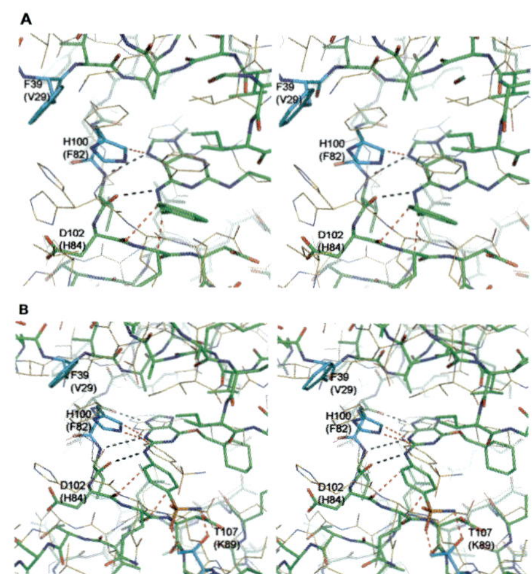

COLOR FIGURE 7.9 Stereo drawings of the ATP-binding pocket of CDK6 with CDK2-specific inhibitors olomoucine (A) and NU6102 (B) modeled into the binding pocket. The structures of the active CDK6 and the modeled inhibitor are shown as stick models in green with CDK4/6-specific sequence changes highlighted in cyan. The structure of the corresponding CDK2/inhibitor complex is drawn with thin lines in dark yellow. Conserved hydrogen bonds, used to place the inhibitor into the binding pocket, are represented by black broken lines. Unfavorably close contacts between the inhibitor and CDK6 are shown as broken red lines. Residue labeling refers to the CDK6 sequence with labels for CDK2 displayed in brackets.

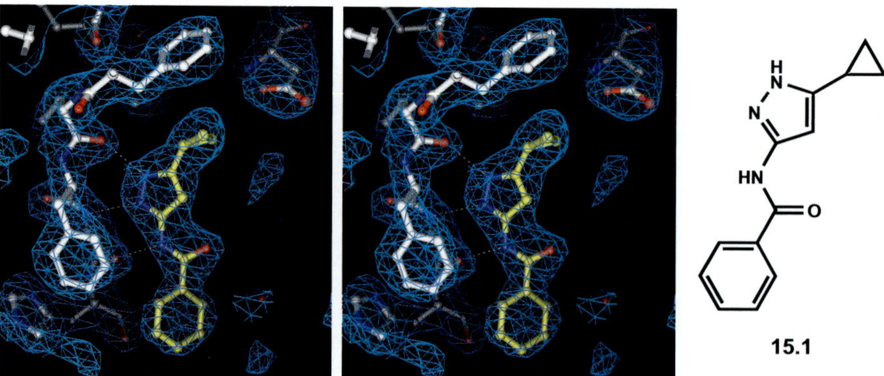

COLOR FIGURE 15.3 ATP-binding pocket residues of the X-ray structure of CDK2 in complex with **15.3**. The three hydrogen bonds between the CDK2 hinge region and the inhibitor are drawn as dashed lines.

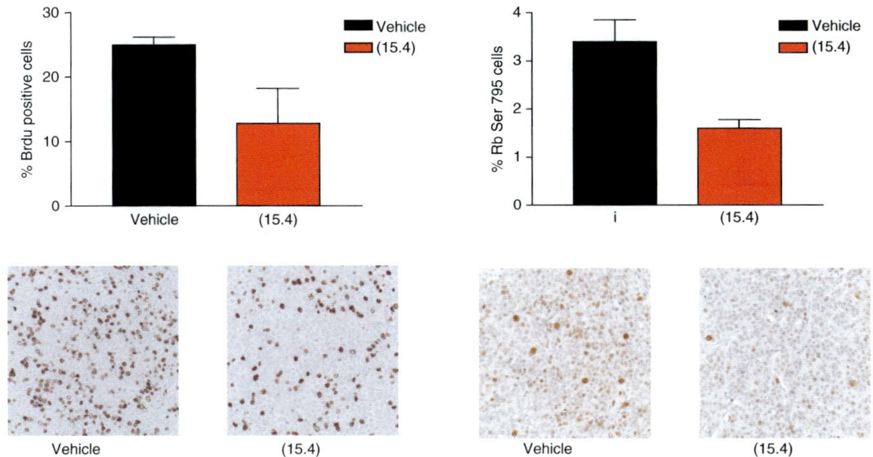

COLOR FIGURE 15.9 Analysis by IHC of the BrdU incorporation (A) and pRb staining (B) of the sections of A2780 xenograft tumors at the end of treatment. Tumor sections were fixed in formalin, paraffin embedded, and stained for BrdU incorporation and phospho-pRb expression. The analysis shows a significant reduction of both BrdU and phospo-pRb positive cells in the treated (**15.4**) vs. control tumors. Representative examples of the stainings are also reported.

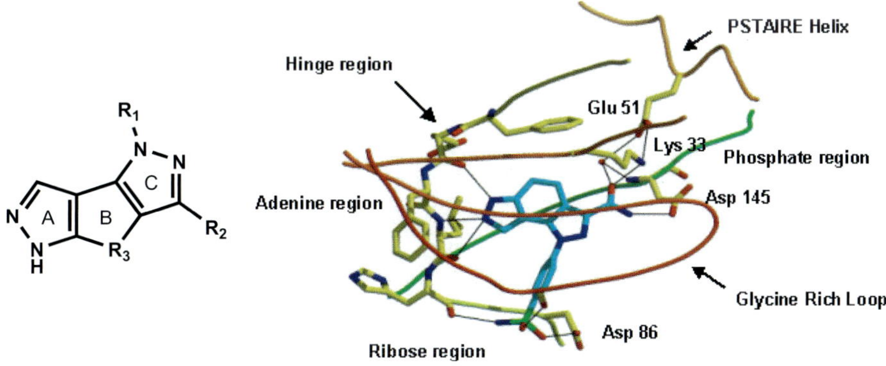

COLOR FIGURE 15.12 General formula of tricyclic benzodipyrazoles (BDP) and binding mode of a derivative into the ATP-binding pocket.

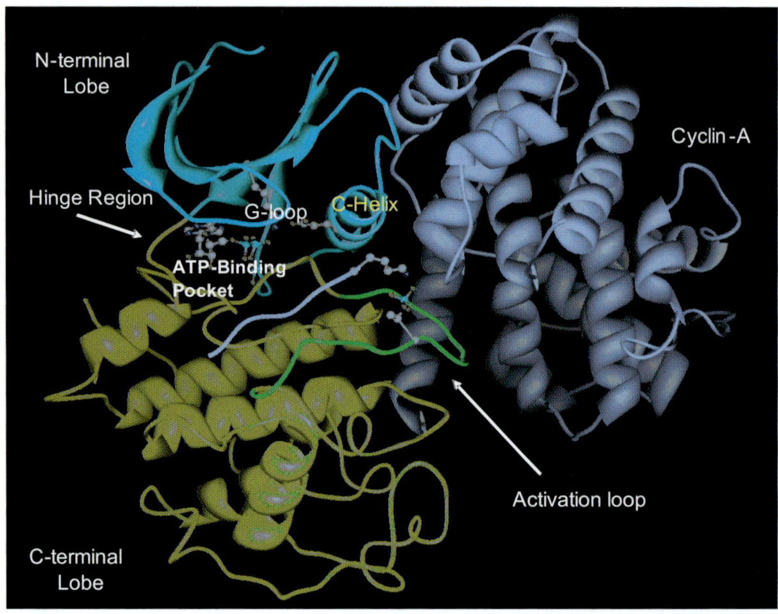

COLOR FIGURE 18.1 Ribbon representation of the crystal structure of CDK2/cyclin A/ATP complex. Cyclin is shown by cornflower blue and CDK2 by cyan, yellow, and green ribbon. The substrate is colored magenta. The critical residues (K33, E51, and T160 in CDK2 and K in the substrate) involved in the activity are depicted by ball-and-stick model. An ATP analog is also shown.

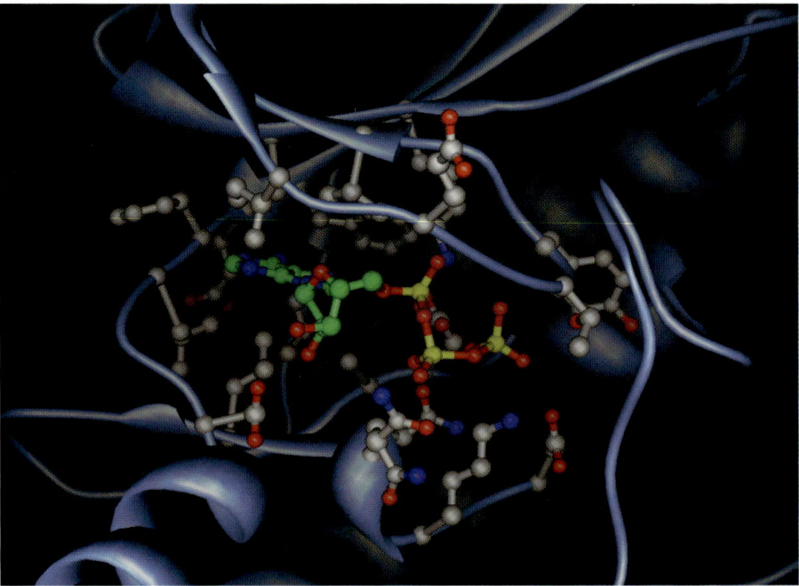

COLOR FIGURE 18.2 The residues of CDK2 in contact with ATP.

7 Three-Dimensional Structures of Cyclin-Dependent Kinases and Their Inhibitor Complexes

Ursula Schulze-Gahmen and Sung-Hou Kim

CONTENTS

7.1 INTRODUCTION

Cyclin-dependent kinases (CDKs) are eukaryotic serine or threonine protein kinases that are closely related to the prototypical Cdc2 protein (Forsburg and Nurse, 1991) and are known to associate with cyclin-related regulatory subunits. In yeast, a single CDK (Cdc28 in the budding yeast *Saccharomyces cerevisiae* and Cdc2 in the fission yeast *Schizosaccharomyces pombe*) regulates all cell cycle transitions by forming complexes with various cyclin proteins (Forsburg and Nurse, 1991). In higher eukaryotes, at least ten CDKs have been identified. However, only a few of them, CDC2 (CDK1) and CDK2, are involved in central cell cycle functions. Other functions include coupling of the cell cycle to extracellular signals (CDK4 and CDK6), phosphorylation of other CDKs (CAK, CDK7) and transcription (CDK7, CDK8, CDK9), and neural differentiation (CDK5) (Morgan, 1997; Sausville, 2002).

Extensive biochemical and structural studies of several of the CDKs have established a clear picture of the activation and regulatory mechanisms that determine the kinase activity in correlation with the cell cycle phases. As the emphasis of this review is on the structural aspects of CDKs and their inhibitor complexes, we will only summarize the common activation and inhibition mechanisms of CDKs to provide a biological background for discussion and comparison of the different activation states of CDKs. These regulatory mechanisms have been previously summarized in excellent reviews (e.g., Jeffrey et al., 2000; Mapelli and Musacchio, 2003; Morgan, 1997; Pavletich, 1999).

7.2 OVERVIEW AND COMPARISON OF CDK STRUCTURES

Four major mechanisms appear to govern the enzymatic activity of CDKs. It seems, however, that not every mechanism is used by all CDKs. All CDK apoenzymes are inactive and require complex formation with cyclins or cyclin-like regulatory subunits as a first step of activation. The CDK5 complexes with p25 and p35 and CDK6 complex with Vcyclin appear to become fully activated by this one-step mechanism (Schulze-Gahmen and Kim, 2002; Tarricone et al., 2001). However, complete activation of other CDKs involved in cell cycle control requires phosphorylation of a conserved threonine in the activation loop (Connell-Crowley et al., 1993; Desai et al., 1992; Solomon et al., 1992). The fully active complexes can be turned off by two mechanisms: CDK inhibitory proteins can bind and inactivate the CDK/cyclin complexes (Brotherton et al., 1998; Russo et al., 1996a; Russo et al., 1998), or regulatory kinases can phosphorylate CDKs on residues in the glycine-rich loop near the N-terminus (McGowan and Russell, 1993; Norbury et al., 1991; Parker et al., 1992). A substantial number of crystal structures of different CDKs in different activation states have been solved over the last 12 years. They revealed an enormous flexibility of the CDK structures and provided a structural basis for the observed activation and inhibition mechanisms. This wealth of structural information is also extremely valuable when designing small molecule inhibitors of CDKs for the development of anti-tumor drugs. For example, if the inhibitors should be CDK specific or specifically directed against the inactive or activated forms of CDKs, the available structural information can define precise structural features for inhibitor design. We compare the structures of different CDK apoenzymes, activated CDK complexes, and CDK-inhibitor complexes here to highlight CDK-specific features and structural changes that are correlated with CDK activation and inhibition. Where specific CDK residues are mentioned in the discussion, residue numbering will be based on the CDK2 sequence (Figure 7.1) unless specifically indicated otherwise.

7.2.1 APOENZYMES

The crystal structures of two apoenzymes, CDK2 and CDK7, have been determined to 1.8 Å and 3.0 Å, respectively (De Bondt et al., 1993; Lolli et al., 2004; Schulze-Gahmen et al., 1996). The kinases provide adjacent binding sites for ATP and protein substrate that orient the reaction partners in such a way that the γ–phosphate of ATP

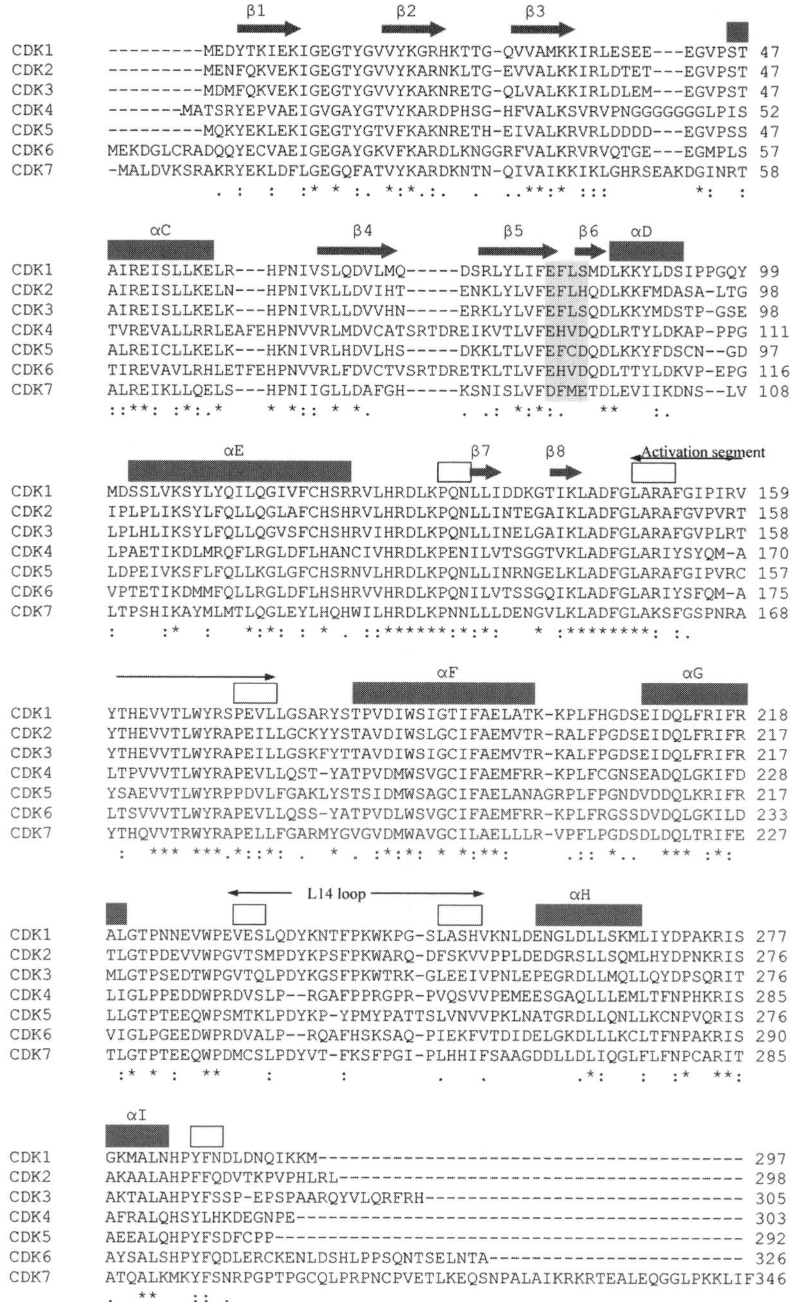

FIGURE 7.1 Sequence alignment of human CDK1 to CDK7. The secondary structures for inactive CDK2 (De Bondt et al., 1993; Schulze-Gahmen et al., 1996) are indicated above the sequences. Residues in the hinge region between the N-terminal and C-terminal domain are highlighted with a gray background.

faces the hydroxylated side chain of the target residue on the protein surface. Catalytic residues then promote the transfer of the phosphate group to the oxygen of the hydroxyl. CDK2 and CDK7 have 44% sequence identity and show the typical protein kinase fold. A smaller N-terminal domain is composed of a beta sheet and the large C-helix (Taylor et al., 1992). It is connected by a linker to the larger, mostly helical C-terminal domain. ATP is bound in the cleft between the two domains so that the hydrophobic purine ring binds in a pocket close to the linker region and the phosphates are oriented outward toward the mouth of the cleft.

The CDK apoenzymes are incapable of catalyzing the phosphotransfer reaction. Their activity is blocked in two ways. First, the activation loop (residues D145 to E172 in CDK2) is folded upward from the C-terminal domain and blocks protein substrate binding. Although the activation loop is flexible in the apoenzymes and adopts somewhat different conformations in CDK2 and CDK7, it prevents substrate peptide binding in both cases (Figure 7.2). Second, essential catalytic residues are located in the catalytic loop (residues 126–132 in CDK2) (Figure 7.2), in β3 and in the αC-helix. The conformation of the activation loop in inactive CDK2 and CDK7 forces a rotation of the αC-helix which leads to the incorrect positioning of several key catalytic residues and the incorrect alignment of β and γ–phosphates in ATP for an efficient in-line phosphotransfer reaction. The structures of CDK2 and CDK7 are similar with a root-mean-square (rms) deviation in Cα positions of 0.9 Å for 134 residues in the C-terminal domain and 1.2 Å for 66 residues in the N-terminal domain. The major

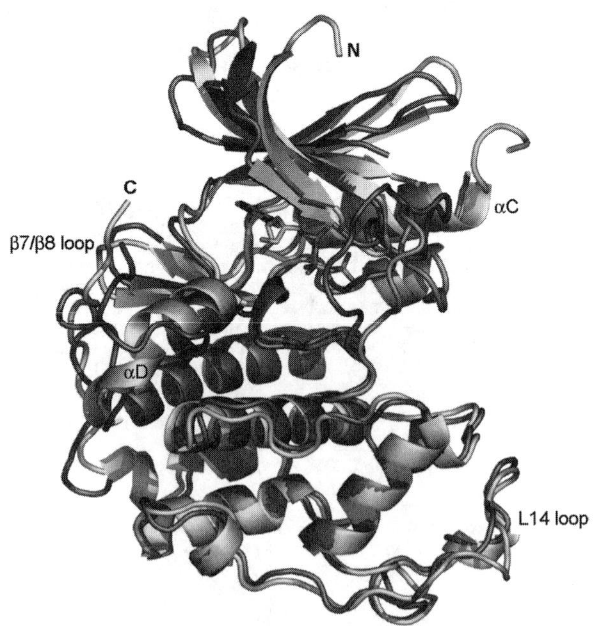

FIGURE 7.2 (See color insert following page 142.) CDK2 and CDK7 apoenzymes superimposed on the C-terminal domain. CDK2 is shown in yellow, CDK7 in blue. The catalytic loop and activation loop in CDK2 are drawn in red and salmon, respectively.

differences between CDK2 and CDK7 occur in three regions: (1) the activation loop; (2) a region where three protein segments pack together comprising αD and the following loop (residues 92–98), the β7/β8 loop (residues 136–139), and the C-terminal region (residues 287–298); and (3) the region of the L14 loop (residues 236–240) which is part of the kinase associated phosphatase (KAP) recognition site (Song et al., 2001) and Cks1 regulatory protein binding site in CDK2 (Bourne et al., 1996). The structural differences in regions 2 and 3 may cause differences in specific interactions of CDKs with substrates or regulatory proteins. For example, members of the MAP kinase family interact with docking site sequences in their protein substrates that are spatially separate from the substrate phosphorylation sites. The structure of p38 in complex with docking peptides from two such interacting proteins show these peptides interacting at a site that is equivalent to the C-terminal region, β7/β8 loop, and αD/αE loop in CDKs (Chang et al., 2002). A similar mechanism might exist for substrate binding to CDK7, or another protein might interact with CDK7 in this region. Structural and sequence differences in region 3 were shown to be important for the regulation of KAP activity, which is active on CDK2 but not on CDK7 (Lolli et al., 2004; Song et al., 2001).

7.2.2 Activated CDK/Cyclin Complexes

Cyclin binding to CDKs activates CDKs partially or completely, depending on the CDK. The CDK2/cyclinA complex requires an additional phosphorylation on T160 for full activation. This activation is correlated with characteristic structural changes that have been observed in the crystal structures of three activated CDKs: CDK2/cyclin A (Jeffrey et al., 1995; Russo et al., 1996b), CDK5/p25 (Tarricone et al., 2001), and CDK6/Vcyclin (Schulze-Gahmen and Kim, 2002). The activation loop that was obstructing the substrate binding cleft in the apoenzyme is moved toward the C-terminal domain where it forms a binding area for the substrate protein (Figure 7.3). Phosphorylation of T160 in CDK2 leads to further conformational changes in the substrate binding site causing an unusual left-handed conformation ($\Phi = 72.5°$, $\Psi = 130.8°$) for residue V164. The valine conformation results in the carbonyl oxygen atom being directed away from the substrate, causing the binding of any residue other than proline at the P+1 site to be unfavorable because of an uncompensated hydrogen bond from the substrate's main-chain nitrogen (Brotherton et al., 1998). The phosphorylation-induced change in the substrate binding pocket in CDK2 is induced in CDK5/p25 and CDK6/Vcyclin by cyclin binding alone.

Simultaneously with the movement of the activation loop, the αC-helix is moved into the binding cleft and rotated so that the side chain of E51 in the αC-helix becomes rotated into the ATP-binding pocket, where it forms ionic interactions with K33. The role of K33 is to bind to the α- and β-phosphates of ATP and align them to enable the transfer of the γ-phosphate to substrate proteins (Taylor et al., 1992).

p25 and Vcyclin induce very similar structural changes in CDK5 and CDK6, respectively, leading to activation of these kinases without requiring additional phosphorylation of the threonine in the activation loop. Figure 7.4 shows the superposition of activated CDK5 and CDK6 on the C-terminal domain of activated CDK2. The structural elements of the C-terminal domain and the αC-helix overlap very well. Structural differences are found in the relative orientation of the N-terminal

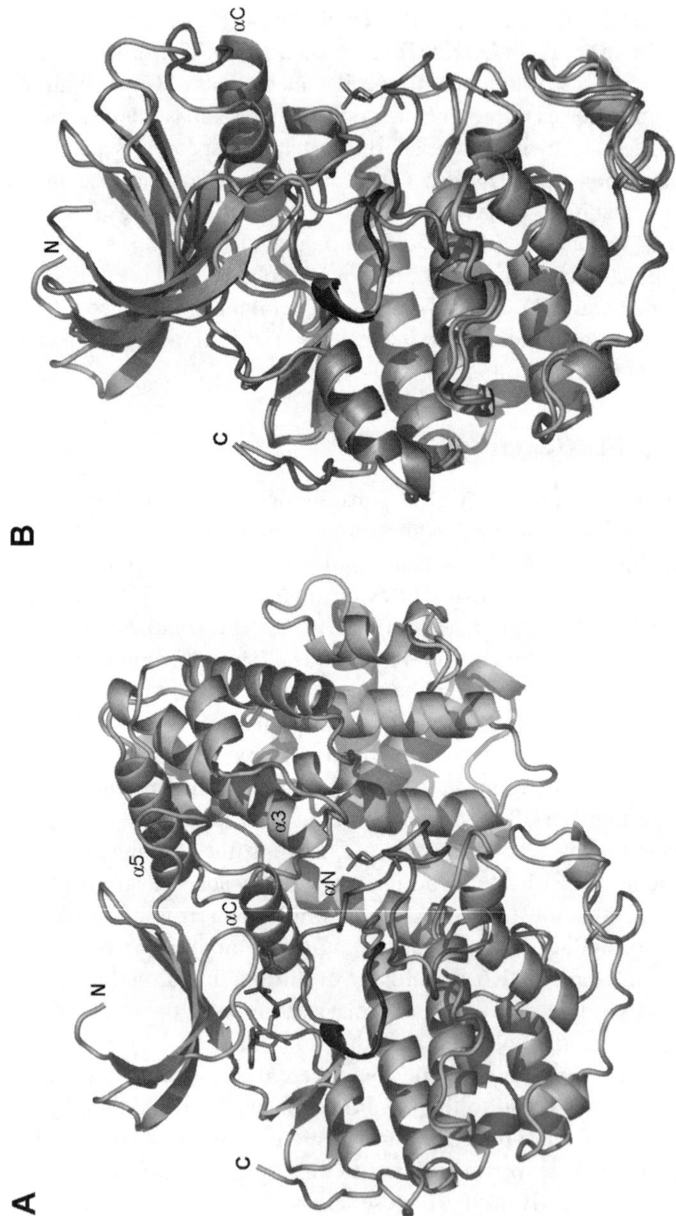

FIGURE 7.3 (See color insert.) (A) Phosphorylated CDK2/cyclinA. (B) CDK2 apoenzyme in gray and activated CDK2 in yellow are superimposed. The catalytic loop and activation loop in the activated CDK2 are drawn in red and salmon (blue in the apoenzyme).

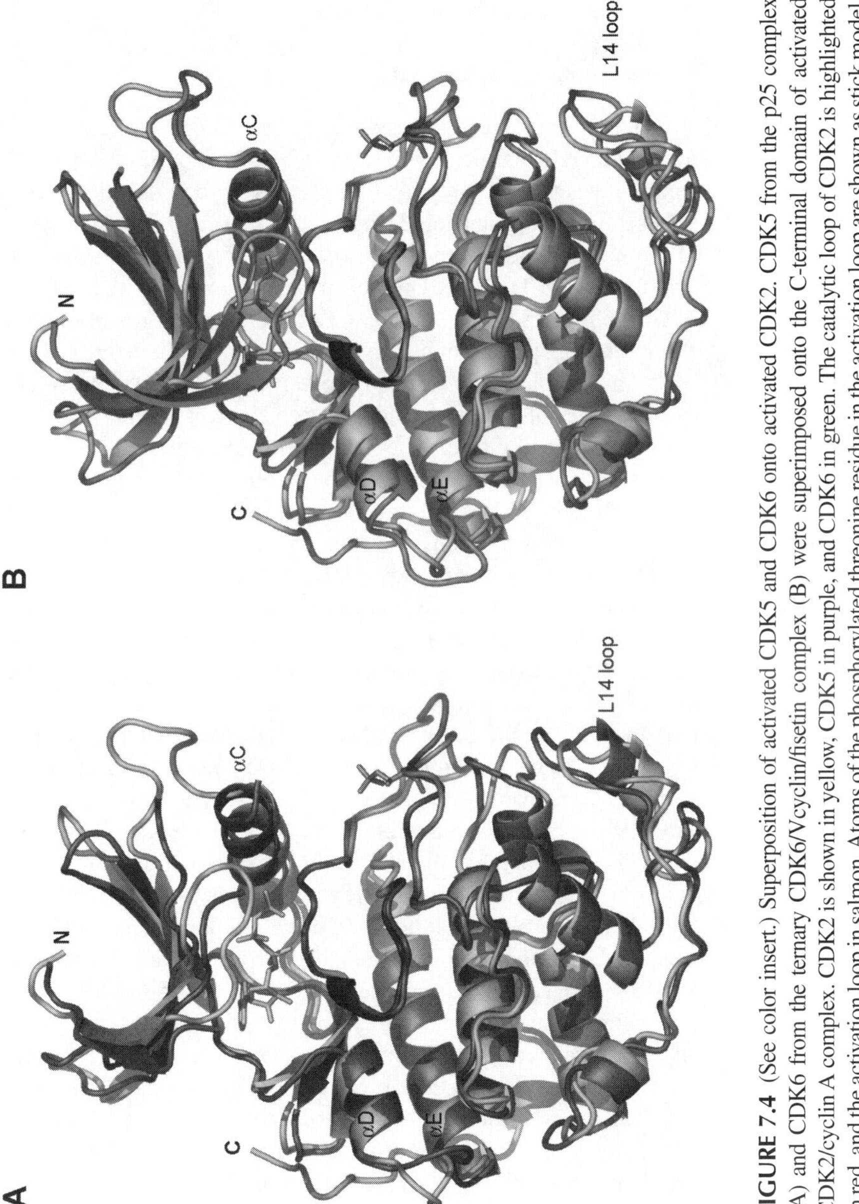

FIGURE 7.4 (See color insert.) Superposition of activated CDK5 and CDK6 onto activated CDK2. CDK5 from the p25 complex (A) and CDK6 from the ternary CDK6/Vcyclin/fisetin complex (B) were superimposed onto the C-terminal domain of activated CDK2/cyclin A complex. CDK2 is shown in yellow, CDK5 in purple, and CDK6 in green. The catalytic loop of CDK2 is highlighted in red, and the activation loop in salmon. Atoms of the phosphorylated threonine residue in the activation loop are shown as stick model.

β-sheet, the tip of the activation loop, the L14-loop, and the αD/αE loop. These structural differences may indicate higher flexibility in these regions or may be the reason for differential interactions with other CDK-interacting proteins, such as KAP or protein substrates. The critical structural changes for enzymatic activation of CDKs are, however, the same for all three structures.

7.2.3 CDK/Protein Inhibitor Complexes

The active CDKs can be counteracted by two families of cell cycle–inhibitory proteins. Members of the CIP family, which include p27[KIP1], p21[CIP1,WAF-1], and p57[KIP2], bind and inhibit the active CDK/cyclin complexes (Sherr and Roberts, 1995). Members of the INK4 family, which include p15, p16, p18, and p19, are specific for CDK4 and CDK6, and can bind to either the isolated CDK subunit or its complex with cyclin D (Serrano, 1997). The INK4 inhibitors are specific for the G1 phase CDKs, whereas the CIP inhibitors have a broader CDK preference. Crystal structures of a 69-amino-acid peptide from p27[KIP1] bound to CDK2/CyclinA (Russo et al., 1996a) and p16[INK4] and p19[INK4] in complex with CDK6 (Brotherton et al., 1998; Russo et al., 1998) revealed that both types of inhibitors induce large changes in the shape of the ATP and substrate binding sites of CDKs. The p27 peptide binds with its N-terminal part to cyclin A, and with its C-terminal part to the N-terminal domain of CDK2. The conserved "LFG" motif in p27 is inserted in a pocket on the face of cyclin A without changing the conformation of cyclin A (Figure 7.5A). The same "LFG" sequence motif is found in tight binding substrates of CDK2-CyclinA (Zhu et al., 1995). The p27-peptide then stretches over to the N-terminal domain of CDK2, where its binding induces large structural changes in the kinase. First, the β-turn segment of p27 flattens the β-sheet of the kinase. Second, the β-strand segment of p27 replaces the first β-strand of the CDK2 sheet, which becomes disordered. Third, the carboxyl-terminal segment of the p27 peptide, the 3_{10}-helix, binds deep within the active site–cleft and prevents ATP binding by mimicking the ATP substrate in its position and the contacts it makes to active site groups (Figure 7.5A).

The p16[INK4] and p19[INK4] inhibitors also take advantage of the kinase flexibility to distort the catalytic cleft and prevent the necessary activating structural changes. p16 and p19 consist of four and five ankyrin repeats, respectively. Each ankyrin repeat has a structure that resembles the letter L (Gorina and Pavletich, 1996), with a pair of antiparallel helices forming the stem and a β-hairpin forming the base. The L-shaped repeats stack to give an extended concave surface (Figure 7.5B). CDK6-bound p16 and p19 have very similar structures (Brotherton et al., 1998; Russo et al., 1998) with p19 having an extra repeat added at the part of the structure furthest away from CDK6. p19 binds to one side of the catalytic cleft opposite to the cyclin binding site. The concave site of the p19 molecule binds to the β-sheet in the N-terminal domain of CDK6, and the ends of the helices of ankyrin repeats 1 to 3 bind to the C-terminal domain where they interact with the activation loop and with structural elements of the catalytic cleft. Like the inactive CDK2 apoenzyme, p19-bound CDK6 does not have any of the activating changes. However, p19 binding induces structural distortions in CDK6 that move it further away from the active structure. The N-terminal and C-terminal domains are twisted away from each other, distorting the cyclin binding site and preventing the translocation of the αC-helix into the

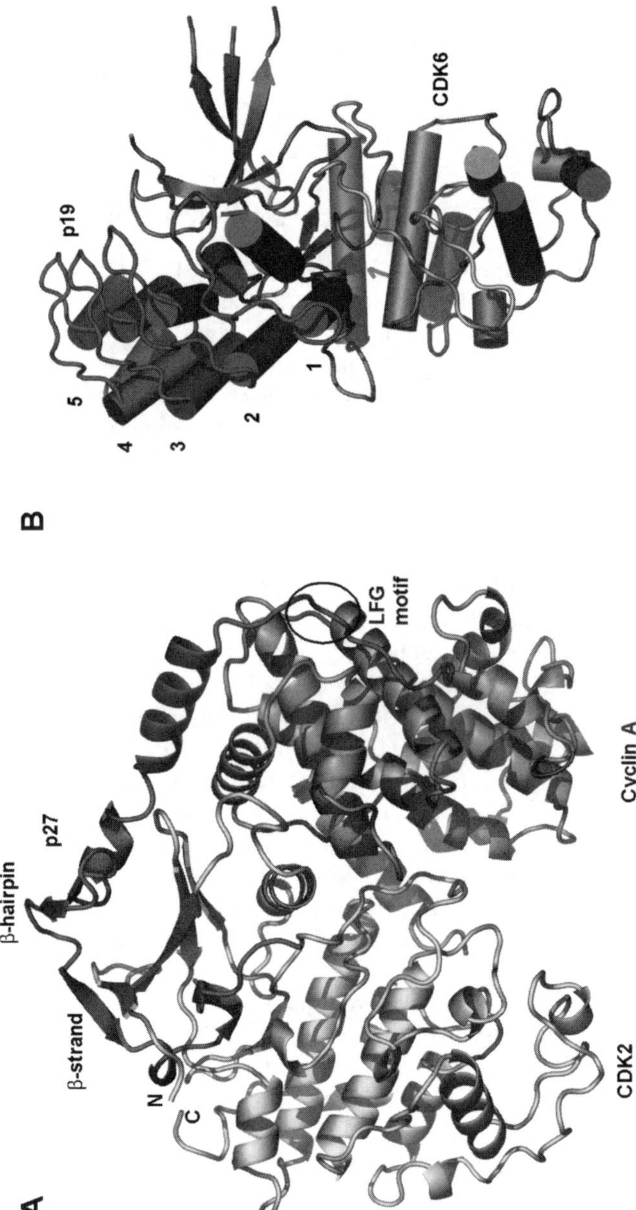

FIGURE 7.5 (See color insert.) Structures of CDK/inhibitor complexes. (A) The p27^{KIP1} inhibitor is bound in a ternary complex to CDK2 and cyclin A. The inhibitor is shown in red, CDK2 in yellow, and cyclin A in green. (B) The p19^{INK4} inhibitor, shown in red, is bound to CDK6, shown in green.

binding cleft. The ATP-binding site is distorted because of p19 contacts with CDK6 residues in the linker region between the N-terminal and C-terminal domain and in the catalytic cleft, and because of the relative domain movements. The p19/CDK6 structures indicate that the inhibitor-kinase interactions result in extensive movements of the N-terminal domain of the kinase relative to the C-terminal domain, inhibiting productive binding with ATP and movement of the αC-helix to its active conformation. The distortions of the cyclin binding site are consistent with the observation that p16 binding weakens the affinity of Cyclin D for CDK4 and CDK6 (Parry et al., 1995; Ragione et al., 1996).

7.2.4 THE ATP-BINDING POCKET IN ACTIVE AND INACTIVE CDKs

All of the known chemical inhibitors of CDKs bind in the ATP-binding pocket located between the two kinase domains. Hence, a detailed structural characterization of the ATP-binding pocket is very valuable for rational drug design efforts. Because residues lining the ATP-binding pocket in different CDKs are quite conserved, we will describe here the ATP-binding pocket in the high-resolution structures of activated CDK2 (Brown et al., 1999b) and the CDK2 apoenzyme (De Bondt et al., 1993; Schulze-Gahmen et al., 1996) (Figure 7.6 and Figure 7.7). In the ternary complex of CDK2 with ATP and a substrate peptide, the purine ring of ATP is binding close to the hinge region connecting the N-terminal and C-terminal domain (residues 81 to 84). The N1 and N6 nitrogens of the adenine base form hydrogen bonds with the peptide bond nitrogen of L83 and oxygen of E81, respectively. Hydrogen bonds with the same hinge residue atoms are conserved for most molecules binding in the kinase-active site, but they can be formed from a diverse range of heterocyclic frameworks. In fact, satisfying the hydrogen bond–donating capacity of the backbone nitrogen of L83 is the only conserved feature of all CDK2/inhibitor complex structures determined to date (Davies et al., 2002b). The adenine ring is also sandwiched between a number of hydrophobic residues, including I10, A31, F82, and L134. The back wall of the ATP-binding cleft is formed by the side chain of F80, the "gatekeeper" residue (Liu et al., 1999; Schindler et al., 1999). Although the adenine ring of ATP does not come close to the F80 side chain, other bound inhibitors pack hydrophobic groups into this site and take advantage of hydrophobic interactions with the phenyl ring. Members of other kinase families have a much larger hydrophobic pocket in this area due to a smaller "gatekeeper" residue. The difference in size of the hydrophobic pocket has been successfully used to create kinase-specific inhibitors (Cohen et al., 2005; Liu et al., 1999; Schindler et al., 1999; Tong et al., 1997). The ribose and phosphate moiety of ATP bind in an open cleft formed in part by the flexible glycine loop (residues 11 to 18), residues from the catalytic loop (residues 126 to 132), K33, and D145 (Figure 7.7). The ribose and phosphate atoms form a network of hydrogen bonds with side-chain and main-chain atoms lining the cleft, as well as with water molecules in the binding pocket. A single Mg ion is coordinated to oxygens from ATP α- and γ-phosphates, N132, and D145 (Figure 7.6). The ATP-binding pockets of the activated CDK2 enzyme and inactive CDK2 apoenzyme are structurally similar. After superposition of the molecules on the C-terminal domain, relatively small shifts are observed in the position of residues

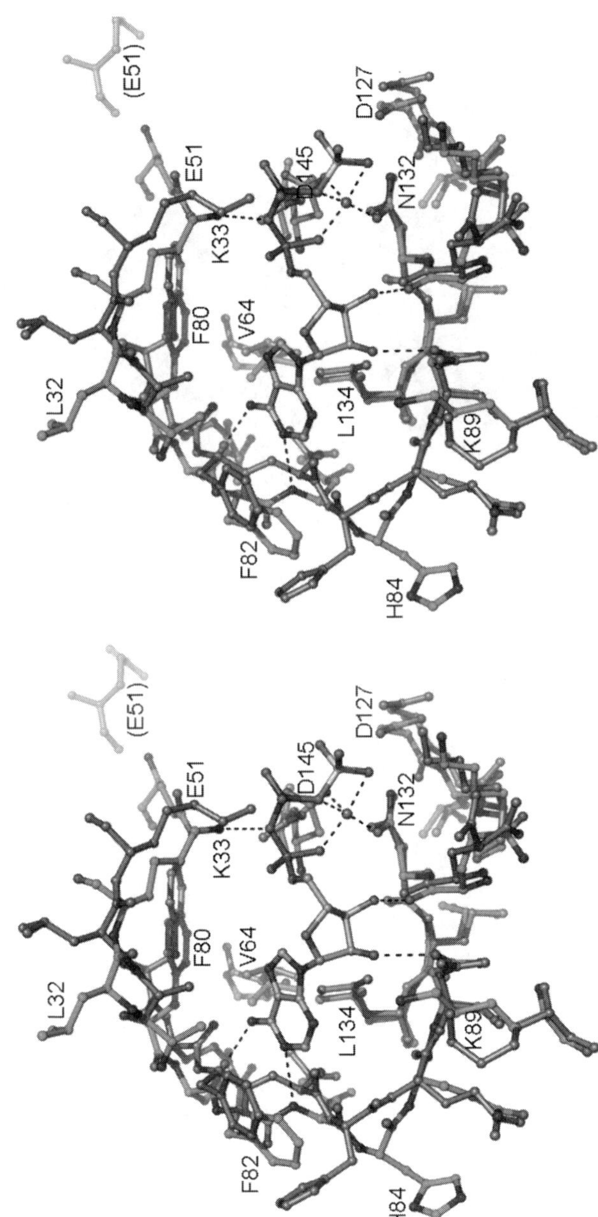

FIGURE 7.6 (See color insert.) Stereo drawing of the ATP-binding pocket in CDK2. Residues 11 to 18 that form the glycine loop were omitted for clarity. The binding pocket in the activated CDK2 is shown in green with bound ATP in yellow and the Mg ion as a gray sphere. The structure of the binding pocket in the CDK2 apoenzyme is shown in gray. Hydrogen bonds are drawn as broken lines.

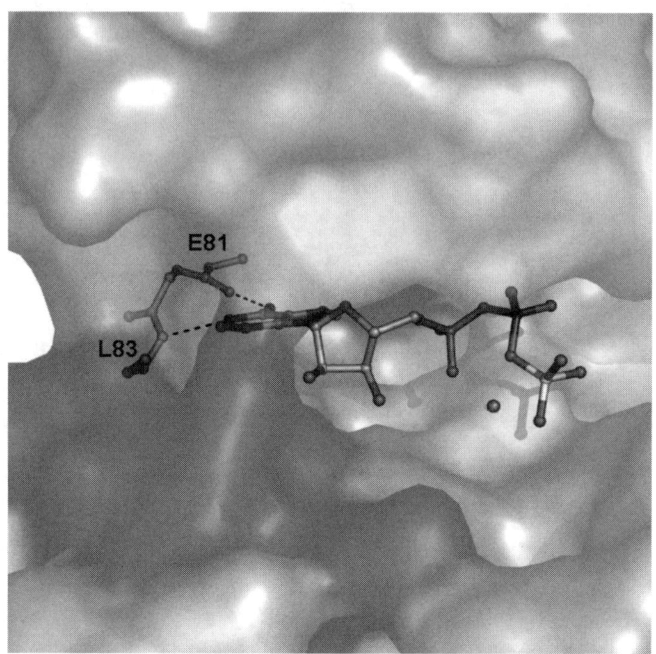

FIGURE 7.7 (See color insert.) Surface representation of the ATP-binding pocket in CDK2 (Brown, N.R. et al. (1999a) *Nat Cell Biol 1*, 438–443). The surface is colored by pocket regions (Davies, T.G. et al. (2002b) *Pharmacol Ther 93*, 125–133). The hinge region is shown in cyan, with the backbone of residues 81 to 83 drawn as a stick model. Conserved hydrogen bonds with the hinge region are shown as broken lines. The F80 pocket is colored blue, the ribose and phosphate binding cleft is yellow, and the specificity region is purple. Bound ATP with one Mg ion is drawn as a stick model.

in the N-terminal domain adjacent to the adenine ring. These shifts originate from the rotation of the N-terminal domain relative to the C-terminal domain after activation of the CDK2 apoenzyme. Much larger shifts are observed for residues 11 to 18 in the glycine loop, which contact the ribose and phosphates of the bound ATP. Larger structural differences are also seen in the position of K33 and E51. E51 is located in the αC-helix which is rotated and moved into the binding pocket after cyclin binding to CDK2 (Jeffrey et al., 1995). Before CDK2 activation, E51 points away from the ATP-binding pocket, but moves into the vicinity of the pocket to form ionic interactions with K33 after CDK2 activation. K33 interacts directly with the ATP phosphates and also with some inhibitor molecules (Davies et al., 2001; Lu et al., 2005; Meijer et al., 2000). E51 does not interact with ATP directly, but it was shown to interact with a bound inhibitor to CDK6 (Lu et al., 2005). Structural differences in the ATP-binding pocket between inactive and activated CDKs may need to be taken into account for drug design studies, especially if the inhibitor molecule extends beyond the adenine binding region of the ATP-binding pocket (Davies et al., 2001). Inhibitor interactions with residues that show structural differences between the two kinase states, such as K33 and E51, can be used to design inhibitors with preference for different kinase states.

7.3 COMPARISON OF INHIBITOR COMPLEX STRUCTURES

The central role that CDKs play in cell cycle control and the high incidence of genetic alterations of CDKs or their regulators in a number of cancers make the CDK family an important target for therapeutic intervention in various proliferative diseases, including cancer (Sielecki et al., 2000). Hence, considerable efforts have been made to screen and design CDK-specific inhibitors. A large number of chemically diverse inhibitors with high affinities for CDK1, CDK2, and CDK5 are available today and several of them have entered clinical trials to treat cancer (Blagden and de Bono, 2005; Meijer et al., 1999; Senderowicz and Sausville, 2000). Drug design of CDK2-specific inhibitors was greatly aided by the high-resolution x-ray structures of CDK2 and its inhibitor complexes. In the following sections, we will highlight the structural features of several CDK/inhibitor complexes and suggest a structural basis for the lower binding affinities of some of these inhibitors for CDK4 and CDK6.

7.3.1 INTERACTIONS OF CDK2-SPECIFIC INHIBITORS IN THE ATP-BINDING SITE

Over the last ten years, a large number of heterocyclic CDK inhibitors with a variety of chemical scaffolds have been identified (Table 7.1). The compounds can be grouped into those that nonspecifically inhibit all CDKs and often other protein kinases, those that are specific for the group including CDK1, CDK2, and CDK5, and those that are specific for CDK4 and CDK6. CDK4 and CDK6 form a subgroup within the CDKs based on sequence similarities. For simplicity, we will call the inhibitors of the second group CDK2-specific inhibitors because most of the structural work was performed on CDK2 complexes. The structures of the first CDK2 complexes with CDK2-specific purine-based inhibitors, such as olomoucine (Schulze-Gahmen et al., 1995), roscovitine (De Azevedo et al., 1997), and purvalanol (Gray et al., 1998), showed that the aromatic ring system is bound in a different orientation than in the natural ligand ATP to make room for the larger substituent groups on the purine ring. The hydrogen bonds from the backbone atoms of L83 and E81 in CDK2 to the adenine base in ATP are conserved (Figure 7.7 and Figure 7.8), although with different partner atoms in the substituted purine rings of the inhibitors. Subsequently, other compounds with a variety of scaffolds were identified as more or less specific inhibitors of CDK2. The binding mode for many of these inhibitors was determined by crystallographic studies of their complexes with the CDK2 apoenzyme. Table 7.1 shows a list of selected CDK inhibitors with structures that were determined in complex with one of the CDKs. Comparative analysis of the binding modes of many of these inhibitors revealed some common features: (1) The hydrophobic core of the inhibitor molecule is sandwiched between the two kinase domains,

TABLE 7.1
IC_{50} Values (μM) for Selected CDK Inhibitors with Three-Dimensional Structures of CDK-Inhibitor Complexes

Inhibitor	CDK1/B	CDK2/A	CDK2/E	CDK4/D	CDK6/ Vcyclin	CDK5/ p25
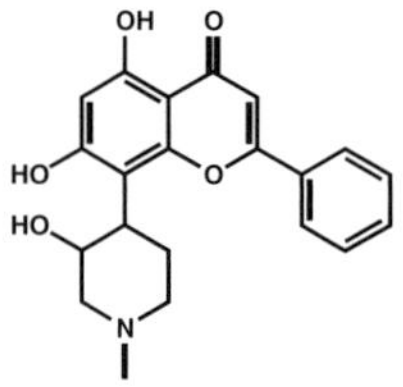 Des-chloro-flavopiridol	0.4	0.4	—	0.4	0.8	—
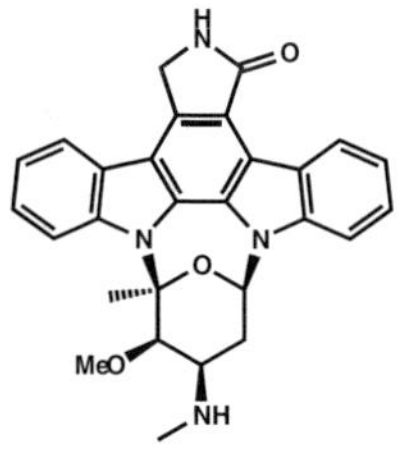 Staurosporine	—	0.007	—	3–10	—	—
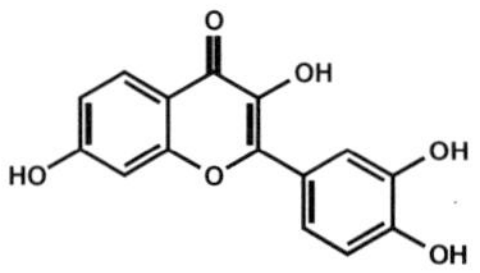 Fisetin	0.79	—	—	—	0.85	0.57
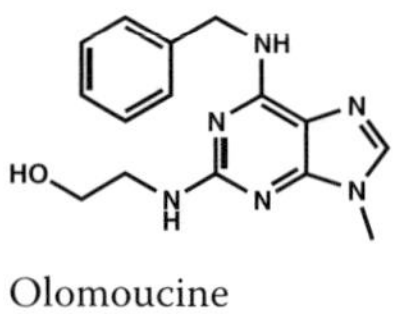 Olomoucine	7	7	7	>10,000	>10,000	3
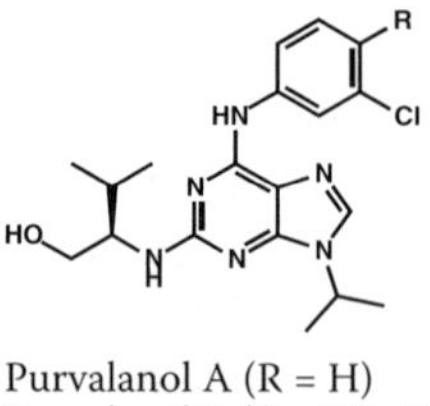 Purvalanol A (R = H)	0.004	0.070	0.035	0.850	—	0.075
Purvalanol B (R = CO$_2$H)	0.006	0.006	0.009	>10	—	0.006

TABLE 7.1 (CONTINUED)
IC_{50} Values (μM) for Selected CDK Inhibitors with Three-Dimensional Structures of CDK-Inhibitor Complexes

Inhibitor	CDK1/B	CDK2/A	CDK2/E	CDK4/D	CDK6/ Vcyclin	CDK5/ p25
NU6102	0.009	0.006	—	1.6	—	—
Hymenialdisine	0.022	0.07	0.04	0.6	0.7	0.028
Indirubin-3'-monoxime	0.18	0.44	0.25	3.33	0.46	0.1
Aloisine A	0.15	0.12	0.40	>100	0.55	0.20

forming contacts with hydrophobic residues such as I10, A31, F82, and L134; (2) High-affinity inhibitors form two or three conserved hydrogen bonds with backbone atoms in the hinge region of CDK2 (Davies et al., 2002b); (3) Many inhibitors fill out the hydrophobic pocket in front of the gate keeper residue F80 with hydrophobic groups; (4) The specificity of CDK-inhibitor interactions is most likely based on inhibitor interactions with the less conserved hinge region and specificity area (residues 84 to 89) that borders the ATP-binding site from the outside of the cleft (Figure 7.7). The amino acid sequence for different CDKs is less conserved in these

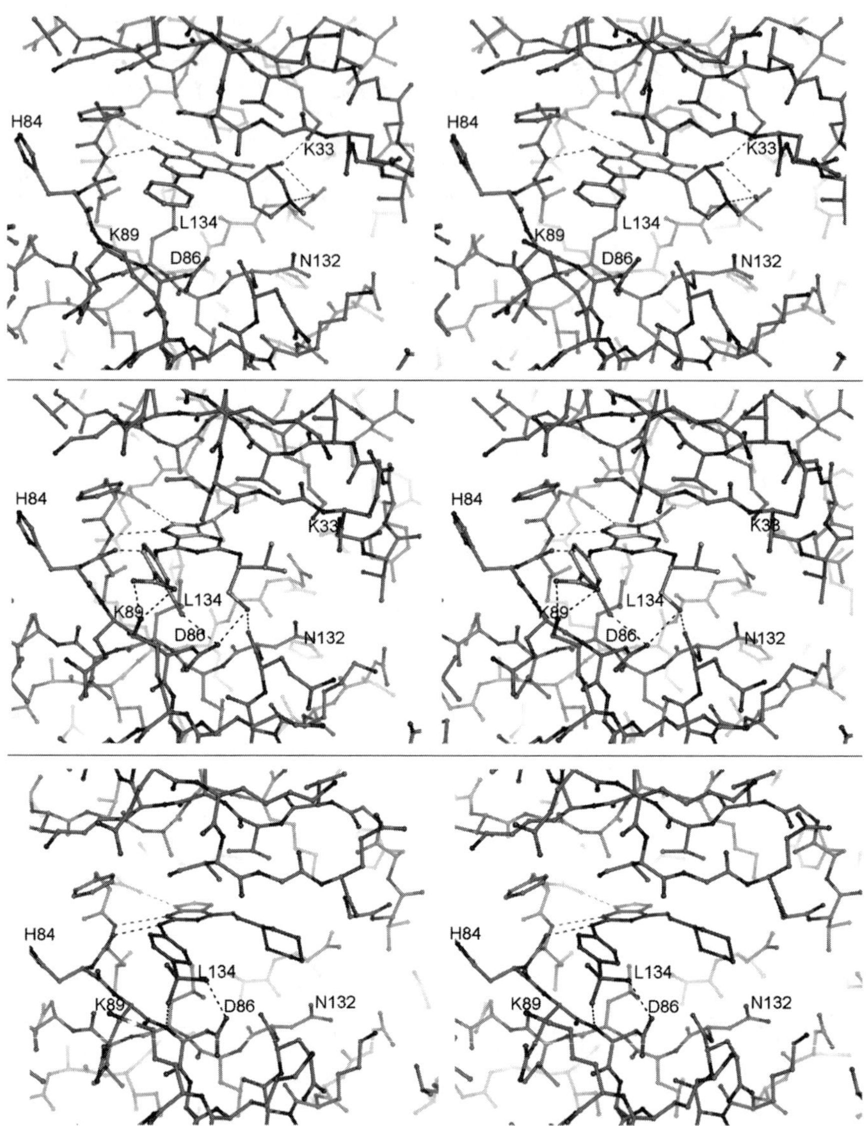

FIGURE 7.8 (See color insert.) Structure of the binding pocket of three CDK2/inhibitor complexes. The top shows the complex with the nonspecific inhibitor des-chloro-flavopiridol. There are few contacts of the inhibitor with residues in the specificity region of the binding pocket. The middle panel shows the interactions of CDK2 with the highly specific inhibitor purvalanol B. There are many interactions with the specificity region. However, the inhibitor fills very little space in the area of the binding pocket that binds the ribose and phosphates of ATP. The bottom panel shows the interactions with NU6102, another CDK2-specific inhibitor. There are several interactions with the specificity region. The inhibitor also fills the ribose-binding pocket with the cyclohexylmethyl group of the inhibitor. Hydrogen bonds between inhibitors and CDK2 are shown in black broken lines.

areas (Figure 7.1) and can contribute to the CDK specificity for a ligand. Examples of the structures of three inhibitor complexes with varying specificity for CDK2 are shown in Figure 7.8.

7.3.2 STRUCTURAL BASIS FOR CDK SPECIFICITY

Analysis of the complex structures of CDK2 with inhibitors led to the suggestion that the kinase specificity of many of these inhibitors is based on their interactions with residues in the specificity region that are less conserved among different CDKs. The availability of a structure of the active CDK6/Vcyclin complex (Lu et al., 2005; Schulze-Gahmen and Kim, 2002) provides the opportunity to model CDK2-specific inhibitors into the CDK6 binding pocket by superimposing the respective kinase structures, followed by optimization of the inhibitor orientation to satisfy the conserved hydrogen bonds between the inhibitor and atoms in the hinge region of CDK6. Using this procedure, we placed the CDK2-specific inhibitors olomoucine (Schulze-Gahmen et al., 1995) and NU6102 (Davies et al., 2002a) into the ATP-binding pocket of CDK6-Vcyclin, as well as the less specific hymenialdisine (Meijer et al., 2000) and aloisine A (Mapelli et al., 2005; Mettey et al., 2003). The placement of the olomoucine ligand results in several close contacts between the N7 atom of the adenine ring and the $H100^{CDK6}$ side chain (F82 in CDK2), and between the benzylamino group of the inhibitor and backbone atoms of residues 102 and 103 (84 and 85 in CDK2) (Figure 7.9). The unfavorable interactions with the H100 side chain cannot be easily relieved by rotating the histidine side chain because it interacts with the side chain of $F39^{CDK6}$. Both H100 and F39 are unique to CDK4 and CDK6 and are replaced by phenylalanine and a small hydrophobic residue in the other CDKs (Figure 7.1).

The docking of NU6102 into the binding pocket of CDK6 leads to an even larger number of unfavorably close contacts (Figure 7.9). Close interactions with the side chain of $H100^{CDK6}$, backbone atoms of residues 102 and 103, and with the side chain of T^{107} would prevent a tight binding of NU6102 to the active CDK6-Vcyclin complex. The underlying reason for the different inhibitor interactions in CDK2 and CDK6 is a conformational difference in the hinge region. If superimposed on atoms of the C-terminal domain, the atom positions start diverging at residue 103 (85 in CDK2) and continue doing so towards the N-terminal domain. The structural differences between CDK6 and CDK2 in the hinge region are very similar in the comparison of CDK6 with the inactive CDK2 complexed with olomoucine, and with active CDK2-Cyclin A complexed with NU6102. Hence, this difference is likely to be caused by sequence differences rather than by varied crystal forms or CDK activation states, which might induce different domain rotations and result in dissimilar hinge conformations. The fact that the hinge residue $H100^{CDK6}$ is unique to CDK4 and CDK6 and forms a close contact with the CDK4 and CDK6 conserved residue $F39^{CDK6}$ also supports the possibility that sequence differences in the hinge region might induce a different conformation in the CDK6 hinge region.

In contrast to the close contacts found in the CDK6-model complexes with CDK2 inhibitors of high specificity, we found few close interactions in the model of hymenialdisine bound to CDK6, and no significant bad contacts in the model

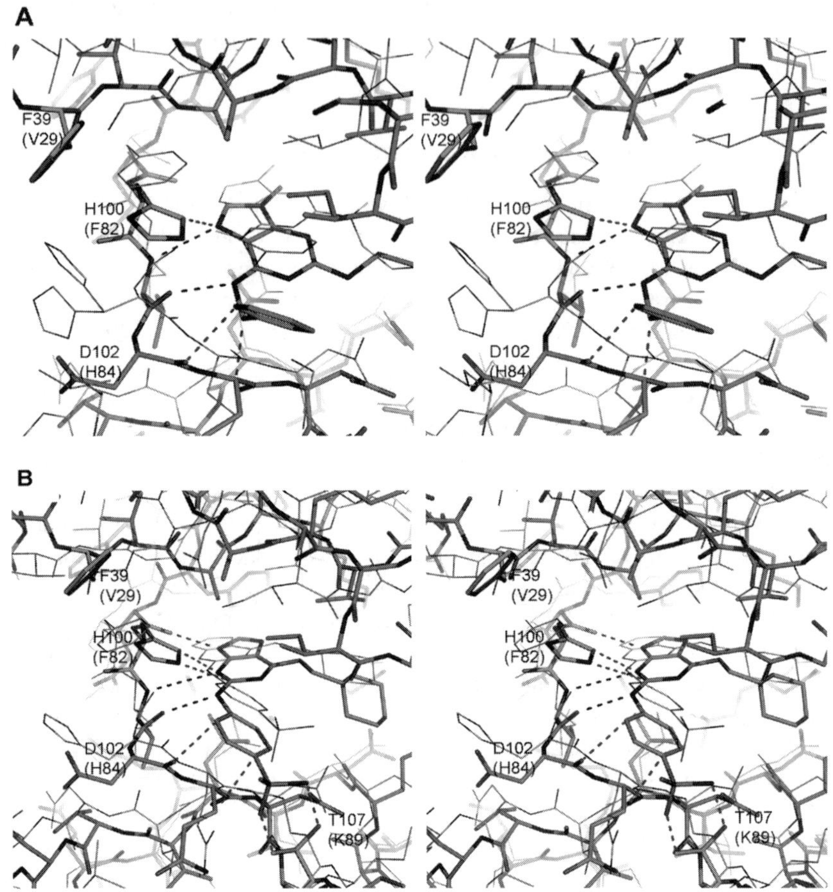

FIGURE 7.9 (See color insert.) Stereo drawings of the ATP-binding pocket of CDK6 with CDK2-specific inhibitors olomoucine (A) and NU6102 (B) modeled into the binding pocket. The structures of the active CDK6 and the modeled inhibitor are shown as stick models in green with CDK4/6-specific sequence changes highlighted in cyan. The structure of the corresponding CDK2/inhibitor complex is drawn with thin lines in dark yellow. Conserved hydrogen bonds, used to place the inhibitor into the binding pocket, are represented by black broken lines. Unfavorably close contacts between the inhibitor and CDK6 are shown as broken red lines. Residue labeling refers to the CDK6 sequence with labels for CDK2 displayed in brackets.

with aloisine A. Only the Br atom in hymenialdisine is positioned too close to Q103$^{\text{CDK6}}$ in our CDK6-complex model, possibly explaining the lower IC$_{50}$ value of hymenialdisine for CDK6 compared to CDK2 (Table 7.1). The observations from these CDK6-inhibitor models agree quite well with the relative affinities of the CDK inhibitors for different CDKs and suggest that residues in the hinge region and the specificity region can contribute to the specificity of certain CDK inhibitors.

Recently, some highly specific inhibitors of CDK4 and CDK6 have been identified that have chemical structures based on the pyrido[2,3-d]pyrimidin-7-one scaffold (Fry et al., 2004; VanderWel et al., 2005). A complex structure of these inhibitors with CDK4 or CDK6 would be extremely valuable in understanding the structural basis of their specificity and for the future design of CDK4 and CDK6 specific inhibitors.

ACKNOWLEDGMENTS

Research was supported by a grant from the National Institute of Health (R29 AI4204101) to USG.

REFERENCES

Blagden, S. and de Bono, J. (2005). Drugging cell cycle kinases in cancer therapy. *Curr Drug Targets 6*, 325–335.

Bourne, Y., Watson, M.H., Hickey, M.J., Holmes, W., Rocque, W., Reed, S.I., and Tainer, J.A. (1996). Crystal structure and mutational analysis of the human CDK2 kinase complex with cell cycle-regulatory protein CksHs1. *Cell 84*, 863–874.

Brotherton, D.H., Dhanaraj, V., Wick, S., Brizuela, L., Domaille, P.J., Volyanik, E., Xu, X., Parisini, E., Smith, B.O., Archer, S.J. et al. (1998). Crystal structure of the complex of the cyclin D-dependent kinase Cdk6 bound to the cell-cycle inhibitor p19INK4d. *Nature 395*, 244–250.

Brown, N.R., Noble, M.E., Endicott, J.A., and Johnson, L.N. (1999a). The structural basis for specificity of substrate and recruitment peptides for cyclin-dependent kinases. *Nat Cell Biol 1*, 438–443.

Brown, N.R., Noble, M.E., Lawrie, A.M., Morris, M.C., Tunnah, P., Divita, G., Johnson, L.N., and Endicott, J.A. (1999b). Effects of phosphorylation of threonine 160 on cyclin-dependent kinase 2 structure and activity. *J Biol Chem 274*, 8746–8756.

Chang, C.I., Xu, B.E., Akella, R., Cobb, M.H., and Goldsmith, E.J. (2002). Crystal structures of MAP kinase p38 complexed to the docking sites on its nuclear substrate MEF2A and activator MKK3b. *Mol Cell 9*, 1241–1249.

Cohen, M.S., Zhang, C., Shokat, K.M., and Taunton, J. (2005). Structural bioinformatics-based design of selective, irreversible kinase inhibitors. *Science 308*, 1318–1321.

Connell-Crowley, L., Solomon, M.J., Wei, N., and Harper, J.W. (1993). Phosphorylation independent activation of human cyclin-dependent kinase 2 by cyclin A in vitro. *Mol Biol Cell 4*, 79–92.

Davies, T.G., Bentley, J., Arris, C.E., Boyle, F.T., Curtin, N.J., Endicott, J.A., Gibson, A.E., Golding, B.T., Griffin, R.J., Hardcastle, I.R. et al. (2002a). Structure-based design of a potent purine-based cyclin-dependent kinase inhibitor. *Nat Struct Biol 9*, 745–749.

Davies, T.G., Pratt, D.J., Endicott, J.A., Johnson, L.N., and Noble, M.E. (2002b). Structure-based design of cyclin-dependent kinase inhibitors. *Pharmacol Ther 93*, 125–133.

Davies, T.G., Tunnah, P., Meijer, L., Marko, D., Eisenbrand, G., Endicott, J.A., and Noble, M.E. (2001). Inhibitor binding to active and inactive CDK2: the crystal structure of CDK2-cyclin A/indirubin-5-sulphonate. *Structure (Camb) 9*, 389–397.

De Azevedo, W.F., Leclerc, S., Meijer, L., Havlicek, L., Strnad, M., and Kim, S.H. (1997). Inhibition of cyclin-dependent kinases by purine analogues: crystal structure of human cdk2 complexed with roscovitine. *Eur J Biochem 243*, 518–526.

De Bondt, H.L., Rosenblatt, J., Jancarik, J., Jones, H.D., Morgan, D.O., and Kim, S.H. (1993). Crystal structure of cyclin-dependent kinase 2. *Nature 363*, 595–602.

Desai, D., Gu, Y., and Morgan, D.O. (1992). Activation of human cyclin-dependent kinases in vitro. *Mol Biol Cell 3*, 571–582.

Forsburg, S.L. and Nurse, P. (1991). Cell cycle regulation in the yeasts Saccharomyces cerevisiae and Schizosaccharomyces pombe. *Annu Rev Cell Biol 7*, 227–256.

Fry, D.W., Harvey, P.J., Keller, P.R., Elliott, W.L., Meade, M., Trachet, E., Albassam, M., Zheng, X., Leopold, W.R., Pryer, N.K., and Toogood, P.L. (2004). Specific inhibition of cyclin-dependent kinase 4/6 by PD 0332991 and associated antitumor activity in human tumor xenografts. *Mol Cancer Ther 3*, 1427–1438.

Gorina, S. and Pavletich, N.P. (1996). Structure of the p53 tumor suppressor bound to the ankyrin and SH3 domains of 53BP2. *Science 274*, 1001–1005.

Gray, N.S., Wodicka, L., Thunnissen, A.M., Norman, T.C., Kwon, S., Espinoza, F.H., Morgan, D.O., Barnes, G., LeClerc, S., Meijer, L. et al. (1998). Exploiting chemical libraries, structure, and genomics in the search for kinase inhibitors. *Science 281*, 533–538.

Jeffrey, P.D., Russo, A.A., Polyak, K., Gibbs, E., Hurwitz, J., Massague, J., and Pavletich, N.P. (1995). Mechanism of CDK activation revealed by the structure of a cyclin A-CDK2 complex. *Nature 376*, 313–320.

Jeffrey, P.D., Tong, L., and Pavletich, N.P. (2000). Structural basis of inhibition of CDK-cyclin complexes by INK4 inhibitors. *Genes Dev 14*, 3115–3125.

Liu, Y., Bishop, A., Witucki, L., Kraybill, B., Shimizu, E., Tsien, J., Ubersax, J., Blethrow, J., Morgan, D.O., and Shokat, K.M. (1999). Structural basis for selective inhibition of Src family kinases by PP1. *Chem Biol 6*, 671–678.

Lolli, G., Lowe, E.D., Brown, N.R., and Johnson, L.N. (2004). The crystal structure of human CDK7 and its protein recognition properties. *Structure (Camb) 12*, 2067–2079.

Lu, H., Chang, D.J., Baratte, B., Meijer, L., and Schulze-Gahmen, U. (2005). Crystal structure of a human cyclin-dependent kinase 6 complex with a flavonol inhibitor, fisetin. *J Med Chem 48*, 737–743.

Mapelli, M., Massimiliano, L., Crovace, C., Seeliger, M.A., Tsai, L.H., Meijer, L., and Musacchio, A. (2005). Mechanism of CDK5/p25 binding by CDK inhibitors. *J Med Chem 48*, 671–679.

Mapelli, M. and Musacchio, A. (2003). The structural perspective on CDK5. *Neurosignals 12*, 164–172.

McGowan, C.H. and Russell, P. (1993). Human Wee1 kinase inhibits cell division by phosphorylating p34cdc2 exclusively on Tyr15. *EMBO J 12*, 75–85.

Meijer, L., Leclerc, S., and Leost, M. (1999). Properties and potential-applications of chemical inhibitors of cyclin-dependent kinases. *Pharmacol Ther 82*, 279–284.

Meijer, L., Thunnissen, A.M., White, A.W., Garnier, M., Nikolic, M., Tsai, L.H., Walter, J., Cleverley, K.E., Salinas, P.C., Wu, Y.Z. et al. (2000). Inhibition of cyclin-dependent kinases, GSK-3beta and CK1 by hymenialdisine, a marine sponge constituent. *Chem Biol 7*, 51–63.

Mettey, Y., Gompel, M., Thomas, V., Garnier, M., Leost, M., Ceballos-Picot, I., Noble, M., Endicott, J., Vierfond, J.M., and Meijer, L. (2003). Aloisines, a new family of CDK/GSK-3 inhibitors. SAR study, crystal structure in complex with CDK2, enzyme selectivity, and cellular effects. *J Med Chem 46*, 222–236.

Morgan, D.O. (1997). Cyclin-dependent kinases: engines, clocks, and microprocessors. *Annu Rev Cell Dev Biol 13*, 261–291.

Norbury, C., Blow, J., and Nurse, P. (1991). Regulatory phosphorylation of the p34cdc2 protein kinase in vertebrates. *EMBO J 10*, 3321–3329.

Parker, L.L., Atherton-Fessler, S., and Piwnica-Worms, H. (1992). p107wee1 is a dual-specificity kinase that phosphorylates p34cdc2 on tyrosine 15. *Proc Natl Acad Sci USA 89*, 2917–2921.

Parry, D., Bates, S., Mann, D.J., and Peters, G. (1995). Lack of cyclin D-Cdk complexes in Rb-negative cells correlates with high levels of p16INK4/MTS1 tumor suppressor gene product. *EMBO J 14*, 503–511.

Pavletich, N. P. (1999). Mechanisms of cyclin-dependent kinase regulation: structures of Cdks, their cyclin activators, and Cip and INK4 inhibitors. *J Mol Biol 287*, 821–828.

Ragione, F.D., Russo, G.L., Oliva, A., Mercurio, C., Mastropietro, S., Pietra, V.D., and Zappia, V. (1996). Biochemical characterization of p16INK4- and p18-containing complexes in human cell lines. *J Biol Chem 271*, 15942–15949.

Russo, A.A., Jeffrey, P.D., Patten, A.K., Massague, J., and Pavletich, N.P. (1996a). Crystal structure of the p27Kip1 cyclin-dependent-kinase inhibitor bound to the cyclin A-Cdk2 complex. *Nature 382*, 325–331.

Russo, A.A., Jeffrey, P.D., and Pavletich, N.P. (1996b). Structural basis of cyclin-dependent kinase activation by phosphorylation. *Nat Struct Biol 3*, 696–700.

Russo, A.A., Tong, L., Lee, J.O., Jeffrey, P.D., and Pavletich, N.P. (1998). Structural basis for inhibition of the cyclin-dependent kinase Cdk6 by the tumour suppressor p16INK4a. *Nature 395*, 237–243.

Sausville, E.A. (2002). Complexities in the development of cyclin-dependent kinase inhibitor drugs. *Trends Mol Med 8*, S32–37.

Schindler, T., Sicheri, F., Pico, A., Gazit, A., Levitzki, A., and Kuriyan, J. (1999). Crystal structure of Hck in complex with a Src family-selective tyrosine kinase inhibitor. *Mol Cell 3*, 639–648.

Schulze-Gahmen, U., Brandsen, J., Jones, H. D., Morgan, D. O., Meijer, L., Vesely, J., and Kim, S. H. (1995). Multiple modes of ligand recognition: crystal structures of cyclin-dependent protein kinase 2 in complex with ATP and two inhibitors, olomoucine and isopentenyladenine. *Proteins 22*, 378–391.

Schulze-Gahmen, U., De Bondt, H.L., and Kim, S.H. (1996). High-resolution crystal structures of human cyclin-dependent kinase 2 with and without ATP: bound waters and natural ligand as guides for inhibitor design. *J Med Chem 39*, 4540–4546.

Schulze-Gahmen, U. and Kim, S. H. (2002). Structural basis for CDK6 activation by a virus-encoded cyclin. *Nat Struct Biol 9*, 177–181.

Senderowicz, A.M. and Sausville, E.A. (2000). Preclinical and clinical development of cyclin-dependent kinase modulators. *J Natl Cancer Inst 92*, 376–387.

Serrano, M. (1997). The tumor suppressor protein p16INK4a. *Exp Cell Res 237*, 7–13.

Sherr, C.J. and Roberts, J.M. (1995). Inhibitors of mammalian G1 cyclin-dependent kinases. *Genes Dev 9*, 1149–1163.

Sielecki, T.M., Boylan, J.F., Benfield, P.A., and Trainor, G.L. (2000). Cyclin-dependent kinase inhibitors: useful targets in cell cycle regulation. *J Med Chem 43*, 1–18.

Solomon, M.J., Lee, T., and Kirschner, M.W. (1992). Role of phosphorylation in p34cdc2 activation: identification of an activating kinase. *Mol Biol Cell 3*, 13–27.

Song, H., Hanlon, N., Brown, N.R., Noble, M.E., Johnson, L.N., and Barford, D. (2001). Phosphoprotein-protein interactions revealed by the crystal structure of kinase-associated phosphatase in complex with phosphoCDK2. *Mol Cell 7*, 615–626.

Tarricone, C., Dhavan, R., Peng, J., Areces, L.B., Tsai, L.H., and Musacchio, A. (2001). Structure and regulation of the CDK5-p25(nck5a) complex. *Mol Cell 8*, 657–669.

Taylor, S.S., Knighton, D.R., Zheng, J., Ten Eyck, L.F., and Sowadski, J.M. (1992). Structural framework for the protein kinase family. *Annu Rev Cell Biol 8*, 429–462.

Tong, L., Pav, S., White, D.M., Rogers, S., Crane, K.M., Cywin, C.L., Brown, M.L., and Pargellis, C.A. (1997). A highly specific inhibitor of human p38 MAP kinase binds in the ATP pocket. *Nat Struct Biol 4*, 311–316.

VanderWel, S.N., Harvey, P.J., McNamara, D.J., Repine, J.T., Keller, P.R., Quin, J., III, Booth, R.J., Elliott, W.L., Dobrusin, E.M., Fry, D.W., and Toogood, P.L. (2005). Pyrido[2,3-d]pyrimidin-7-ones as specific inhibitors of cyclin-dependent kinase 4. *J Med Chem 48*, 2371–2387.

Zhu, L., Harlow, E., and Dynlacht, B.D. (1995). p107 uses a p21CIP1-related domain to bind cyclin/cdk2 and regulate interactions with E2F. *Genes Dev 9*, 1740–1752.

Section III

CDK Inhibitors: Chemistry Focus

8 Cyclin-Dependent Kinase Small Molecule Modulators for Cancer Therapy

Adrian M. Senderowicz

CONTENTS

8.1 BRIEF OVERVIEW OF CELL CYCLE REGULATION

Upon activation of mitogenic signaling cascades, cells commit to entry into a series of regulated steps allowing traverse of the cell cycle. Synthesis of DNA (genome duplication), also known as S phase, is followed by separation into two daughter cells (chromatid separation) or M phase. During the G_2 phase (the time between the S and M phases), cells can repair errors that occur during DNA duplication, preventing the propagation of these errors to daughter cells. In contrast, the G_1 phase (the time between the M and S phases) represents the period of commitment to cell cycle progression. For cells to continue cycling to the next phase, the prior phase has to be properly completed; otherwise, "fail-safe" mechanisms, also known as *cell cycle checkpoints* are elicited.[1]

The cell cycle machinery is governed by the cyclical activation of the cyclin-dependent kinases (CDKs), serine-threonine kinases composed of the CDK catalytic subunit, and cofactors such as cyclins and endogenous CDK inhibitors (CKIs) such as p21[CIP1/WAF1]. The main function of CDKs is the phosphorylation of substrates required for cell cycle progression.[2] One crucial substrate of CDKs is the gene product of the retinoblastoma gene (Rb), a tumor suppressor gene that is dysfunctional in the majority of human neoplasms due to "overactive" CDKs.[3,4] Thus, manipulation of CDKs and cofactors is a potentially valuable strategy in cancer therapeutics.[3,4]

The fact that most tumors are aneuploid, reflecting abnormal sister chromatid separation, has motivated increasing interest in the understanding of the mitotic checkpoints.[5,6] There are at least two serine-threonine kinases relevant to mitotic checkpoints that are being targeted by small molecules: aurora- and polo-like kinases.[5,7–10] Depletion of several mitotic components (including aurora- and polo-like kinases) by either small molecules, intracellular antibodies, dominant negative alleles, or siRNA promotes cell death in *in vitro* cancer models.[8,11–13] This novel concept is being investigated intensely and several molecules are approaching Phase I or II trials.[14]

Another gene relevant to cell cycle regulation (and also to apoptosis) is p53, a tumor-suppressor gene frequently inactivated in human cancer cells.[15] Transformed cells with inactivated p53 are unable to undergo apoptosis, which leads to growth imbalance and genomic instability.[15] Because most tumor cells have lost the G_1 (because of p53 mutations) but not the G_2 checkpoint, they would arrest in G_2 upon DNA damage. Thus, the use of combination therapy of DNA-damaging agents (radiation or chemotherapy) and small molecules that selectively abrogate G_2 checkpoint represents an attractive approach to cancer therapy. This approach could lead to tumor cell death due to accelerated mitosis and unrepaired DNA lesions and spare normal cells from some of the cytotoxic effects.[4,16,17]

8.2 CELL CYCLE ALTERATIONS IN HUMAN NEOPLASMS

In the last few years, it became clear that cyclins, CDK complexes, and other cell cycle regulators are mechanistically involved in the development of human tumors.[18–20] This is consistent with a large body of literature showing the importance of inactivation of the Rb pathway in tumor development.[3,21] The inactivation of Rb can be produced by direct mutation of the Rb protein, but this is a relatively rare event occurring only in Rb's, osteosarcomas, and a minority of breast and some other tumors.[22,23] More frequent alterations of this pathway occur by functional inactivation of Rb by hyperphosphorylation. This is normally the result of elevated

CDK activities caused by overexpression of cyclins or CDKs. For example, several laboratories have reported that some tumors show loss of Rb or, alternatively, overexpression of cyclin D1.[24–26] Similarly, in other tumors, loss of p16[INK4a] and Rb are mutually exclusive.[27,28] This observation led to the hypothesis that inactivation of the cyclin D/CDK/p16/pRb pathway can promote tumor development and that either loss of the suppressor activity of Rb or p16[INK4a], or overexpression of cyclin D1 can override this checkpoint.[22,23] Several reports have implicated D-type cyclins in neoplastic development, although limited information is available on the participation of its partner, CDK4, in these events. The involvement of CDK4 in the neoplastic process was suggested by the fact that CDK4 amplification and overexpression were detected in human glioblastomas, but overexpression and amplification of D-type cyclins were not detected in these tumors.[29] In addition, CDK4 mutations were identified in patients with familial melanoma,[30] and amplification and overexpression of CDK4 were also recently detected in sporadic breast carcinomas,[31] ovarian carcinomas,[32] and sarcomas.[33] Taken together, proteins that govern cell cycle control are reasonable targets for cancer therapy.[17,34]

8.3 MANIPULATION OF CDK ACTIVITY FOR THERAPEUTIC PURPOSES

Several strategies could be considered to modulate CDK activity. These are divided into direct effects on the catalytic CDK subunit or indirect modulation of regulatory pathways that govern CDK activity.[17,34,35] Small molecular endogenous CDK inhibitors (SCDKI) are compounds that directly target the catalytic CDK subunit. Most of these compounds modulate CDK activity by interacting specifically with the ATP-binding site of CDKs.[17,34–37] Examples of this class include flavopiridol, roscovitine, aminothiazole (BMS 387032), UCN-01 (7-hydroxystaurosporine), and alsterpaullone (Figure 8.1). The second class consists of compounds that modulate CDK activity by targeting the regulatory upstream pathways that govern CDK activity: by altering the expression and synthesis of the CDK/cyclin subunits or the CDK inhibitory proteins; by modulating the phosphorylation of CDKs; by targeting CDK-activating kinase (CAK), cdc25, and wee1/myt1; or by manipulating the proteolytic machinery that regulates the catabolism of CDK/cyclin complexes or their regulators.[3,4,35] Examples of this class of compounds include perifosine and UCN-01, among others.

8.4 SMALL MOLECULE CDK MODULATORS

As mentioned previously, CDKs can be modulated by direct effects on the catalytic subunit or by disruption of upstream regulatory pathways. Several examples and mechanisms are described elsewhere[3,4,34–41] and in the following chapters in this section.

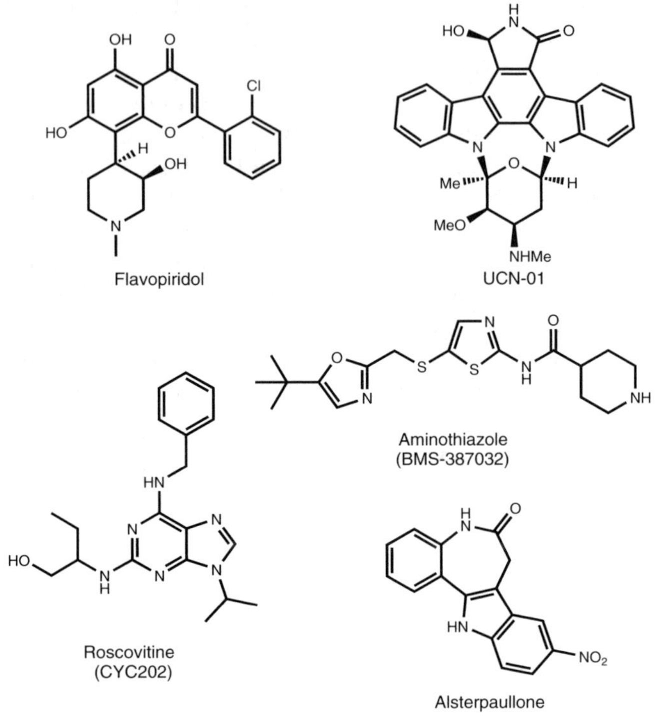

Flavopiridol

UCN-01

Aminothiazole
(BMS-387032)

Roscovitine
(CYC202)

Alsterpaullone

FIGURE 8.1 Examples of small molecule CDK inhibitors.

8.5 CDK MODULATORS IN CLINICAL TRIALS

8.5.1 FLAVOPIRIDOL

8.5.1.1 Mechanism of Antiproliferative Effects

Flavopiridol (L86-8275 or HMR 1275) is a semisynthetic flavonoid derived from the stem bark of *Dysoxylum binectariferum*, an indigenous plant from India. Initial studies with this flavonoid revealed clear evidence of G_1/S or G_2/M arrest, because of loss in CDK1 and CDK2.[42,43] Studies using purified CDKs showed that the inhibition observed is reversible and competitively blocked by ATP, with a K_i of 41 nM.[42–44] Furthermore, the crystal structure of the complex of deschloroflavopiridol and CDK2 showed that flavopiridol binds to the ATP-binding pocket, with the benzopyran occupying the same region as the purine ring of ATP,[45] confirming the earlier biochemical studies with flavopiridol.[46] Flavopiridol inhibits all CDKs thus far examined (IC_{50}~100 nM), but it inhibits CDK7 (CAK) less potently (IC_{50} ~ 300 nM).[46,47]

In addition to directly inhibiting CDKs, flavopiridol promotes a decrease in the level of cyclin D1, an oncogene that is overexpressed in many human neoplasias. It is noteworthy that neoplasms that overexpress cyclin D1 have a poor prognosis.[48] Depletion of cyclin D1 appears to lead to the loss of CDK activity.[49] Cyclin D1 decrease is caused by depletion of cyclin D1 mRNA and was associated with a specific decline in cyclin D1 promoter, measured by a luciferase reporter assay.[49] The transcriptional repression of cyclin D1 observed after treatment with flavopiridol is consistent with the effects of flavopiridol on yeast cells (see preceding text) and underscores the conserved effect of flavopiridol on eucaryotic cyclin transcription.[50] In summary, flavopiridol can induce cell cycle arrest by at least three mechanisms: (1) direct inhibition of CDK activities by binding to the ATP-binding site, (2) prevention of the phosphorylation of CDKs at threonine-160/161 by inhibition of CDK7/cyclin H[42,44], and (3) decrease in the amount of cyclin D1, an important cofactor for CDK4 and CDK6 activation (G_1/S arrest only).

In part, flavopiridol regulates transcription due to potent inhibition of P-TEFb (also known as CDK9/cyclin T), with a K_i of 3 nM, leading to inhibition of transcription by RNA polymerase II by blocking the transition into productive elongation. Interestingly, in contrast with all CDKs tested so far, flavopiridol was not competitive with ATP in this reaction. P-TEFb is a required cellular cofactor for the human immunodeficiency virus (HIV-1) transactivator, Tat. Consistent with its ability to inhibit P-TEFb, flavopiridol blocked Tat transactivation of the viral promoter *in vitro*. Furthermore, flavopiridol blocked HIV-1 replication in both single-round and viral spread assays with an IC_{50} of less than 10 nM.[51] These actions of the drug led to the testing of flavopiridol through clinical trials for patients with HIV-related malignancies.[52]

An important biochemical effect involved in the antiproliferative activity of flavopiridol is the induction of apoptotic cell death. Hematopoietic cell lines are often quite sensitive to flavopiridol-induced apoptotic cell death, [53–57] but the mechanisms by which flavopiridol induces apoptosis have not yet been elucidated. Flavopiridol does not modulate topoisomerase I/II activity.[56] In certain hematopoietic cell lines, neither BCL-2/BAX nor p53 appeared to be affected,[54,56] whereas, BCL-2 may be inhibited in other systems.[55,58] It is still unclear whether the putative flavopiridol-induced inhibition of CDK activity is required for induction of apoptosis.

Clear evidence of cell cycle arrest along with apoptosis was observed in a panel of squamous head and neck cancer cell lines, including a cell line (HN30) that is refractory to several DNA-damaging agents, such as italic *gamma*-irradiation and bleomycin.[59] Again, the apoptotic effect was independent of p53 status and was associated with the depletion of cyclin D1.[59] These findings have been corroborated in other preclinical models.[49,60–62]

Flavopiridol targets not only tumor cells but also angiogenesis pathways. Brusselbach et al.[63] incubated primary human umbilical vein endothelial cells (HUVECs) with flavopiridol and observed apoptotic cell death even in cells that were not cycling, leading to the notion that flavopiridol may have antiangiogenic properties due to endothelial cytotoxicity. In other model systems, Kerr et al.[64] tested flavopiridol in an *in vivo* Matrigel model of angiogenesis and found that flavopiridol decreased blood vessel formation, a surrogate marker for the antiangiogenic effect of this compound. Furthermore, as mentioned earlier, Melillo et al.[65] demonstrated that, at low nanomolar concentrations, flavopiridol prevented the induction of vascular endothelial growth factor (VEGF) by hypoxic conditions in human monocytes owing to decreased VEGF mRNA stability. Similar antiangiogenic effects were observed in zebrafish *in vivo* models.[66] Thus, the anti-tumor activity of flavopiridol observed may be in part due to antiangiogenic effects. Whether the various antiangiogenic actions of flavopiridol result from its interaction with a CDK target or other targets requires further study.

The anti-tumor effect observed with flavopiridol can also be explained by activation of differentiation pathways. It became clear recently that cells become differentiated when exit of the cell cycle (G_0) and loss of CDK2 activity occur. Based on this information, Lee et al.[67] tested flavopiridol and roscovitine, both known CDK2 inhibitors, to determine if they induce a differentiated phenotype. For this purpose, NCI-H358 lung carcinoma cell lines were exposed to CDK2 antisense construct, flavopiridol, or roscovitine. Clear evidence of mucinous differentiation along with loss in CDK2 activity was observed in this lung carcinoma model. Thus, it is plausible that the anti-tumor effect of flavopiridol in lung carcinoma models may be due to induction of differentiation, among others.[67]

Several investigators determined that flavopiridol has synergistic effects with standard chemotherapeutic agents in several *in vitro* models. Synergistic effects were observed in A549 lung carcinoma cells when treatment with flavopiridol followed treatment with paclitaxel, cytarabine, topotecan, doxorubicin, and/or etoposide.[68–71] In contrast, a synergistic effect was observed with 5-fluorouracil only when cells were treated with flavopiridol for 24 h before the addition of 5-fluorouracil. Furthermore, synergistic effects with cisplatin were not schedule dependent.[69] However, Chien et al.[72] failed to demonstrate a synergistic effect between flavopiridol and cisplatin or italic *gamma*-irradiation in bladder carcinoma models. One important issue to mention is that most of these studies were performed in *in vitro* models. Thus, confirmatory studies in *in vivo* animal models are needed.

Experiments using colorectal (Colo205) and prostate (LnCaP/DU-145) carcinoma xenograft models in which flavopiridol was administered frequently over a protracted period demonstrated that flavopiridol is cytostatic.[73,74] These demonstrations led to human clinical trials of flavopiridol administered as a 72 h continuous infusion every 2 weeks[75] (see following text). Subsequent studies in human leukemia and lymphoma xenografts demonstrated that flavopiridol administered intravenously as a bolus rendered animals tumorfree, whereas flavopiridol administered as an infusion only delayed tumor growth.[53] Moreover, in HN12 head and neck cancer xenografts, flavopiridol administered intraperitoneally for 5 d demonstrated a substantial growth delay.[59] Again, apoptotic cell death and cyclin D1 depletion were observed in tissues from xenografts treated with flavopiridol.[53] Based on these results, a Phase I trial of 1 h daily infusional flavopiridol every 3 weeks has been conducted at the NCI.[76]

8.5.1.2   Clinical Experience with Flavopiridol

Two Phase I clinical trials of flavopiridol administered as a 72 h continuous infusion every 2 weeks have been completed.[75,78] In the NCI Phase I trial (N = 76) of infusional flavopiridol, dose-limiting toxicity (DLT) was secretory diarrhea with a maximal tolerated dose (MTD) of 50 mg/m^2/d for 3 d. In the presence of antidiarrheal prophylaxis (a combination of cholestyramine and loperamide), patients tolerated higher doses, defining a second MTD, 78 mg/m^2/d for 3 d. The DLT observed at the higher dose level was a substantial proinflammatory syndrome that is associated with induction of plasma IL-6.[75,77] Minor responses were observed in patients with non-Hodgkin's lymphoma, colon, and kidney cancer for more than 6 months. Moreover, one patient with refractory renal cancer achieved a partial response for more than 8 months.[75] Plasma concentrations of 300 to 500 nM flavopiridol, which inhibit CDK activity *in vitro*, were safely achieved during this trial.[75]

In a complementary Phase I trial also exploring the same schedule (72 h continuous infusion every 2 weeks), Thomas et al.[78] found that the DLT was diarrhea, corroborating the NCI experience. Moreover, plasma concentrations of 300 to 500 nM flavopiridol were also observed. Interestingly, there was one patient in this trial with refractory metastatic gastric cancer that progressed after a treatment regimen containing 5-fluorouracil. When treated with flavopiridol, this patient achieved a sustained complete response without any evidence of disease for more than 2 years after treatment was completed.

The first Phase I trial of a daily 1 h infusion of flavopiridol was recently completed.[76] This schedule was based on anti-tumor results observed in leukemia and lymphoma and head and neck cancer xenografts treated with flavopiridol.[53,59] A total of 55 patients were treated in this trial. The recommended Phase II dose is 37.5 mg/m^2/d for five consecutive days. DLTs observed at 52.5 mg/m^2/d are nausea and vomiting, neutropenia, fatigue, and diarrhea.[76] Other side effects are local tumor pain and anorexia. To reach higher flavopiridol concentrations, the protocol was amended to administer flavopiridol for 3 d and then for 1 d only. With these protocol modifications, we were able to achieve concentrations (~4 μm) necessary to induce apoptosis in xenograft models.[53,59] It is noteworthy that the half-life observed in this trial is much shorter (~3 h) than the infusional trial (~10 h). Thus, the high micromolar concentrations achieved in the 1 h infusional trial could be maintained only for short periods of time.

Several Phase II trials using the continuous infusion schedule (50 mg/m^2/d over 72 h) were recently conducted for several malignancies including melanoma, lung, kidney, and prostate, in patients with refractory head and neck cancer, chronic lymphocytic leukemia (CLL), and mantle cell lymphoma (MCL), among others. Unfortuntately, at this dose and schedule, flavopiridol monotherapy did not show significant anti-tumor activity.[79–83]

Based on the interesting preclinical data in combination of cytotoxics and the feasibility of flavopiridol administration as a short infusion, several Phase I combination trials have been performed.[84–87] Activity was observed in patients with taxane-refractory disease. However, results of Phase III trials with these combinations are needed before concluding that combinations are active in refractory cases.

8.5.2 UCN-01

8.5.2.1 Mechanism of Antiproliferative Effects

Staurosporine is a potent nonspecific protein and tyrosine kinase inhibitor with a very low therapeutic index in animals.[88] Thus, efforts to find staurosporine analogs of staurosporine have identified compounds specific for protein kinases. One staurosporine analog, UCN-01, has potent activity against several protein kinase C isoenzymes, particularly the Ca^{2+}-dependent protein kinase C with an $IC_{50} \sim 30$ nM.[89,90] In addition to its effects on protein kinase C, UCN-01 has antiproliferative activity in several human tumor cell lines.[90–92] These effects appear not to be related to the effects of UCN-01 in PKC signaling.[93]

Another interesting feature, again unrelated to PKC, is "inappropriate activation" of CDK kinases in intact cells.[93] This phenomenon correlates with the G_2 abrogation checkpoint observed with this agent. Experimental evidence suggests that DNA damage leads to cell cycle arrest to allow DNA repair. In the presence of UCN-01, irradiated cells are unable to accumulate in the G_2 phase with subsequent early mitosis accompanying the onset of apoptotic cell death.[92] The accelerated mitosis is due to activation of cdc2 kinase. These activations could be partially explained by the inactivation of Wee1, the kinase that negatively regulates the G_2/M phase transition.[94] Moreover, UCN-01 can have a direct effect on chk1, the protein kinase that regulates the G_2 checkpoint.[95–97] Thus, although UCN-01 at high concentrations can directly inhibit CDKs *in vitro*, UCN-01 can modulate cellular "upstream" regulators at much lower concentrations, leading to inappropriate cdc2 activation. Studies from other groups suggest that not only is UCN-01 able to abrogate the G_2 checkpoint induced by DNA-damaging agents but also, in some circumstances, UCN-01 is able to abrogate the DNA-damage-induced S phase checkpoint.[98,99]

Another interesting property of UCN-01 is its ability to arrest cells in the G_1 phase of the cell cycle.[90,100–107] When human epidermoid carcinoma A431 cells (mutated p53) or HN12 head and neck carcinoma cell lines are incubated with UCN-01, these cells were arrested in the G_1 phase with Rb hypophosphorylation and p21waf1/p27kip1 accumulation.[101,105] Chen et al.[102] suggest that Rb, but not p53, function is essential for UCN-01-mediated G_1 arrest. However, Shimizu et al.[106] demonstrated that lung carcinoma cell lines with either absent, mutant, or wild-type Rb exposed to UCN-01 displayed G_1 arrest and antiproliferative effects irrespective of Rb function. Thus, the exact role of Rb or p53 in the G_1 arrest induced by UCN-01 is still unknown. Further studies on the putative targets for UCN-01 in the G_1 phase arrest of cells are warranted.

Recently, Facchinetti et al. demonstrated that the G_1/S arrest induced by UCN-01 is due to the transcriptional upregulation of p21.[103] This effect is due to activation of the MAPK/ERK pathway, leading to p53-independent transactivation of p21. Further studies are needed to understand the mechanism by which UCN-01 activates MAPK.

As shown in several *in vitro* models, lack of functional p53 does not preclude the cell cycle arrest and cytotoxicity induced by this agent.[90,92,93,100–107] Thus, a common feature observed in more than 50% of human neoplasias, associated with poor outcome and refractoriness to standard chemotherapies,[108,109] may render tumor cells more sensitive to UCN-01.

A very exciting recent development is the discovery that UCN-01 can modulate the PI3 kinase/AKT survival pathway.[110] UCN-01 displays potent inhibition *in vitro* of the PDK1 serine/threonine kinase, leading to dephosphorylation and inactivation of AKT.[110] Of note, induction of p21 is not due to inhibition of PDK1/AKT but occurs by activation of MAPK.[103] Although this is an exciting novel feature of

UCN-01, it is of utmost importance to demonstrate whether the anti-tumor effects of UCN-01 are mediated by this action. Moreover, demonstration that these effects also occur in *in vivo* settings is crucial.

As previously mentioned, synergistic effects of UCN-01 have been observed with many signal transduction and chemotherapeutic agents, including mitomycin C, 5-fluorouracil, carmustine, and camptothecin.[98,99,111–119] Therefore, it is possible that combining UCN-01 with these or other agents could improve its therapeutic index. Moreover, UCN-01 has demonstrated synergistic effects with *gamma*-irradiation.[92,120] Clinical trials exploring these possibilities are currently being developed.

8.5.2.2 Clinical Trials of UCN-01

In the first Phase I trial UCN-01 was initially administered as a 72-h continuous infusion every 2 weeks based on data from *in vitro* and xenograft preclinical models.[121,122] However, it became apparent in the first few patients that the drug had an unexpectedly long half-life (~30 days). This half-life was 100 times longer than the half-life observed in preclinical models, most likely due to the avid binding of UCN-01 to alpha1-acid glycoprotein.[123,124] Thus, the protocol was modified to administer UCN-01 every 4 weeks (one half-life) and, in subsequent courses, the duration of infusion was decreased by half (total 36 h). Thus, it was possible to reach similar peak plasma concentrations in subsequent courses with no evidence of drug accumulation. There was no evidence of myelotoxicity or gastrointestinal toxicity (prominent side effects observed in animal models), despite very high plasma concentrations achieved (35 to 50 μM).[121–124] Major toxicities were nausea and vomiting (amenable to standard antiemetic treatments), symptomatic hyperglycemia associated with an insulin-resistance state (increase in insulin and c-peptide levels while receiving UCN-01), and pulmonary toxicity characterized by substantial hypoxemia without obvious radiologic changes. The recommended Phase II dose of UCN-01 given on a 72 h continuous infusion schedule was 42.5 mg/m^2/d.[122] One patient with refractory metastatic melanoma developed a partial response that lasted 8 months. Another patient with refractory anaplastic large-cell lymphoma that had failed multiple chemotherapeutic regimens including high-dose chemotherapy has no evidence of disease more than 4 years after the initiation of UCN-01. Moreover, a few patients with leiomyosarcoma, non-Hodgkin's lymphoma, and lung cancer demonstrated stable disease for 6 months.[122,125]

In order to estimate "free UCN-01 concentrations" in body fluids, several efforts were considered. Plasma ultracentrifugation and salivary determination of UCN-01 revealed similar results. At the recommended Phase II dose (37.5 mg/m^2/d over 72 h), concentrations of "free-salivary" UCN-01 (~100 nM) that may cause G$_2$ checkpoint abrogation can be achieved. As mentioned earlier, UCN-01 is a potent PKC inhibitor. In order to determine the putative signaling effects of UCN-01 in tissues, bone marrow aspirates and tumor cells were obtained from patients before and during the first cycle of UCN-01 administration. Western blot studies were performed in those samples against phosphorylated adducin, a cytoskeletal membrane protein, a specific substrate phosphorylated by PKC. A clear loss in phospho-adducin content in the post-treatment samples was observed in all tumor and bone marrow samples tested, concluding that UCN-01 can modulate PKC activity in tissues from patients in this trial.[121,122]

Several groups have conducted shorter duration (3 h) infusional trials of UCN-01.[126,127] A recently published report by Dees et al.[127] presented the experience with 1 h (and then 3 h) infusional experience with UCN-01. A total of 24 patients participated in this trial. The study started as a 1 h infusion; however, it appeared too toxic. MTD for the 3 h infusion was 95 mg/m^2 over 3 h for the first course and 47.5 mg/m^2 over 3 h for the second and subsequent courses. Dose-limiting toxicity was hypotension. Other toxicities observed were similar to the 72-h infusion trial. However, in the 3 h infusion UCN-01 trial, CNS toxicities including seizures and changes in mental status occurred. No objective responses were observed. Mean (SD) pharmacokinetic variable values in nine patients treated at 95 mg/m^2 over 3 h were volume of distribution at steady state, 14 (5.4) L; ß half-life, 406 (151) h; systemic clearance, 0.028 (0.017) L/h; C_{max}, 51(16) µM; and area under the curve, 19,732(12,195) µM/L h. When compared with the 72-h infusional trial, the 3 h infusional trial has some similarities and differences. Unfortunately, based on the accelerated dose escalation design,[128] they were not able to have "robust" PK parameters. Despite these limitations, PK parameters are similar; it appears that, at MTD, the 3-h trials demonstrated a predictable slightly higher peak plasma concentration than the 72 h. However, AUC in the former was slightly lower. In the 72 h infusion, anti-tumor activity was demonstrated in a patient with metastatic melanoma and in a patient with anaplastic non-Hodgkin's lymphoma whereas the 3 h infusion had no evidence of anti-tumor effects. Finally, evidence of "free" UCN-01 concentrations were demonstrated in the 72 h infusion trial by salivary UCN-01 concentration, by plasma centrifugation, by plasma G$_2$ checkpoint *ex vivo* assay and by downmodulation of phospho-adducin, a known substrate of PKC. In contrast, the 3 h infusion showed salivary data in some patients at the MTD dose. In summary, the best schedule to administer UCN-01 is still unknown.

Recently, a Phase I combination study of 72 h infusion UCN-01 and 24 h weekly flourouracil (FU) was reported.[129] The protocol schedule consisted of increasing doses of weekly 24 h infusion of FU and were followed by a UCN-01 dose of 135 mg/m^2 over 72 h in cycle 1 and 67.5 mg/m^2 over 36 h in subsequent cycles, based on the initial Phase I trial.[121,122] The authors were able to escalate FU up to 2,600 mg/m^2 in combination with monthly UCN-01. Dose-limiting toxicity included arrhythmia and syncope. Other toxicities included hyperglycemia, headache, and nausea and vomiting. The highest mean peak plasma concentration of UCN-01 was obtained in cohort 5 (1.265 mg/m^2 FU) of 48.5 µM. Of note, the lowest mean peak plasma UCN-01 concentration was observed in the highest FU dose administered (cohort 8, 2.600 mg/m^2) of 17.6 µM. The investigators proposed cohort 8 as the recommended Phase II dose. Unfortunately, there were no objective responses.

A Phase II trial of 3 h infusional UCN-01 in patients with progressive, metastatic RCC was recently reported[130] A total of 21 patients received 90 mg/m^2 over 3 h based on prior Phase I trials using this schedule.[126,127] Accrual was halted after failure to reach a predetermined efficacy requirement with 7 patients remaining disease-progression free after 4 months. The median TTP for all patients was 2.67 months (range, 0.4 to 7.6 months). There were no objective responses and UCN-01 using this schedule was well tolerated.[130]

8.6 OUTSTANDING ISSUES RELATED TO THE MODULATION OF CELL CYCLE FOR CANCER THERAPY

The role of CDKs as targets for cancer therapy, especially with respect to CDK2, has been recently challenged.[131–137] Briefly, loss of CDK2 function in some colon carcinoma cell lines failed to arrest at the G_1/S entry.[135] Moreover, ablation of CDK2 and cyclin E in mice demonstrated normal development with significant meiotic perturbations.[131,133–137] The lack of significant somatic effects in these models may be explained by the redundancy of CDKs.[136] In order to have significant anti-tumor activity, small molecule CDK modulators may need to target more than one CDK to avoid its rescue by other redundant CDKs.

8.7 SUMMARY

Most human malignancies have an aberration in the Rb pathway due to "CDK hyperactivation." Several small molecule CDK modulators are being discovered and tested in the clinic. The first ATP competitive CDK inhibitors tested in clinical trials, flavopiridol and UCN-01, showed promising results with evidence of anti-tumor activity and plasma concentrations sufficient to inhibit CDK-related functions. The best schedule to be administered, combination with standard chemotherapeutic agents, best tumor types to be targeted, and demonstration of CDK modulation in tumor samples from patients in these trials are important issues that need to be answered in order to advance these agents to the clinical arena.

REFERENCES

1. Paulovich, A., Toczyski, D., and Hartwell, L. When checkpoints fail. *Cell, 88:* 315–321, 1997.
2. Morgan, D. O. Cyclin-dependent kinases: engines, clocks, and microprocessors. *Annu Rev Cell Dev Biol, 13:* 261–291, 1997.
3. Senderowicz, A.M. Small-molecule cyclin-dependent kinase modulators. *Oncogene, 22:* 6609–6620, 2003.
4. Senderowicz, A.M. Novel small molecule cyclin-dependent kinases modulators in human clinical trials. *Cancer Biol Ther, 2:* S84–95, 2003.
5. Bharadwaj, R. and Yu, H. The spindle checkpoint, aneuploidy, and cancer. *Oncogene, 23:* 2016–2027, 2004.
6. Lew, D.J. and Burke, D.J. The spindle assembly and spindle position checkpoints. *Annu Rev Genet, 37:* 251–282, 2003.
7. Chan, G.K. and Yen, T.J. The mitotic checkpoint: a signaling pathway that allows a single unattached kinetochore to inhibit mitotic exit. Prog Cell Cycle Res, 5: 431–439, 2003.

8. Kops, G.J.P.L., Foltz, D.R., and Cleveland, D.W. Lethality to human cancer cells through massive chromosome loss by inhibition of the mitotic checkpoint. *Proc Natl Acad Sci U S A, 101:* 8699–8704, 2004.

9. Meraldi, P., Honda, R., and Nigg, E.A. Aurora kinases link chromosome segregation and cell division to cancer susceptibility. *Curr Opin Genet Dev, 14:* 29–36, 2004.

10. Jiang, Y., Zhang, Y., Lees, E., and Seghezzi, W. AuroraA overexpression overrides the mitotic spindle checkpoint triggered by nocodazole, a microtubule destabilizer. *Oncogene, 22:* 8293–8301, 2003.

11. Harrington, E.A., Bebbington, D., Moore, J., Rasmussen, R.K., Ajose-Adeogun, A.O., Nakayama, T., Graham, J.A., Demur, C., Hercend, T., Diu-Hercend, A., Su, M., Golec, J.M.C., and Miller, K.M. VX-680, a potent and selective small-molecule inhibitor of the Aurora kinases, suppresses tumor growth in vivo. *Nat Med, 10:* 262–267, 2004.

12. Hauf, S., Cole, R.W., LaTerra, S., Zimmer, C., Schnapp, G., Walter, R., Heckel, A., van Meel, J., Rieder, C.L., and Peters, J.M. The small molecule Hesperadin reveals a role for Aurora B in correcting kinetochore-microtubule attachment and in maintaining the spindle assembly checkpoint. *J Cell Biol, 161:* 281–294, 2003.

13. Ditchfield, C., Johnson, V.L., Tighe, A., Ellston, R., Haworth, C., Johnson, T., Mortlock, A., Keen, N., and Taylor, S.S. Aurora B couples chromosome alignment with anaphase by targeting BubR1, Mad2, and Cenp-E to kinetochores. *J Cell Biol, 161:* 267–280, 2003.

14. Sillje, H.H. and Nigg, E.A. Signal transduction: capturing polo kinase. *Science, 299:* 1190–1191, 2003.

15. Vousden, K.H. and Lu, X. Live or let die: the cell's response to p53. *Nat Rev Cancer, 2:* 594–604, 2002.

16. Kawabe, T. G_2 checkpoint abrogators as anticancer drugs. *Mol Cancer Ther, 3:* 513–519, 2004.

17. Senderowicz, A.M. Targeting cell cycle and apoptosis for the treatment of human malignancies. *Curr Opin Cell Biol, 16:* 670–678, 2004.

18. Motokura, T., Bloom, T., Kim, H.G., Juppner, H., Ruderman, J.V., Kronenberg, H.M., and Arnold, A. A novel cyclin encoded by a bcl1-linked candidate oncogene [see comments]. *Nature, 350:* 512–515, 1991.

19. Weinberg, R.A. The integration of molecular genetics into cancer management. *Cancer, 70:* 1653–1658, 1992.

20. Weinberg, R.A. The molecular basis of carcinogenesis: understanding the cell cycle clock. *Cytokines Mol Ther, 2:* 105–110, 1996.

21. Hatakeyama, M., Brill, J.A., Fink, G.R., and Weinberg, R.A. Collaboration of G_1 cyclins in the functional inactivation of the retinoblastoma protein. *Genes Dev, 8:* 1759–1771, 1994.

22. Sherr, C.J. Cell cycle control and cancer. *Harvey Lect, 96:* 73–92, 2000.

23. Pines, J. Cyclins and cyclin-dependent kinases: a biochemical view. *Biochem. J, 308:* 697–711, 1995.

24. Bartek, J., Staskova, Z., Draetta, G., and Lukas, J. Molecular pathology of the cell cycle in human cancer cells. *Stem Cells (Dayt), 11(Suppl. 1):* 51–58, 1993.

25. Lukas, J., Aagaard, L., Strauss, M., and Bartek, J. Oncogenic aberrations of p16INK4/CDKN2 and cyclin D1 cooperate to deregulate G_1 control. *Cancer Res, 55:* 4818–4823, 1995.

26. Bartek, J., Bartkova, J., and Lukas, J. The retinoblastoma protein pathway and the restriction point. *Curr Opin Cell Biol, 8:* 805–814, 1996.

27. Aagaard, L., Lukas, J., Bartkova, J., Kjerulff, A.A., Strauss, M., and Bartek, J. Aberrations of p16Ink4 and retinoblastoma tumour-suppressor genes occur in distinct sub-sets of human cancer cell lines. *Int J Cancer, 61:* 115–120, 1995.

28. Lukas, J., Bartkova, J., and Bartek, J. Convergence of mitogenic signalling cascades from diverse classes of receptors at the cyclin D-cyclin-dependent kinase-pRb-controlled G_1 checkpoint. *Mol Cell Biol, 16:* 6917–6925, 1996.

29. Sonoda, Y., Yoshimoto, T., and Sekiya, T. Homozygous deletion of the MTS1/p16 and MTS2/p15 genes and amplification of the CDK4 gene in glioma. *Oncogene, 11:* 2145–2149, 1995.

30. Wolfel, T., Hauer, M., Schneider, J., Serrano, M., Wolfel, C., Klehmann-Hieb, E., De Plaen, E., Hankeln, T., Meyer zum Buschenfelde, K.H., and Beach, D. A p16INK4a-insensitive CDK4 mutant targeted by cytolytic T lymphocytes in a human melanoma. *Science, 269:* 1281–1284, 1995.

31. An, H.X., Beckmann, M.W., Reifenberger, G., Bender, H.G., and Niederacher, D. Gene amplification and overexpression of CDK4 in sporadic breast carcinomas is associated with high tumor cell proliferation. *Am J Pathol, 154:* 113–118, 1999.

32. Masciullo, V., Scambia, G., Marone, M., Giannitelli, C., Ferrandina, G., Bellacosa, A., Benedetti Panici, P., and Mancuso, S. Altered expression of cyclin D1 and CDK4 genes in ovarian carcinomas. *Int J Cancer, 74:* 390–395, 1997.

33. Kanoe, H., Nakayama, T., Murakami, H., Hosaka, T., Yamamoto, H., Nakashima, Y., Tsuboyama, T., Nakamura, T., Sasaki, M. S., and Toguchida, J. Amplification of the CDK4 gene in sarcomas: tumor specificity and relationship with the RB gene mutation. *Anticancer Res, 18:* 2317–2321, 1998.

34. Meijer, L. Cyclin-dependent kinases inhibitors as potential anticancer, antineurodegenerative, antiviral and antiparasitic agents. *Drug Resist Update, 3:* 83–88, 2000.

35. Senderowicz, A.M. and Sausville, E.A. Preclinical and clinical development of cyclin-dependent kinase modulators. *J Natl Cancer Inst, 92:* 376–387, 2000.

36. Zaharevitz, D.W., Gussio, R., Leost, M., Senderowicz, A.M., Lahusen, T., Kunick, C., Meijer, L., and Sausville, E.A. Discovery and initial characterization of the paullones, a novel class of small-molecule inhibitors of cyclin-dependent kinases. *Cancer Res, 59:* 2566–2569, 1999.

37. De Azevedo, W.F., Leclerc, S., Meijer, L., Havlicek, L., Strnad, M., and Kim, S.H. Inhibition of cyclin-dependent kinases by purine analogues: crystal structure of human cdk2 complexed with roscovitine. *Eur J Biochem, 243:* 518–526, 1997.

38. Meijer, L., Borgne, A., Mulner, O., Chong, J.P., Blow, J.J., Inagaki, N., Inagaki, M., Delcros, J.G., and Moulinoux, J.P. Biochemical and cellular effects of roscovitine, a potent and selective inhibitor of the cyclin-dependent kinases cdc2, cdk2 and cdk5. *Eur J Biochem, 243:* 527–536, 1997.

39. Misra, R.N., Xiao, H.Y., Kim, K.S., Lu, S., Han, W.C., Barbosa, S.A., Hunt, J.T., Rawlins, D.B., Shan, W., Ahmed, S.Z., Qian, L., Chen, B.C., Zhao, R., Bednarz, M.S., Kellar, K.A., Mulheron, J.G., Batorsky, R., Roongta, U., Kamath, A., Marathe, P., Ranadive, S.A., Sack, J.S., Tokarski, J.S., Pavletich, N.P., Lee, F.Y., Webster, K.R., and Kimball, S.D. N-(cycloalkylamino)acyl-2-aminothiazole inhibitors of cyclin-dependent kinase 2. N-[5-[[[5-(1,1-dimethylethyl)-2-oxazolyl]methyl]thio]-2-thiazolyl]-4-piperidinecarboxamide (BMS-387032), a highly efficacious and selective antitumor agent. *J Med Chem, 47:* 1719–1728, 2004.

40. Mettey, Y., Gompel, M., Thomas, V., Garnier, M., Leost, M., Ceballos-Picot, I., Noble, M., Endicott, J., Vierfond, J.M., and Meijer, L. Aloisines, a new family of CDK/GSK-3 inhibitors. SAR study, crystal structure in complex with CDK2, enzyme selectivity, and cellular effects. *J Med Chem, 46:* 222–236, 2003.

41. Ortega, M.A., Montoya, M.E., Zarranz, B., Jaso, A., Aldana, I., Leclerc, S., Meijer, L., and Monge, A. Pyrazolo[3,4-b]quinoxalines: a new class of cyclin-dependent kinases inhibitors. *Bioorg Med Chem, 10:* 2177–2184, 2002.

42. Worland, P.J., Kaur, G., Stetler-Stevenson, M., Sebers, S., Sartor, O., and Sausville, E.A. Alteration of the phosphorylation state of p34cdc2 kinase by the flavone L86-8275 in breast carcinoma cells: correlation with decreased H1 kinase activity. *Biochem Pharmacol, 46:* 1831–1840, 1993.

43. Kaur, G., Stetler-Stevenson, M., Sebers, S., Worland, P., Sedlacek, H., Myers, C., Czech, J., Naik, R., and Sausville, E. Growth inhibition with reversible cell cycle arrest of carcinoma cells by flavone L86-8275. *J Natl Cancer Inst, 84:* 1736–1740, 1992.

44. Carlson, B.A., Dubay, M.M., Sausville, E.A., Brizuela, L., and Worland, P.J. Flavopiridol induces G_1 arrest with inhibition of cyclin-dependent kinase (CDK) 2 and CDK4 in human breast carcinoma cells. *Cancer Res, 56:* 2973–2978, 1996.

45. De Azevedo, W.F., Jr., Mueller-Dieckmann, H.J., Schulze-Gahmen, U., Worland, P.J., Sausville, E., and Kim, S.H. Structural basis for specificity and potency of a flavonoid inhibitor of human CDK2, a cell cycle kinase. Proc *Natl Acad Sci U S A, 93:* 2735–2740, 1996.

46. Losiewicz, M.D., Carlson, B.A., Kaur, G., Sausville, E.A., and Worland, P.J. Potent inhibition of CDC2 kinase activity by the flavonoid L86-8275. *Biochem Biophys Res Commun, 201:* 589–595, 1994.

47. Carlson, B., Pearlstein, R., Naik, R., Sedlacek, H., Sausville, E., and Worland, P. Inhibition of CDK2, CDK4 and CDK7 by flavopiridol and structural analogs. In *Proceedings of the American Association for Cancer Research,* 1996, p. 424.

48. Fredersdorf, S., Burns, J., Milne, A.M., Packham, G., Fallis, L., Gillett, C.E., Royds, J.A., Peston, D., Hall, P.A., Hanby, A.M., Barnes, D.M., Shousha, S.O., Hare, M.J., and Lu, X. High level expression of p27(kip1) and cyclin D1 in some human breast cancer cells: inverse correlation between the expression of p27(kip1) and degree of malignancy in human breast and colorectal cancers. *Proc Natl Acad Sci U S A, 94:* 6380–6385, 1997.

49. Carlson, B., Lahusen, T., Singh, S., Loaiza-Perez, A., Worland, P.J., Pestell, R., Albanese, C., Sausville, E.A., and Senderowicz, A.M. Down-regulation of cyclin D1 by transcriptional repression in MCF-7 human breast carcinoma cells induced by flavopiridol. *Cancer Res, 59:* 4634–4641, 1999.

50. Gray, N.S., Wodicka, L., Thunnissen, A.M., Norman, T.C., Kwon, S., Espinoza, F.H., Morgan, D.O., Barnes, G., LeClerc, S., Meijer, L., Kim, S.H., Lockhart, D.J., and Schultz, P.G. Exploiting chemical libraries, structure, and genomics in the search for kinase inhibitors. *Science, 281:* 533–538, 1998.

51. Chao, S.H., Fujinaga, K., Marion, J.E., Taube, R., Sausville, E.A., Senderowicz, A.M., Peterlin, B.M., and Price, D.H. Flavopiridol inhibits P-TEFb and blocks HIV-1 replication. *J Biol Chem, 275:* 28345–28348, 2000.

52. Wright, J., Blatner, G.L., and Cheson, B.D. Clinical trials referral resource. Clinical trials of flavopiridol. *Oncology (Huntingt), 12:* 1018, 1023–1014, 1998.

53. Arguello, F., Alexander, M., Sterry, J., Tudor, G., Smith, E., Kalavar, N., Greene, J., Koss, W., Morgan, D., Stinson, S., Siford, T., Alvord, W., Labansky, R., and Sausville, E. Flavopiridol induces apoptosis of normal lymphoid cells, causes immunosuppresion, and has potent antitumor activity in vivo against human and leukemia xenografts. *Blood, 91:* 2482–2490, 1998.

54. Byrd, J.C., Shinn, C., Waselenko, J.K., Fuchs, E.J., Lehman, T.A., Nguyen, P.L., Flinn, I.W., Diehl, L.F., Sausville, E., and Grever, M.R. Flavopiridol induces apoptosis in chronic lymphocytic leukemia cells via activation of caspase-3 without evidence of bcl-2 modulation or dependence on functional p53. *Blood, 92:* 3804–3816, 1998.

55. Konig, A., Schwartz, G.K., Mohammad, R.M., Al-Katib, A., and Gabrilove, J.L. The novel cyclin-dependent kinase inhibitor flavopiridol downregulates Bcl-2 and induces growth arrest and apoptosis in chronic B-cell leukemia lines. *Blood, 90:* 4307–4312, 1997.

56. Parker, B.W., Kaur, G., Nieves-Neira, W., Taimi, M., Kohlhagen, G., Shimizu, T., Losiewicz, M.D., Pommier, Y., Sausville, E.A., and Senderowicz, A.M. Early induction of apoptosis in hematopoietic cell lines after exposure to flavopiridol. *Blood, 91:* 458–465, 1998.

57. Decker, R.H., Dai, Y., and Grant, S. The cyclin-dependent kinase inhibitor flavopiridol induces apoptosis in human leukemia cells (U937) through the mitochondrial rather than the receptor-mediated pathway. *Cell Death Differ, 8:* 715–724, 2001.

58. Kitada, S., Zapata, J.M., Andreeff, M., and Reed, J.C. Protein kinase inhibitors flavopiridol and 7-hydroxy-staurosporine down-regulate antiapoptosis proteins in B-cell chronic lymphocytic leukemia [in process citation]. *Blood, 96:* 393–397, 2000.

59. Patel, V., Senderowicz, A.M., Pinto, D., Igishi, T., Raffeld, M., Quintanilla-Martinez, L., Ensley, J.F., Sausville, E.A., and Gutkind, J.S. Flavopiridol, a novel cyclin-dependent kinase inhibitor, suppresses the growth of head and neck squamous cell carcinomas by inducing apoptosis. *J Clin Invest, 102:* 1674–1681, 1998.

60. Guedez, L., Quintanilla-Martinez, L., Lahusen, T., Davies, T., Singh, S.S., Barotto, N., Vistica, D., Raffeld, M., Sausville, E.A., and Senderowicz, A.M. Flavopiridol-induced apoptosis is associated with a decrease in Cyclin D1 in mantle lymphoma cell lines. In *Proceedings of the 9th Annual Meeting of the American Association of Cancer Research*, Philadelphia, PA, 1999.

61. Schrump, D.S., Matthews, W., Chen, G.A., Mixon, A., and Altorki, N.K. Flavopiridol mediates cell cycle arrest and apoptosis in esophageal cancer cells. *Clin Cancer Res, 4:* 2885–2890, 1998.

62. Wu, K., Wang, C., D'Amico, M., Lee, R.J., Albanese, C., Pestell, R.G., and Mani, S. Flavopiridol and trastuzumab synergistically inhibit proliferation of breast cancer cells: association with selective cooperative inhibition of cyclin D1-dependent kinase and Akt signaling pathways. *Mol Cancer Ther, 1:* 695–706, 2002.

63. Brusselbach, S., Nettelbeck, D.M., Sedlacek, H.H., and Muller, R. Cell cycle-independent induction of apoptosis by the anti-tumor drug Flavopiridol in endothelial cells. *Int J Cancer, 77:* 146–152, 1998.

64. Kerr, J.S., Wexler, R.S., Mousa, S.A., Robinson, C.S., Wexler, E.J., Mohamed, S., Voss, M.E., Devenny, J.J., Czerniak, P.M., Gudzelak, A., Jr., and Slee, A.M. Novel small molecule alpha v integrin antagonists: comparative anti-cancer efficacy with known angiogenesis inhibitors [in process citation]. *Anticancer Res, 19:* 959–968, 1999.

65. Melillo, G., Sausville, E. A., Cloud, K., Lahusen, T., Varesio, L., and Senderowicz, A.M. Flavopiridol, a protein kinase inhibitor, down-regulates hypoxic induction of vascular endothelial growth factor expression in human monocytes. *Cancer Res, 59:* 5433–5437, 1999.

66. Parng, C., Seng, W.L., Semino, C., and McGrath, P. Zebrafish: a preclinical model for drug screening. *Assay Drug Dev Technol, 1:* 41–48, 2002.

67. Lee, H.R., Chang, T.H., Tebalt, M.J., III, Senderowicz, A.M., and Szabo, E. Induction of differentiation accompanies inhibition of Cdk2 in a non-small cell lung cancer cell line. *Int J Oncol, 15:* 161–166, 1999.

68. Schwartz, G., Farsi, K., Maslak, P., Kelsen, D., and Spriggs, D. Potentiation of apoptosis by flavopiridol in mitomycin-C-treated gastric and breast cancer cells. *Clin Cancer Res, 3:* 1467–1472, 1997.

69. Bible, K.C. and Kaufmann, S.H. Cytotoxic synergy between flavopiridol (NSC 649890, L86-8275) and various antineoplastic agents: the importance of sequence of administration. *Cancer Res, 57:* 3375–3380, 1997.

70. Yu, C., Krystal, G., Dent, P., and Grant, S. Flavopiridol potentiates STI571-induced mitochondrial damage and apoptosis in BCR-ABL-positive human leukemia cells. *Clin Cancer Res, 8:* 2976–2984, 2002.

71. Dai, Y., Rahmani, M., Pei, X.Y., Dent, P., and Grant, S. Bortezomib and Flavopiridol interact synergistically to induce apoptosis in chronic myeloid leukemia cells resistant to imatinib mesylate through both Bcr/Abl-dependent and -independent mechanisms. *Blood, 104:* 509–518, 2004.

72. Chien, M., Astumian, M., Liebowitz, D., Rinker-Schaeffer, C., and Stadler, W.M. In vitro evaluation of flavopiridol, a novel cell cycle inhibitor, in bladder cancer. *Cancer Chemother Pharmacol, 44:* 81–87, 1999.

73. Sedlacek, H.H., Czech, J., Naik, R., Kaur, G., Worland, P., Losiewicz, M., Parker, B., Carlson, B., Smith, A., Senderowicz, A., and Sausville, E. Flavopiridol (L86-8275, NSC-649890), a new kinase inhibitor for tumor therapy. *Int J Oncol, 9:* 1143–1168, 1996.

74. Drees, M., Dengler, W., Roth, T., Labonte, H., Mayo, J., Malspeis, L., Grever, M., Sausville, E., and Fiebig, H. Flavopiridol (L86-8275): Selective antitumor activity in vitro and activity in vivo for prostate carcinoma cells. *Clin Cancer Res, 32:* 273–279, 1997.

75. Senderowicz, A.M., Headlee, D., Stinson, S.F., Lush, R.M., Kalil, N., Villalba, L., Hill, K., Steinberg, S.M., Figg, W.D., Tompkins, A., Arbuck, S.G., and Sausville, E.A. Phase I trial of continuous infusion flavopiridol, a novel cyclin-dependent kinase inhibitor, in patients with refractory neoplasms. *J Clin Oncol, 16:* 2986–2999, 1998.

76. Tan, A.R., Headlee, D., Messmann, R., Sausville, E.A., Arbuck, S.G., Murgo, A.J., Melillo, G., Zhai, S., Figg, W.D., Swain, S.M., and Senderowicz, A.M. Phase I clinical and pharmacokinetic study of flavopiridol administered as a daily 1-h infusion in patients with advanced neoplasms. *J Clin Oncol, 20:* 4074–4082, 2002.

77. Messmann, R.A., Ullmann, C.D., Lahusen, T., Kalehua, A., Wasfy, J., Melillo, G., Ding, I., Headlee, D., Figg, W.D., Sausville, E.A., and Senderowicz, A.M. Flavopiridol-related proinflammatory syndrome is associated with induction of interleukin-6. *Clin Cancer Res, 9:* 562–570, 2003.

78. Thomas, J.P., Tutsch, K.D., Cleary, J.F., Bailey, H.H., Arzoomanian, R., Alberti, D., Simon, K., Feierabend, C., Binger, K., Marnocha, R., Dresen, A., and Wilding, G. Phase I clinical and pharmacokinetic trial of the cyclin-dependent kinase inhibitor flavopiridol. *Cancer Chemother Pharmacol, 50:* 465–472, 2002.

79. Stadler, W.M., Vogelzang, N.J., Amato, R., Sosman, J., Taber, D., Liebowitz, D., and Vokes, E.E. Flavopiridol, a novel cyclin-dependent kinase inhibitor, in metastatic renal cancer: a University of Chicago Phase II Consortium study. *J Clin Oncol, 18:* 371–375, 2000.

80. Schwartz, G.K., Ilson, D., Saltz, L., O'Reilly, E., Tong, W., Maslak, P., Werner, J., Perkins, P., Stoltz, M., and Kelsen, D. Phase II study of the cyclin-dependent kinase inhibitor flavopiridol administered to patients with advanced gastric carcinoma. *J Clin Oncol, 19:* 1985–1992, 2001.

81. Shapiro, G.I., Supko, J.G., Patterson, A., Lynch, C., Lucca, J., Zacarola, P.F., Muzikansky, A., Wright, J.J., Lynch, T.J., Jr., and Rollins, B.J. A phase II trial of the cyclin-dependent kinase inhibitor flavopiridol in patients with previously untreated stage IV non-small cell lung cancer. *Clin Cancer Res, 7:* 1590–1599, 2001.

82. Aklilu, M., Kindler, H.L., Donehower, R.C., Mani, S., and Vokes, E.E. Phase II study of flavopiridol in patients with advanced colorectal cancer. *Ann Oncol, 14:* 1270–1273, 2003.

83. Liu, G., Gandara, D.R., Lara, P.N., Jr., Raghavan, D., Doroshow, J.H., Twardowski, P., Kantoff, P., Oh, W., Kim, K., and Wilding, G. A phase II trial of flavopiridol (NSC #649890) in patients with previously untreated metastatic androgen-independent prostate cancer. *Clin Cancer Res, 10:* 924–928, 2004.

84. Bible, K., Lensing, J., Nelson, S., Atherton, P., Sloan, J., and Erlichman, C. A phase 1 trial of flavopiridol combined with 5-fluorouracil (5-FU) and leucovorin (CF) in patients with advanced malignancies. In *Proceedings of the Annual Meeting of the American Society of Clinical Oncology*, Chicago, IL, 2003.

85. Schwartz, G.K., O'Reilly, E., Ilson, D., Saltz, L., Sharma, S., Tong, W., Maslak, P., Stoltz, M., Eden, L., Perkins, P., Endres, S., Barazzoul, J., Spriggs, D., and Kelsen, D. Phase I study of the cyclin-dependent kinase inhibitor flavopiridol in combination with paclitaxel in patients with advanced solid tumors. *J Clin Oncol, 20:* 2157–2170, 2002.

86. Gries, J., Kasimis, B., Schwarzenberger, P., Shapiro, G., Fidias, P., Rodrigues, L., Cogswell, J., and Bukowski, R. Phase I study of flavopiridol (HMR1275) in combination with paclitaxel and carboplatin in non-small cell lung cancer (NCSLC) patients. *Eur J Cancer, 38:* S49–S50, 2002.

87. Tan, A.R., Yang, X., Berman, A., Zhai, S., Sparreboom, A., Parr, A.L., Chow, C., Brahim, J.S., Steinberg, S.M., Figg, W.D., and Swain, S.M. Phase I trial of the cyclin-dependent kinase inhibitor flavopiridol in combination with docetaxel in patients with metastatic breast cancer. *Clin Cancer Res, 10:* 5038–5047, 2004.

88. Tamaoki, T. Use and specificity of staurosporine, UCN-01, and calphostin C as protein kinase inhibitors. *Methods Enzymol, 201:* 340–347, 1991.

89. Takahashi, I., Kobayashi, E., Asano, K., Yoshida, M., and Nakano, H. UCN-01, a selective inhibitor of protein kinase C from Streptomyces. *J Antibiot (Tokyo), 40:* 1782–1784, 1987.

90. Seynaeve, C.M., Stetler-Stevenson, M., Sebers, S., Kaur, G., Sausville, E.A., and Worland, P.J. Cell cycle arrest and growth inhibition by the protein kinase antagonist UCN-01 in human breast carcinoma cells. *Cancer Res, 53:* 2081–2086, 1993.

91. Akinaga, S., Gomi, K., Morimoto, M., Tamaoki, T., and Okabe, M. Antitumor activity of UCN-01, a selective inhibitor of protein kinase C, in murine and human tumor models. *Cancer Res, 51:* 4888–4892, 1991.

92. Wang, Q., Fan, S., Eastman, A., Worland, P.J., Sausville, E.A., and O'Connor, P. UCN-01: a potent abrogator of G_2 checkpoint function in cancer cells with disrupted p53. *J Natl Cancer Inst, 88:* 956–965, 1996.

93. Wang, Q., Worland, P.J., Clark, J.L., Carlson, B.A., and Sausville, E.A. Apoptosis in 7-hydroxystaurosporine-treated T lymphoblasts correlates with activation of cyclin-dependent kinases 1 and 2. *Cell Growth Differ, 6:* 927–936, 1995.

94. Yu, L., Orlandi, L., Wang, P., Orr, M.S., Senderowicz, A.M., Sausville, E.A., Silvestrini, R., Watanabe, N., Piwnica-Worms, H., and O'Connor, P.M. UCN-01 abrogates G_2 arrest through a Cdc2-dependent pathway that is associated with inactivation of the Wee1Hu kinase and activation of the Cdc25C phosphatase. *J Biol Chem, 273:* 33455–33464, 1998.

95. Sarkaria, J.N., Busby, E.C., Tibbetts, R.S., Roos, P., Taya, Y., Karnitz, L.M., and Abraham, R.T. Inhibition of ATM and ATR kinase activities by the radiosensitizing agent, caffeine. *Cancer Res, 59:* 4375–4382, 1999.

96. Graves, P.R., Yu, L., Schwarz, J.K., Gales, J., Sausville, E.A., O'Connor, P.M., and Piwnica-Worms, H. The Chk1 protein kinase and the Cdc25C regulatory pathways are targets of the anticancer agent UCN-01. *J Biol Chem, 275:* 5600–5605, 2000.

97. Busby, E.C., Leistritz, D.F., Abraham, R.T., Karnitz, L.M., and Sarkaria, J.N. The radiosensitizing agent 7-hydroxystaurosporine (UCN-01) inhibits the DNA damage checkpoint kinase hChk1. *Cancer Res, 60:* 2108–2112, 2000.

98. Shao, R.G., Cao, C.X., Shimizu, T., O'Connor, P.M., Kohn, K.W., and Pommier, Y. Abrogation of an S-phase checkpoint and potentiation of camptothecin cytotoxicity by 7-hydroxystaurosporine (UCN-01) in human cancer cell lines, possibly influenced by p53 function. *Cancer Res, 57:* 4029–4035, 1997.

99. Bunch, R.T. and Eastman, A. Enhancement of cisplatin-induced cytotoxicity by 7-hydroxystaurosporine (UCN-01), a new G_2-checkpoint inhibitor. *Clin Cancer Res, 2:* 791–797, 1996.

100. Akinaga, S., Nomura, K., Gomi, K., and Okabe, M. Effect of UCN-01, a selective inhibitor of protein kinase C, on the cell-cycle distribution of human epidermoid carcinoma, A431 cells. *Cancer Chemother Pharmacol, 33:* 273–280, 1994.

101. Akiyama, T., Sugiyama, K., Shimizu, M., Tamaoki, T., and Akinaga, S. G_1-checkpoint function including a cyclin-dependent kinase 2 regulatory pathway as potential determinant of 7-hydroxystaurosporine (UCN-01)-induced apoptosis and G_1-phase accumulation. *Jpn J Cancer Res, 90:* 1364–1372, 1999.

102. Chen, X., Lowe, M., and Keyomarsi, K. UCN-01-mediated G_1 arrest in normal but not tumor breast cells is pRb-dependent and p53-independent. *Oncogene, 18:* 5691–5702, 1999.

103. Facchinetti, M.M., De Siervi, A., Toskos, D., and Senderowicz, A.M. UCN-01-induced cell cycle arrest requires the transcriptional induction of p21(waf1/cip1) by activation of mitogen-activated protein/extracellular signal-regulated kinase kinase/extracellular signal-regulated kinase pathway. *Cancer Res, 64:* 3629–3637, 2004.

104. Kawakami, K., Futami, H., Takahara, J., and Yamaguchi, K. UCN-01, 7-hydroxyl-staurosporine, inhibits kinase activity of cyclin- dependent kinases and reduces the phosphorylation of the retinoblastoma susceptibility gene product in A549 human lung cancer cell line. *Biochem Biophys Res Commun, 219:* 778–783, 1996.

105. Patel, V., Lahusen, T., Leethanakul, C., Igishi, T., Kremer, M., Quintanilla-Martinez, L., Ensley, J. F., Sausville, E.A., Gutkind, J.S., and Senderowicz, A.M. Antitumor activity of UCN-01 in carcinomas of the head and neck is associated with altered expression of cyclin D3 and p27(KIP1). *Clin Cancer Res, 8:* 3549–3560, 2002.

106. Shimizu, E., Zhao, M. R., Nakanishi, H., Yamamoto, A., Yoshida, S., Takada, M., Ogura, T., and Sone, S. Differing effects of staurosporine and UCN-01 on RB protein phosphorylation and expression of lung cancer cell lines. *Oncology, 53:* 494–504, 1996.

107. Usuda, J., Saijo, N., Fukuoka, K., Fukumoto, H., Kuh, H.J., Nakamura, T., Koh, Y., Suzuki, T., Koizumi, F., Tamura, T., Kato, H., and Nishio, K. Molecular determinants of UCN-01-induced growth inhibition in human lung cancer cells. *Int J Cancer, 85:* 275–280, 2000.

108. Marchetti, A., Buttitta, F., Merlo, G., Diella, F., Pellegrini, S., Pepe, S., Macchiarini, P., Chella, A., Angeletti, C.A., Callahan, R., and et al. p53 alterations in non-small cell lung cancers correlate with metastatic involvement of hilar and mediastinal lymph nodes. *Cancer Res, 53:* 2846–2851, 1993.

109. Lowe, S.W., Bodis, S., Bardeesy, N., McClatchey, A., Remington, L., Ruley, H.E., Fisher, D.E., Jacks, T., Pelletier, J., and Housman, D.E. Apoptosis and the prognostic significance of p53 mutation. *Cold Spring Harb Symp Quant Biol, 59:* 419–426, 1994.

110. Sato, S., Fujita, N., and Tsuruo, T. Interference with PDK1-Akt survival signaling pathway by UCN-01 (7-hydroxystaurosporine). *Oncogene, 21:* 1727–1738, 2002.

111. Tsuchida, E., Tsuchida, M., and Urano, M. Synergistic cytotoxicity between a protein kinase C inhibitor, UCN-01, and monoclonal antibody to the epidermal growth factor receptor on MDA-468 cells. *Cancer Biother Radiopharm, 12:* 117–121, 1997.

112. Sugiyama, K., Shimizu, M., Akiyama, T., Tamaoki, T., Yamaguchi, K., Takahashi, R., Eastman, A., and Akinaga, S. UCN-01 selectively enhances mitomycin C cytotoxicity in p53 defective cells which is mediated through S and/or G(2) checkpoint abrogation. *Int J Cancer, 85:* 703–709, 2000.

113. Pollack, I.F., Kawecki, S., and Lazo, J.S. Blocking of glioma proliferation in vitro and in vivo and potentiating the effects of BCNU and cisplatin: UCN-01, a selective protein kinase C inhibitor. *J Neurosurg, 84:* 1024–1032, 1996.

114. Jones, C.B., Clements, M.K., Redkar, A., and Daoud, S.S. UCN-01 and camptothecin induce DNA double-strand breaks in p53 mutant tumor cells, but not in normal or p53 negative epithelial cells. Int J Oncol, *17:* 1043–1051, 2000.

115. Husain, A., Yan, X.J., Rosales, N., Aghajanian, C., Schwartz, G.K., and Spriggs, D.R. UCN-01 in ovary cancer cells: effective as a single agent and in combination with cis-diamminedichloroplatinum (II) independent of p53 status. *Clin Cancer Res, 3:* 2089–2097, 1997.

116. Hsueh, C.T., Kelsen, D., and Schwartz, G.K. UCN-01 suppresses thymidylate synthase gene expression and enhances 5- fluorouracil-induced apoptosis in a sequence-dependent manner [in process citation]. *Clin Cancer Res, 4:* 2201–2206, 1998.

117. Akinaga, S., Nomura, K., Gomi, K., and Okabe, M. Enhancement of antitumor activity of mitomycin C in vitro and in vivo by UCN-01, a selective inhibitor of protein kinase C. Cancer Chemother Pharmacol, *32:* 183–189, 1993.

118. Hahn, M., Li, W., Yu, C., Rahmani, M., Dent, P., and Grant, S. Rapamycin and UCN-01 synergistically induce apoptosis in human leukemia cells through a process that is regulated by the Raf-1/MEK/ERK, Akt, and JNK signal transduction pathways. *Mol Cancer Ther, 4:* 457–470, 2005.

119. Dasmahapatra, G.P., Didolkar, P., Alley, M.C., Ghosh, S., Sausville, E.A., and Roy, K.K. In vitro combination treatment with perifosine and UCN-01 demonstrates synergism against prostate (PC-3) and lung (A549) epithelial adenocarcinoma cell lines. *Clin Cancer Res, 10:* 5242–5252, 2004.

120. Tsuchida, E. and Urano, M. The effect of UCN-01 (7-hydroxystaurosporine), a potent inhibitor of protein kinase C, on fractionated radiotherapy or daily chemotherapy of a murine fibrosarcoma. *Int J Radiat Oncol Biol Phys, 39:* 1153–1161, 1997.

121. Senderowicz, A. M., Headlee, D., Lush, R., Bauer, K., Figg, W., Murgo, A., Arbuck, S., Inoue, K., Kobashi, S., Kuwabara, T., and Sausville, E. Phase I trial of infusional UCN-01, a novel protein kinase inhibitor, in patients with refractory neoplasms. In *Proceedings of. 35th Annual Meeting of the American Society of Clinical Oncology*, Atlanta, GA, May 15–18, 1999.

122. Sausville, E.A., Arbuck, S.G., Messmann, R., Headlee, D., Bauer, K.S., Lush, R.M., Murgo, A., Figg, W.D., Lahusen, T., Jaken, S., Jing, X., Roberge, M., Fuse, E., Kuwabara, T., and Senderowicz, A.M. Phase I trial of 72-h continuous infusion UCN-01 in patients with refractory neoplasms. *J Clin Oncol, 19:* 2319–2333, 2001.

123. Fuse, E., Tanii, H., Kurata, N., Kobayashi, H., Shimada, Y., Tamura, T., Sasaki, Y., Tanigawara, Y., Lush, R.D., Headlee, D., Figg, W.D., Arbuck, S.G., Senderowicz, A.M., Sausville, E.A., Akinaga, S., Kuwabara, T., and Kobayashi, S. Unpredicted clinical pharmacology of UCN-01 caused by specific binding to human alpha1-acid glycoprotein. *Cancer Res, 58:* 3248–3253, 1998.

124. Sparreboom, A., Chen, H., Acharya, M.R., Senderowicz, A.M., Messmann, R.A., Kuwabara, T., Venzon, D.J., Murgo, A.J., Headlee, D., Sausville, E.A., and Figg, W.D. Effects of alpha1-acid glycoprotein on the clinical pharmacokinetics of 7-hydroxystaurosporine. *Clin Cancer Res, 10:* 6840–6846, 2004.

125. Wilson, W.H., Sorbara, L., Figg, W.D., Mont, E.K., Sausville, E., Warren, K.E., Balis, F.M., Bauer, K., Raffeld, M., Senderowicz, A.M., and Monks, A. Modulation of clinical drug resistance in a B cell lymphoma patient by the protein kinase inhibitor 7-hydroxystaurosporine: presentation of a novel therapeutic paradigm. *Clin Cancer Res, 6:* 415–421, 2000.

126. Tamura, T., Sasaki, Y., Minami, H., Fujii, H., Ito, K., Igarashi, T., Kamiya, Y., Kurata, T., Ohtsu, T., Onozawa, Y., Yamamoto, N., Yamamoto, N., Watanabe, Y., Tanigaara, Y., Fuse, E., Kuwabara, T., Kobayahsi, S., and Shimada, Y. Phase I study of UCN-01 by 3-h infusion. In *Proceedings of the American Society of Clinical Oncology,* Atlanta, GA, 1999, p. 159.

127. Dees, E.C., Baker, S.D., O'Reilly, S., Rudek, M.A., Davidson, S.B., Aylesworth, C., Elza-Brown, K., Carducci, M.A., and Donehower, R.C. A phase I and pharmacokinetic study of short infusions of UCN-01 in patients with refractory solid tumors. *Clin Cancer Res, 11:* 664–671, 2005.

128. Simon, R., Freidlin, B., Rubinstein, L., Arbuck, S.G., Collins, J., and Christian, M.C. Accelerated titration designs for phase I clinical trials in oncology. *J Natl Cancer Inst, 89:* 1138–1147, 1997.

129. Kortmansky, J., Shah, M.A., Kaubisch, A., Weyerbacher, A., Yi, S., Tong, W., Sowers, R., Gonen, M., O'Reilly, E., Kemeny, N., Ilson, D.I., Saltz, L.B., Maki, R.G., Kelsen, D.P., and Schwartz, G.K. Phase I trial of the cyclin-dependent kinase inhibitor and protein kinase C inhibitor 7-hydroxystaurosporine in combination with Fluorouracil in patients with advanced solid tumors. *J Clin Oncol, 23:* 1875–1884, 2005.

130. Rini, B.I., Weinberg, V., Shaw, V., Scott, J., Bok, R., Park, J.W., and Small, E.J. Time to disease progression to evaluate a novel protein kinase C inhibitor, UCN-01, in renal cell carcinoma. *Cancer, 101:* 90–95, 2004.

131. Mendez, J. Cell proliferation without cyclin E-CDK2. *Cell, 114:* 398–399, 2003.

132. Geng, Y., Yu, Q., Sicinska, E., Das, M., Schneider, J.E., Bhattacharya, S., Rideout, W.M., Bronson, R.T., Gardner, H., and Sicinski, P. Cyclin E ablation in the mouse. *Cell, 114:* 431–443, 2003.

133. Ortega, S., Prieto, I., Odajima, J., Martin, A., Dubus, P., Sotillo, R., Barbero, J.L., Malumbres, M., and Barbacid, M. Cyclin-dependent kinase 2 is essential for meiosis but not for mitotic cell division in mice. *Nat Genet, 35:* 25–31, 2003.

134. Lents, N.H. and Baldassare, J.J. CDK2 and cyclin E knockout mice: lessons from breast cancer. *Trends Endocrinol Metab, 15:* 1–3, 2004.

135. Tetsu, O. and McCormick, F. Proliferation of cancer cells despite CDK2 inhibition. *Cancer Cell, 3:* 233–245, 2003.

136. Sherr, C.J. and Roberts, J.M. Living with or without cyclins and cyclin-dependent kinases. *Genes Dev., 18:* 2699–2711, 2004.

137. Berthet, C., Aleem, E., Coppola, V., Tessarollo, L., and Kaldis, P. Cdk2 knockout mice are viable. *Curr Biol, 13:* 1775–1785, 2003.

9 (R)-Roscovitine (CYC202, Seliciclib)

*Laurent Meijer, Karima Bettayeb,
and Hervé Galons*

CONTENTS

9.1 INTRODUCTION

Phosphorylation of serine, threonine, and tyrosine residues represents one of the most common post-translational mechanisms used by cells to regulate their enzymatic and structural proteins. Phosphorylation is catalyzed by protein kinases, whereas dephosphorylation is carried out by protein phosphatases. Among the 518 human protein kinases, cyclin-dependent kinases (CDKs) (Malumbres and Barbacid, 2005) have aroused considerable interest because of their essential involvement in cell cycle control (Malumbres and Barbacid, 2001; Meijer, 2003), neuronal cell physiology (Cruz and Tsai, 2004), pain signaling (Pareek et al., 2006), apoptosis (Borgne and Golsteyn, 2003), transcription, and RNA splicing (Garriga and Grana, 2004; Loyer et al., 2005). CDKs are regulated in four different ways: (1) transient association with a regulatory partner (cyclin), (2) various post-translational modifications (phosphorylation, ubiquitin-dependent degradation), (3) transient association with a natural inhibitory protein (CIP1, KIP1/2, or INK4A-D), and (4) intracellular localization. Although the human genome sequencing program has resulted in the detection of about 20 CDKs and 25 cyclins, a more limited number of active CDK/cyclin complexes have been identified (Figure 9.1).

Alterations in the phosphorylation of proteins represent a frequent feature associated with human disease. This is the reason for an exponentially growing investment in the discovery, optimization, and therapeutic evaluation of small molecular weight, pharmacological inhibitors of protein kinases (reviews in Cohen, 2002; Fischer, 2004; Weinmann and Metternich, 2005). It is estimated that 30 to 35% of drug discovery programs in the pharmaceutical industry currently target a protein kinase. Presently, 55 kinase inhibitors are undergoing clinical evaluation against diseases such as cancer, inflammation, diabetes, and neurodegeneration.

Abnormalities in CDK activity and regulation in cancers (Vermeulen et al., 2003), viral infections (Schang, 2004), proliferative renal diseases (Nelson and Shankland, 2005), and neurodegenerative disorders such as Alzheimer's (Tsai et al., 2004), Parkinson's (Smith et al., 2003, 2004), and Nieman–Pick's diseases (Zhang et al., 2004a), ischemia (Wang et al., 2003; Rashidian et al., 2005), or traumatic brain injury (Di Giovanni et al., 2005) have encouraged an intensive search for potent and selective pharmacological inhibitors of these kinases (reviews in Knockaert et al., 2002a; Fischer et al., 2003; Benson et al., 2005; Fischer and Gianella-Borradori, 2005; Shapiro, 2006). Over 100 small-molecular-weight inhibitors of CDKs have been characterized, most of which appear to act by direct competition with ATP for binding to the catalytic site of the kinase. Over 30 of these compounds have been cocrystallized with CDK2 (Noble et al., 2004) or CDK5 (Mapelli et al., 2005), demonstrating their binding in the ATP-binding pocket of CDKs.

The family of 2,6,9-trisubstituted purines encompasses some of the first CDK inhibitors that have been described (review in Haesslein et al., 2002; Meijer and Raymond, 2003). Among these purines, the (*R*)-stereoisomer of roscovitine is one of the most frequently studied and used CDK inhibitors (Figure 9.2). Also referred to as CYC202 or Seliciclib, (*R*)-roscovitine is developed by Cyclacel Pharmaceuticals (http: www.cyclacel.com) (Guzi, 2004). It has now reached Phase 2 clinical trials for B-cell malignancies, lung and breast cancer, Phase 1 trials for glomerulonephritis, and Phase 2 trials in IgA nephropathy. The properties and development of roscovitine constitute the object of this review chapter.

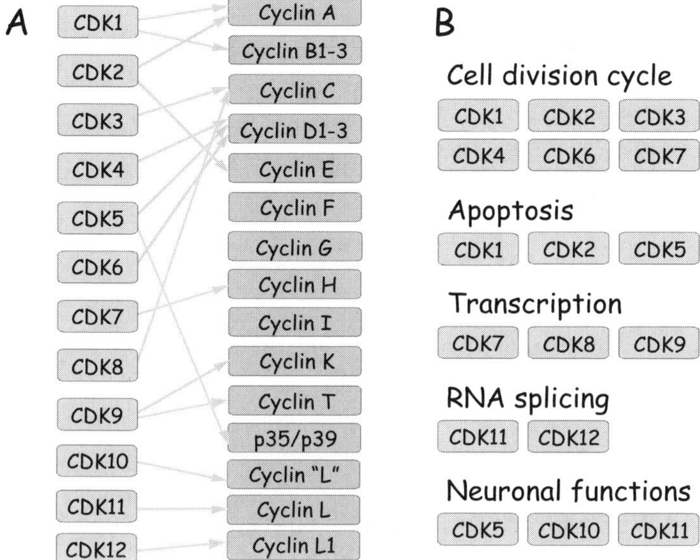

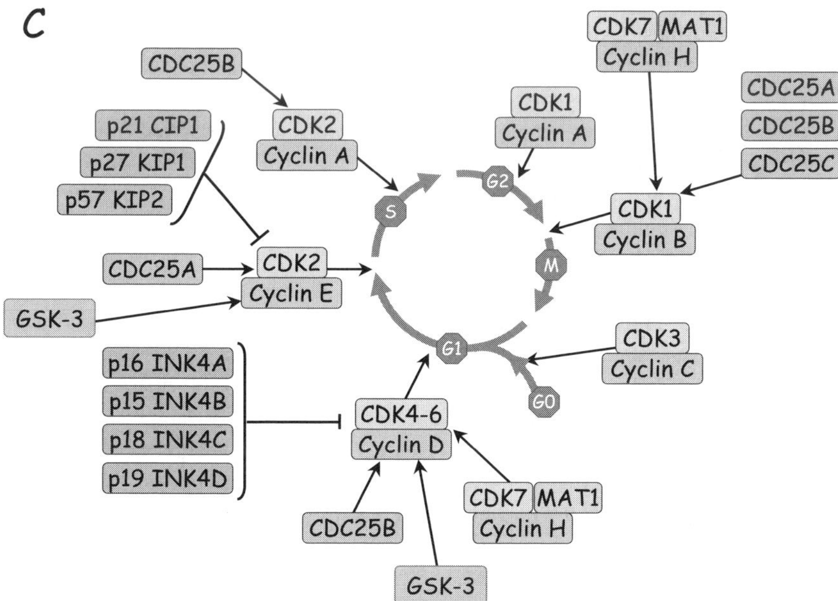

FIGURE 9.1 Cyclin-dependent kinases. CDKs (catalytic subunit) associate with regulatory subunits (cyclins and other) to constitute active protein kinase complexes (A). Different CDKs are involved in various physiological processes (B), including the cell division cycle (C), many regulators of which are altered in human cancer.

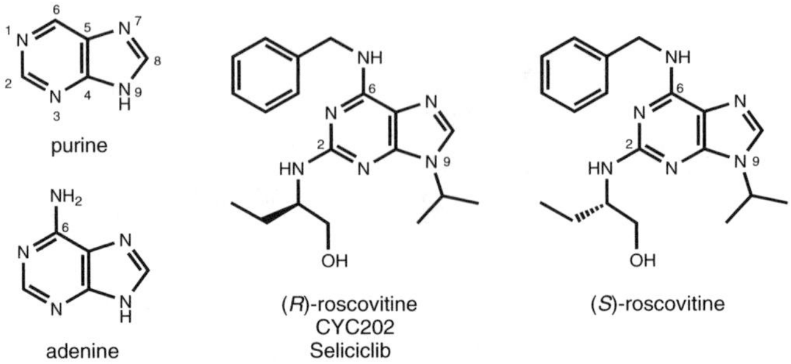

FIGURE 9.2 Structure and atom numbering of the purine ring. Structure of 6-aminopurine (adenine), and the two isomers of 2-(1-ethyl-2-hydroxyethylamino)-6-benzylamino-9-isopropylpurine (roscovitine).

9.2 DISCOVERY OF ROSCOVITINE AND OTHER 2,6,9-TRISUBSTITUTED PURINES

Roscovitine belongs to the family of purines (Figure 9.2), one of the most widely distributed heterocycles in nature, and has the basic ring structure of biologically important molecules such as ATP, cyclic AMP, NAD, FAD, acetyl-coenzyme A, caffeine, and theophylline, to name a few (reviews in Rosemeyer, 2004; Legraverend and Grierson, 2006). Among purines, adenine (6-aminopurine) and guanine constitute two of the four nucleotide building blocks of DNA. The history of the discovery of roscovitine has been reviewed in detail previously (Meijer and Raymond, 2003). Briefly, its development stems from initial studies performed by Lionel Rebhun, who identified 6-dimethylaminopurine (6-DMAP) (Figure 9.3) as an analog of puromycin that was able to prevent cell division of sea urchin embryos, although it had lost puromycin's ability to block protein synthesis (Rebhun et al., 1973). 6-DMAP was later found to inhibit the activity of the so-called "M-phase-specific histone H1 kinase" (Meijer and Pondaven, 1988; Néant and Guerrier, 1988), later to be identified as an equimolar complex between CDK1 (Arion et al., 1988) and cyclin B (Meijer et al., 1989). Following this finding, a small screening assay was established to search for other inhibitors of CDK1/cyclin B (Rialet and Meijer, 1991). Isopentenyladenine (Figure 9.3) was one of the first inhibitors to be identified. However, both 6-DMAP and isopentenyladenine were rather unselective and weakly active. A more extensive screen carried out with Jaroslav Vesely led to the discovery of olomoucine (Figure 9.3), a 2,6,9-trisubstituted purine that displayed promising selectivity toward some of the CDKs, among a panel of over 35 kinases (Vesely et al., 1994). In fact, olomoucine had been initially synthesized by David Letham as an antagonist of plant cytokinin 7-glucosyltransferase. A classical medicinal chemistry and structure–activity relationship study led to the synthesis and extensive characterization of roscovitine (Azevedo et al., 1997; Meijer et al., 1997). Combinatorial chemistry from this lead structure allowed the identification of purvalanols (Gray et al., 1998; Chang et al., 1999). Since then, the family of 2,6,9-trisubstituted purines has been the subject of numerous studies (reviews in Haesslein et al., 2002; Meijer and Raymond, 2003).

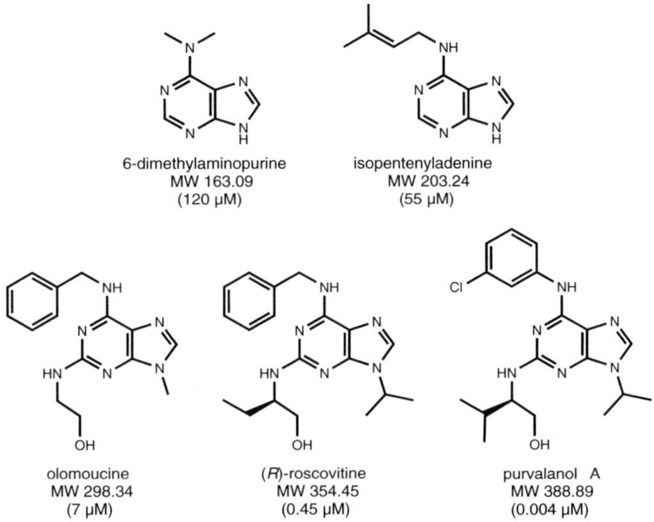

FIGURE 9.3 Structure of 6-dimethylaminopurine, isopentenyladenine, olomoucine, (*R*)-roscovitine, and purvalanol A, with molecular weight and, in parentheses, IC_{50} values for *in vitro* inhibition of CDK1/cyclin B.

9.3 SYNTHESIS OF (*R*)-ROSCOVITINE

(*R*)-Roscovitine is prepared by a simple and inexpensive three-step procedure, starting from commercially available 2,6-dichloropurine (Figure 9.4) (Havlicek et al., 1997; Wang et al., 2001b). The overall yield is 50%. The reactive 6-chloro is first substituted by benzylamine upon heating in butanol. Alkylation with 2-bromo or 2-iodopropane, using K_2CO_3 as a base, is then achieved at 20°C in DMSO. Finally, the less reactive 2-chloro is displaced upon heating with (*R*)-2-amino-butan-1-ol. This last step is improved when DMSO is used as a solvent. The first two steps of

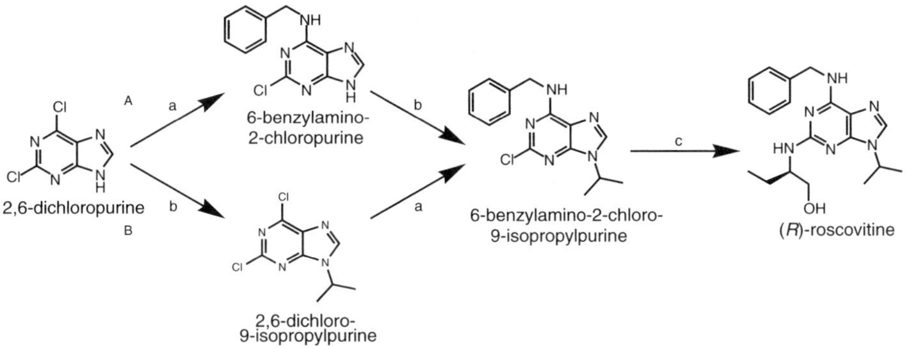

FIGURE 9.4 Chemical synthesis of (*R*)-roscovitine.

TABLE 9.1
Nomenclature and Physicochemical Properties of (*R*)-Roscovitine

Chemical names	2-(1-ethyl-2-hydroxyethylamino)-6-benzylamino-9-isopropylpurine
	1-butanol, 2-[[9-(1-methylethyl)-6-[(phenylmethyl)amino]-9*H*-purin-2yl]amino], (2*R*)-(*R*)-2-(6-benzylamino-9-isopropyl-9*H*-purin-2-ylamino)-butan-1-ol
Other names	(*R*)-roscovitine, CYC202, Seliciclib
CAS registry number	186692-46-6
NCI number	NSC-701554
Atomic composition	$C_{19}H_{26}N_6O$
	C, 64.38%; H, 7.39%; N, 23.71%; O, 4.51%
Molecular weight	354.45
Rotation values	(*R*)-roscovitine: $[\alpha]_D^{20}$ + 56.3 (Wang et al., 2001b)
	(*S*)-roscovitine: $[\alpha]_D^{20}$ − 56.3
Melting point	106–108°C (Wang et al., 2001b)
pKa	4.4 (Vita et al., 2005a)
Absorption	λ max: 230 nm and 292 nm
Chromatographic analysis	HPLC/UV detection (Vita et al., 2004)
	LC-MS/MS (Vita et al., 2005c; Raynaud et al., 2005)
Crystal structure	Orthorhombic, space group $P2_12_12_1$ (Wang et al., 2001b)
	Coordinates available at Cambridge Crystallographic Data Centre (deposit@ccdc.cam.ac.uk)
	CDC 157779, 157780

the synthesis can be switched (route B). However, alkylation of 2,6-dichloropurine leads to the formation of a mixture (82/18) of the 9/7 regioisomers which need to be separated by column chromatography.

9.4 CHEMICAL PROPERTIES OF (*R*)-ROSCOVITINE

(*R*)-Roscovitine is a white powder that is soluble in DMSO (up to 50 mM) and in 50 mM HCl with the pH adjusted to 2.5. Its nomenclature and physicochemical properties are summarized in Table 9.1.

9.5 SELECTIVITY AND BIOCHEMICAL PROPERTIES OF (*R*)-ROSCOVITINE

9.5.1 SELECTIVITY

(*R*)-Roscovitine has been optimized from the related purine olomoucine using an *in vitro* CDK1/cyclin B kinase assay (Meijer et al., 1997). During this initial work, it was realized that (*R*)-roscovitine displayed rather good selectivity toward CDK1, CDK2, and CDK5 compared to other kinases among a panel of 24 kinases (Meijer et al., 1997). Since then, the selectivity has been extensively investigated by various methods.

First, (*R*)-roscovitine has been run on other kinase selectivity panels such as Sir Philip Cohen's laboratory kinase selectivity panel (28 kinases) (Bain et al., 2003), ProQinase's selectivity panel (85 kinases), Invitrogen's SelectScreen™ Kinase Profiling panel (70 kinases), and Cerep's kinase selectivity panel (50 kinases) (see Bach et al., 2005 (supplementary material) for a compilation of all available data). A total of 151 protein kinases have been tested for their sensitivity to roscovitine. IC_{50} values are below 1 µM for CDK1, CDK2, CDK5, CDK7, and CDK9 only, whereas CDK4, CDK6, and CDK8 are poorly, if at all, sensitive to roscovitine (Table 9.2). Only a few kinases are sensitive to roscovitine in the 1 to 40 µM range (CaM Kinase 2, CK1α, CK1δ, DYRK1A, EPHB2, ERK1, ERK2, FAK, and IRAK4), but most other kinases are insensitive to roscovitine. Based on these data, roscovitine appears to be

---

TABLE 9.2
Selectivity of (*R*)-Roscovitine Toward CDKs

Protein Kinase	IC_{50} (µM)
CDK1/cyclin B	0.65[a], 2.69[b], 23[c], 0.45/0.95 (*R/S*)[d], >80%[e] and 98%[f] inhibition at 10 µM, 14.1[g], 1.9[l], 0.67[m]
CDK2/cyclin A	0.7[a], 0.25[h], 0.71[b], 1.2/1.8[c], >80% inhibition at 10 µM [e], 2.2[g], 2.1[m]
CDK2/cyclin E	0.7[a], 0.95/1.4[c], 0.10/0.24 (*R/S*)[b], 98% inhibition at 10 µM [f], 0.13[g], 0.05[l], 0.19[m]
CDK3/cyclin E	1.4/1.5[c]
CDK4/cyclin D1	>100[a], 14.2[b], 75[c], 14.7[g], 14.6[l], 10[m]
CDK5/p25	0.16[a], >80% inhibition at 10 µM [e]
CDK6/cyclin D1	51[c]
CDK6/cyclin D3	>100[a], 50[g]
CDK7/cyclin H	0.5–0.6[i,j], 0.49[b], <5[k], 0.46[g], 0.51[m]
CDK8/cyclin C	>100[i], >50[k]
CDK9/cyclin T1	0.6[j], <5[k], 0.78[g]

Note: The purified protein kinases were assayed in the presence of an increasing concentration of (*R*)-roscovitine. IC_{50} values are presented in µM.

[a] From Meijer L. et al. *Eur. J. Biochem.*, 243, 527, 1997.

[b] From McClue, S.J. et al. *Int. J. Cancer.*, 102, 463, 2002.

[c] Courtesy of ProQinase.

[d] From Azevedo, W.F. et al. *Eur. J. Biochem.*, 243, 518, 1997.

[e] Courtesy of Invitrogen SelectScreen™ Kinase Profiling Service.

[f] Courtesy of Cerep Kinase Selectivity profiling service.

[g] From Raynaud, F.I. et al. *Clin. Cancer Res.*, 11, 4875, 2005.

[h] From Bain, J. et al. *Biochem. J.*, 371, 199, 2003.

[i] From Schang, L.M. et al. *J. Virol.*, 76, 7874, 2002.

[j] From Wang, D. et al. *J. Virol.*, 75, 7266, 2001a.

[k] From Pinhero, R. et al. *Biol. Procedures Online*, 6, 163, 2004.

[l] From Nutley, B.P. et al. *Mol. Cancer Ther.*, 4, 125, 2005.

[m] From Byth, K.F. et al. *Mol. Cancer Ther.*, 5, 655, 2006.

a reasonably selective kinase inhibitor. However, this panel only reflects 29.2% of the reported 518+ kinases of the human kinome.

The second method used to address the selectivity of (R)-roscovitine is based on the identification by mass spectrometry of the roscovitine-binding proteins that can be purified by affinity chromatography on sepharose-immobilized roscovitine from various tissue and cell extracts (Bach et al., 2005). This method has been successfully applied to purvalanol (Knockaert et al., 2000) and other kinase inhibitors (review in Knockaert and Meijer, 2002; Valsasina et al., 2004). Roscovitine beads allowed the identification of expected targets such as CDKs, but also various CaM Kinase 2 isoforms, ERK1, ERK2, and CK1α. Surprisingly, pyridoxal kinase (PDXK), the enzyme responsible for the phosphorylation and activation of vitamin B_6, a cofactor of many enzymes, was identified as a roscovitine-binding protein in all biological materials tested. This interaction was investigated in detail and further confirmed by the cocrystallization of (R)-roscovitine with sheep brain PDXK (Bach et al., 2005; Tang et al., 2005).

The third method that has been used is a yeast three-hybrid screen (Becker et al., 2005) based on the reconstitution of an active transcription factor from the close association of a DNA-binding domain (DBD) and the activation domain (AD) of a transcriptional activator (*GAL4*) expressed separately. The DBD is fused to dihydro-folate reductase (DHFR), and the AD is fused to a library of potential kinase targets. The kinase inhibitor is attached to methotrexate through a polyethylene-glycol linker. The binding of methotrexate to DHFR, on one hand, and the inhibitor to its target, on the other hand, reconstitutes a functional DBD/AD transcription factor, allowing the detection of the inhibitor's targets and their identification (Becker et al., 2004). This elegant method was used with purvalanol and roscovitine as proofs of principle. The results showed that (R)-roscovitine interacts with its known target CDK2, but also with CK1δ, CK1ϵ, and the CDK-like kinase PCTK1, and more weakly with CLK1, PAK4, PCTK3, PKWA, and GSK3α (Becker et al., 2004).

A fourth approach that has been used to investigate the selectivity of (R)-roscovitine is a quantitative competition assay carried out in the absence of ATP or protein substrate in contrast to the classical kinase inhibition assays. It is based on the interaction of a given inhibitor immobilized to biotin with a library of protein kinases expressed as T7 bacteriophage capsid protein fusion proteins (Fabian et al., 2005). This method has been applied to 20 known, ATP-competitive, clinical kinase inhibitors and to 113 kinases. It confirmed the rather good selectivity of (R)-roscovitine, which was found to bind to CDK2, CDK5, PCTK1, CK1γ1, CK1γ2, CK1ϵ, CLK1, CLK2, CLK4, TTK, and RPS6KA2 (Kinase Domain 1).

These and other methods used to identify the targets of inhibitors of CDKs (Bach et al., 2006) and other kinases (Daub et al., 2004) have been extensively reviewed.

9.5.2 BIOCHEMICAL PROPERTIES: (*R*)-ROSCOVITINE/TARGET CO-CRYSTAL STRUCTURES

Classical enzymology has shown that (R)-roscovitine acts by competing with ATP for binding at the ATP-binding site of CDK1/cyclin B (Azevedo et al., 1997). This binding at the catalytic site was confirmed by direct cocrystallization of (R)-roscovitine with CDK2 (coordinates available from Dr. S.H. Kim, <shkim@lbl.gov>)

(Azevedo et al., 1997; Dobes et al., 2006; Otyepka et al., 2002), and later with CDK5/p25 (1UNL) (Mapelli et al., 2005) and CDK2/cyclin A (Echallier, Endicott, and Meijer, unpublished). A CDK1/roscovitine model has also been described (Canduri et al., 2004). These crystal structures reveal the interaction between (*R*)-roscovitine and the amino acids that line up the ATP-binding pocket of the CDK catalytic subunit (Figure 9.5). Briefly, the interaction involves mostly hydrophobic and van der Waals contacts and two hydrogen bonds (involving N^7 and N^6 of the purine) with backbone atoms of Leu83 (CDK2). In addition, a weak hydrogen bond is formed between O^1 and a water molecule. A similar binding mode is observed with CDK5 (involving Cys83) (Mapelli et al., 2005). The binding mode suggested that N^6-methyl-(*R*)-roscovitine or O^6-benzyl-(*R*)-roscovitine (Figure 9.6) would be unable to interact at the ATP site, and would therefore constitute useful kinase-inactive controls. This was confirmed experimentally by kinase assays and also by affinity chromatography on immobilized N^6-methyl-(*R*)-roscovitine (Tang et al., 2005).

The cocrystal structures also reveal that the benzyl ring is facing the outside of the ATP-binding pocket. This property has been used to select the place where a linker can be tethered to roscovitine to immobilize it on sepharose beads while still maintaining the potential interaction with its protein kinase targets. A control

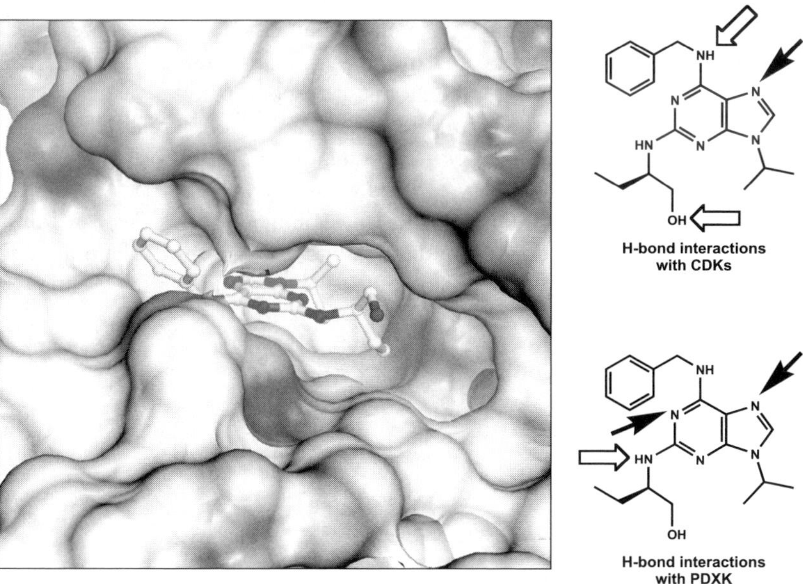

FIGURE 9.5 Interactions of (*R*)-roscovitine with its targets. Left, crystal structure of CDK2 in complex with (*R*)-roscovitine, illustrating the position of (*R*)-roscovitine in the ATP-binding pocket and how its benzyl ring is facing the outside of the kinase. Right, roscovitine and its atoms involved in H-bonds with either CDKs or PDXK (hydrogen acceptors are shown by solid arrows and hydrogen donors by empty arrows).

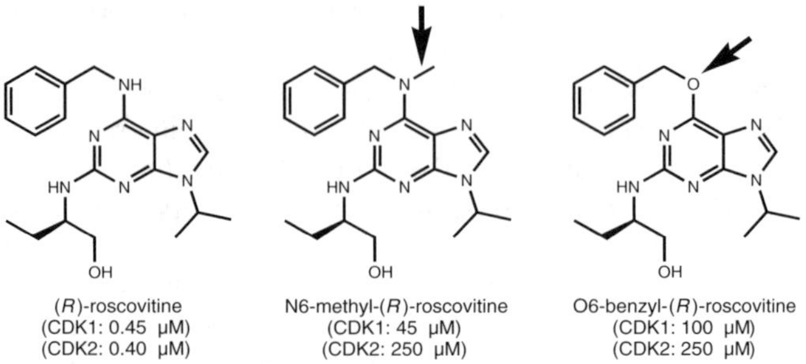

FIGURE 9.6 (*R*)-roscovitine and its control, protein kinase inactive analogs, *R*-2-[6-(benzyl-methyl-amino)-9-isopropyl-9H-purin-2-ylamino]-butan-1-ol) (*N*[6]-methyl-(*R*)-roscovitine), and *R*-2-(6-benzyloxy-9-isopropyl-9H-purin-2-ylamino)-butan-1-ol (O6- benzyl-(*R*)-roscovitine). Arrows point to the chemical changes introduced in the roscovitine analogs. The IC_{50} values for *in vitro* inhibition of CDK1/cyclin B and CDK2/cyclin E are shown in parentheses.

matrix is obtained when N^6-methyl-(*R*)-roscovitine is immobilized to sepharose beads (Figure 9.7).

Affinity chromatography with sepharose-immobilized roscovitine revealed that roscovitine interacts with PDXK from all species and tissues that have been tested (Bach et al., 2005). (*R*)-Roscovitine has been cocrystallized with sheep brain PDXK (coordinates: 1YGK) (Tang et al., 2005) and the interaction investigated in detail. Surprisingly, (*R*)-roscovitine is located in the pyridoxal-binding site, rather than at the ATP-binding site. Furthermore, the atoms of roscovitine involved in the binding to PDXK are not the same as those implicated in the binding to CDKs (Figure 9.5). This has allowed us to synthesize two compounds (Figure 9.6) that interact with PDXK — they were actually cocrystallized with PDXK (coordinates: 1YGJ and 1YHJ) (Tang et al., 2005) — but that do not bind to CDKs. We are currently synthesizing roscovitine analogs that should bind CDKs without interacting with PDXK. The two sets of molecules should allow us to distinguish the cellular effects

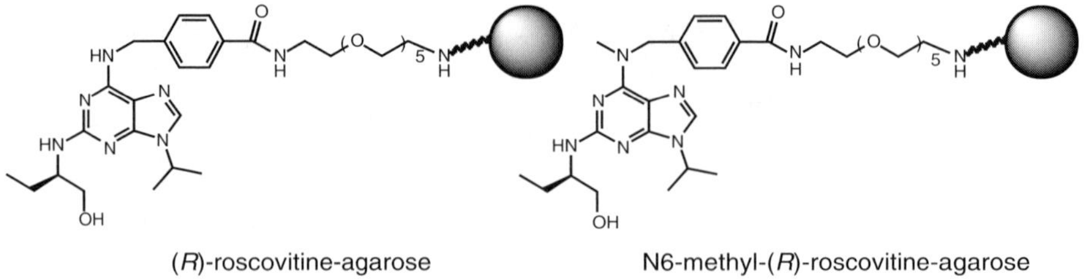

FIGURE 9.7 (*R*)-Roscovitine affinity chromatography matrices. (*R*)-Roscovitine and its control, protein kinase inactive analog, N^6-methyl-(*R*)-roscovitine, are covalently immobilized on sepharose beads through a polyethylene-glycol linker attached to the benzyl ring of the ligand. These matrices can be used for affinity purification of the biological targets of (*R*)-roscovitine.

of roscovitine due to interaction with protein kinases from those due to interaction with PDXK.

These examples illustrate the extraordinary diversity of information that can be drawn from the structure of kinase/inhibitor complexes (Noble et al., 2004).

9.6 CELLULAR EFFECTS OF (*R*)-ROSCOVITINE

(*R*)-Roscovitine has been evaluated for its effects on a wide variety of cultured cells. Table 9.3 summarizes most of the available data published using mammalian (mostly human) cells. Broadly speaking, two major events have been described: an arrest in cell cycle progression and an induction of cell death.

9.6.1 ANTIMITOTIC EFFECTS

(*R*)-Roscovitine arrests the cell division cycle in most, if not all, cell lines. Many methods have been used to monitor the effects of the drug on cell proliferation from direct enumeration to FACS analysis, estimation of the number of viable cells, and direct assay of DNA synthesis. The average IC_{50} value for inhibition of proliferation from all available cell line data is about 17 μM (Table 9.3). This comprises the NCI 60 cell line panel (Meijer et al., 1997) (average IC_{50} = 16 μM), the McClue et al. (2002) panel (19 cell lines; average IC_{50} = 15.2 μM), and the Raynaud et al. (2005) panel (24 cell lines; average IC_{50} = 14.6 μM).

(*R*)-Roscovitine blocks cell proliferation in G_0, G_1, S, G_2/M, or a combination of these, depending on the cell line studied, the duration of treatment, and the dose of (*R*)-roscovitine. These cell cycle stage arrests can presumably be attributed to a direct inhibition of CDKs (Figure 9.8A): CDK3/cyclin C (inhibition of exit from G0), CDK2/cyclin E (inhibition of G_1/S transition), CDK2/cyclin A (inhibition of S phase progression), CDK1/cyclin A (inhibition of G_2 phase), CDK1/cyclin B (inhibition of the prophase-to-metaphase transition in M phase), and, under some conditions, CDK1/cyclin E (inhibition of G_1/S transition), as recently suggested by Aleem et al. (2005).

In addition there are two indirect mechanisms by which (*R*)-roscovitine can block the cell division cycle (Figure 9.8B). First, by inhibiting the CDK7/cyclin H/MAT1 complex, (*R*)-roscovitine prevents phosphorylation of the T-loop threonine of various CDKs, a step that is essential for activity of the kinases. Consequently, CDK1, CDK2, and CDK4 catalytic activities are expected to decrease. Second, the CDK2/CDK4 natural peptidic inhibitor p27[KIP1] is phosphorylated by CDK2/cyclin E (Vlach et al., 1997; Bloom and Pagano, 2003). This phosphorylation targets p27[KIP1] for rapid, ubiquitin-dependent degradation by the proteasome. The presence of (*R*)-roscovitine prevents p27[KIP1] phosphorylation, leads to its stabilization, accumulation (Zhang et al., 2004b), and enhanced inhibition of its CDK targets and, consequently, to an arrest in G_1 (Figure 9.8B). A p27[KIP1] luciferase fusion protein has been constructed that provides a convenient and elegant *in vivo* and *in vitro* reporter system for the induction of p27[KIP1] by CDK inhibitors, including (*R*)-roscovitine (Zhang et al., 2004b).

TABLE 9.3
Cellular Effects of (*R*)-Roscovitine in Mammalian Cells

Cell Line (Cancer Type)	IC$_{50}$ (μM)	Cell Cycle Effect	Cell Proliferation and Death Assay	Reference
501mel (melanoma)	29		15	Du et al., 2004
2008 (ovary) (+ FTI)			9	Edamatsu et al., 2000
A172 (glioma) (+ TRAIL)			3	Kim et al., 2004
A375 (melanoma)	17.5		15	Du et al., 2004
A375 (melanoma)	19.9	θ DNA synthesis		Payton et al., 2006
A431 (epidermoid)	10	θ DNA synthesis	3	Atanasova et al., 2005
A4573 (Ewing's sarcoma)			16	Tirado et al., 2005
A498 (renal)	13.2		3	McClue et al., 2002
A549 (lung)	15.9		3	McClue et al., 2002
A549 (lung)	9.3		2	Raynaud et al., 2005
A2780 (ovarian)	4.9		2	Raynaud et al., 2005
A2780Cis^R (ovarian)	8.4		2	Raynaud et al., 2005
ACHN (renal)	11.9		3	McClue et al., 2002
AN3CA (endometrial)	14.1		3	McClue et al., 2002
BE (colon)	17.5		2	Raynaud et al., 2005
Bon-1 (carcinoid cell line)	60		3	Goke et al., 2004
CH1 (ovarian)	7.7		2	Raynaud et al., 2005
CH1Cis^R (ovarian)	9.3		2	Raynaud et al., 2005
CH1Dox^R (ovarian)	7.4		2	Raynaud et al., 2005
CHAGO-K1 (lung)	30.2		3	McClue et al., 2002
CHP212 (neuroblastoma)			11	van Engeland et al., 1997
Chronic lymphocytic leukemia cells				Hahntow et al., 2004
Chronic lymphocytic leukemia cells				Alvi et al., 2005
COLO-205 (colon)	8.5		2	Raynaud et al., 2005
COLO-205 (colon)	18.6	θ DNA synthesis		Payton et al., 2006
COLO-320 (colon)	15.9	θ DNA synthesis		Payton et al., 2006
Colon 26 (murine colon)	42.5		2	Raynaud et al., 2005
CORL23 (lung)	10.5		2	Raynaud et al., 2005
Daudi (Burkitt's lymphoma)	19.2	θ DNA synthesis		Payton et al., 2006
Dox40 (multiple myeloma)	15		14	Raje et al., 2005
DU145 (prostate)	18.8		3	McClue et al., 2002
Fibroblasts (human fetal lung)		G$_1$, θ S, θ Rb-P		Alessi et al., 1998
GCT27 (testicular)	5.2		2	Raynaud et al., 2005
Granta-519 (mantle cell lymphoma)	25	G$_2$/M	3, 4	Lacrima et al., 2005
H929 (multiple myeloma)	8.84		12	MacCallum et al., 2005
HS27 (foreskin fibroblast)	22.2		2	McClue et al., 2002

TABLE 9.3 (CONTINUED)
Cellular Effects of (R)-Roscovitine in Mammalian Cells

Cell Line (Cancer Type)	IC$_{50}$ (μM)	Cell Cycle Effect	Cell Proliferation and Death Assay	Reference
HCT15 (colon)	18.3	NCC	3, 4	McClue et al., 2002
HCT116 (colon)	10.7		3	McClue et al., 2002
HCT116 (colon)		θ Rb S608-P		Barrie et al., 2003
HCT116 (colon)	6.9	G$_2$/M	2	Raynaud et al., 2005
HCT116 p53+ (colon)	13.7	θ DNA synthesis		Payton et al., 2006
HCT116 p53- (colon)	23.7	θ DNA synthesis		Payton et al., 2006
HEK293 (embryonic kidney)	48.9		13	Bettayeb and Meijer, unpubl.
HepG2 (hepatic)	11.3		3	McClue et al., 2002
HL-60 (leukemia)	20.5	θ DNA synthesis		Payton et al., 2006
HL-60 (leukemia) (+ FTI)			9, 10	Edamatsu et al., 2000
HNSCC cell lines	9.8 – 25.0	G$_2$/M	6	Mihara et al., 2002
HS294 (melanoma)	16.6	θ DNA synthesis		Payton et al., 2006
HT29 (colon)	14.6		3	McClue et al., 2002
HT29 (colon)		θ Rb S608-P		Barrie et al., 2003
HT29 (colon)	20.3		2	Raynaud et al., 2005
HT29 (colon)	16		2	Whittaker et al., 2004
HT29 (colon) (+ FTI)			9	Edamatsu et al., 2000
HT1376 (bladder)	18.4		3	McClue et al., 2002
HX147 (lung)	19		2	Raynaud et al., 2005
IMR-90 (fetal lung fibroblast)	>100		2	McClue et al., 2002
INR1-G9 (glucagonoma)	80		3	Goke et al., 2004
INS-1 (insulinoma)	90		3	Goke et al., 2004
JeKo-1 (mantle cell lymphoma)	25	G$_2$/M	3, 4	Lacrima et al., 2005
Jurkat (acute T-cell leukemia)	27.8	θ DNA synthesis		Payton et al., 2006
K562 (chronic myelogenous leukemia)		G$_2$/M, θ S		Penuelas et al., 1998
K562 (chronic myelogenous leukemia)	35	θ DNA synthesis		Payton et al., 2006
K562 (leukemia) (+ FTI)			9	Edamatsu et al., 2000
KM12 (colon)	17		2	Whittaker et al., 2004
KM12 (colon)	15		2	Raynaud et al., 2005
Keratinocytes (confluent)	35		3	Atanasova et al., 2005
Keratinocytes (subconfluent)	17		3	Atanasova et al., 2005
L1210 (mouse leukemia)	40	G$_2$/M	1	Meijer et al., 1997
L1210 (mouse leukemia)	47	G$_2$/M		Somerville and Cory, 2000

(continued)

TABLE 9.3 (CONTINUED)
Cellular Effects of (*R*)-Roscovitine in Mammalian Cells

Cell Line (Cancer Type)	IC$_{50}$ (μM)	Cell Cycle Effect	Cell Proliferation and Death Assay	Reference
L1210 (Y8) (mouse leukemia)	52	G$_2$/M	8	Somerville and Cory, 2000
LNcaP (pancreas)			17	Mohapatra et al., 2005
LNcaP (pancreas) (+ FTI)			9	Edamatsu et al., 2000
LP-1 (multiple myeloma)	12.68		12	McCallum et al., 2005
LoVo (colon)	9.3		3	McClue et al., 2002
LoVo (colon)	20		2	Raynaud et al., 2005
LR5 (multiple myeloma)	25		14	Raje et al., 2005
M14 (melanoma)	17		15	Du et al., 2004
MALME-3M (melanoma)	30		15	Du et al., 2004
Mawi (colon)	18		2	Raynaud et al., 2005
MCF-7 (breast)	10.9		3	McClue et al., 2002
MCF-7 (breast)			5	Mgbonyebi et al., 1998
MCF-7 (breast)	14.7	G$_2$/M, θ DNA synthesis	7	Wojciechowski et al., 2003
MCF-7 (breast)	14	G$_2$/M	7	Wesierska-Gadek et al., 2003
MCF-7 (breast)	7.8		2	Raynaud et al., 2005
MCF-10F (breast)			5	Mgbonyebi et al., 1998
MDA-MB-231 (breast)		θ DNA synthesis	5	Mgbonyebi et al., 1998, 1999
MDA-MB-231 (breast)		G$_2$/M (+ irradiation)		Maggiorella et al., 2003
MDA-MB-231 (breast)	20.8	θ DNA synthesis		Payton et al., 2006
MDA-MB-231 (breast)	15		2	Raynaud et al., 2005
MDA-MB-435S (breast)	17.5		3	McClue et al., 2002
MES-SA (uterine)	11.9		3	McClue et al., 2002
MES-SA/Dx5 (MDR+) (uterine)	7.9		3	McClue et al., 2002
MES-SA (uterine)	25	θ DNA synthesis		Payton et al., 2006
MES-SA/Dx5 (MDR+) (uterine)	12.5	θ DNA synthesis		Payton et al., 2006
MiaPaCa2 (pancreatic)	13.9		3	McClue et al., 2002
MiaPaCa2 (pancreatic)	12.6	θ DNA synthesis		Payton et al., 2006
MKN45 (stomach) (+ FTI)			9	Edamatsu et al., 2000
MM.1r (multiple myeloma)	25		14	Raje et al., 2005
MM.1s (multiple myeloma)	23		14	Raje et al., 2005
MOR (lung)	12.5		2	Raynaud et al., 2005
MR65 (NSC-lung)			11	van Engeland et al., 1997

TABLE 9.3 (CONTINUED)
Cellular Effects of (*R*)-Roscovitine in Mammalian Cells

Cell Line (Cancer Type)	IC$_{50}$ (μM)	Cell Cycle Effect	Cell Proliferation and Death Assay	Reference
MT-2 (leukemia)			8	Mohapatra et al., 2003
NCI 60 cell line panel	Mean: 16		2	Meijer et al., 1997
NCI-H69 (lung)	26.0		3	McClue et al., 2002
NCI-H460 (lung)	13.1		3	McClue et al., 2002
NCEB-1 (mantle cell lymphoma)	50	G$_2$/M	3, 4	Lacrima et al., 2005
OPM2 (multiple myeloma)	18.46		12	MacCallum et al., 2005
OPM2 (multiple myeloma)	15		14	Raje et al., 2005
PC-3 (prostate) (+ FTI)			9	Edamatsu et al., 2000
PC-3 (prostate)	12	θ DNA synthesis		Payton et al., 2006
REC (mantle cell lymphoma)	25	G$_2$/M	3, 4	Lacrima et al., 2005
RPMI (multiple myeloma)	23		14	Raje et al., 2005
RPMI 8226 (multiple myeloma)	19.5		12	MacCallum et al., 2005
Sa-SO2 (osteosarcoma)	16.5		2	Raynaud et al., 2005
Sa-SO2 (osteosarcoma)	17.1	θ DNA synthesis		Payton et al., 2006
SH-SY5Y (neuroblastoma)	25		13	Ribas and Boix, 2004
SH-SY5Y (neuroblastoma)	16.1		13	Bettayeb and Meijer, unpubl.
SKMEL2 (melanoma)	21		15	Du et al., 2004
SKMEL5 (melanoma)	23		15	Du et al., 2004
SKMEL28 (melanoma)	25		15	Du et al., 2004
SKOV-3 (ovarian)	31		2	Raynaud et al., 2005
SW480 (colon carcinoma)	30		15	Du et al., 2004
SW620 (colon)	23		2	Raynaud et al., 2005
TC-71 (Ewing's sarcoma)			16	Tirado et al., 2005
T98 (glioma) (+ TRAIL)			3	Kim et al., 2004
U2-OS (osteosarcoma)	15		2	Raynaud et al., 2005
U2-OS (osteosarcoma)		G$_2$/M	15	Maude and Enders, 2005
U-2 OS (osteosarcoma)	28.3	θ DNA synthesis		Payton et al., 2006
WI38 (fetal lung)	24.0		2	McClue et al., 2002
U87MG (glioma) (+ TRAIL)			3	Kim et al., 2004
U266 (multiple myeloma)	17.93		12	MacCallum et al., 2005
U266 (multiple myeloma)	25		14	Raje et al., 2005
U251 (glioma) (+ TRAIL)			3	Kim et al., 2004

(*continued*)

TABLE 9.3 (CONTINUED)
Cellular Effects of (*R*)-Roscovitine in Mammalian Cells

Cell Line (Cancer Type)	IC_{50} (μM)	Cell Cycle Effect	Cell Proliferation and Death Assay	Reference
U937 (leukemia)			11	Rosato et al., 2006
UACC62 (melanoma)	12		15	Du et al., 2004
UACC257 (melanoma)	25		15	Du et al., 2004

Note: The IC_{50} values for inhibition of proliferation are provided in μM. Cell cycle effects are monitored by FACS analysis, [³H]-thymidine uptake, or BrdU incorporation. Cell proliferation and cell death are monitored by various methods.

Abbreviations: LDH = lactate dehydrogenase; MTS = 3-(4,5-dimethylthiazol-2-yl)-5-(3-carboxy methoxyphenyl)-2-(4-sulfophenyl)-2*H*-tetrazolium salt; MTT = 3-(4,5-dimethylthiazol-2-yl)-2,5-diphenyl tetrazolium bromide; NCC = no apparent increase in a specific cell cycle phase; θ Rb S608-P = inhibition of retinoblastoma protein phosphorylation on Serine 608; θ DNA synthesis = inhibition of BrdUrd or [³H]-thymidine incorporation; θ S = reduction in S phase cells.

Cell proliferation and cell death assays: 1 = microculture tetrazolium assay; 2 = sulforhodamine B assay; 3 = MTT assay; 4 = sub-G_1 + TUNEL assay; 5 = Cell proliferation reagent WST-1 (Boehringer); 6 = DNA fragmentation, PARP cleavage; 7 = CellTiter-Glo Luminescent Viability assay (Promega); Trypan blue dye exclusion; TUNEL assay; PARP cleavage; 8 = annexin V, caspase-3, PARP cleavage; 9 = caspase-3 activation; 10 = sub-G_1, cytochrome C release; 11 = annexin-V/PI; 12 = Alamar blue, TUNEL, PARP cleavage; 13 = MTS assay; DNA fragmentation, caspase activation, TUNEL assay, LDH release; 14 = MTT assay, [³H]-thymidine incorporation, sub-G_1, caspase-3, PARP cleavage; 15 = WST-1 cell proliferation assay (Roche); 16 = Trypan blue dye exclusion, TUNEL, caspase-3 activation; 17 = DNA fragmentation, PARP and keratin 18 cleavage.

A decrease in the phosphorylation of substrates at sites that are specifically phosphorylated by CDKs is a clear demonstration of the direct or indirect inhibition of the CDK catalytic activity by (*R*)-roscovitine (Table 9.4). Furthermore, these molecular events constitute tools that can be used as surrogate markers to monitor the efficacy of (*R*)-roscovitine in animal models and during human clinical trials.

Besides these direct targets, there are many biochemical and morphological events that rely on CDK activity, and are, therefore, sensitive to (*R*)-roscovitine. Among those, DNA replication initiation depends on CDK2. *In vitro* initiation of DNA replication is inhibited by (*R*)-roscovitine (Krude, 2000) as monitored in cell cultures by inhibition of [³H] thymidine uptake and incorporation in DNA. Nucleolus formation and functions rely on active CDKs and are inhibited by (*R*)-roscovitine (Sirri et al., 2002). Centrosome duplication depends on CDK2 and is also prevented by (*R*)-roscovitine (Matsumoto et al., 1999). Golgi fragmentation is inhibited by (*R*)-roscovitine through dephosphorylation of GM130 on Ser25 (Lowe et al., 1998). Nuclear envelope breakdown at the end of prophase is a consequence of lamins phosphorylation by CDK1/cyclin B, and is consequently inhibited by (*R*)-roscovitine treatment. (*R*)-Roscovitine treatment induces Erk1/2 activation in HT29 and KM12 colon carcinoma cell lines (Whittaker et al., 2004).

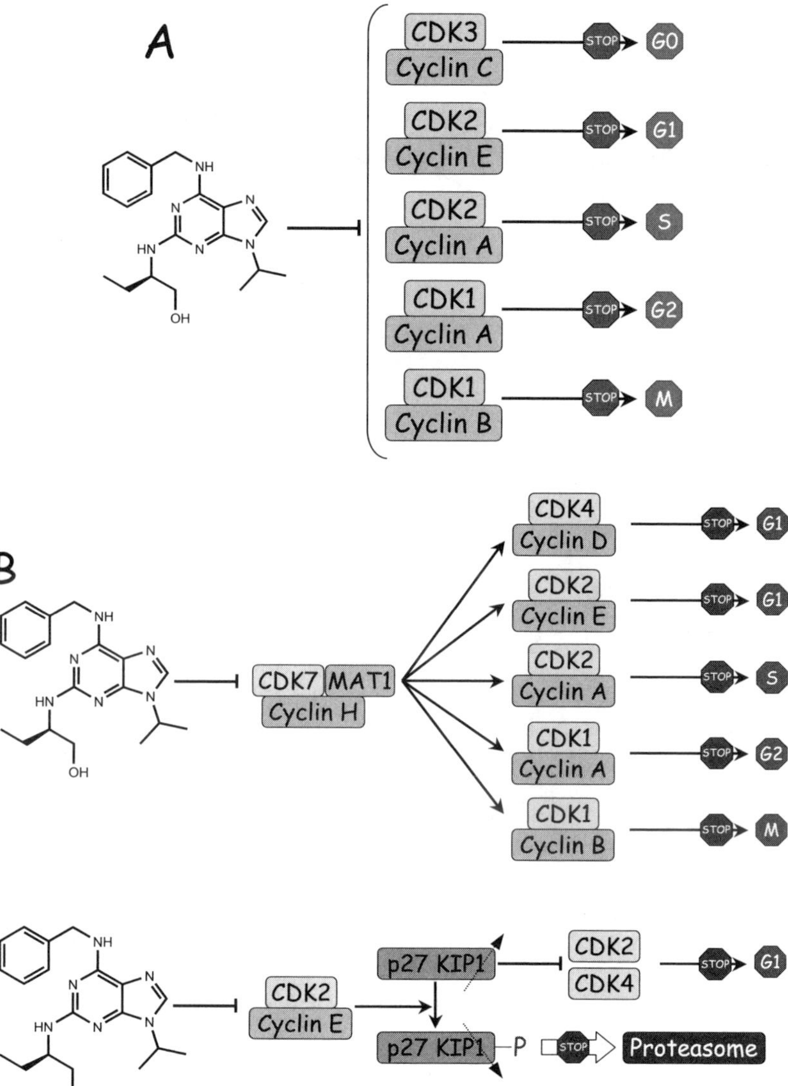

FIGURE 9.8 Multiple mechanisms of action of (*R*)-roscovitine and their cellular consequences. (A) Direct stoichiometric interaction with CDKs leads to inhibition of the catalytic activity of various CDK/cyclin complexes with a direct effect on various cell cycle phases (indicated by a "STOP" sign). (B) Indirect inhibition of cell cycle progression: (i) interaction with CDK7/cyclin H/MAT1 prevents the phosphorylation of a key activating threonine residue located on the T-loop of the substrate CDKs. Consequently, the activity of various CDK/cyclin complexes are reduced; (ii) inhibition of CDK2/cyclin E prevents phosphorylation and subsequent proteolytic degradation of p27^{KIP1}, a natural CDK2/CDK4 inhibitor. p27^{KIP1} Accumulation thus contributes to an arrest in G_1.

C

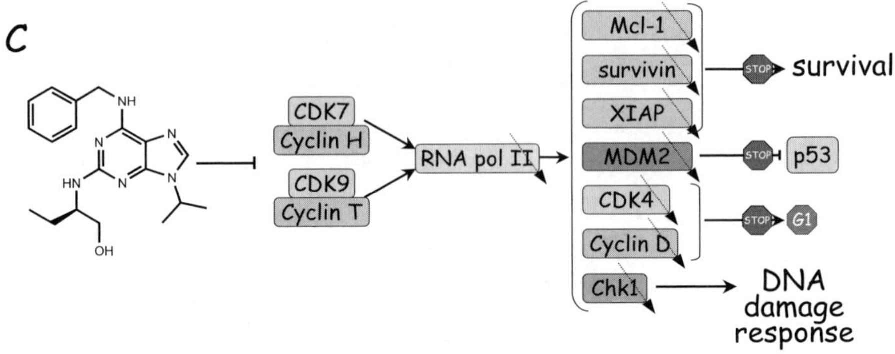

D

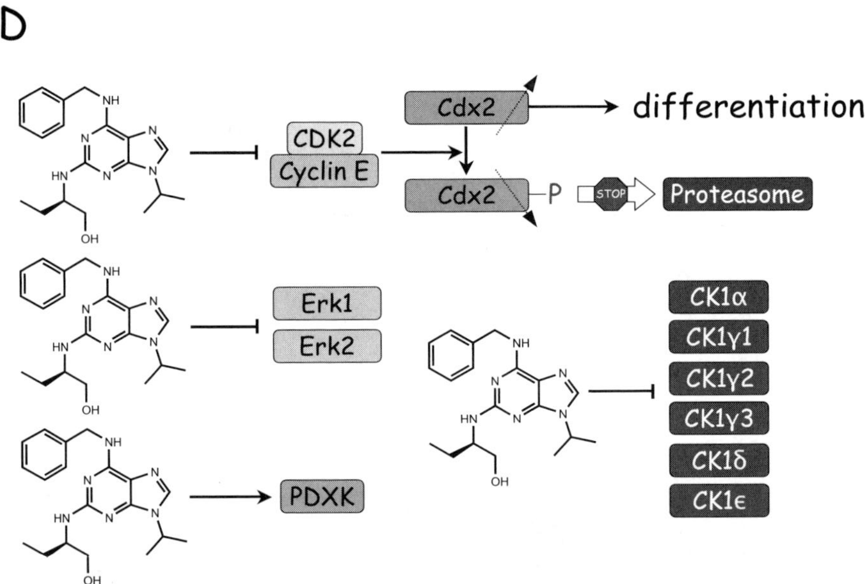

FIGURE 9.8 (CONTINUED) (C) Direct interaction with CDK7/cyclin H and CDK9/cyclin T leads to inhibition of RNA polymerase II (by lack of serine 2 and 5 phosphorylation). Consequently, transcription is reduced and short-lived proteins are rapidly downregulated. In particular, the reduction of survival factors such as Mcl-1, survivin, and XIAP contributes to cell death, the reduction of MDM2 level upregulates p53 level, reduced CDK4 and cyclin D contribute to a G_1 arrest, and downregulation of Chk1 leads to activation of a DNA damage response. Furthermore, CDK7/CDK9 inhibition by (*R*)-roscovitine leads to a downregulation of RNA polymerase II expression. (D) Other possible mechanisms of action. Inhibition of CDK2/cyclin E prevents phosphorylation and subsequent proteolytic degradation of Cdx2, a transcription factor involved in intestinal cell differentiation. (*R*)-Roscovitine also inhibits the MAP-kinases Erk1 and Erk2, and members of the casein kinase 1 (CK1) family, contributing to its antiproliferative effects. Finally, (*R*)-roscovitine binds to pyridoxal kinase (PDXK).

TABLE 9.4
Phosphorylation Sites That Are Sensitive to (*R*)-Roscovitine in Mammalian Cells

Phosphorylated Substrate	CDK-Specific Site	P-Specific Antibody	Reference
Vimentin	Ser55 (CDK1)	4A4 antibody	Meijer et al., 1997
Nucleolin	—	TG3 antibody	Knockaert et al., 2002b
GM130 Golgi protein	Ser25 (CDK1)	—	Lowe et al., 1998
Survivin	Thr34 (CDK1)	—	O'Connor et al., 2000, 2002; Wall et al., 2003
Retinoblastoma protein	Ser608	51B7 antibody	Barrie et al., 2003 Raynaud et al., 2005
Retinoblastoma protein	Ser249/Thr252	Antibody 44-584 (Biosource)	MacCallum et al., 2005
p27^{KIP1}	Thr187 (CDK2)	—	Vlach et al., 1997 Bloom and Pagano, 2003 Gherardi et al., 2004 Zhang et al., 2004b
PP-1 phosphatase	Thr320 (CDK1)	P-Thr320 antibody (Cell Signaling Technology)	Kwon et al., 1997 Payton et al., 2006
Inhibitor-2 (at centrosomes)	Thr72 (CDK1)	P-Thr72 antibody	Leach et al., 2000
Peroxiredoxin I	Thr90 (CDK1)	P-Thr90 PrxI antibody	Chang et al., 2002

9.6.2 CELL DEATH EFFECTS

(*R*)-Roscovitine induces cell death in many cell lines (Table 9.3). Roscovitine-induced cell death can occur at all phases of the cell cycle (McClue et al., 2002). It displays all the characteristics of apoptosis: chromatin condensation, nuclear DNA fragmentation, accumulation in the sub-G_1 compartment as detected by FACS analysis, DNA laddering, release of cytochrome C, activation of caspases, PARP cleavage, positive TUNEL staining, and LDH release (see, for example, Ribas and Boix, 2004).

Functional p53 does not appear to be necessary for cell death induction, but roscovitine has a slightly greater potency in p53 wild-type cells compared to cells bearing mutant p53 (Meijer et al., 1997; McClue et al., 2002; Raynaud et al., 2005). Little (Payton et al., 2006) or no (Raynaud et al., 2005) difference in sensitivity to roscovitine was observed between HCT-116 (wild-type p53) and an HCT-116 isogenic p53$^{-/-}$ variant. Nevertheless, roscovitine induces an increased expression (David-Pfeuty, 1999; Blaydes et al., 2000; Kotala et al., 2001; Lu et al., 2001; Wojciechowski et al., 2003; Wesierska-Gadek et al., 2003; Mohapatra et al., 2005), Ser46 phosphorylation (in MCF-7 cells) (Wesierska-Gadek et al., 2005), and nuclear accumulation (Ljungman and Paulsen, 2001; David-Pfeuty, 1999; David-Pfeuty et al.,

2001) of wild-type p53. An increase of p53-dependent transcription is therefore observed, leading, for example, to the accumulation of p21[CIP1] (MCF-7 cells) (Kotala et al., 2001; Lu et al., 2001) or p53AIP1 (Wesierska-Gadek et al., 2005). The effect may be due to downregulation of MDM2, which would contribute to stabilize p53 and induce p53-dependent transcription as monitored in a p53-responsive BP100-luciferase reporter (Lu et al., 2001). This effect is not observed with expression of p16, p21, p27, p57, or dominant negative mutants of CDK1, CDK2, CDK3, CDK4, and CDK6, suggesting that another target of roscovitine is involved. Interestingly, the removal of roscovitine results in superactivation of p53 (Lu et al., 2001).

Although roscovitine-induced cell death was initially thought to result directly from cell cycle arrest, it is increasingly considered to be the consequence of an inhibition of CDK7/CDK9-dependent transcription (Figure 9.8C). Indeed, transcription largely rests on the activity of RNA polymerase II, the activity of which relies on phosphorylation of its C-terminal domain by CDK7/cyclin H, CDK8/cyclin C, and CDK9/cyclin T. Both CDK7 and CDK9 are potently inhibited by roscovitine (Table 9.2). (R)-Roscovitine inhibits the phosphorylation of the C-terminal domain of RNA polymerase II (by CDK7), reducing mRNA synthesis in both human fibroblasts and HCT-116 colon cancer cells (Ljungman and Paulsen, 2001). In multiple myeloma cells, (R)-roscovitine induces the dephosphorylation of RNA polymerase II (Ser2, Ser5) and consequently decreases transcription (MacCallum et al., 2005). In contrast to flavopiridol, which inhibits gene expression globally, the effects of (R)-roscovitine on transcription are rather limited (Lam et al., 2001). The expression of only a small number of proteins is found to be severely reduced. One would predict the greatest effect to be observed on gene products with mRNA and protein short half-lives. This would result in a rapid decline in the level of these proteins. This appears to be true for important survival factors such as Mcl-1, a member of the antiapoptotic Bcl-2 family, XIAP, and survivin. Induction of cell death by roscovitine thus seems to correlate rather well with inhibition of transcription of essential cell survival factors (Figure 9.8C). Furthermore, among the genes that are downregulated by treatment of cells with (R)-Roscovitine is RNA polymerase II, as well as cyclin D1 and CDK4 in HCT-116 (Raynaud et al., 2005), and cyclin D1 in some mantle cell lymphoma cell lines (Lacrima et al., 2005). Treatment with (R)-roscovitine also leads to the downregulation of Chk1 and activation of a DNA damage response marked by an activation of ATM and Chk2 (Maude and Enders, 2005).

Mcl-1 expression is strongly reduced in mantle cell lymphoma cell lines exposed to (R)-roscovitine (Lacrima et al., 2005). Mcl-1 downregulation is sufficient by itself to induce apoptosis in multiple myeloma cells, as demonstrated by the use of siRNA (McCallum et al., 2005). Mcl-1 downregulation is also observed in multiple myeloma cells (Raje et al., 2005) and in U937 leukemia cells (Rosato et al., 2005).

(R)-Roscovitine reduces the level of the antiapoptotic protein XIAP by downregulating XIAP mRNA expression (Mohapatra et al., 2003, 2005). It also decreases the tyrosine phosphorylation and consequent activation of STAT5a, an upstream regulator of XIAP. (R)-Roscovitine downregulates survivin and XIAP, which contributes to the activation of caspase cascades, overcoming glioma cell resistance to TRAIL-mediated apoptosis (Kim et al., 2004). Combined treatment of glioma cells with TRAIL and (R)-roscovitine leads to a decrease in expression of XIAP and survivin, and activation of caspases and cell apoptosis (Kim et al., 2004).

Chronic lymphocytic leukemia (CLL) B-lymphocytes are noncycling, G_0-arrested cells in which CDK2 is expressed but inactive. Nevertheless, (R)-roscovitine induces caspase-dependent cell death in these CCL cells to a significantly much higher level than in peripheral blood mononuclear cells, purified normal B-lymphocytes (Hahntow et al., 2004; Alvi et al., 2005), or normal T-lymphocytes (Alvi et al., 2005). Again, this effect is associated with downregulation of Mcl-1 and XIAP proteins (Hahntow et al., 2004; Alvi et al., 2005) and other proteins such as RNA polymerase II (Alvi et al., 2005). In CCL cells, roscovitine does not trigger an increase in p53 or its nuclear translocation, and roscovitine-induced cell death is independent of the p53 and ATM status (Alvi et al., 2005).

Other mechanisms may be involved in the effects of (R)-roscovitine on cell proliferation and cell death (Figure 9.8D). Inhibition of the MAP-kinases Erk1 and Erk2 (Meijer et al., 1997), and of several members of the casein kinase 1 (CK1) family (Ferandin et al., unpublished), certainly contributes to the antiproliferative effects of (R)-roscovitine. As discussed previously, (R)-roscovitine binds to pyridoxal kinase (PDXK) (Bach et al., 2005; Tang et al., 2005) with undetermined effects on the cell cycle. Finally, inhibition of CDK2/cyclin E prevents Ser281 phosphorylation and subsequent proteolytic degradation of Cdx2, a transcription factor involved in the balance between proliferation and differentiation of intestinal cells (Boulanger et al., 2005; Gross et al., 2005; Gespach, 2005).

9.6.3 ANTI-TUMOR EFFECTS: CONVERGENCE OF DIFFERENT MECHANISMS OF ACTION

We have seen how (R)-roscovitine acts in different ways, all of which converge towards cell cycle arrest and cell death, thereby providing the observed anti-tumor effects. Induction of cell cycle arrest originates from both a direct inhibition of cell-cycle-regulating CDKs (Figure 9.8A) and an indirect effect by inhibition of the upstream CDK-activating CDK7 and an increased level of the CDK inhibitory p27[KIP1] (Figure 9.8B). Induction of cell death originates from a transient reduction in transcription due to direct inhibition of CDK7 and CDK9, leading to the down-regulation of essential, short-lived survival factors that are typically expressed in cancer cells (such as Mcl-1, XIAP, survivin) (Figure 9.8C). Furthermore, we believe that the short half-life of (R)-roscovitine and the lack of activity of its metabolites together prevent a long-term and massive inhibition of transcription, which is likely to be deleterious to normal cells. We hypothesize that a brief inhibition of transcription selectively affects tumor cells that are highly dependent on short-lived survival factors. The transient downregulation of these survival factors then triggers an irreversible activation of apoptosis that can proceed even after roscovitine has been metabolized away. In contrast, normal cells, which do not rely on these survival factors, are only transiently and reversibly arrested in their cell cycle progression.

In addition to these direct anti-tumor effects, roscovitine appears to display synergistic properties with a number of anti-tumor treatments (Table 9.5). More examples are available in the patent literature and should be published soon. These effects are

TABLE 9.5
Additive and Synergistic Effects of Various Treatments with (*R*)-Roscovitine in Mammalian Cells

Treatment	Target	Cells	Reference
Irradiation	DNA	Breast carcinoma	Maggiorella et al., 2003
SCH56582	Farnesyltransferase	Leukemic and prostate cancer	Edamatsu et al., 2000
Camptothecin	Topoisomerase I	MCF-7 breast tumor	Lu et al., 2001
Irinotecan	Topoisomerase I	p53-mutated colon cancer	Abal et al., 2004
Doxorubicin	Topoisomerase II	A549 and HEC1B adenocarcinoma HCT116 and H1299	Crescenzi et al., 2005
Etoposide	Topoisomerase II	U2-OS osteosarcoma	Maude and Enders, 2005
LY294002	Phophatidylinositol 3-kinase	U937 monocytic leukemia	Yu et al., 2003
TRAIL[a]		U87MG, T98, A172, U251 glioma cells	Kim et al., 2004
LAQ824	Histone deacetylase	U937, Jurkat and HL-60 leukemia	Rosato et al., 2005
nutlin-3	MDM2/p53 binding	SH-SY5Y neuroblastoma	Ribas et al., 2006

[a] Tumor-necrosis-factor-related apoptosis-inducing ligand.

highly dependent on the sequence of drug treatment. These additive or synergistic effects have strong implications for the use of (*R*)-roscovitine in chemotherapy.

In contrast with the cell-death-inducing properties described earlier, (*R*)-roscovitine, similar to other CDK inhibitors, has well-established antiapoptotic properties, mostly but not exclusively (Borgne et al., 2006), in nondividing cells such as neural cells (review in Borgne and Golsteyn, 2003). These properties are being extensively investigated for their applications in the neurodegeneration diseases field (Knockaert et al., 2002a). It is still largely unknown how CDK inhibitors are able to protect cells from apoptosis induced by various factors. Depending on the model, CDK1 (Konishi et al., 2002; Park et al., 2005; Borgne et al., 2006), CDK2 (Gil-Gomez et al., 1998; Hakem et al., 1999), or CDK5 (Sandal et al., 2002) is involved. The contribution of CDKs varies according to cell type, conditions, and the nature and concentration of the apoptosis-inducing drug. The antiapoptotic properties of CDK inhibitors may reduce the use of these compounds as anti-tumor drugs. Nevertheless, CDK inhibitors might diminish their own side effects and even be used to counteract damaging effects of other anti-tumor drugs. For example, roscovitine could be used as therapy for cisplatin-induced nephrotoxicity (Price et al., 2004). In addition, these antiapoptotic properties of roscovitine might provide some protection to normal cells. Understanding the paradoxical apoptosis-inducing and apoptosis-preventing properties of roscovitine and other CDK inhibitors is a major challenge for current research.

CDK2 has been closely linked with melanoma growth (Du et al., 2004). The CDK2 gene overlaps with the melanocyte-specific gene SILVER/PMEL17, which encodes an antigen commonly used for melanoma diagnosis and immune therapy.

Both genes are regulated by the melanocyte lineage transcription factor MITF. CDK2 appears to be an essential target gene for MITF, which is important for survival of melanocytes and melanoma. Mutations in MITF lead to melanocyte defects. Expression of MITF and CDK2 are tightly correlated in human melanoma samples and melanoma cell lines, and their levels predict sensitivity to (*R*)-roscovitine (Du et al., 2004).

Finally, interesting studies have shown that cyclin E is expressed as low molecular weight (LMW) forms in breast and melanoma cancers (review in Akli and Keyomarsi, 2004; Hunt and Keyomarsi, 2005). Expression of LMW cyclin E strongly correlates with poor prognosis. CDK2 associated with these LMW forms is quite active and resistant to inhibition by protein inhibitors such as p27[KIP1] and, therefore, constitutes an attractive target in these clinical settings.

9.7 PHARMACOLOGY OF (*R*)-ROSCOVITINE

A correlation has been established between some chemical parameters of a molecule and its absorption or permeation properties (Lipinski et al., 1997). (*R*)-roscovitine appears to meet all required parameters for favorable absorption (Table 9.6).

9.7.1 QUANTIFICATION OF (*R*)-ROSCOVITINE

Two methods have been developed to quantify (*R*)-roscovitine. The first is high-performance liquid chromatography, associated with detection at 292 nm, and provides a working linear range of detection between 100 ng/ml to 5000 ng/ml (i.e., 0.28 to 14 µM) (Vita et al., 2004). The second is liquid chromatography coupled to tandem mass spectrometry (LC-MS/MS) and allows the detection and quantification of (*R*)-roscovitine over a range of 0.5 to 2000 ng/ml (i.e., 0.0014 to 5.6 µM) (Vita et al., 2005c; Raynaud et al., 2004, 2005).

9.7.2 PHARMACOKINETICS

Pharmacokinetic studies of (*R*)-roscovitine injected in rat (25 mg/kg body weight) show a rapid, biphasic elimination of the drug with a 5 min and a <30 min half-life, in accordance with a two-compartment open model (Vita et al., 2004, 2005b).

TABLE 9.6
(*R*)-Roscovitine and the Lipinski Rule of 5

Favorable Parameters According to the Lipinski Rule of 5	(*R*)-Roscovitine
Molecular mass: ≤ 500	354.45
Log P[a]: < 5	3.244
Number of H-bond donors (OH and NH): ≤ 5	3
Number of H-bond acceptors (O and N): ≤ 10	4

[a] P is the octanol-water partition coefficient.

In mouse, the plasma level of roscovitine also rapidly drops within 30 min to <1% of the I.V. injected dose (40 nmol/g, i.e., about 15 mg/kg) (Chmela et al., 2003). Detailed pharmacokinetic studies performed in BALB/c mice (Nutley et al., 2005; Raynaud et al., 2004, 2005) and Tg26 mice (Gherardi et al., 2004) have been recently published. They show rapid, biexponential clearance of (R)-roscovitine from plasma following I.V., I.P., or oral administration (Nutley et al., 2005; Raynaud et al., 2004, 2005; Gherardi et al., 2004). (R)-Roscovitine uptake into the general circulation was fast and its plasma half-life was 1.19 h. Plasma concentrations could be maintained above 15 μM (the average IC_{50} values obtained with various tumor cell lines) for 4, 12, and 24 h following oral administration at 50, 500, and 2000 mg/kg, respectively (Raynaud et al., 2005).

The pharmacokinetics of (R)-roscovitine were recently studied in humans (de la Motte and Gianella-Borradori, 2004). Following oral administration of a single dose (50, 100, 200, 400, and 800 mg) in healthy men, (R)-roscovitine and its carboxylated metabolite were measured in plasma and urine. (R)-Roscovitine undergoes rapid passage into the blood, distribution in tissues, and metabolism.

9.7.3 METABOLITES

When (R)-roscovitine was injected I.V. at 100 mg/kg in mouse (Nutley et al., 2005) or at 25 mg/kg in rat (Vita et al., 2005b), several metabolites were identified in the plasma (Figure 9.9). (R)-Roscovitine undergoes a rapid loss of the isopropyl group (M1), several oxidations (M2–M7), or conjugation of a glucose residue (M8). M3 is the most abundantly produced metabolite, which is then excreted in urine (Chmela et al., 2003; Nutley et al., 2005; Vita et al., 2005b).

When (R)-roscovitine was incubated in microsomal preparations, M1 to M6 metabolites were generated, the COOH-(R)-roscovitine M3 being the most abundant. Sensitivity to the absence of NADPH and to SKF-525A demonstrates that this main metabolite is produced through an NADPH- and cytochrome-P450-dependent process. Glycosidation is also a major pathway observed in rodent and primate microsomes (Cervenkova et al., 2003).

Both M3 and M6 were synthesized and found to be much less potent than the parent (R)-roscovitine at inhibiting CDK2 (Nutley et al., 2005).

In humans, the carboxylated derivative was also the main metabolite formed following oral administration of (R)-roscovitine (de la Motte and Gianella-Borradori, 2004).

9.8 ANTI-TUMOR EFFECTS OF (R)-ROSCOVITINE

9.8.1 TOXICITY

(R)-Roscovitine appears to be well tolerated. The maximum tolerated dose (MTD) in mice could not be reached when (R)-roscovitine was delivered intravenously, owing to poor solubility (maximal achievable dose was 20 mg/kg) (McClue et al., 2002). Nor could MTD be reached when (R)-roscovitine was given orally (maximal achievable dose was 2000 mg/kg) (McClue et al., 2002). When administered intraperitoneally, three doses of 100 mg/kg were well tolerated (McClue et al., 2002). In BALB/c mice,

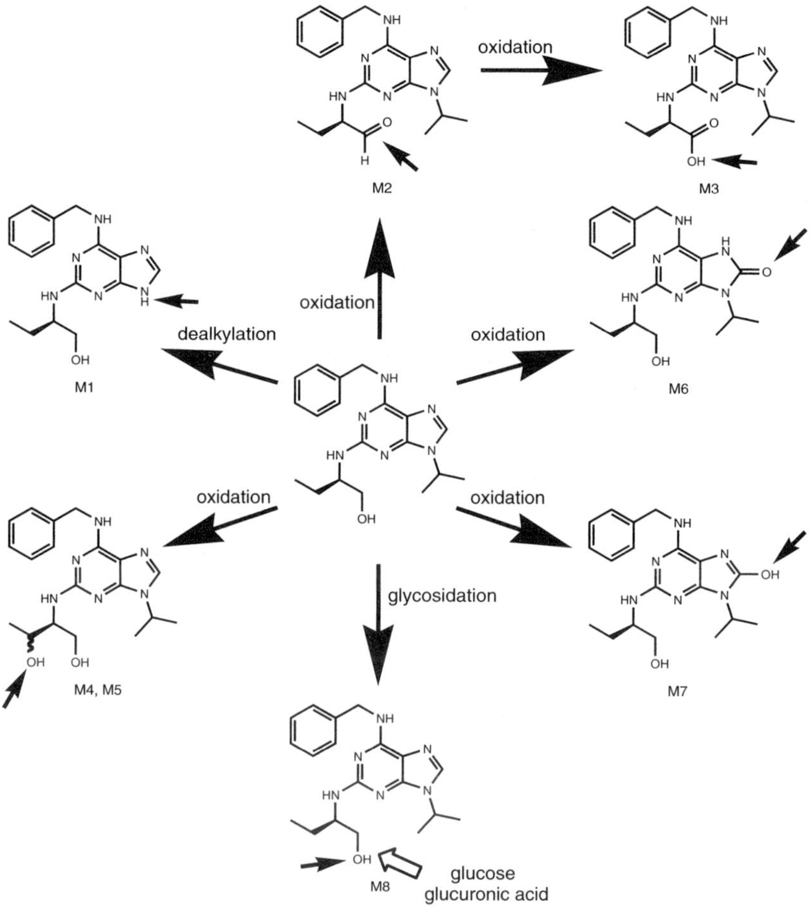

FIGURE 9.9 (*R*)-Roscovitine and its metabolites. Small arrows point to the chemical changes introduced in the initial roscovitine structure.

the MTD was 100 mg/kg for intravenous injection. For intraperitoneal injections, 150 mg/kg was well tolerated. Finally, (*R*)-roscovitine was well tolerated up to 2000 mg/kg when administered as a single oral dose (Raynaud et al., 2005).

9.8.2 ANIMAL MODELS

The anti-tumor activity of (*R*)-roscovitine has been investigated in various human tumor xenografts (summarized in Table 9.7) (review in Meijer and Raymond, 2003; Guzi, 2004). An interesting additivity of irradiation and (*R*)-roscovitine treatments of breast carcinoma *in vitro* and *in vivo* was reported (Maggiorella et al., 2003). Irinotecan and (*R*)-roscovitine treatments also showed some additivity in a colon cancer xenograft model (Abal et al., 2004). Other results were reported in meeting abstracts and have not been published in detail at the time of writing this chapter (Guzi, 2004; Fischer and Gianella-Borradori, 2005).

TABLE 9.7
***In Vivo* Anti-Tumor Activity of (*R*)-Roscovitine in Nude Mice Models**

Tumor Cells	Delivery Mode	Concentration	Frequency	Results	Reference
LoVo (colon carcinoma)	Intraperitoneal	100 mg/kg	3×/d/5 d	44.8% reduction in tumor size T/C: 58.3%	McClue et al., 2002
MES-SA/ Dx5 (uterine carcinoma)	Oral	200 mg/kg	3×/d/10 d	62% reduction in tumor size T/C: 47.3%	McClue et al., 2002
		500 mg/kg	3×/d/4 d	35% reduction in tumor size T/C: 85.7%	
MDA-MB 231 (breast cancer)	Intraperitoneal	100 mg/kg (± 7.5 Gy irradiation)	Single dose	54% (irradiation) or 73% (irradiation + roscovitine) reduction in tumor size	Maggiorella et al., 2003
HT29 (colon cancer)	Oral	200 mg/kg (± irinotecan)	2×/d/3–4 d	Additivity	Abal et al., 2004
HCT116 (colon cancer)	Oral	500 mg/kg	2×/d/5 d	T/C: 65%	Raynaud et al., 2005
PC-3	Subcutaneous infusion with osmotic minipump	10 mg/ml	2 mg/week for 3 weeks	35% tumor growth inhibition	Payton et al., 2006
A4573 (Ewing's sarcoma)	Intraperitoneal	50 mg/kg	5 d or two 5-d series with a 2-d break	11.5-fold difference in tumor volume of treated vs. control animals 1 d after 5-d treatment Overall 5-fold reduction in tumor growth	Tirado et al., 2005

9.8.3 HUMAN CANCERS

(*R*)-Roscovitine has been evaluated in several Phase I clinical trials. A total of 70 patients have been evaluated for anti-tumor activity. One partial response (hepatocellular carcinoma) and 10 stable diseases lasting for more than 4 months were reported. Two of these patients had non-small-cell lung cancers that remained stable for more than a year.

 213

(R)-Roscovitine is currently under investigation in several Phase 2 clinical trials in four different clinical situations as a monotherapy against B-cell malignancies (chronic lymphocytic leukemia, mantle cell lymphoma, and multiple myeloma) and in combination trials against non-small-cell lung cancer (+ gemcitabine/cisplatin or docetaxel) and metastatic breast cancer (+ capecitabine).

Unfortunately, the results of the Phase 1 and 2 clinical trials of (R)-roscovitine in human patients are only available as meeting abstracts and have not been published in detail at the time of writing this chapter. Overviews of the clinical data are available in recent reviews (Guzi, 2004; Fischer and Gianella-Borradori, 2005).

9.9 USING ROSCOVITINE

9.9.1 STORAGE, DILUTION, CONCENTRATIONS, AND AFFINITY CHROMATOGRAPHY

Roscovitine is a rather stable compound as raw material or in nonaqueous solutions. For long-term storage, we recommend storage as a dry powder at −20°C. Roscovitine is usually dissolved in dimethylsulfoxide (DMSO) as a stock solution of 10 to 50 mM. It can be aliquoted and stored at −20°C. However, some precipitation is sometimes observed, probably because of the absorption of water by DMSO.

When tested in *in vitro* enzymatic assays, we suggest the following range of roscovitine concentrations for initial testing: 0–0.01–0.025–0.05–0.1–0.25–0.5–1–2.5–5–10–25–50–100 μM (a final ATP concentration of 15 μM ATP has been routinely used in our kinase assays). In initial cellular tests, we recommend the following range of concentration: 0–0.1–0.25–0.5–1–2.5–5–10–25–50–100 μM.

For *in vivo* testing, several routes have been used to deliver roscovitine. These methods are summarized in Table 9.8. Experimental data show that roscovitine displays high bioavailability when delivered intraperitoneally or orally.

For affinity chromatography on immobilized roscovitine (Bach et al., 2005), extracts are first prepared in a homogenization buffer and centrifuged for 10 min at 14,000 g at 4°C. The supernatant is assayed for protein content and immediately loaded batchwise on the affinity matrix. Just before use, packed roscovitine beads are washed with bead buffer and resuspended in this buffer. The cell or tissue extract supernatant (up to 3 mg total protein) or purified protein is then added. The tubes are rotated at 4°C for 30 min. After a brief spin and removal of the supernatant, the beads are washed four times with bead buffer before addition of 2X Laemmli sample buffer. Following heat denaturation, the bound proteins are analyzed by SDS-PAGE and Western blotting or silver staining. Protein bands can then be excised and digested in gel with trypsin. The resulting peptides are purified and concentrated prior to mass spectrometry analysis.

9.9.2 USING ROSCOVITINE: A "ROSCOVITINE USER'S KIT"

With an assumed selectivity for CDKs, (R)-roscovitine is frequently used to evaluate the contribution of CDKs to cellular processes. Although this assumption of selectivity is partially confirmed, the use of (R)-roscovitine and the interpretation of its

TABLE 9.8
Methods Used to Deliver Roscovitine to Animals and Humans

Animal	Delivery Mode	Concentration	Frequency	Vehicle	Reference
Mouse	Intravenous	20 mg/kg		PEG400:10% Kollidon 17PF (7:13)	McClue et al., 2002
Mouse	Oral	500 mg/kg 200 mg/kg	3x/d/4 d 3x/d/10 d	50 mM HCl	McClue et al., 2002
Mouse	Intraperitoneal	100 mg/kg	3x/d/5 d	50 mM HCl pH 2.5	McClue et al., 2002
Mouse	Intraperitoneal	100 mg/kg		0.25% Tween 20 in 0.9% NaCl	Maggiorella et al., 2003
Mouse	Intraperitoneal	20–75 mg/kg	2x/d/20 d	DMSO:PBS (4:6)	Gherardi et al., 2004
Mouse	Oral	500 mg/kg	2x/d/5 d	DMSO: Tween 20:50 mM HCl/saline (10:5:85)	Raynaud et al., 2005
Mouse	Intravenous	100 mg/kg	—	—	Nutley et al., 2005
Mouse	Intraperitoneal	2.8 mg/kg	1x/d/5 or 14 d	DMSO:PBS (2:8)	Griffin et al., 2005
Mouse	Intraperitoneal	50 mg/kg	1x/d/5 d, 2-d interval, then 1x/d/5 d	Dissolve in ethanol or DMSO. Dilute in Tween 80, *N,N*-dimethylacetamide, polyethylene glycol 400 (Fisher Scientific) (1:2:7)	Tirado et al., 2005
Rat	Intraperitoneal	2.8 mg/kg	1x/d/3–10 d	DMSO	Pippin et al., 1997
Rat	Oral	100 mg/kg	1x/d/10 d then daily I.P.	30 mM HCl, then DMSO	Milovanceva-Popovska et al., 2002
Human	Oral	50–800 mg	1 dose	50 mg capsules	de la Motte and Gianella-Borradori, 2004

cellular effects has limitations. First, (*R*)-roscovitine inhibits CDK1, CDK2, CDK5, CDK7, and CDK9 almost equally. Second, the effects on Erk1/2, CK1, and CaM kinase 2 cannot be neglected as they may significantly contribute to the cellular response to (*R*)-roscovitine. Third, interference with PDXK might, under some circumstances, lead to a significant reduction in intracellular vitamin B_6 level, and this is expected to alter a large number of metabolic enzymes, with numerous cellular consequences. In addition, high levels of PDXK in some tissues may reduce the availability of CDKs to roscovitine.

Nevertheless, (*R*)-roscovitine can provide valuable and convincing information if used in conjunction with a "kit" of other molecules:

1. *(S)-Roscovitine*, apparently much less active on PDXK than (*R*)-roscovitine, in most models (except primates); in addition, (*S*)-roscovitine is slightly less efficient on CDKs than (*R*)-roscovitine, and this difference should be observed in CDK-dependent cellular events (Bach et al., 2005).

2. *O^6-Benzyl-(R)-roscovitine* or *N^6-methyl-(R)-roscovitine*, two compounds closely related to (*R*)-roscovitine, which are essentially inactive on CDKs but still interact with PDXK (Tang et al., 2005).

3. *Olomoucine* (Vesely et al., 1994) and *aminopurvalanol* (Chang et al., 1999), which are less and more active on CDKs, respectively.

4. *U0126, PD184352,* or *PD98059*, three MEK1 inhibitors that prevent the activation of Erk1/2 and thus mimic the Erk1/2 inhibiting properties of (*R*)-roscovitine, but do not inhibit CDKs.

5. *IC261, D4476,* and *SB203580*, three CK1 inhibitors that mimic the CK1-inhibiting properties of (*R*)-roscovitine, but do not inhibit CDKs.

6. At least another structurally unrelated CDK inhibitor. For example, the combined use of both kenpaullone and roscovitine has been recommended to help identify substrates and physiological roles of CDKs, whereas the combined use of kenpaullone and lithium could be useful for identifying substrates and physiological roles of glycogen synthase kinase-3 (GSK-3) (Bain et al., 2003).

In addition, it is highly recommended to complement the use of (*R*)-roscovitine with molecular biology methods that prevent or reduce the expression and activity of specific CDKs or cyclins such as siRNAs, antisense oligonucleotides, peptidic nucleic acids, expression of dominant negative mutants, and CDK or cyclin knockouts.

Finally, the use of specific antibodies that cross-react with sites that are specifically phosphorylated by CDKs (Table 9.4) provides further support for specific inhibition of CDKs.

9.10 FUTURE PROSPECTS

Considerable work has been carried out since the early discovery of roscovitine starting from the isopentenyladenine and olomoucine structures and using native CDK1/cyclin B kinase purified from the highly synchronous starfish M phase oocytes. But how can roscovitine be improved?

(*R*)-Roscovitine appears to be a rather selective inhibitor of a few CDKs, but it interacts with a few other protein kinases and pyridoxal kinase. The antimitotic and proapoptotic properties of (*R*)-roscovitine are to be accounted for by a favorable combination of effects converging both to multiple cell cycle arrest points and to induction of cell death by several parallel mechanisms. This multitarget effect of (*R*)-roscovitine constitutes a weakness when the drug is used as a pharmacological tool in cell biology studies, and results should be interpreted with care (see Subsection 9.9.2). In contrast, this diversity of molecular actions becomes an advantage when (R)-roscovitine is investigated as a potential drug for the treatment of cancer. Indeed, with such complexity of cellular targets, we do not expect rapid resistance to develop following (R)-roscovitine treatment. This is supported by the fact that, despite several serious efforts, no (R)-roscovitine-resistant cell lines have been reported. Also, despite the genetic demonstration that CDK2 is dispensable for mitotic division (Ortega et al., 2003), the sensitivity of $CDK2^{-/-}$ cells to (R)-roscovitine is only slightly lower than that of $CDK2^{+/+}$ wild-type cells (Bach et al., 2005). The artificial absence of CDK2 appears to be compensated by CDK1 (Aleem et al., 2005), another target of (R)-roscovitine. The contribution of the interaction between (R)-roscovitine and PDXK to its anti-tumor effects remains an open question. To address this issue, we are currently designing roscovitine derivatives deprived of interaction with PDXK but still inhibiting CDKs. Protein kinase-inactive but PDXK-binding (R)-roscovitine derivatives (Figure 9.6) are not completely devoid of anti-proliferative effects, but these effects may be unspecific as they require very high doses (Meijer et al., unpublished).

The pharmacological parameters of (R)-roscovitine certainly could be improved by slight modifications of the parent structure. Whether the relatively short half-life of (R)-roscovitine constitutes an advantage or a disadvantage is still to be determined. One way to address this issue would be to generate more metabolically stable (R)-roscovitine derivatives with identical biochemical properties. The oral bioavailability of (R)-roscovitine certainly constitutes a great advantage. Dosing, frequency of administration, and circadian optimization of drug delivery need to be investigated further to obtain the most suitable drug exposure. Furthermore, the combination of (R)-roscovitine with currently used cancer treatments represents a promising field of investigation. Finally, the identification of the cancer subtypes that are most sensitive to (R)-roscovitine remains an open field. Currently, B-cell malignancies, lung cancer, breast cancer, and melanoma seem to be promising clinical targets.

ACKNOWLEDGMENTS

The work presented here was supported by a grant from the EEC (FP6-2002-Life Sciences & Health, PRO-KINASE Research Project) (L.M.) and the Canceropole Grand-Ouest. Many thanks to Dr. Martin Noble for providing the representation of the roscovitine/CDK2 cocrystal structure in Figure 9.5.

REFERENCES

Abal, M., Bras-Goncalves, R., Judde, J.G., Fsihi, H., De Cremoux, P., Louvard, D., Magdelenat, H., Robine, S., and Poupon, M.F. Enhanced sensitivity to irinotecan by Cdk1 inhibition in the p53-deficient HT29 human colon cancer cell line. *Oncogene* 23, 1737, 2004.

Akli, S. and Keyomarsi, K. Low-molecular-weight cyclin E: the missing link between biology and clinical outcome. *Breast Cancer Res.*, 6, 188, 2004.

Aleem, E., Kiyokawa, H., and Kaldis, P. Cdc2-cyclin E complexes regulate the G1/S phase transition. *Nat. Cell Biol.*, 7, 831, 2005.

Alessi, F., Quarta, S., Savio, M., Riva, F., Stivala, L.A., Scovassi, A.I., Meijer, L., and Prosperi, E. The cyclin-dependent kinase inhibitors olomoucine and roscovitine arrest human fibroblasts in G1 phase by inhibition of CDK2 kinase activity. *Exp. Cell Res.*, 245, 8, 1998.

Alvi, A.J., Austen, B., Weston, V.J., Fegan, C., MacCallum, D., Gianella-Borradori, A., Lane, D.P., Hubank, M., Powell, J.E., Wei, W., Taylor, A.M., Moss, P.A., and Stankovic, T. A novel CDK inhibitor, CYC202 (*R*-roscovitine), overcomes the defect in p53-dependent apoptosis in B-CLL by down-regulation of genes involved in transcription regulation and survival. *Blood*, 105, 4484, 2005.

Arion, D., Meijer, L., Brizuela, L., and Beach, D. Cdc2 is a component of the M-phase specific histone H_1 kinase: evidence for identity with MPF. *Cell*, 55, 371, 1988.

Atanasova, G., Jans, R., Zhelev, N., Mitev, V., and Poumay, Y. Effects of the cyclin-dependent kinase inhibitor CYC202 (*R*)-roscovitine) on the physiology of cultured human keratinocytes. *Biochem. Pharmacol.*,70, 824, 2005.

Azevedo, W.F., Leclerc, S., Meijer, L., Havlicek, L., Strnad, M., and Kim, S.H. Inhibition of cyclin-dependent kinases by purine analogues: crystal structure of human cdk2 complexed with roscovitine. *Eur. J. Biochem.*, 243, 518, 1997.

Bach, S., Knockaert, M., Lozach, O., Reinhardt, J., Baratte, B., Schmitt, S., Coburn, S.P., Tang, L., Jiang. T., Liang, D.C., Galons, H., Dierick, J.F., Totzke, F., Schächtele, C., Lerman, A.S., Carnero, A., Wan, Y., Gray, N., and Meijer, L. Roscovitine targets: protein kinases and pyridoxal kinase. *J. Biol. Chem.*, 280, 31208, 2005.

Bach, S., Blondel, M., and Meijer, L. Evaluation of CDK inhibitors' selectivity: from affinity chromatography to yeast genetics. In *Monographs on Enzyme Inhibitors. Vol. 2. CDK Inhibitors and Their Potential as Anti-Tumor Agents*, E. Yue and P.J. Smith, Eds. CRC Press, submitted, 2006.

Bain, J., McLauchlan, H., Elliott, M., and Cohen, P. The specificities of protein kinase inhibitors: an update. *Biochem. J.*, 371, 199, 2003.

Barrie, S.E., Eno-Amooquaye, E., Hardcastle, A., Platt, G., Richards, J., Bedford, D., Workman, P., Aherne, W., Mittnacht, S., and Garrett, M.D. High-throughput screening for the identification of small-molecule inhibitors of retinoblastoma protein phosphorylation in cells. *Anal. Biochem.*, 320, 66, 2003.

Becker, F., Murthi, K., Smith, C., Come, J., Costa-Roldan, N., Kaufmann, C., Hanke, U., Degenhart, C., Baumann, S., Wallner, W., Huber, A., Dedier, S., Dill, S., Kinsman, D., Hediger, M., Bockovich, N., Meier-Ewert, S., Kluge, A.F., and Kley, N. A three hybrid approach to scanning the proteome for targets of small molecule kinase inhibitors. *Chem. Biol.*, 11, 211, 2004.

Benson, C., Kaye, S., Workman, P., Garrett, M., Walton, M., and de Bono, J. Clinical anticancer drug development: targeting the cyclin-dependent kinases. *Br. J. Cancer*, 92, 7, 2005.

Blaydes, J.P., Craig, A.L., Wallace, M., Ball, H.M., Traynor, N.J., Gibbs, N.K., and Hupp, T.R. Synergistic activation of p53-dependent transcription by two cooperating damage recognition pathways. *Oncogene*, 19, 3829, 2000.

Bloom, J. and Pagano, M. Deregulated degradation of the cdk inhibitor p27 and malignant transformation. *Semin. Cancer Biol.*, 13, 41, 2003.

Borgne, A. and Golsteyn, R.M. The role of cyclin-dependent kinases in apoptosis. In *Progress in Cell Cycle Research*, L. Meijer, A. Jézéquel, and M. Roberge, Eds., Life in Progress Editions, Roscoff, France, 2003, p. 453.

Borgne, A., Versteege, I., Mahé, M., Studeny, A., Léonce, S., Rodriguez, M., Hickman, J.A., Meijer, L., and Golsteyn, R.M., 2006. Analysis of cyclin B1 in apoptosis and mitotic catastrophe induced by camptothecin treatment in HT29 cells. *Oncogene*, in press, 2006.

Boulanger, J., Vezina, A., Mongrain, S., Boudreau, F., Perreault, N., Auclair, B.A., Laine, J., Asselin, C., and Rivard, N. Cdk2-dependent phosphorylation of homeobox transcription factor CDX2 regulates its nuclear translocation and proteasome-mediated degradation in human intestinal epithelial cells. *J. Biol. Chem.*, 280, 18095, 2005.

Byth, K.F., Geh, C., Forder, C.L., Oakes, S.E., and Thomas, A.P. The cellular phenotype of AZ703, a novel selective imidazo[1,2-a]pyridine cyclin-dependent kinase inhibitor. *Mol. Cancer Ther.*, 5, 655, 2006.

Canduri, F., Uchoa, H.B., and de Azevedo, W.F. Molecular models of cyclin-dependent kinase 1 complexed with inhibitors. *Biochem. Biophys. Res. Commun.*, 324, 661, 2004.

Cervenkova, K., Belejova, M., Chmela, Z., Rypka, M., Riegrova, D., Michnova, K., Michalikova, K., Surova, I., Brejcha, A., Hanus, J., Cerny, B., Fuksova, K., Havlícek, L., and Vesely, J. In vitro glycosidation potential towards olomoucine-type cyclin-dependent kinase inhibitors in rodent and primate microsomes. *Physiol. Res.*, 52, 467, 2003.

Chang, T.S., Jeong, W., Choi, S.Y., Yu, S., Kang, S.W., and Rhee, S.G. Regulation of peroxiredoxin I activity by Cdc2-mediated phosphorylation. *J. Biol. Chem.*, 277, 25370, 2002.

Chang, Y.T., Gray, N.G., Rosania, G.R., Sutherlin, D.P., Kwon, S., Norman, T.C., Sarohia, R., Leost, M., Meijer, L., and Schultz, P.G. Synthesis and application of functionally diverse 2,6,9-trisubstituted purine libraries as CDK inhibitors. *Chem. Biol.*, 6, 361, 1999.

Chmela, Z., Rypka, M., Cervenkova, K., Lemr, K., Riegrova, D., and Vesely, J. Carboxylic acid is the main product of roscovitine biotransformation in mice in vivo: comparison with bohemine. *Proceedings of the 13th International Conference on Cytochromes P450,*. Prague, June 29–July 3, 2003. D629C0041, 163, 2003.

Cohen, P. Protein kinases — the major drug targets of the twenty-first century? *Nat. Rev. Drug Discovery*, 1, 309, 2002.

Crescenzi, E., Palumbo, G., and Brady, H.J. Roscovitine modulates DNA repair and senescence: implications for combination chemotherapy. *Clin. Cancer Res.*, 11, 8158, 2005.

Cruz, J.C. and Tsai, L.H. A Jekyll and Hyde kinase: roles for Cdk5 in brain development and disease. *Curr. Opin. Neurobiol.*, 14, 390, 2004.

David-Pfeuty T. Potent inhibitors of cyclin-dependent kinase 2 induce nuclear accumulation of wild-type p53 and nucleolar fragmentation in human untransformed and tumor-derived cells. *Oncogene*, 18, 7409, 1999.

David-Pfeuty, T., Nouvian-Dooghe, Y., Sirri ,V., Roussel, P., and Hernandez-Verdun, D. Common and reversible regulation of wild-type p53 function and of ribosomal biogenesis by protein kinases in human cells. *Oncogene*, 20, 5951, 2001.

Daub, H., Godl, K., Brehmer, D., Klebl, B. and Müller, G. Evaluation of kinase inhibitor selectivity by chemical proteomics. *Assay Drug Dev. Technol.* 2, 215–224, 2004.

de la Motte, S. and Gianella-Borradori, A. Pharmacokinetic model of *R*-roscovitine and its metabolite in healthy male subjects. *Int. J. Clin. Pharmacol. Ther.*, 42, 232, 2004.

Di Giovanni, S., Movsesyan, V., Ahmed, F., Cernak, I., Schinelli, S., Stoica, B., and Faden, A.I. Cell cycle inhibition provides neuroprotection and reduces glial proliferation and scar formation after traumatic brain injury. *Proc. Natl. Acad. Sci. U.S.A.*, 102, 8333, 2005.

Dobes, P., Otyepka, M., Strnad, M., and Hobza, P. Interaction energies for the purine inhibitor roscovitine with cyclin-dependent kinase 2: correlated Ab initio quantum-chemical, DFT and empirical calculations. *Chemistry* [Epub ahead of print], 2006.

Du, J., Widlund, H.R., Horstmann, M.A., Ramaswamy, S., Ross, K., Huber, W.E., Nishimura, E.K., Golub, T.R., and Fisher, D.E. Critical role of CDK2 for melanoma growth linked to its melanocyte-specific transcriptional regulation by MITF. *Cancer Cell*, 6, 565, 2004.

Edamatsu, H., Gau, C.L., Nemoto, T., Guo, L., and Tamanoi, F. Cdk inhibitors, roscovitine and olomoucine, synergize with farnesyltransferase inhibitor (FTI) to induce efficient apoptosis of human cancer cell lines. *Oncogene*, 19, 3059, 2000.

Fabian, M.A., Biggs, W.H., Treiber, D.K., Atteridge, C.E., Azimioara, M.D., Benedetti, M.G., Carter, T.A., Ciceri, P., Edeen, P.T., Floyd, M., Ford, J.M., Galvin, M., Gerlach, J.L., Grotzfeld, R.M., Herrgard, S., Insko, D.E., Insko, M.A., Lai, A.G., Lelias, J.M., Mehta, S.A., Milanov, Z.V., Velasco, A.M., Wodicka, L.M., Patel, H.K., Zarrinkar, P.P., and Lockhart, D.J. A small molecule-kinase interaction map for clinical kinase inhibitors. *Nat. Biotechnol.*, 23, 329, 2005.

Fischer, P.M. The design of drug candidate molecules as selective inhibitors of therapeutically relevant protein kinases. *Curr. Med. Chem.*, 11, 1563, 2004.

Fischer, L., Endicott, J., and Meijer, L. Cyclin-dependent kinase inhibitors. In *Cell Cycle Regulators as Therapeutic Targets*, L. Meijer, A. Jézéquel and M. Roberge, Eds., Progr. Cell Cycle Res., Vol. 5, Life in Progress Editions, Station Biologique de Roscoff, France, 2003, p. 248.

Fischer, P.M. and Gianella-Borradori, A. Recent progress in the discovery and development of cyclin-dependent kinase inhibitors. *Exp. Opin. Invest. Drugs*, 14, 457, 2005.

Garriga, J. and Grana, X. Cellular control of gene expression by T-type cyclin/CDK9 complexes. *Gene*, 337, 15, 2004.

Gespach, C. Crosstalk between the cell cycle regulator Cdk2 and the Cdx2 tumor suppressor at the proliferation-differentiation interface. *Oncogene*, 24, 7953-, 2005.

Gherardi, D., D'Agati, V., Chu, T.H.T., Barnett, A., Gianella-Borradori, A., Gelman, I.H., and Nelson, P.J. Reversal of collapsing glomerulopathy in mice with the cyclin-dependent kinase inhibitor CYC202. *J. Am. Soc. Nephrol.*, 15, 1212, 2004.

Gil-Gomez, G., Berns, A., and Brady, H.J. A link between cell cycle and cell death: Bax and Bcl-2 modulate Cdk2 activation during thymocyte apoptosis. *EMBO J.*, 17, 7209, 1998.

Goke, R., Barth, P., Schmidt, A., Samans, B., and Lankat-Buttgereit B. Programmed cell death protein 4 suppresses CDK1/cdc2 via induction of p21(Waf1/Cip1). *Am. J. Physiol. Cell Physiol.*, 287, C1541, 2004.

Gray, N., Wodicka, L., Thunnissen, A.M., Norman, T., Kwon, S., Espinoza, F.H., Morgan, D.O., Barnes, G., Leclerc, S., Meijer, L., Kim, S.H., Lockhart, D.J., and Schultz, P.G. Exploiting chemical libraries, structure, and genomics in the search for new kinase inhibitors. *Science*, 281, 533, 1998.

Griffin, S.V., Krofft, R.D., Pippin, J.W., and Shankland S.J. Limitation of podocyte proliferation improves renal function in experimental crescentic glomerulonephritis. *Kidney Int.*, 67, 977, 2005.

Gross, I., Lhermitte, B., Domon-Dell, C., Duluc, I., Martin, E., Gaiddon, C., Kedinger, M., and Freund, J.N. Phosphorylation of the homeotic tumor suppressor Cdx2 mediates its ubiquitin-dependent proteasome degradation. *Oncogene*, 24, 7955, 2005.

Guzi, T. Cyc-202 Cyclacel. *Curr. Opin. Invest. Drugs*, 5, 1311, 2004.

Haesslein, J.L. and Jullian, N. Recent advances in cyclin-dependent kinase inhibition. Purine-based derivatives as anti-cancer agents. Roles and perspectives for the future. *Curr. Top. Med. Chem.*, 2, 1035, 2002.

Hahntow, I.N., Schneller, F., Oelsner, M., Weick, K., Ringshausen, I., Fend, F., Peschel, C., and Decker, T. Cyclin-dependent kinase inhibitor Roscovitine induces apoptosis in chronic lymphocytic leukemia cells. *Leukemia*, 18, 747, 2004.

Hakem, A., Sasaki, T., Kozieradzki, I., and Penninger, J.M. The cyclin-dependent kinase Cdk2 regulates thymocyte apoptosis. *J. Exp. Med.*, 189, 957, 1999.

Havlícek, L., Hanus, J., Vesely, J., Leclerc, S., Meijer L., Shaw, G., and Strnad, M. Cytokinin-derived cyclin-dependent kinase inhibitors: Synthesis of cdc2 inhibitory activity of olomoucine and related compounds. *J. Med. Chem.*, 40, 408, 1997.

Hunt, K.K. and Keyomarsi, K. Cyclin E as a prognostic and predictive marker in breast cancer. *Semin. Cancer Biol.*, 15, 319, 2005.

Kim, E.H., Kim, S.U., Shin, D.Y., and Choi, K.S. Roscovitine sensitizes glioma cells to TRAIL-mediated apoptosis by downregulation of survivin and XIAP. *Oncogene*, 23, 446, 2004.

Knockaert, M., Gray, N., Damiens, E., Chang, Y.T., Grellier, P., Grant, K., Fergusson, D., Mottram, J., Soete, M., Dubremetz, J.F., LeRoch, K., Doerig, C., Schultz, P.G., and Meijer, L. Intracellular targets of cyclin-dependent kinase inhibitors: identification by affinity chromatography using immobilized inhibitors. *Chem. Biol.*, 7, 411, 2000.

Knockaert, M., Greengard, P., and Meijer, L. Pharmacological inhibitors of cyclin-dependent kinases. *Trends Pharmacol. Sci.*, 23, 417, 2002a.

Knockaert, M., Lenormand, P., Gray, N., Schultz, P., Pouyssegur, J., and Meijer, L. p42/p44MAPKS are intracellular targets of the CDK inhibitor purvalanol. *Oncogene*, 21, 6413, 2002b.

Knockaert, M. and Meijer, L. Identifying *in vivo* targets of cyclin-dependent kinase inhibitors by affinity chromatography. *Biochem. Pharmacol.*, 64, 819, 2002.

Konishi, Y., Lehtinen, M., Donovan, N., and Bonni, A. Cdc2 phosphorylation of BAD links the cell cycle to the cell death machinery. *Mol. Cell*, 9, 1005, 2002.

Kotala, V., Uldrijan, S., Horky, M., Trbusek, M., Strnad, M., and Vojtesek, B. Potent induction of wild-type p53-dependent transcription in tumour cells by a synthetic inhibitor of cyclin-dependent kinases. *Cell Mol Life Sci.*, 58, 1333, 2001.

Krude, T. Initiation of human DNA replication in vitro using nuclei from cells arrested at an initiation-competent state. *J. Biol. Chem.*, 275, 13699, 2000.

Kwon, Y.G., Lee, S.Y., Choi, Y., Greengard, P., and Nairn, A.C. Cell cycle-dependent phosphorylation of mammalian protein phosphatase 1 by cdc2 kinase. *Proc. Natl. Acad. Sci. U.S.A.*, 94, 2168, 1997.

Lacrima, K., Valentini, A., Lambertini, C., Taborelli, M., Rinaldi, A., Zucca, E., Catapano, C., Cavalli, F., Gianella-Borradori, A., Maccallum, D.E., and Bertoni, F. In vitro activity of cyclin-dependent kinase inhibitor CYC202 (Seliciclib, *R*-roscovitine) in mantle cell lymphomas. *Ann. Oncol.*, 16, 1169, 2005.

Lam, L.T., Pickeral, O.K., Peng, A.C., Rosenwald, A., Hurt, E.M., Giltnane, J.M., Averett, L.M., Zhao H., Davis, R.E., Sathyamoorthy, M., Wahl, L.M., Harris, E.D., Mikovits, J.A., Monks, A.P., Hollingshead, M.G., Sausville, E.A., and Staudt, L.M. Genomic-scale measurement of mRNA turnover and the mechanisms of action of the anti-cancer drug flavopiridol. *Genome Biol.*, 2, RESEARCH0041, 2001.

Leach, C., Shenolikar, S., and Brautigan, D.L. Phosphorylation of phosphatase inhibitor-2 at centrosomes during mitosis. *J. Biol. Chem.*, 278, 26015, 2003.

Legraverend, M. and Grierson, D.S. The purines: potent and versatile small molecule inhibitors and modulators of key biological targets. *Bioorg. Med. Chem.*, 14, 3987, 2006.

Lipinski, C.A., Lombardo, F., Dominy, B.W., and Feeney, P.J. Experimental and computational approaches to estimate solubility and permeability in drug discovery and development settings. *Adv. Drug Delivery Rev.*, 23, 3, 1997.

Ljungman, M. and Paulsen, M.T. The cyclin-dependent kinase inhibitor roscovitine inhibits RNA synthesis and triggers nuclear accumulation of p53 that is unmodified at Ser15 and Lys382. *Mol. Pharmacol.*, 60, 785, 2001.

Lowe, M., Rabouille, C., Nakamura, N., Watson, R., Jackman, M., Jamsa, E., Rahman, D., Pappin, D.J., and Warren, G. Cdc2 kinase directly phosphorylates the cis-Golgi matrix protein GM130 and is required for Golgi fragmentation in mitosis. *Cell*, 94, 783, 1998.

Loyer, P., Trembley, J.H., Katona, R., Kidd, V.J., and Lahti, J.M. Role of CDK/cyclin complexes in transcription and RNA splicing. *Cell Signal.*, 17, 1033, 2005.

Lu, W., Chen, L., Peng, Y., and Chen, J. Activation of p53 by roscovitine-mediated suppression of MDM2 expression. *Oncogene*, 20, 3206, 2001.

MacCallum, D.E., Melville, J., Frame, S., Watt, K., Anderson, S., Gianella-Borradori, A., Lane, D.P., and Green, S.R. Seliciclib (CYC202, *R*-Roscovitine) induces cell death in multiple myeloma cells by inhibition of RNA polymerase II-dependent transcription and down-regulation of Mcl-1. *Cancer Res.*, 65, 5399, 2005.

Maggiorella, L., Deutsch, E., Frascogna, V., Chavaudra, N., Jeanson, L., Milliat, F., Eschwege, F., and Bourhis, J. Enhancement of radiation response by roscovitine in human breast carcinoma in vitro and in vivo. *Cancer Res.*, 63, 2513, 2003.

Malumbres, M. and Barbacid, M. To cycle or not to cycle: a critical decision in cancer. *Nat. Rev. Cancer*, 1, 222, 2001.

Malumbres, M. and Barbacid, M. Mammalian cyclin-dependent kinases. *Trends Biochem. Sci.*, 30, 630, 2005.

Mapelli, M., Massimiliano, L., Crovace, C., Seeliger, M., Tsai, L. H., Meijer, L., and Musacchio, A. Mechanism of CDK5/p25 binding by CDK inhibitors. *J. Med. Chem.*, 48, 671, 2005.

Matsumoto, Y., Hayashi, K., and Nishida, E. Cyclin-dependent kinase 2 (Cdk2) is required for centrosome duplication in mammalian cells. *Curr. Biol.*, 9, 429, 1999.

Maude, S.L. and Enders, G.H. Cdk inhibition in human cells compromises chk1 function and activates a DNA damage response. *Cancer Res.*, 65, 780, 2005.

McClue, S.J., Blake, B., Clarke, R., Cowan, A., Cummings, L., Fischer, P.M., MacKenzie, M., Melville, J., Stewart, K., Wang, S., Zhelev, N., Zheleva, N., and Lane, D.P. *In vitro* and *in vivo* antitumor properties of the cyclin dependent kinase inhibitor CYC202 (*R*)-roscovitine). *Int. J. Cancer.*, 102, 463, 2002.

Meijer, L. Le cycle cellulaire et sa régulation. *Oncologie*, 5, 1, 2003.

Meijer, L., Arion, D., Golsteyn, R., Pines, J., Brizuela, L., Hunt, T., and Beach, D. Cyclin is a component of the sea urchin egg M-phase specific histone H1 kinase. *EMBO J.*, 8, 2275, 1989.

Meijer, L., Borgne, A., Mulner, O., Chong, J.P.J., Blow, J.J., Inagaki, N., Inagaki, M., Delcros, J.G., and Moulinoux, J.P. Biochemical and cellular effects of roscovitine, a potent and selective inhibitor of the cyclin-dependent kinases cdc2, cdk2 and cdk5. *Eur. J. Biochem.*, 243, 527, 1997.

Meijer, L. and Pondaven, P. Cyclic activation of histone H1 kinase during the sea urchin egg mitotic divisions. *Exp. Cell Res.*, 174, 116, 1988.

Meijer, L. and Raymond, E. Roscovitine and other purines as kinase inhibitors: from starfish oocytes to clinical trials. *Acc. Chem. Res.*, 36, 417, 2003.

Mgbonyebi, O.P., Russo, J., Russo, I.H. Roscovitine induces cell death and morphological changes indicative of apoptosis in MDA-MB-231 breast cancer cells. *Cancer Res.*, 59, 1903, 1999.

Mgbonyebi, O.P., Russo, J., Russo, I.H. Roscovitine inhibits the proliferative activity of immortal and neoplastic human breast epithelial cells. *Anticancer Res.*, 18, 751, 1998.

Mihara, M., Shintani, S., Kiyota, A., Matsumura, T., and Wong, D.T. Cyclin-dependent kinase inhibitor (roscovitine) suppresses growth and induces apoptosis by regulating Bcl-x in head and neck squamous cell carcinoma cells. *Int. J. Oncol.*, 21, 95, 2002.

Milovanceva-Popovska, M., Kunter, U., Ostendorf, T., Petermann, A., Rong, S., Eitner, F., Kerjaschki, D., Barnett, A., and Floege, J. *R*-roscovitine (CYC202) alleviates renal cell proliferation in nephritis without aggravating podocyte injury. *Kidney Int.*, 67, 1362, 2005.

Mohapatra, S., Chu, B., Wei, S., Djeu, J., Epling-Burnette, P.K., Loughran, T., Jove, R., and Pledger, W.J. Roscovitine inhibits STAT5 activity and induces apoptosis in the human leukemia virus type 1-transformed cell line MT-2. *Cancer Res.*, 63, 8523, 2003.

Mohapatra, S., Chu, B., Zhao, X., and Pledger, W.J. Accumulation of p53 and reductions in XIAP abundance promote the apoptosis of prostate cancer cells. *Cancer Res.*, 65, 7717, 2005. Erratum in *Cancer Res.*, 65, 10635, 2005.

Neant, I., and Guerrier, P. 6-Dimethylaminopurine blocks starfish oocyte maturation by inhibiting a relevant protein kinase activity. *Exp. Cell Res.*, 176, 68, 1988.

Nelson, P.J. and Shankland, S.J. Therapeutics in renal disease: the road ahead for antiproliferative targets. *Nephron Exp. Nephrol.*, 103, e6, 2005.

Noble, M.E., Endicott, J.A., and Johnson, L.N. Protein kinase inhibitors: insights into drug design from structure. *Science*, 303, 1800, 2004.

Nutley, B.P., Raynaud, F.I., Wilson, S.C., Fischer, P.M., Hayes, A., Goddard, P.M., McClue, S.J., Jarman, M., Lane, D.P., and Workman, P. Metabolism and pharmacokinetics of the cyclin-dependent kinase inhibitor *R*-roscovitine in the mouse. *Mol. Cancer Ther.*, 4, 125, 2005.

O'Connor, D.S., Grossman, D., Plescia, J., Li, F., Zhang, H., Villa, A., Tognin, S., Marchisio, P.C., and Altieri, D.C. Regulation of apoptosis at cell division by p34cdc2 phosphorylation of survivin. *Proc. Natl. Acad. Sci. U.S.A.*, 97, 13103, 2000.

O'Connor, D.S., Wall, N.R., Porter, A.C., and Altieri, D.C. A p34(cdc2) survival checkpoint in cancer. *Cancer Cell.*, 2, 43, 2002.

Ortega, S., Prieto, I., Odajima, J., Martin, A., Dubus, P., Sotillo, R., Barbero, J.L., Malumbres, M., and Barbacid, M. Cyclin-dependent kinase 2 is essential for meiosis but not for mitotic cell division in mice. *Nat. Genet.*, 35, 25, 2003.

Otyepka, M., Kriz, Z., and Koca, J. Dynamics and binding modes of free cdk2 and its two complexes with inhibitors studied by computer simulations. *J. Biomol. Struct. Dyn.*, 2, 141, 2002.

Pareek, T.K., Keller, J., Kesavapany, S., Pant, H.C., Iadarola, M.J., Brady, R.O., and Kulkarni, A.B. Cyclin-dependent kinase 5 activity regulates pain signaling. *Proc. Natl. Acad. Sci. U.S.A.*, 103, 791, 2006.

Park, S.S., Eom, Y.W., and Choi, K.S. Cdc2 and Cdk2 play critical roles in low dose doxorubicin-induced cell death through mitotic catastrophe but not in high dose doxorubicin-induced apoptosis. *Biochem. Biophys. Res. Commun.*, 334, 1014, 2005.

Payton, M., Chung, G., Yakowec, P., Wong, A., Powers, D., Xiong, L., Zhang, N., Leal, J., Bush, T.L., Santora, V., Askew, B., Tasker, A., Radinsky, R., Kendall, R., and Coats, S. Discovery and evaluation of dual CDK1 and CDK2 inhibitors. *Cancer Res.*, 66, 4299, 2006.

Price, P.M., Safirstein, R.L., and Megyesi, J. Protection of renal cells from cisplatin toxicity by cell cycle inhibitors. *Am. J. Physiol. Renal Physiol.*, 286, F378, 2004.

Penuelas, S., Alemany, C., Noe, V., and Ciudad, C.J. The expression of retinoblastoma and Sp1 is increased by low concentrations of cyclin-dependent kinase inhibitors. *Eur. J. Biochem.*, 24, 4809, 2003.

Pinhero, R., Liaw, P., and Yankulov, K. A uniform procedure for the purification of CDK7/CycH/MAT1, CDK8/CycC and CDK9/CycT1. *Biol. Procedures Online*, 6, 163, 2004.

Pippin, J.W., Qu, Q., Meijer, L., and Shankland, S.J. Direct in vivo inhibition of the nuclear cell cycle cascade in experimental mesangial proliferative glomerulonephritis with roscovitine, a novel CDK2 antagonist. *J. Clin. Invest.*, 100, 2512, 1997.

Raje, N., Kumar, S., Hideshima, T., Roccaro, A., Ishitsuka, K., Yasui, H., Shiraishi, N., Chauhan, D., Munshi, N.C., Green, S.R., and Anderson, K.C. Seliciclib (CYC202 or *R*-roscovitine), a small-molecule cyclin-dependent kinase inhibitor, mediates activity via down-regulation of Mcl-1 in multiple myeloma. *Blood*, 106, 1042, 2005.

Rashidian, J., Iyirhiaro, G., Aleyasin, H., Rios, M., Vincent, I., Callaghan, S., Bland, R.J., Slack, R.S., During, M.J., and Park, D.S. Multiple cyclin-dependent kinases signals are critical mediators of ischemia/hypoxic neuronal death in vitro and in vivo. *Proc. Natl. Acad. Sci. U.S.A.*, 102, 14080, 2005.

Raynaud, F.I., Fischer, P.M., Nutley, B.P., Goddard, P.M., Lane, D.P., and Workman, P. Cassette dosing pharmacokinetics of a library of 2,6,9-trisubstituted purine cyclin-dependent kinase 2 inhibitors prepared by parallel synthesis. *Mol. Cancer Ther.*, 3, 353, 2004.

Raynaud, F.I., Whittaker, S.R., Fischer, P.M., McClue, S., Walton, M.I., Barrie, S.E., Garrett, M.D., Rogers, P., Clarke, S.J., Kelland, L.R., Valenti, M., Brunton, L., Eccles, S., Lane, D.P., and Workman, P. In vitro and in vivo pharmacokinetic-pharmacodynamic relationships for the trisubstituted aminopurine cyclin-dependent kinase inhibitors olomoucine, bohemine and CYC202. *Clin. Cancer Res.*, 11, 4875, 2005.

Rebhun, L.I., White, D., Sander, G., and Ivy, N. Cleavage inhibition in marine eggs by puromycin and 6-dimethylaminopurine. *Exp. Cell Res.*, 77, 312, 1973.

Ribas, J., Boix, J., and Meijer, L. (R)-Roscovitine (CYC202, Seliciclib) sensitizes SH-SY5Y neuroblastoma cells to Nutlin-3 induced apoptosis. *Exp. Cell Res.*, in press, 2006.

Rialet, V. and Meijer, L. A new screening test for antimitotic compounds using the universal M phase-specific protein kinase, p34^{cdc2}/cyclin B^{cdc13}, affinity-immobilized on p13^{suc1}-coated microtitration plates. *Anticancer Res.*, 11, 1581, 1991.

Ribas, J. and Boix, J. Cell differentiation, caspase inhibition, and macromolecular synthesis blockage, but not BCL-2 or BCL-XL proteins, protect SH-SY5Y cells from apoptosis triggered by two CDK inhibitory drugs. *Exp. Cell Res.*, 295, 9, 2004.

Rosato, R.R., Almenara, J.A., Maggio, S.C., Atadja, P., Craig, R., Vrana, J., Dent, P., and Grant, S. Potentiation of the lethality of the histone deacetylase inhibitor LAQ824 by the cyclin-dependent kinase inhibitor roscovitine in human leukemia cells. *Mol. Cancer Ther.*, 4, 1772, 2005.

Rosemeyer, H. The chemodiversity of purine as a constituent of natural products. *Chem. Div.*, 1, 361, 2004.

Sandal, T., Stapnes, C., Kleivdal, H., Hedin, L., and Doskeland, S.O. A novel, extraneuronal role for cyclin-dependent protein kinase 5 (CDK5): modulation of cAMP-induced apoptosis in rat leukemia cells. *J. Biol. Chem.*, 277, 20783, 2002.

Schang, L.M. Effects of pharmacological cyclin-dependent kinase inhibitors on viral transcription and replication. *Biochim. Biophys. Acta*, 1697, 197, 2004.

Schang, L.M., Bantly, A., Knockaert, M., Shaheen, F., Meijer, L., Malim, M.H., Gray, N.S., and Schaffer, P.A. Pharmacological cyclin-dependent kinase inhibitors inhibit replication of wild-type and drug-resistant strains of herpes simplex virus and human immunodeficiency virus type 1 by targeting cellular, not viral, proteins. *J. Virol.*, 76, 7874, 2002.

Shapiro, G.I. Cyclin-dependent kinase pathways as targets for cancer treatment. *J. Clin. Oncol.*, 24, 1770, 2006.

Sirri, V., Hernandez-Verdun, D., and Roussel, P. Cyclin-dependent kinases govern formation and maintenance of the nucleolus. *J. Cell Biol.*, 156, 969, 2002.

Smith, P.D., Crocker, S.J., Jackson-Lewis, V., Jordan-Sciutto, K.L., Hayley, S., Mount, M.P., O'Hare, M.J., Callaghan, S., Slack, R.S., Przedborski, S., Anisman, H., and Park, D.S. Cyclin-dependent kinase 5 is a mediator of dopaminergic neuron loss in a mouse model of Parkinson's disease. *Proc. Natl. Acad. Sci. U.S.A.*, 100, 13650, 2003.

Smith, P.D., O'Hare, M.J. and Park, D.S. CDKs: taking on a role as mediators of dopaminergic loss in Parkinson's disease. *Trends Mol. Med.*, 10, 445, 2004.

Somerville, L. and Cory, J.G. Enhanced roscovitine-induced apoptosis is mediated by a caspase-3-like activity in deoxyadenosine-resistant mouse leukemia L1210 cells. *Anticancer Res.*, 20, 3347, 2000.

Tang, L., Li, M.H., Cao, P., Wang, F., Chang, W.R., Bach, S., Reinhardt, J., Koken, M., Galons, H., Wan, Y., Gray, N., Meijer, L., Jiang. T., and Liang, D.C. Crystal structure of pyridoxal kinase in complex with roscovitine and derivatives. *J. Biol. Chem.* 280, 31220, 2005.

Tirado, O.M., Mateo-Lozano, S., and Notario, V. Roscovitine is an effective inducer of apoptosis of Ewing's sarcoma family tumor cells in vitro and in vivo. *Cancer Res.*, 65, 9320, 2005.

Tsai, L.H., Lee, M.S., and Cruz, J. Cdk5, a therapeutic target for Alzheimer's disease? *Biochim. Biophys. Acta*, 1697, 137, 2004.

Valsasina, B., Kalisz, H.M., and Isacchi, A. Kinase selectivity profiling by inhibitor affinity chromatography. *Future Drugs*, 1, 303, 2004.

van Engeland, M., Kuijpers, H.J., Ramaekers, F.C., Reutelingsperger, C.P., and Schutte, B. Plasma membrane alterations and cytoskeletal changes in apoptosis. *Exp. Cell Res.*, 235, 421, 1997.

Vermeulen, K., Van Bockstaele, D.R., and Berneman, Z.N. The cell cycle: a review of regulation, deregulation and therapeutic targets in cancer. *Cell Prolif.*, 36, 131, 2003.

Vesely, J., Havlicek, L., Strnad, M., Blow, J. J., Donella-Deana, A., Pinna, L., Letham, D. S., Kato, J. Y., Détivaud, L., Leclerc, S., and Meijer, L. Inhibition of cyclin-dependent kinases by purine derivatives. *Eur. J. Biochem.*, 224, 771, 1994.

Vita, M., Abdel-Rehim, M., Nilsson, C., Hassan, Z., Skansen, P., Wan, H., Meurling, L., and Hassan, M. Stability, pKa and plasma protein binding of roscovitine *J. Chromatogr. B,* 821, 75, 2005a.

Vita, M., Abdel-Rehim, M., Olofsson, S., Hassan, Z, Meurling, L., Siden, A., Siden, M., Pettersson T., and Hassan, M. Tissue distribution, pharmacokinetics and identification of roscovitine metabolites in rat. *Eur. J. Pharm. Sci.*, 25, 91, 2005b.

Vita, M., Meurling, L., Pettersson, T., Cruz-Siden, M., Siden, A., and Hassan, M. Analysis of roscovitine using novel high performance liquid chromatography and UV-detection method: pharmacokinetics of roscovitine in rat. *J. Pharm. Biomed. Anal.*, 34, 425, 2004.

Vita, M., Skansen, P., Hassan, M., and Abdel-Rehim, M. Development and validation of a liquid chromatography and tandem mass spectrometry method for determination of roscovitine in plasma and urine samples utilizing on-line sample preparation. *J. Chromatogr. B*, 817, 303, 2005c.

Vlach, J., Hennecke, S., and Amati, B. Phosphorylation-dependent degradation of the cyclin-dependent kinase inhibitor p27. *EMBO J.*, 16, 5334, 1997.

Wall, N.R., O'Connor, D.S., Plescia, J., Pommier, Y., and Altieri, D.C. Suppression of survivin phosphorylation on Thr34 by flavopiridol enhances tumor cell apoptosis. *Cancer Res.*, 63, 230, 2003.

Wang, D., de La Fuente, C., Deng, L., Wang, L., Zilberman, I., Eadie, C., Healey, M., Stein, D., Denny, T., Harrison, L.E., Meijer, L., and Kashanchi, F. Inhibition of human immunodeficiency virus type 1 transcription by chemical cyclin-dependent kinase inhibitors. *J. Virol.*, 75, 7266, 2001a.

Wang, J., Liu, S.H., Fu, Y.P., Wang, J.H., and Lu, Y.M. Cdk5 activation induces hippocampal CA1 cell death by directly phosphorylating NMDA receptors. *Nat. Neurosci.*, 6, 1039, 2003.

Wang, S., McClue, S.J., Ferguson, J.R., Hull, J.D., Stokes, S., Parsons, S., Westwood, R., and Fischer, P.M. Synthesis and configuration of the cyclin-dependant kinase inhibitor roscovitine and its enantiomer. *Tetrahedron: Asymmetry*, 12, 2891, 2001b.

Weinmann, H. and Metternich, R. Drug discovery process for kinase inhibitors. *ChemBio-Chemistry*, 6, 455, 2005.

Wesierska-Gadek, J., Gueorguieva, M., and Horky, M. Dual action of cyclin-dependent kinase inhibitors: induction of cell cycle arrest and apoptosis: a comparison of the effects exerted by roscovitine and cisplatin. *Pol. J. Pharmacol.*, 55, 895, 2003.

Wesierska-Gadek, J., Gueorguieva, M., and Horky, M. Roscovitine-induced up-regulation of p53AIP1 protein precedes the onset of apoptosis in human MCF-7 breast cancer cells. *Mol. Cancer Ther.*, 4, 113, 2005. Erratum in *Mol. Cancer Ther.*, 4, 503, 2005.

Whittaker, S.R., Walton, M.I., Garrett, M.D., and Workman, P. The cyclin-dependent kinase inhibitor CYC202 (*R*-roscovitine) inhibits retinoblastoma protein phosphorylation, causes loss of Cyclin D1, and activates the mitogen-activated protein kinase pathway. *Cancer Res.*, 64, 262, 2004.

Wojciechowski, J., Horky, M., Gueorguieva, M., and Wesierska-Gadek, J. Rapid onset of nucleolar disintegration preceding cell cycle arrest in roscovitine-induced apoptosis of human MCF-7 breast cancer cells. *Int. J. Cancer*, 106, 486, 2003.

Yu, C., Rahmani, M., Dai, Y., Conrad, D., Krystal, G., Dent, P., and Grant, S. The lethal effects of pharmacological cyclin-dependent kinase inhibitors in human leukemia cells proceed through a phosphatidylinositol 3-kinase/Akt-dependent process. *Cancer Res.*, 63, 1822, 2003.

Zhang, M., Li, J., Chakrabarty, P., Bu, B., and Vincent, I. Cyclin-dependent kinase inhibitors attenuate protein hyperphosphorylation, cytoskeletal lesion formation, and motor defects in Niemann-Pick Type C mice. *Am. J. Pathol.*, 165, 843, 2004a.

Zhang, G.J., Safran, M., Wei, W., Sorensen, E., Lassota, P., Zhelev, N., and Neuberg, D.S., Shapiro, G., and Kaelin, W.G. Jr. Bioluminescent imaging of Cdk2 inhibition in vivo. *Nat. Med.*, 10, 643, 2004b.

10 Development, Selectivity, and Application of Paullones, a Family of CDK Inhibitors

Conrad Kunick, Thomas Lemcke, and Laurent Meijer

CONTENTS

10.1 INTRODUCTION

Among the manifold chemotypes described as cyclin-dependent kinase (CDK) inhibitors, the paullones are an exceptional compound family for at least two reasons. First, the series was discovered by systematic data mining in the National Cancer Institute's (NCI's) database of selectivity patterns from the Antitumor Drug Screening program. Therefore, the discovery of paullones is an example of a successful drug discovery program based on bioinformatics beyond virtual screening. Second, several groups have used the commercially available paullones as biochemical tools; consequently, a variety of interesting results from diverse research fields have been reported. As with the many other CDK inhibitor families, a number of qualitative and quantitative structure–activity relationship (QSAR) investigations and structure optimization programs were carried out on the paullone series, which will be summarized in this chapter. Short survey articles concentrating on special aspects of the paullone kinase inhibitor family have already been published elsewhere (Kunick and Lemcke, 2002; Meijer et al., 2005).

10.2 DISCOVERY OF KENPAULLONE, THE PARENT STRUCTURE OF THE PAULLONES

In many human cancers, hyperactivity of CDKs has been associated with the pathological mechanisms underlying hyperproliferation (Malumbres and Barbacid, 2001). Consequently, the inhibition of CDKs has been suggested as a therapeutic option for cancer patients (Buolamwini, 2000; Knockaert et al., 2002a; Grant and Roberts, 2003; Sausville, 2003). The NCI has participated in the development of flavopiridol and UCN-01, two anticancer drugs exhibiting CDK-inhibitory activity that are currently undergoing clinical trials (Monks et al., 1997). In the mid-1990s, scientists at the NCI were interested in additional CDK inhibitors structurally unrelated to both flavopiridol and UCN-01. A project was therefore initiated at the NCI to extract such a novel CDK inhibitor chemotype from the database containing the selectivity patterns of compounds tested in the NCI's Antitumor Drug Screen (Sausville et al., 1999). In this screening system, test compounds are evaluated *in vitro* for growth inhibitory activity on a panel of 60 human tumor cell lines from nine different tumor entities (melanoma, leukemia, breast, prostate, lung, colon, ovary, kidney, and central nervous system origin) (Monks et al., 1991). This disease-oriented screening strategy was originally based on the hypothesis that selective activity against tumor cells from a particular organ could predict selective activity against corresponding tumors in humans. After several years of application, it became obvious that antiproliferative compounds with organ selectivity are rare. However, the selectivity pattern generated with an antiproliferative agent over the 60 cell lines has turned out to be characteristic of the biochemical mechanism underlying growth inhibition (Weinstein et al., 1997). Therefore, chemically different structures with similar biological targets in tumor cells can be identified by a comparison of the described selectivity patterns or "fingerprints." The COMPARE algorithm has been developed as a useful computerized tool for the discovery of chemotypes that act with a cellular mechanism similar to that of a reference or "seed compound" (Paull et al., 1989). Employing this strategy, novel chemotypes of agents interacting with

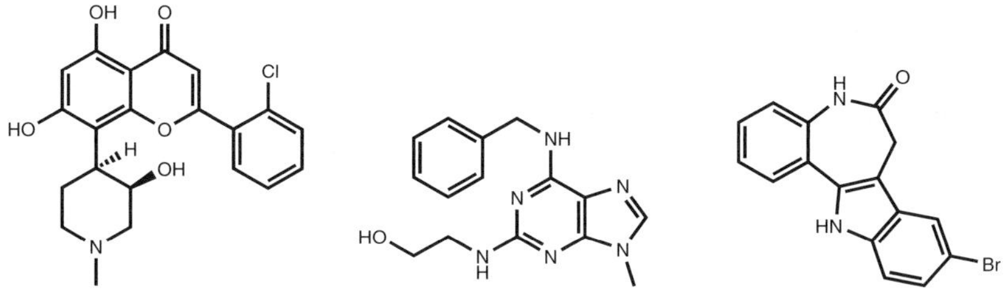

Flavopiridol (10.1)   Olomoucine (10.2)   NSC-664704
                                                                      (Kenpaullone) (10.3)

FIGURE 10.1 Structures of the CDK inhibitors flavopiridol (**10.1**), olomoucine (**10.2**), and kenpaullone (NSC-664704, **10.3**).

topoisomerase I (Kohlhagen et al., 1998), dihydroorotate dehydrogenase (Cleaveland et al., 1995), tubulin (Paull et al., 1992), and inosine monophosphate dehydrogenase (Gharehbaghi et al., 1994) have been discovered. Using the selectivity pattern of flavopiridol **10.1** (Figure 10.1) as a seed, the COMPARE analysis carried out on a database comprising the selectivity patterns of approximately 35,000 nonproprietary compounds revealed 11 structures with Pearson correlation coefficients >0.6. Highest ranked among these structures was the established CDK inhibitor olomoucine **10.2** (NSC-666069, Pearson correlation coefficient = 0.67). One of the remaining ten compounds was the 7,12-dihydroindolo[3,2-*d*][1]benzazepin-6(5*H*)-one NSC-664704 **10.3**. Although structurally unrelated to both flavopiridol and olomoucine, NSC-664704 turned out to be a moderately potent inhibitor of CDK1/cyclin B (IC$_{50}$ = 0.4 µM @ 15 µM ATP, apparent K$_i$ = 2.5 µM). Kinetic studies at different ATP concentrations showed that NSC-664704 inhibits CDK1/cyclin B by competition with ATP (Zaharevitz et al., 1999). Despite the high similarity of the selectivity patterns, the absolute averaged antiproliferative activity of NSC-664704 in the cell line screening (GI$_{50}$ ~ 42 µM) is clearly inferior compared to flavopiridol, which exhibits growth inhibition in the nanomolar concentration range (GI$_{50}$ ~ 75 nM) (Sausville et al., 1999).

NSC-664704 was later named kenpaullone to honor the memory of the late Dr. Ken Paull, the scientist at the National Cancer Institute who invented the COMPARE algorithm (Paull et al., 1989). Eventually, the unsubstituted derivative **10.13** was named "paullone," and the whole family of structures based on the dihydroindolo[3,2-*d*][1]benzazepine scaffold was designated paullones (Zaharevitz et al., 1999). At the time of their discovery, CDK inhibitors with structural features resembling the paullone scaffold were not known. Several projects were initiated at that point aimed at the development of optimized compounds and the characterization of the effect of the paullones in biological and biochemical models.

10.3 SYNTHESIS OF PAULLONES

For the synthesis of the paullone scaffold, two general strategies have been realized. The majority of the compounds published so far have been prepared by a sequence employing a Fischer indole ring closure as the key step, which is outlined in Figure 10.2.

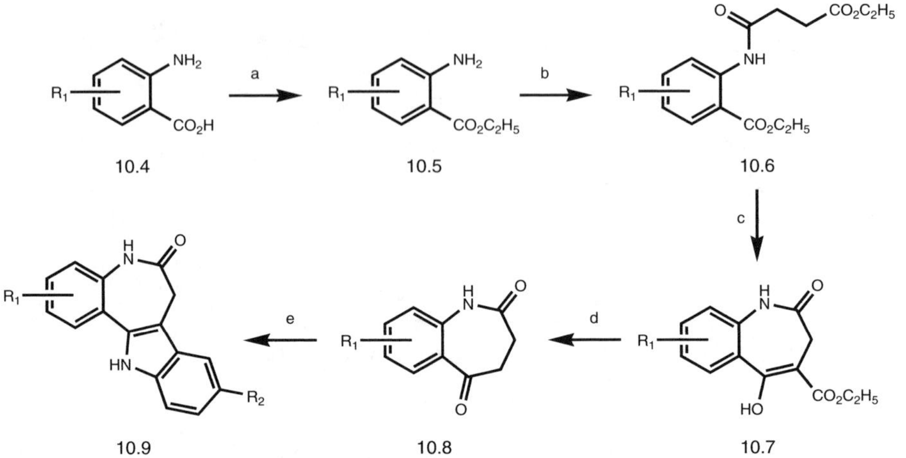

FIGURE 10.2 Synthesis of paullone derivatives. Reagents and conditions: (a) iodoethane, DMF, K_2CO_3; (b) ethyl succinyl chloride, toluene, pyridine, reflux; (c) KH, toluene, DMF, 70°C; (d) DMSO, H_2O, 150°C; (e) 1. 4-R_2-Ph-NH-NH_2, AcOH, 70°C; 2. H_2SO_4, AcOH, 70°C or diphenyl ether, reflux. (From Schultz, C. et al. (1999) *Journal of Medicinal Chemistry* **42**, 2909–2919; Kunick, C. et al. (2000) *Bioorganic and Medicinal Chemistry Letters* **10**, 567–569; Kunick, C. et al. (2004b) *Journal of Medicinal Chemistry* **47**, 22–36.)

The sequence starts with an anthranilic acid derivative **10.4**, which is esterified in the first step to yield the ester **10.5**. Acylation at the nitrogen with ethyl succinyl chloride furnishes the diester **10.6**, which is subsequently cyclized by treatment with potassium hydride in a Dieckmann ring closure reaction. The resulting enolized β-oxocarboxylic ester **10.7** is dealkoxycarbonylated by heating in wet DMSO. Eventually, the cyclic ketone **10.8** undergoes a Fischer indole reaction with a suitable phenyl hydrazine in the presence of sulfuric acid (Kunick, 1992; Schultz et al., 1999). The Fischer indole ring closure fails or gives only poor yields with electron-deficient

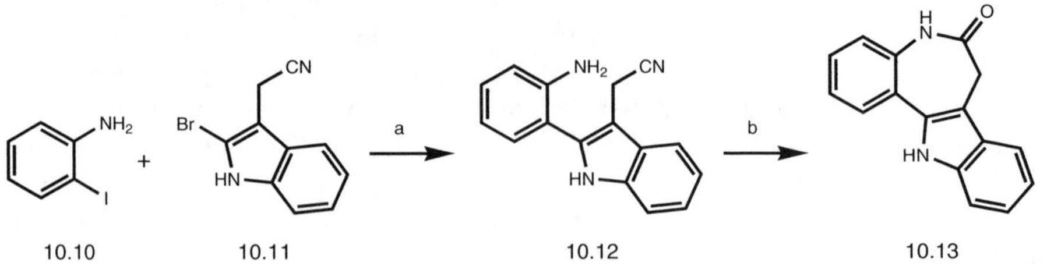

FIGURE 10.3 Synthesis of unsubstituted paullone **10.13** based on borylation/Suzuki coupling (BSC) reaction as key step. Reagents and conditions: (a) **10.10**, Et_3N, $Pd(OAc)_2$, PCy_2(o-biph), pinacolborane, dioxane, 80°C, then H_2O, **10.11**, $Ba(OH)_2$ • 8 H_2O, 100°C (62%); (b) NaOH, MeOH, H_2O, reflux (51%). (From Baudoin, O. et al. *Journal of Organic Chemistry* **67**, 1199–1207.)

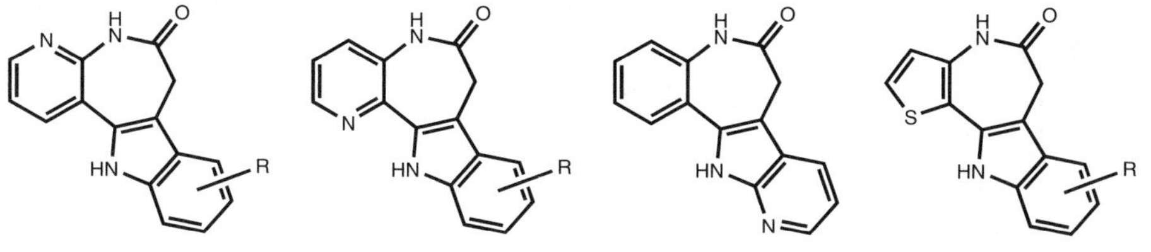

10.14   10.15   10.16   10.17

FIGURE 10.4 Structures of heterocyclic paullone analogs.

phenyl hydrazine derivatives. In these cases, a thermal indole ring closure can be induced by heating the isolated phenylhydrazone precursor in diphenyl ether (Kunick, et al., 2004b). In addition to the dealkoxycarbonylation of precursor **10.7**, several different methods have been reported for the preparation of intermediates **10.8** from other starting materials (Rees, 1959; Jones, 1967; Moriconi and Maniscalco, 1972; Witte and Boekelheide, 1972; Rivett and Stewart, 1978; Chen and Gilman, 1983; Wieking et al., 2002). An alternative route leading to the paullone scaffold via a palladium-catalyzed borylation/Suzuki coupling (BSC) reaction as outlined in Figure 10.3 has been carried out, yielding the unsubstituted paullone **10.13** (Baudoin et al., 2002). The method outlined in Figure 10.2 is also useful for the preparation of paullone analogs with modified heterocyclic scaffolds starting either from heterocyclic aminocarboxylic acids instead of the anthranilic acids **10.4,** or by applying heterocyclic hydrazides in the final indole ring closure step. Employing such procedures, heterocyclic paullone analogs represented by general structures **10.14–10.17** (Figure 10.4) have been synthesized (Kunick, 1992; Kunick et al., 2004a, 2004b). The nitrogen atoms of the lactam and the indole substructure of paullones can be alkylated regioselectively. While reaction of a paullone with alkyl halide in the presence of sodium hydride in DMF leads to substitution at the lactam nitrogen, the use of potassium hydroxide in acetone leads to alkylation at the indole nitrogen (Schultz et al., 1999).

10.4 BIOLOGICAL PROPERTIES OF PAULLONES

10.4.1 INHIBITION OF CDKs

10.4.1.1 Structure–Activity Relationships for CDK1 Inhibition

Soon after the discovery of the paullone scaffold as a kinase inhibitor, a variety of derivatives were prepared following the synthetic strategies outlined earlier and tested for CDK1/cyclin B inhibitory activity. The corresponding data for representative paullones are summarized in Table 10.1 (Schultz et al., 1999; Zaharevitz et al., 1999; Leost et al., 2000; Wieking et al., 2002; Kunick et al., 2004b).

TABLE 10.1

CDK1/Cyclin B Inhibition[a] and *In Vitro* Growth Inhibition of Cancer Cell Lines[b] by Paullones

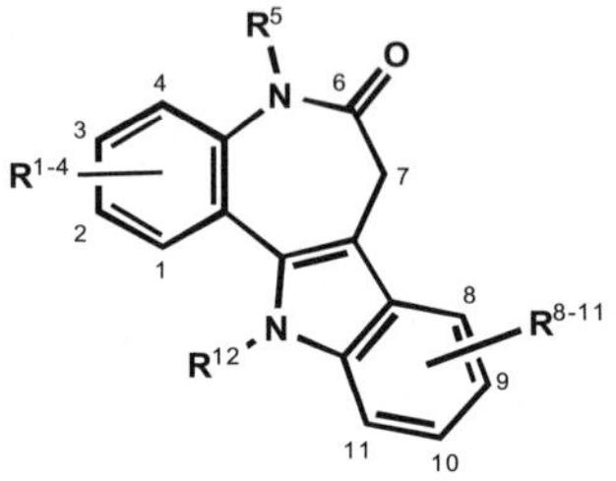

Structure 10.18

Entry	R	IC_{50} CDK1/Cyclin B [μM][a]	Log GI_{50} HCT 116 [c] [M]	Log GI_{50} MG MID [d] [M]
10.3	9-Br (Kenpaullone)	0.4	−5.7	−4.3
10.13	R = H (unsubstituted Paullone)	7	−4.8	−4.5
10.19	9-Cl	0.6	−5.5	−4.1
10.20	9-F	1.6	−5.1	−4.6
10.21	$9\text{-}OCH_3$	0.9	−4.6	−4.6
10.22	$9\text{-}CH_3$	2.0	−4.0	−4.0
10.23	$9\text{-}CF_3$	0.4	−5.4	−4.1
10.24	$9\text{-}NH_2$	20	−4.1[e]	−4.1[e]
10.25	9-NH-Ac	1.3	>−4.0[e]	−4.0[e]
10.26	9-CN	0.024	−4.7	−4.1
10.27	$9\text{-}NO_2$ (Alsterpaullone)	0.035	−6.7	−6.4
10.28	$2,3\text{-di-}OCH_3$; 9-Br	0.2	−5.6	−5.1
10.29	2,3-di-OH; 9-Br	3.0	>−4.0	−4.0
10.30	$2,3\text{-di-}OCH_3$; 9-CN	0.044	−6.3	−5.3[e]
10.31	$2,3\text{-di-}OCH_3$; $9\text{-}NO_2$	0.024	−6.3	−5.7[e]
10.32	$2\text{-}OCH_3$; 9-Br	0.4	−5.4	−4.5
10.33	$2\text{-}OCH_3$; $9\text{-}NO_2$	0.08	n.a.	n.a.
10.34	$2\text{-}OCH_3$; $9\text{-}SCH_3$	0.34	>−4.0	−4.1[e]
10.35	$2\text{-}OCH_3$; $9\text{-}SOCH_3$	0.08	−5.0	−4.6[e]
10.36	$2\text{-}OCH_3$; $9\text{-}SO_2NH_2$	0.18	−4.6	−4.2[e]
10.37	$2\text{-}OCH_3$; $9\text{-}SO_2CH_3$	0.009	>−4.0	−4.1[e]
10.38	10-Br	1.3	−5.5	−4.8
10.39	11-Br	0.9	n.a.	−4.3
10.40	$4\text{-}OCH_3$, 9-Br	250	−5.0	−4.7
10.41	$5\text{-}CH_3$, 9-Br	20	−4.7	−4.6
10.42	9-Br, $12\text{-}CH_3$	6.2	−5.1	−4.6
10.43	8,10-di-Cl	2.5	>−4.0	−4.0
10.44	Thiolactam analog of kenpaullone (see Figure 10.5)	2.3	n.a.	n.a.
10.45	Hydroxyamidine analog of kenpaullone (see Figure 10.5)	1.0	−5.2	−5.0
10.46	Methylthioimidate analog of kenpaullone (see Figure 10.5)	43	−5.8	−5.6
10.47	Nonlactam analog of kenpaullone (see Figure 10.5)	51	n.a.	n.a.
10.48	2-Substituted paullone (see Figure 10.6)	0.3	−5.5	−4.8

TABLE 10.1 (CONTINUED)
CDK1/Cyclin B Inhibition[a] and *In Vitro* Growth Inhibition of Cancer Cell Lines[b] by Paullones

Entry	R	IC_{50} CDK1/Cyclin B $[\mu M]^a$	Log GI_{50} HCT 116 [c] [M]	Log GI_{50} MG MID [d] [M]
10.49	2-Substituted paullone (see Figure 10.6)	0.32	−5.8	−5.4
10.50	2-Substituted paullone (see Figure 10.6)	0.047	−5.9	−5.6
10.51	4-Azakenpaullone (see Figure 10.6)	8.4	n.a.	n.a.
10.52	1-Azakenpaullone (see Figure 10.6)	2.0	n.a.	n.a.
10.53	Thiophene analog of kenpaullone (see Figure 10.6)	0.6	>−4.0[e]	−4.0[e]

Note: n.a. = not available.

[a] Kinase assays were carried out in the presence of 15 µM ATP. (From Schultz, C. et al. (1999) *Journal of Medicinal Chemistry* **42**, 2909–2919 and Leost, M. et al. (2000) *European Journal of Biochemistry* **267**, 5983–5994; Kunick, C. et al. (2000). *Bioorganic and Medicinal Chemistry Letters* **10**, 567–569 and Wieking, K. et al. (2002) *Archiv der Pharmazie — Pharmaceutical and Medicinal Chemistry* **335**, 311–317; Pies, T. et al. (2004) *Archiv der Pharmazie — Pharmaceutical and Medicinal Chemistry* **337**, 486–492; Kunick, C. et al. (2004a) *Bioorganic and Medicinal Chemistry Letters* **14**, 413–416 and Kunick, C. et al. (2004b) *Journal of Medicinal Chemistry* **47**, 22–36).

[b] Data obtained from the NCIs *in vitro* Antitumor Drug Screen (ADS). Indicated are $\log_{10}$ values of molar concentrations. (From Schultz, C. et al. (1999) *Journal of Medicinal Chemistry* **42**, 2909–2919; Kunick, C. et al. (2000) *Bioorganic and Medicinal Chemistry Letters* **10**, 567–569 and Wieking, K.et al. (2002). *Archiv der Pharmazie — Pharmaceutical and Medicinal Chemistry* **335**, 311–317; Pies, T. et al. (2004). *Archiv der Pharmazie — Pharmaceutical and Medicinal Chemistry* **337**, 486–492).

[c] Colon cancer cell line.

[d] MG MID = Mean graph Midpoint = mean value for all tested cancer cell lines. If the indicated effect was not attainable for distinct cell lines within the used concentration interval, the highest tested concentration (= 100 µM) was used for the calculation.

[e] Previously unpublished results.

Evaluation of the CDK1/cyclin B kinase inhibition assay results showed that a substitution at the 9-position with an electron-withdrawing group leads to relatively strong kinase inhibition; namely, the 9-cyano derivative **10.26**, the 9-nitro derivative **10.27** (alsterpaullone), and the 9-sulfonyl derivative **10.37** are considerably more potent compared to the lead structure kenpaullone. In contrast, electron-donating groups at position 9 are detrimental to kinase inhibition as exemplified by the methoxy, methyl, and amino derivatives **10.21, 10.22,** and **10.24**. The bromo substituent can be replaced by a trifluoromethyl group without loss of activity as shown by **10.23**. The position of the electron-withdrawing group seems to be optimal in the 9-position because a shift of the bromo substituent of kenpaullone to the 10- (cf. **10.38**) or the 11-position causes decreased activity. A similar decrease is observed after exchange of the 9-bromo-substituent for a 8,10-dichlorosubstitution (cf. **10.43**). Substitutions at the 4-position, or at either the lactam or the indole nitrogens, are clearly detrimental as is demonstrated by the reduced kinase inhibition exhibited by **10.40, 10.41,** and **10.42**, respectively. In contrast, a substitution with methoxy

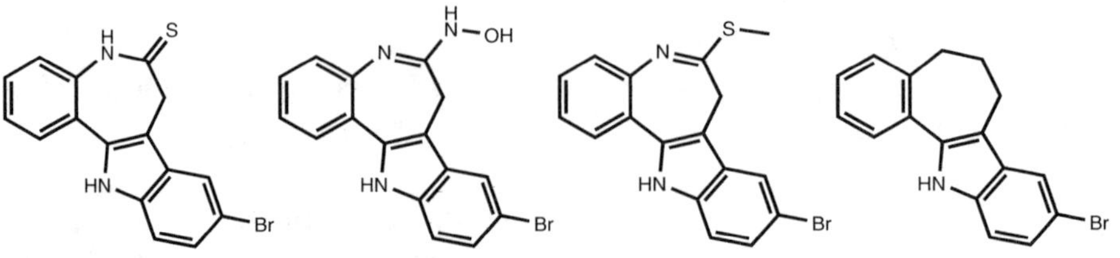

10.44   10.45   10.46   10.47

FIGURE 10.5 Kenpaullone derivatives modified at the lactam moiety.

groups in positions 2 and 3 is tolerated (cf. **10.28** and **10.30–10.37**). However, the 2,3-dihydroxy derivative **10.29** of kenpaullone is clearly inferior to the parent structure. The importance of the unchanged secondary lactam group is demonstrated by a comparison of kenpaullone with its derivatives depicted in Figure 10.5. Neither a thiolactam (cf. **10.44**), a methylthioimidate (cf. **10.46**), nor an ethylene bridge, as in **10.47**, can replace the lactam without considerable loss of activity. The only group that might be similar to the secondary lactam is the hydroxyamidine structure displayed by compound **10.45**.

On the basis of preliminary docking studies (Zaharevitz et al., 1999) with the paullones in a homology model of CDK1, it was postulated that even larger substituents than methoxy would be tolerated in the 2-position because the 2,3-edge of the paullone molecules is directed toward the entrance of the ATP-binding pocket. This assumption was verified by the results of kinase inhibition tests with the derivatives **10.48–10.50** (Figure 10.6), which proved to be equipotent or superior to the parent structure 9-trifluoromethylpaullone **10.23**.

Although the modification of the basic paullone scaffold by introduction of pyridine nitrogen atoms in the 1- or 4-position (**10.52** and **10.51**, respectively) leads to a decrease in CDK-inhibitory activity of roughly one order of magnitude, the thiophene analog **10.53** of kenpaullone is basically equipotent with regard to the lead structure (cf. Figure 10.7).

Besides qualitative structure–activity considerations, QSAR evaluations have also been reported for CDK1 inhibition by the paullones. In a classical approach, a multiple linear regression analysis was applied on a small series of 17 9-substituted paullones (Pies et al., 2004). While the electronic properties of the 9-substituents proved to be of high relevance for CDK1 inhibition, both lipophilic and steric parameters could

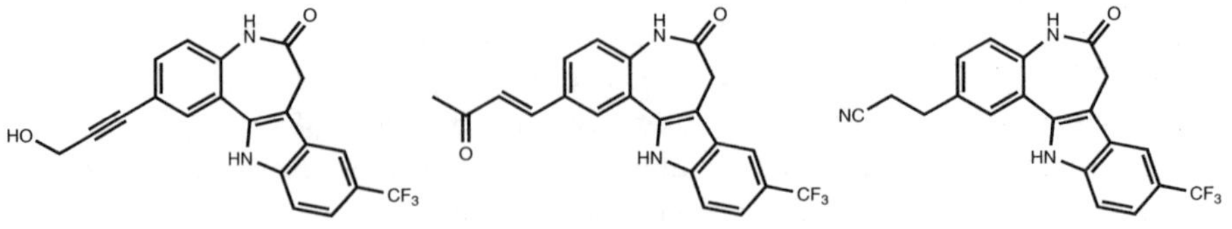

10.48   10.49   10.50

FIGURE 10.6 2-Substituted paullones.

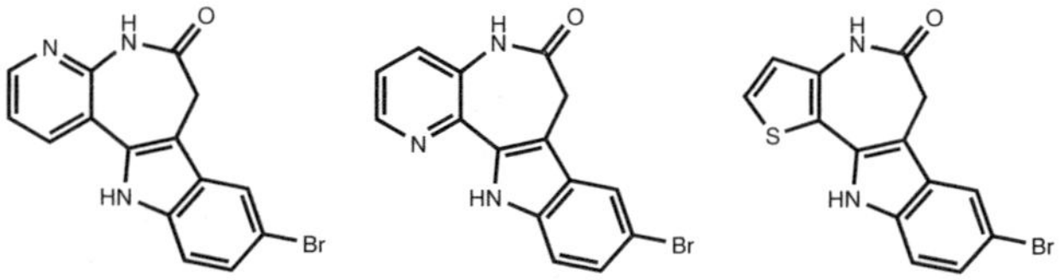

10.51   10.52   10.53

FIGURE 10.7 Heterocyclic analogs of kenpaullone.

not be included in a statistically significant manner for the calculation of biological properties. Equation 10.1, which is based exclusively on the electronic parameter σ_m of the 9-substituent, demonstrated satisfactory predictive power when it was challenged by a small test set of novel paullone derivatives that were synthesized and tested for this purpose.

$$pIC_{50} \; (CDK1/cyclin \; B) = 5.35(\pm 0.26) + 3.06(\pm 0.64)\sigma_m$$

$$n = 17; \; t(\sigma_m) = 10.21; \; r = 0.935; \; s = 0.314; \; F = 104.28; \; q^2 = 0.838 \quad (10.1)$$

A 3D-QSAR model has been reported in which the CDK1/cyclin B inhibition data of 26 paullones substituted on different positions were included (Gussio et al., 2000). To realize this approach, a CDK1/cyclin B homology model was constructed based on the crystal structure of CDK2/cyclin A (McGrath et al., 2005). After docking the paullones into the ATP-binding pocket, hydropathic analyses of the complex were performed based on the electronic properties of the paullones calculated by quantum mechanics. The hydropathic descriptors thus obtained served as predictors in a multiple regression analysis with the CDK1/cyclin B IC_{50} data serving as outcome measure. The striking result of this effort was the accuracy in predicting the high inhibitory activity of the 9-nitro and 9-cyano paullones.

In another 3D-QSAR approach, the CDK1/cyclin B kinase inhibition data of 52 paullone entities were included (Kunick et al., 2004b). Docking the paullone structures into the ATP-binding cleft of McGrath's CDK1/cyclin B homology model (McGrath et al., 2005) nicely reproduced the three hydrogen bonds (azepine NH–Leu83 backbone carbonyl, azepine carbonyl–Leu83 backbone NH, indole NH–Asp86 side-chain carbonyl) of the model reported by Gussio (Gussio et al., 2000). The resulting relative conformations of the docked paullones were used as the alignment for subsequent CoMSIA calculations. Grid spacing (2 Å) and column filtering (0.5) resulted in a five component model with a q^2 value of 0.699 and an r^2 value of 0.929. The performance of the resulting model was demonstrated by the prediction of an independent set of test compounds that also comprised heterocyclic paullone analogs of the general structures **10.14**, **10.16,** and **10.17**. The analysis of the field distribution of the five CoMSIA fields showed a relatively low influence of the steric field, whereas the electrostatic field added high contributions to the component composition for CDK1, an observation that is in agreement with the results of the two other QSAR reports mentioned previously.

10.4.1.2 Antiproliferative Activity on Cancer Cells

The majority of compounds from the paullone series have been tested for antiproliferative activity in the Antitumor Drug Screen of the NCI (Schultz et al., 1999; Kunick
et al., 2000; Wieking et al., 2002; Pies et al., 2004). The evaluation of the selected
data presented in Table 10.1 shows that the CDK1/cyclin B inhibitory activity does
not necessarily parallel the antiproliferative activity on cancer cells. However, a characteristic feature found in this class of antiproliferative agents is the high sensitivity of
the colon cancer cell line HCT-116, which is often inhibited by lower drug concentrations than the average cell line as expressed by the mean graph midpoint. Exceptions
to this rule are usually derivatives with lower CDK inhibitory activity, e.g., **10.21**,
10.22, and **10.29**. Interestingly, some paullones that are very potent CDK1 inhibitors
(cf. 9-cyanopaullone **10.26**: IC_{50} CDK1/cyclin B = 24 nM and 2-methoxy-9-
methylsulfonylpaullone **10.37**: IC_{50} CDK1/cyclin B = 9 nM) are practically devoid of
cancer cell growth inhibitory activity, a phenomenon that might be due to poor cell
uptake. On the other hand, the 9-nitropaullone (alsterpaullone, **10.27**) exhibits out-
standing tumor cell line inhibition within the series (Schultz et al., 1999). Because a
correlation between cancer cell growth inhibition and inhibition of CDKs is not
apparent, a question regarding the identity of the other mechanisms contributing to
the anti-tumor activity of the paullones arises. It has been reported that alsterpaullone
10.27 induces apoptosis in Jurkat cell lines (Lahusen et al., 2003). Apoptosis by
alsterpaullone was accompanied by activation of caspase-8 and caspase-9, which
resulted in cleavage of poly(ADP-ribose)polymerase (PARP) and caspase-3. The
involvement of caspases was shown by the application of the general caspase inhibitor
ZVAD, which prevented alsterpaullone-induced apoptosis. On the other hand, it was
demonstrated that neither a modulation of pro- or antiapoptotic proteins nor an activation of the extrinsic pathway by production of the extracellular CD-95 ligand are
pivotal for alsterpaullone-induced apoptosis. In fact, it was shown that a disruption of the
inner mitochondrial membrane potential by alsterpaullone results in subsequent activation
of caspase 9 and apoptosis induction, a mechanism independent of the cell cycle effects
of alsterpaullone (Lahusen et al., 2003). According to their CDK inhibitory activity,
paullones lead to a delay in cell cycle progression, as was shown for alsterpaullone **10.27**
in Jurkat cells (Lahusen et al., 2003) and for kenpaullone **10.3** in serum-starved MCF-10
A cells (Zaharevitz et al., 1999). In a panel of eight colon cancer cell lines, alsterpaullone
exhibited growth inhibition in the submicromolar concentration range. The activity was
similar in all cell lines with no evident relationship to p53, cyclin, or CDK expression
(Price et al., 2000). Using DNA microarrays, it has been shown that, in contrast to
flavopiridol, alsterpaullone **10.27** is not a global gene expression inhibitor (Lam et al.,
2001). In lung cancer cells, alsterpaullone **10.27** concomitantly decreased the antiapoptotic factor survivin and enhanced the cytotoxicity induced by sodium nitroprusside (Chao
et al., 2004). A recent study on thiophene analogs of kenpaullone showed that high
concentrations of **10.53** (100 µM) increased both the p53 and the p21 mRNA level in
MCF7 breast cancer cells. In terminal deoxynucleotidyl-transferase-mediated dUTP nick
end-labeling (TUNEL) experiments, a 24 h treatment of MCF-7 cells with 50 µM **10.53**
revealed a significant induction of apoptosis. Introduction of aryl or alkyl substituents at
the thiophene ring diminished the antiproliferative activity of **10.53** for breast cancer

cells (Brault et al., 2005). In contrast, the attachment of epoxide-bearing side chains into the 2-position of either kenpaullone **10.3** or 9-trifluoromethylpaullone **10.23** led to increased *in vitro* antitumor activity in a panel of cancer cell lines (Xie et al., 2005).

10.4.2 KINASE SELECTIVITY DISPLAYED BY PAULLONES

Similar to most other known small synthetic kinase inhibitors, the paullones occupy the ATP binding pocket of the target kinase (Zaharevitz et al., 1999). In this regard, the question of selectivity against the various kinases present in living organisms is of major interest. The comparison of the kinase selectivity profile of kenpaullone **10.3**, 10-bromopaullone **10.38**, and alsterpaullone **10.27** is summarized in Table 10.2.

TABLE 10.2
Kinase Selectivity of Kenpaullone 10.3, 10-Bromopaullone 10.38, and Alsterpaullone 10.27[a]

Kinase	Kenpaullone 10.3 IC_{50} [μM]	10-Bromopaullone 10.38 IC_{50} [μM]	Alsterpaullone 10.27 IC_{50} [μM]
CDK1/cyclin B	0.4	1.3	0.035
CDK2/cyclin A	0.68	3.0	0.015
CDK2/cyclin E	7.5	4.0	0.20
CDK4/cyclin D1	>100	n.a.	>10
CDK5/p25	0.85	2.7	0.04
GSK-3α	n.a.	n.a.	0.004
GSK-3β	0.023	0.14	0.004
erk1	20	88	22
erk2	9	100	4.5
c-raf	38	>100	>10
MAPKK	>100	n.a.	>100
c-jun N-terminal kinase	>100	n.a.	>10
Protein kinase Cα	>100	>100	>100
Protein kinase Cβ1	>100	>100	>100
Protein kinase Cβ2	>100	0.8	>100
Protein kinase Cγ	>100	4.4	>100
Protein kinase Cδ	>100	>100	>100
Protein kinase Cϵ	>100	0.75	>100
Protein kinase Cη	>100	>100	>100
Protein kinase Cζ	>100	>100	>100
cAMP-dependent protein kinase	>1000	>100	7.0
cGMP-dependent protein kinase	>1000	>100	>100
Casein kinase 1	>100	>100	>100
Casein kinase 2	20	0.8	>100
Insulin receptor tyrosine kinase	>1000	>100	>100

Note: n.a. = not available.

[a]ATP concentration 15 μM. (From Zaharevitz, D.W. et al. (1999). *Cancer Research* **59**, 2566–2569; Leost, M. et al. (2000). *European Journal of Biochemistry* **267**, 5983–5994.)

Both kenpaullone and alsterpaullone display a high degree of selectivity, inhibiting only CDK1, CDK2, and CDK5 within the CDK family and also the α and β isoforms of GSK-3 in submicromolar concentrations. In contrast to the pan-CDK inhibitor flavopiridol, the paullones do not inhibit CDK4. The position of the substituent at the indole substructure of the heterocyclic scaffold is apparently of major importance for kinase selectivity. The shift of the 9-bromo substituent to the 10-position leads to a significant decrease in selectivity, as is illustrated by 10-bromopaullone **10.38**. This isomer of kenpaullone is rather an inhibitor of the protein kinase C isoforms β and ε, as well as of casein kinase 2, than of the tested CDKs. Besides the CDKs, glycogen synthase kinase-3β was found to be another protein kinase inhibited by paullones (Leost et al., 2000). This observation was not surprising, because GSK-3 is phylogenetically closely related to the CDKs (Frame and Cohen, 2001). Similar to the paullones, at least three other kinase inhibitor classes have been characterized to act on both GSK-3 and distinct CDKs, namely, the indirubins (Hoessel et al., 1999; Damiens et al., 2001; Leclerc et al., 2001; Marko et al., 2001), hymenialdisine (Meijer et al., 2000), and the aloisines (Mettey et al., 2003).

A comparative investigation of 15 commercially available kinase inhibitors from a panel of 28 kinases showed that kenpaullone and alsterpaullone exhibited selectivity toward CDK2/cyclinA, GSK-3β, and the lymphocyte kinase Lck, a nonreceptor protein tyrosine kinase of the src family. As a result of their study, the authors suggested the combined use of roscovitine and kenpaullone for biochemical investigations regarding the physiological role of CDKs. The combination of kenpaullone and lithium chloride was recommended to study the relevance of GSK-3 dependent processes (Bain et al., 2003). GSK-3 is a serine/threonine kinase ubiquitous in mammalian tissues. Occurring in two isoforms (GSK-3α and GSK-3β), it is involved in several physiological processes, e.g., Wnt signaling, dorsoventral patterning during development, glucose metabolism, and cell cycle regulation (Ali et al., 2001; Frame and Cohen, 2001; Doble and Woodgett, 2003). GSK-3 inhibition might be a rewarding strategy to develop novel treatments for many diseases such as diabetes, muscle hypertrophy, cancer, bipolar mood disorder, schizophrenia, and Alzheimer's disease (AD) (Jope and Johnson, 2004; Meijer et al., 2004). Particularly, with regard to AD, GSK-3 has recently attracted interest as a potential drug target (Bhat et al., 2004). In AD, two characteristic neuropathological features are observed: intracellular neurofibrillary tangles (NFT) and extracellular deposited senile plaques. Together with CDK-5, GSK-3 is responsible for hyperphosphorylation of the microtubule-associated tau-protein, which is the main constituent of the neurofibrillary tangles. The senile plaques are aggregates of the Aβ peptide, which itself is an activator of GSK-3. The neurotoxicity induced by Aβ can be diminished by GSK-3 inhibition (Grimes and Jope, 2001). In light of these findings, GSK-3β inhibitors such as the paullones might attenuate the formation of both the neurofibrillary tangles and the senile plaques. A systematic study on the GSK-3β inhibition in the paullone series revealed that the majority of paullones are roughly twofold to tenfold selective for GSK-3β versus CDK1(Leost et al., 2000; Kunick et al., 2004b). Inspection of the representative results summarized in Table 10.3 reveals two prominent exceptions.

TABLE 10.3
GSK-3 Inhibition by Paullones[a]

10.18

Entry	R	IC_{50} GSK-3 [μM]	IC_{50} CDK1/ Cyclin B [μM]
10.3	9-Br (Kenpaullone)	0.023	0.4
10.24	9-NH$_2$	12	20
10.26	9-CN	0.010	0.024
10.27	9-NO$_2$ (Alsterpaullone)	0.004	0.035
10.37	2-OCH$_3$; 9-SO$_2$CH$_3$	2.0[b]	0.009
10.52	1-Azakenpaullone (see Figure 10.7)	0.018	2.0

[a] Kinase assays were carried out in the presence of 15 μM ATP. (From Leost, M. et al. (2000) *European Journal of Biochemistry* **267**, 5983–5994; Kunick, C. et al. (2004a) *Bioorganic and Medicinal Chemistry Letters* **14**, 413–416; Kunick, C. et al. (2004b) *Journal of Medicinal Chemistry* **47**, 22–36; Pies, T. et al. (2004). *Archiv der Pharmazie — Pharmaceutical and Medicinal Chemistry* **337**, 486–492.)

[b] Previously unpublished result.

While the 2-methoxy-9-methylsulfonylpaullone **10.37** is a very potent inhibitor of CDK1, it shows only weak inhibition for GSK-3β and *vice versa*, and the 1-azakenpaullone **10.52** exhibits a 100-fold selectivity for GSK-3β (Kunick et al., 2004a).

Recently, a crystal structure analysis has been reported that revealed details of the alsterpaullone binding to the ATP-binding pocket of GSK-3β. As illustrated in Figure 10.8, the lactam group of alsterpaullone makes a pair of hydrogen bonds with the carbonyl oxygen and the nitrogen of Val135 of the hinge area. An additional hydrogen bond is observed between the nitro group and the side-chain amino group of Lys85. The indole nitrogen of alsterpaullone is hydrogen-bonded to a water molecule that is part of a hydrogen bond network bridging the backbone carbonyl oxygen of Gln185 and the side-chain oxygen of Thr138 (Bertrand et al., 2003).

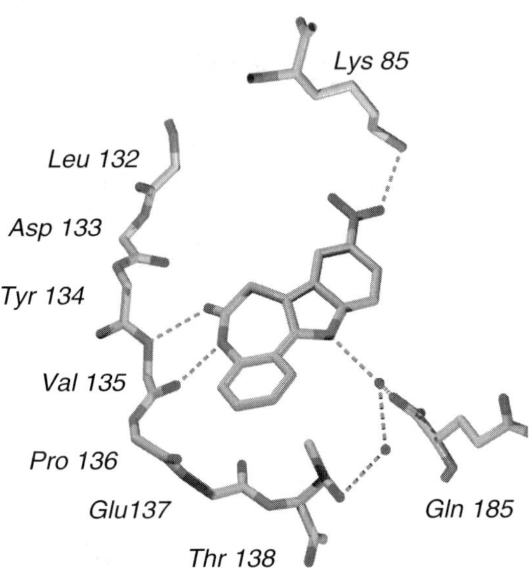

FIGURE 10.8 Alsterpaullone bound to the ATP-binding pocket of GSK-3. For clarity, the hinge-region amino acids side chains (Leu132–Glu137) have been omitted. The illustration was constructed based on data deposited in pdb-file 1Q3W. (From Bertrand, J.A. et al. (2003) *Journal of Molecular Biology* **333**, 393–407.)

10.4.3 NON-KINASE TARGETS OF PAULLONES

Based on the lack of a correlation between antiproliferative activity and CDK-inhibitory activity in the paullone series, it was speculated that the inhibition of other targets might contribute to their growth-inhibitory activity. To identify such additional targets, paullone moieties were immobilized on the surface of agarose beads. Extracts from various tissues were then incubated with the matrices and proteins having affinity toward the paullone moiety, were purified by washing procedures, and isolated by gel electrophoresis. This affinity approach verified the GSK-3 isoforms as major intracellular targets and also revealed mitochondrial malate dehydrogenase (mMDH) as a protein with paullone affinity. Cytosolic and mitochondrial malate dehydrogenases (MDHs) catalyze the interconversion of malate and oxaloacetate using NAD/NADH as a cofactor system. The enzymes play a role in several important metabolic events, comprising gluconeogenesis, the tricarboxylic acid cycle, amino acid synthesis, and the malate-aspartate shuttle. Interestingly, MDH has recently been addressed as a putative biological target of the anticancer drug E7070 (Oda et al., 2003). It was shown that mMDH is indeed inhibited by micromolar concentrations of paullones and that the paullones compete directly with NAD/NADH (Knockaert et al., 2002b). However, it is yet to be investigated whether the inhibition of mMDH by paullones is relevant to their growth inhibitory activity toward cancer cells and whether structure–activity relationships can be established for this additional paullone target.

10.4.4 Paullones as Potential Drugs against Neurodegenerative Disorders

Paullones inhibit GSK-3 and CDK5, both of which are relevant for neurodegenerative disorders. In this regard, some studies have been published in which paullones were used as tools in experiments aimed either at the elucidation of neurological patho-biochemistry or at the development of novel therapeutic approaches. Among other substrates in the brain, CDK5 phosphorylates the DARPP-32 protein at Thr75, which then acts as an inhibitor of protein kinase A (PKA) (Bibb et al., 1999). Being a part of the intracellular signaling cascade downstream, the D_1 receptor DARPP-32 plays a central role in regulating the efficacy of dopaminergic neurotransmission. Drugs acting on DARPP-32 have been suggested as agents for the treatment of neuropsychiatric disorders (Greengard, 2001). To assess the effect of CDK5 inhibition on DARPP-32, slices of mouse striatum (the brain area producing DARPP-32) were treated with alsterpaullone. Using an antibody that interacts selectively with DARPP-32 phosphorylated at Thr75, it was shown that 10 µM alsterpaullone inhibits the phosphorylation at this site by 50% (Leost et al., 2000). Regarding potential use for a treatment of Parkinson's disease, it was reported that kenpaullone stimulated neuronal differentiation in ventral mesancephalon precursor cultures and increased the size of the dopamine neuron population. These effects were interpreted as being due to the GSK-3β inhibition by kenpaullone, resulting in a modified Wnt signaling pathway. The use of GSK-3 inhibitors is suggested as an approach to improve stem cell or precursor cell therapy approaches in Parkinson's disease (Castelo-Branco et al., 2004).

GSK-3 and CDK5 are the main kinases responsible for hyperphosphorylation of the microtubule-associated tau-protein that is the cardinal constituent of the neurofibrillary tangles found in the brain cells of patients suffering from Alzheimer's disease. Consequently, both kinases are suggested as targets for potential anti-Alzheimer's drugs (Michaelis, 2003). Alsterpaullone has been used as a dual CDK5/GSK-3 inhibitor to study the effects of reduced tau phosphorylation in differentiating P12 cells. In this situation, alsterpaullone led to an inhibition of anterograde mitochondria transport and clustering of mitochondria around the cell nuclei (Tatebayashi et al., 2004). It was shown that alsterpaullone inhibits the GSK-3-dependent phosphorylation of bacterially expressed recombinant human tau *in vitro* with an IC_{50} of 33 nM (Leost et al., 2000). Moreover, 20 µM alsterpaullone inhibited the tau-phosphorylation in cultured tau-expressing Sf9-cells at epitopes characteristic of Alzheimer's disease (Leost et al., 2000). Recently, it was reported that in AD GSK-3 is also relevant for the production of the Aβ-peptide, the monomer forming the senile plaques, from its precursor, the carboxyterminal fragment of the amyloid precursor protein (APP). Although the detailed mechanism of the interaction between GSK-3 and Aβ production is still a matter of debate, it was demonstrated that kenpaullone inhibits Aβ formation in a model of Chinese hamster ovary cells stably expressing APP (CHO-APP$_{695}$ cells). It was concluded that this was due to GSK-3 inhibition, because lithium, another GSK-3 inhibitor, similarly inhibited Aβ formation in the cellular model. GSK-3 inhibitors, like the paullones might consequently bear the potential to address both the neurofibrillary tangles and the amyloid plaques, the two pathological hallmarks of Alzheimer's disease (Phiel et al., 2003).

10.4.5 PAULLONES AS PUTATIVE ANTI-INFECTIOUS AGENTS

The metabolism of pathogenic microorganisms involves enzymes that show homology to mammalian kinases. These enzymes might be suitable targets for treatment of infectious diseases by paullones. In the protozoan parasite *Leishmania*, the kinase CRK3 is the most likely homologue to the mammalian CDK1, playing an important role in cell division. During the course of a systematic search for CRK3 inhibitors, 9-cyanopaullone **10.26** was identified as an *in vitro* inhibitor of CRK3 (IC$_{50}$ = 2.9 µM). On testing against *Leishmania donovani*-infected mouse peritoneal macrophages, **10.26** exhibited an ED$_{50}$ of 5.77 µM. However, the paullone showed toxicity against the macrophages at 10 µM (Grant et al., 2004). The inhibition of nonkinase targets, namely, mitochondrial malate dehydrogenase (mMDH), might contribute to the antiparasitic activity by paullones. When a *Leishmania mexicana* extract was incubated with agarose beads bearing immobilized paullone moieties, mMDH was identified as a major paullone-binding protein. In agreement with the mMDH inhibitory properties of paullones, 3 µM alsterpaullone suppressed the growth of *Leishmania mexicana* promastigotes (Knockaert et al., 2002b). The malaria-causing parasite *Plasmodium falciparum* possesses the kinase PfGSK-3, a homologue of the mammalian GSK-3. The search for selective PfGSK-3 inhibitors revealed that paullones are roughly two orders of magnitude less potent in the parasite enzyme compared to the mammalian GSK-3β (Droucheau et al., 2004). Pfmrk, a CDK of *Plasmodium falciparum*, was inhibited by high concentrations of alsterpaullone (IC$_{50}$ = 51 µM) and kenpaullone (IC$_{50}$ = 15 µM) (Bhattacharjee et al., 2004).

10.4.6 INTERACTION WITH INVERTEBRATE DEVELOPMENT

Alsterpaullone has been used to study the role of GSK-3 homologues in the Wnt signaling pathway during the development of invertebrates, namely, sea- and freshwater polyps. In *Hydractinia*, a colonial marine hydroid, alsterpaullone accelerates the formation of ectopic head structures and the budding of new polyps. In this regard, alsterpaullone acts antagonistically to the natural endogenous stolon-inducing factor SIF (Müller et al., 2004). Treatment of female individuals of the freshwater polyp *Hydra* with either 2.5 µM alsterpaullone or 1 mM lithium chloride prevented the differentiation of nurse cells and subsequent oocyte formation. This observation was interpreted as a hint that GSK-3 is necessary for the induction of apoptosis in germline cells, a step previous to the gametogenesis in *Hydra* (Rentzsch et al., 2005).

10.5 STRUCTURE OPTIMIZATION OF PAULLONES

10.5.1 DEVELOPMENT OF WATER-SOLUBLE PAULLONES

In spite of the promising anticancer activity exhibited by alsterpaullone toward cultured cancer cell lines, *in vivo* tests with the compounds in animal models for anticancer activity were compromised by the poor water solubility of the compound. To overcome this problem, alsterpaullone derivatives with ionizable side chains in the 2-position were prepared, among which the water-soluble hydrochlorides of the 2-aminoalkylalsterpaullones **10.54** and **10.55** (Figure 10.9) exhibited CDK1/cyclin B inhibitory activity in the range of alsterpaullone (IC_{50} values 80 nM and 27 nM, respectively) (Kunick et al., 2005). Scientists of Enzon Pharmaceuticals have designed and synthesized water-soluble prodrugs of alsterpaullone (e.g., **10.56**) with polyethylene glycol side chains attached to the indole nitrogen. The prodrug **10.56** (Mr~21012) proved to be stable in saline for 24 h but was cleaved in rat plasma to release alsterpaullone with a $t_{1/2}$ of 1 h. Both I.P. and I.V. application of **10.56** resulted in favorable pharmacokinetic parameters, namely, high relative bioavailability and extended half-life of the released alsterpaullone (Greenwald et al., 2004). Pharmacological results from animal studies with regard to the anti-tumor efficacy of paullones have not yet been published.

10.5.2 STRUCTURE-BASED APPROACHES FOR ACTIVITY IMPROVEMENT

For a structure-based approach to enhance the CDK1-inhibitory activity of alsterpaullone, the molecule was docked into McGrath's (McGrath et al., 2005) CDK1/cyclin B homology model. Subsequently, substituents were attached to the 2-position of the parent scaffold to enable additional hydrophobic contacts to the hydrocarbon side chains of Ile10 and Leu298/Met85 positioned in the entrance area to the ATP-binding cleft. This concept was realized by the construction and synthesis of 2-cyanoethylalsterpaullone **10.57**, in which the nitrile group forms a hydrogen bond to Lys89, resulting in an alignment that directs the ethylene group in such a way as to enable the hydrophobic contacts mentioned earlier (see Figure 10.10). Actually, **10.57** proved to be a very potent kinase inhibitor for both CDK1/cyclin B (IC_{50} = 0.23 nM) and GSK-3β (IC_{50} = 0.8 nM) (Kunick et al., 2005).

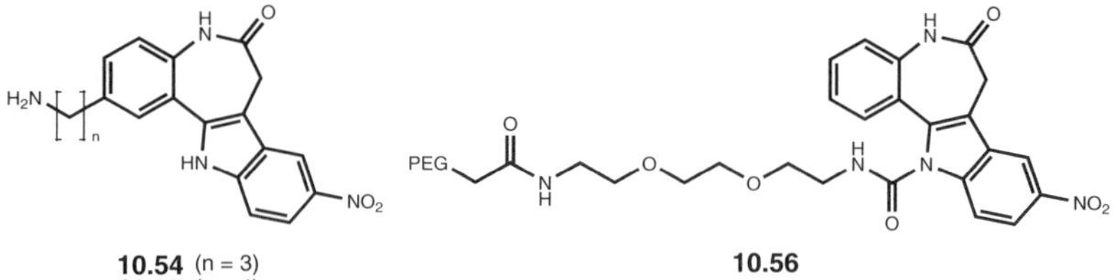

10.54 (n = 3)
10.55 (n = 4)

10.56

FIGURE 10.9 Water-soluble paullone derivatives.

FIGURE 10.10 Structure of 2-cyanoethylalsterpaullone **10.57** (left) and 2-cyanoethylalsterpaullone **10.57** docked into the ATP-binding pocket of CDK1/cyclin B (right) depicted from outside to the gate area of the ATP-binding pocket. The nitrile group of the ligand is directed toward the protonated side chain nitrogen of Lys89 forming a hydrogen bond, enabling hydrophobic contacts between the carbons of the ethylene bridge of the ligand and the side chains of Ile10 and Leu298.

10.6 FUTURE PROSPECTS

The paullones are a good example of the unforeseen possibilities in the development and applications of a compound family. Discovered and developed to a certain extent as CDK inhibitors and anticancer agents, GSK-3 and mMDH have been identified as additional biological paullone targets. Owing to their commercial availability, kenpaullone and alsterpaullone have been employed in various research fields, e.g., neurobiochemistry and development biology. The most promising area of research apparently lies in the development of agents against Alzheimer's disease. Because of the dual inhibition of CDK5 and GSK-3, the paullones might be useful in suppressing both tau-hyperphosphorylation and Aβ-peptide production, two main malfunctions in Alzheimer's disease. Evidence that paullones cross the blood–brain barrier and show results in suitable animal models of neurodegenerative diseases would be a major boost for ongoing research.

ACKNOWLEDGMENT

The authors are grateful for funding by the European Commission (Contract No LSHB-CT-2004-503467, to C.K. and L.M.).

REFERENCES

Ali, A., Hoeflich, K.P., and Woodgett, J.R. (2001). Glycogen synthase kinase-3: properties, function and regulation. *Chemical Reviews* (Washington, D.C.) **101**, 2527–2540.

Bain, J., McLauchlan, H., Elliott, M., and Cohen, P. (2003). The specificities of protein kinase inhibitors: an update. *Biochemical Journal* **371**, 199–204.

Baudoin, O., Cesario, M., Guénard, D., and Guéritte, F. (2002). Application of the palladium-catalyzed borylation/Suzuki coupling (BSC) reaction to the synthesis of biologically active biaryl lactams. *Journal of Organic Chemistry* **67**, 1199–1207.

Bertrand, J.A., Thieffine, S., Vulpetti, A., Cristiani, C., Valsasina, B., Knapp, S., Kalisz, H.M., and Flocco, M. (2003). Structural characterization of the GSK-3beta active site using selective and non-selective ATP-mimetic inhibitors. *Journal of Molecular Biology* **333**, 393–407.

Bhat, R.V., Budd Haeberlein, S.L., and Avila, J. (2004). Glycogen synthase kinase-3: a drug target for CNS therapies. *Journal of Neurochemistry* **89**, 1313–1317.

Bhattacharjee, A., Geyer, J., Woodard, C., Kathcart, A., Nichols, D., Prigge, S., Li, Z., Mott, B., and Waters, N. (2004). A three-dimensional in silico pharmacophore model for inhibition of Plasmodium falciparum cyclin-dependent kinases and discovery of different classes of novel Pfmrk specific inhibitors. *Journal of Medicinal Chemistry* **47**, 5418–5426.

Bibb, J.A., Snyder, G.L., Nishi, A., Yan, Z., Meijer, L., Fienberg, A.A., Tsai, L.-H., Kwon, Y.T., Girault, J.-A., Czernik, A.J., Huganir, R.L., Hemmings, H.C.J., Nairn, A.C., and Greengard, P. (1999). Phosphorylation of DARPP-32 by CDK5 modulates dopamine signalling in neurons. *Nature* **402**, 669–671.

Brault, L., Migianu, E., Néguesque, A., Battaglia, E., Bagrel, D., and Kirsch, G. (2005). New thiophene analogues of kenpaullone: synthesis and biological evaluation in breast cancer cells. *European Journal of Medicinal Chemistry* **40**, 757–763.

Buolamwini, J.K. (2000). Cell cycle molecular targets in novel anticancer drug discovery. *Current Pharmaceutical Design* **6**, 379–392.

Castelo-Branco, G., Rawal, N., and Arenas, E. (2004). GSK-3beta inhibition/beta-catenin stabilization in ventral midbrain precursors increases differentiation into dopamine neurons. *Journal of Cell Science* **117**, 5731–5737.

Chao, J., Kuo, P., and Hsu, T. (2004). Down regulation of survivin in nitric oxide-induced cell growth inhibition and apoptosis of the human lung carcinoma cells. *Journal of Biological Chemistry* **279**, 20267–20276.

Chen, W.-Y. and Gilman, N.W. (1983). Synthesis of 7-phenylpyrimido[5,4-*d*][1]benzazepin-2-ones. *Journal of Heterocyclic Chemistry* **20**, 663–666.

Cleaveland, E.S., Monks, A., Vaigro-Wolff, A., Zaharevitz, D.W., Paull, K.D., Ardalan, K., Cooney, D.A., and Ford, H.F.J. (1995). Site of action of two novel pyrimidine biosynthesis inhibitors accurately predicted by the COMPARE program. *Biochemical Pharmacology* **49**, 947–954.

Damiens, E., Baratte, B., Marie, D., Eisenbrand, G., and Meijer, L. (2001). Anti-mitotic properties of indirubin-3-monoxime, a CDK/GSK-3 inhibitor: induction of endoreplication following prophase arrest. *Oncogene* **20**, 3786–3797.

Doble, B.W. and Woodgett, J.R. (2003). GSK-3: tricks of the trade for a multi-tasking kinase. *Journal of Cell Science* **116**, 1175–1186.

Droucheau, E., Primot, A., Thomas, V., Mattei, D., Knockaert, M., Richardson, C., Sallicandro, P., Alano, P., Jafarshad, A., Baratte, B., Kunick, C., Parzy, D., Pearl, L., Doerig, C., and Meijer, L. (2004). Plasmodium falciparum glycogen synthase kinase-3: molecular model, expression, intracellular localisation and selective inhibitors. *Biochimica et Biophysica Acta* **1697**, 181–196.

Frame, S. and Cohen, P. (2001). GSK-3 takes centre stage more than 20 years after its discovery. *Biochemical Journal* **359**, 1–16.

Gharehbaghi, K., Paull, K.D., Kelley, J.A., Barchi, J.J.J., Marquez, V.E., Cooney, D.A., Monks, A., Scudiero, D., Kohn, K., and Jayaram, H.N. (1994). Cytotoxicity and characterization of an active metabolite of benzamide ribose, a novel inhibitor of IMP dehydrogenase. *International Journal of Cancer* **56**, 892–899.

Grant, K., Dunion, M., Yardley, V., Skaltsounis, A., Marko, D., Eisenbrand, G., Croft, S., Meijer, L., and Mottram, J. (2004). Inhibitors of Leishmania mexicana CRK3 cyclin-dependent kinase: chemical library screen and antileishmanial activity. *Antimicrobial Agents and Chemotherapy* **48**, 3033–3042.

Grant, S. and Roberts, J. (2003). The use of cyclin-dependent kinase inhibitors alone or in combination with established cytotoxic drugs in cancer chemotherapy. *Drug Resistance Updates* **6**, 15–26.

Greengard, P. (2001). The neurobiology of slow synaptic transmission. *Science* **294**, 1024–1030.

Greenwald, R., Zhao, H., Xia, J., Wu, D., Nervi, S., Stinson, S., Majerova, E., Bramhall, C., and Zaharevitz, D. (2004). Poly(ethylene glycol) prodrugs of the CDK inhibitor, alsterpaullone (NSC 705701): synthesis and pharmacokinetic studies. *Bioconjugate Chemistry* **15**, 1076–1078.

Grimes, C. and Jope, R. (2001). The multifaceted roles of glycogen synthase kinase 3beta in cellular signaling. *Progress in Neurobiology* **65**, 391–426.

Gussio, R., Zaharevitz, D.W., McGrath, C.F., Pattabiraman, N., Kellogg, G.E., Schultz, C., Link, A., Kunick, C., Leost, M., Meijer, L., and Sausville, E.A. (2000). Structure-based design modifications of the paullone molecular scaffold for cyclin-dependent kinase inhibition. *Anti-Cancer Drug Design* **15**, 53–66.

Hoessel, R., Leclerc, S., Endicott, J., Nobel, M.E.M., Lawrie, A., Tunnah, P., Leost, M., Damiens, E., Marie, D., Marko, D., Niederberger, E., Tang, W., Eisenbrand, G., and Meijer, L. (1999). Indirubin, the active constituent of a Chinese antileukemia medicine, inhibits cyclin-dependent kinases. *Nature Cell Biology* **1**, 60–67.

Jones, G. (1967). Some basic and acidic derivatives of 2,5-dihydro-1*H*-1-benzazepine as potential therapeutic agents. *Journal of Chemical Society* **19**, 1808–1813.

Jope, R. and Johnson, G. (2004). The glamour and gloom of glycogen synthase kinase-3. *Trends in Biochemical Sciences* **29**, 95–102.

Knockaert, M., Greengard, P., and Meijer, L. (2002a). Pharmacological inhibitors of cyclin-dependent kinases. *Trends in Pharmacological Sciences* **23**, 417–425.

Knockaert, M., Wieking, K., Schmitt, S., Leost, M., Grant, K.M., Mottram, J.C., Kunick, C., and Meijer, L. (2002b). Intracellular targets of paullones: Identification following affinity purification on immobilized inhibitor. *Journal of Biological Chemistry* **277**, 25493–25501.

Kohlhagen, G., Paull, K.D., Cushman, M., Nagafuji, P., and Pommier, Y. (1998). Protein-linked DNA strand breaks induced by NSC 314622, a novel noncamptothecin topoisomerase I poison. *Molecular Pharmacology* **54**, 50–58.

Kunick, C. (1992). Synthese von 7,12-Dihydro-indolo[3,2-*d*][1]benzazepin-6(5*H*)-onen und 6,11-Dihydro-thieno[3,2:2,3]azepino[4,5-*b*]indol-5(4*H*)-on. *Archiv der Pharmazie (Weinheim)* **325**, 297–299.

Kunick, C., Lauenroth, K., Leost, M., Meijer, L., and Lemcke, T. (2004a). 1-Azakenpaullone is a selective inhibitor of glycogen synthase kinase-3. *Bioorganic and Medicinal Chemistry Letters* **14**, 413–416.

Kunick, C., Lauenroth, K., Wieking, K., Xie, X., Schultz, C., Gussio, R., Zaharevitz, D., Leost, M., Meijer, L., Weber, A., Jorgensen, F.S., and Lemcke, T. (2004b). Evaluation and comparison of 3D-QSAR CoMSIA models for CDK1, CDK5, and GSK-3 inhibition by paullones. *Journal of Medicinal Chemistry* **47**, 22–36.

Kunick, C. and Lemcke, T. (2002). Paullone als selektive Kinase-Inhibitoren. *Pharmazeutische Zeitung* **147**, 2428–2435.

Kunick, C., Schultz, C., Lemcke, T., Zaharevitz, D.W., Gussio, R., Jalluri, R.K., Sausville, E.A., Leost, M., and Meijer, L. (2000). 2-Substituted paullones: CDK1/cyclin B-inhibiting property and in vitro antiproliferative activity. *Bioorganic and Medicinal Chemistry Letters* **10**, 567–569.

Kunick, C., Zeng, Z., Gussio, R., Zaharevitz, D., Leost, M., Totzke, F., Schächtele, C., Kubbutat, M., Meijer, L., and Lemcke, T. (2005). Structure-aided optimization of kinase inhibitors derived from alsterpaullone. *ChemBioChem* **6**, 541–549.

Lahusen, T., De Siervi, A., Kunick, C., and Senderowicz, A. (2003). Alsterpaullone, a novel cyclin-dependent kinase inhibitor, induces apoptosis by activation of caspase 9 due to perturbation in mitochondrial membrane potential. *Molecular Carcinogenesis* **36**, 183–194.

Lam, T.L., Pickeral, O.K., Peng, A.C., Rosenwald, A., Hurt, E.M., Giltnane, J.M., Averett, L.M., Zhao, H., Davis, R.E., Sathyamoorthy, M., Wahl, L.M., Harris, E.D., Mikovits, J.A., Monks, A.P., Hollingshead, M.G., Sausville, E.A., and Staudt, L.M. (2001). Genomic-scale measurement of mRNA turnover and the mechanism of action of the anti-cancer drug flavopiridol. *Genome Biology* **2**, research 0041.1-0041.11.

Leclerc, S., Garnier, M., Hoessel, R., Marko, D., Bibb, J.A., Snyder, G.L., Greengard, P., Biernat, J., Wu, Y.-Z., Mandelkow, E.-M., Eisenbrand, G., and Meijer, L. (2001). Indirubins inhibit glycogen kinase-3β and CDK5/p25, two protein kinases involved in abnormal tau phosphorylation in Alzheimer's disease. *Journal of Biological Chemistry* **276**, 251–260.

Leost, M., Schultz, C., Link, A., Wu, Y.-Z., Biernat, J., Mandelkow, E.-M., Bibb, J.A., Snyder, G.L., Greengard, P., Zaharevitz, D.W., Gussio, R., Senderowicz, A.M., Sausville, E.A., Kunick, C., and Meijer, L. (2000). Paullones are potent inhibitors of glycogen synthase kinase-3β and cyclin-dependent kinase 5/p25. *European Journal of Biochemistry* **267**, 5983–5994.

Malumbres, M. and Barbacid, M. (2001). To cycle or not to cycle: a critical decision in cancer. *Nature Reviews Cancer* **1**, 222–231.

Marko, D., Schätzle, S., Friedel, A., Genzlinger, A., Zankl, H., Meijer, L., and Eisenbrand, G. (2001). Inhibition of cyclin-dependent kinase 1 (CDK1) by indirubin derivatives in human tumour cells. *British Journal of Cancer* **84**, 283–289.

McGrath, C.F., Pattabiraman, N., Kellogg, G.E., Lemcke, T., Kunick, C., Sausville, E.A., Zaharevitz, D.W., and Gussio, R. (2005). Homology model of the CDK1/cyclin B complex. *Journal of Biomolecular Structure and Dynamics* **22**, 493–502.

Meijer, L., Flajolet, M., and Greengard, P. (2004). Pharmacological inhibitors of glycogen synthase kinase-3. *Trends in Pharmacological Sciences* **25**, 471–480.

Meijer, L., Leost, M., Lozach, O., Schmitt, S., and Kunick, C. (2005). The paullones, a family of pharmacological inhibitors of cyclin-dependent kinases and glycogen synthase kinase-3. In *Handbook of Experimental Pharmacology (HEP)*, L.A. Pinna and P.W. Cohen, Eds., *Inhibitors of Protein Kinases and Protein Phosphatases.* Springer-Verlag, Berlin, 2005, pp. 48–64.

Meijer, L., Thunissen, A.-M.W.H., White, A.W., Garnier, M., Nikolic, M., Tsai, L.-H., Walter, J., Cleverley, K.E., Salinas, P.C., Wu, Y.-Z., Biernat, J., Mandelkow, E.M., Kim, S.-H. and Pettit, G.R. (2000). Inhibition of cyclin-dependent kinases, GSK-3beta and CK1 by hymenialdisine, a marine sponge constituent. *Chemistry and Biology* **7**, 51–63.

Mettey, Y., Gompel, M., Thomas, V., Garnier, M., Leost, M., Ceballos-Picot, I., Noble, M., Endicott, J., Vierfond, J.-M., and Meijer, L. (2003). Aloisines, a new family of CDK/GSK-3 inhibitors. SAR study, crystal structure in complex with CDK2, enzyme selectivity, and cellular effects. *Journal of Medicinal Chemistry* **46**, 222–236.

Michaelis, M. (2003). Drugs targeting Alzheimer's disease: some things old and some things new. *Journal of Pharmacology and Experimental Therapeutics* **304**, 897–904.

Monks, A., Scudiero, D., Skehan, P., Shoemaker, R., Paull, K., Vistica, D., Hose, C., Langley, J., Cronise, P., Vaigro-Wolff, A., Gray-Goodrich, M., Campbell, H., Mayo, J., and Boyd, M. (1991). Feasibility of a high-flux anticancer drug screen using a diverse panel of cultured human tumor cell lines. *Journal of the National Cancer Institute* **83**, 757–766.

Monks, A., Scudiero, D.A., Johnson, G.S., Paull, K.D., and Sausville, E.A. (1997). The NCI anti-cancer drug screen: a smart screen to identify effectors of novel targets. *Anti-Cancer Drug Design* **12**, 533–541.

Moriconi, E.J. and Maniscalco, I.A. (1972). π-Equivalent heterocyclic congeners of tropone. Azatropones. *Journal of Organic Chemistry* **37**, 208–215.

Müller, W., Teo, R. and Möhrlen, F. (2004). Patterning a multi-headed mutant in Hydractinia: enhancement of head formation and its phenotypic normalization. *International Journal of Developmental Biology* **48**, 9–15.

Oda, Y., Owa, T., Sato, T., Boucher, B., Daniels, S., Yamanaka, H., Shinohara, Y., Yokoi, A., Kuromitsu, J., and Nagasu, T. (2003). Quantitative chemical proteomics for identifying candidate drug targets. *Analytical Chemistry* **75**, 2159–2165.

Paull, K.D., Lin, C.M., Malspeis, L., and Hamel, E. (1992). Identification of novel antimitotic agents acting at the tubulin level by computer-assisted evaluation of differential cytotoxicity data. *Cancer Research* **52**, 3892–3900.

Paull, K.D., Shoemaker, R.H., Hodes, L., Monks, A., Scudiero, D.A., Rubinstein, L., Plowman, J., and Boyd, M.R. (1989). Display and analysis of patterns of differential activity of drugs against human tumor cell lines: development of meangraph and COMPARE algorithm. *Journal of the National Cancer Institute* **81**, 1088–1092.

Phiel, C.J., Wilson, C.A., Lee, V.M.-Y., and Klein, P.S. (2003). GSK-3 regulates production of Alzheimer's disease amyloid-β peptides. *Nature* **423**, 435–439.

Pies, T., Schaper, K.-J., Leost, M., Zaharevitz, D.W., Gussio, R., Meijer, L., and Kunick, C. (2004). CDK1-inhibitory activity of paullones depends on electronic properties of 9-substituents. *Archiv der Pharmazie — Pharmaceutical and Medicinal Chemistry* **337**, 486–492.

Price, T., Walton, M., Cunningham, D., and Workmann, P. (2000). Activity of paullones in colon cell lines does not correlate to cell cycle protein expression. *Annals of Oncology* **11(Suppl. 4)**, 13.

Rees, A.H. (1959). Azatropolones. *Journal of the Chemical Society* 3111–3116.

Rentzsch, F., Hobmayer, B., and Holstein, T. (2005). Glycogen synthase kinase 3 has a proapoptotic function in Hydra gametogenesis. *Developmental Biology* **278**, 1–12.

Rivett, D.E. and Stewart, F.H.C. (1978). Formation of substituted 1-benzazepine-2,5-diones. *Australian Journal of Chemistry* **31**, 439–443.

Sausville, E. (2003). Cyclin-dependent kinase modulators studied at the NCI: pre-clinical and clinical studies. *Current Medicinal Chemistry — Anti-Cancer Agents* **3**, 47–56.

Sausville, E.A., Zaharevitz, D., Gussio, R., Meijer, L., Louarn-Leost, M., Kunick, C., Schultz, R., Lahusen, T., Headlee, D., Stinson, S., Arbuck, S.G. and Senderowicz, A. (1999). Cyclin-dependent kinases: initial approaches to exploit a novel therapeutic target. *Pharmacology and Therapeutics* **82**, 285–292.

Schultz, C., Link, A., Leost, M., Zaharevitz, D.W., Gussio, R., Sausville, E.A., Meijer, L., and Kunick, C. (1999). Paullones, a series of cyclin-dependent kinase inhibitors: synthesis, evaluation of CDK1/ cyclin B inhibition, and in vitro antitumor activity. *Journal of Medicinal Chemistry* **42**, 2909–2919.

Tatebayashi, Y., Haque, N., Tung, Y.C., Iqbal, K., and Grundke-Iqbal, I. (2004). Role of tau phosphorylation by glycogen synthase kinase-3{beta} in the regulation of organelle transport. *Journal of Cell Science* **117**,1653–1663.

Weinstein, J.N., Myers, T.G., O'Connor, P.M., Friend, S.H., Fornace Jr., A.J., Kohn, K.W., Fojo, T., Bates, S.E., Rubinstein, L.V., Anderson, N.L., Buolamwini, J.K., van Osdol, W.W., Monks, A.P., Scudiero, D.A., Sausville, E.A., Zaharevitz, D.W., Bunow, B., Viswanadhan, V.N., Johnson, G.S., Wittes, R.E. and Paull, K.D. (1997). An information-intensive approach to the molecular pharmacology of cancer. *Science* **275**, 343–349.

Wieking, K., Knockaert, M., Leost, M., Zaharevitz, D.W., Meijer, L. and Kunick, C. (2002). Synthesis of paullones with aminoalkyl side chains. *Archiv der Pharmazie — Pharmaceutical and Medicinal Chemistry* **335**, 311–317.

Witte, J. and Boekelheide, V. (1972). Stereoselective syntheses of isoquinuclidones. II. *Journal of Organic Chemistry* **37**, 2849–2853.

Xie, X., Lemcke, T., Gussio, R., Zaharevitz, D.W., Leost, M., Meijer, L., and Kunick, C. (2005). Epoxide-containing side chains enhance antiproliferative activity of paullones. *European Journal of Medicinal Chemistry* **40**, 655–661.

Zaharevitz, D.W., Gussio, R., Leost, M., Senderowicz, A., Lahusen, T., Kunick, C., Meijer, L., and Sausville, E.A. (1999). Discovery and initial characterization of the paullones, a novel class of small-molecule inhibitors of cyclin-dependent kinases. *Cancer Research* **59**, 2566–2569.

11 Discovery of BMS-387032, a Potent Cyclin-Dependent Kinase Inhibitor in Clinical Development

John T. Hunt

CONTENTS

11.1 INTRODUCTION

Cell division is a highly ordered process that requires the sequential completion of certain molecular processes before subsequent steps are undertaken. The process by which the orderly progression is monitored and maintained is termed *checkpoint control*. Whether revealed as a failure to undergo growth arrest, apoptosis, or differentiation in response to an appropriate signal, the loss of checkpoint control is a hallmark of cancer. With the initial elucidation of the involvement of cyclins and cyclin-dependent kinases (CDKs) in cell cycle control, it was natural to propose that interfering with these drivers of cell cycle progression might restore checkpoint control to inappropriately proliferating cells.

At Bristol-Myers Squibb, our interest in developing cancer therapeutics that target the CDKs began in the mid-1990s. Cyclin D/CDK4 had been shown to play an important role in control of the restriction point, the time at which a cell becomes committed to cell division.[1,2] Cyclin E/CDK2 and cyclin A had been shown to be important regulators of the S phase transition.[3–5] Numerous reports were appearing that linked these and other cell cycle components to cancer, including the identification of the PRAD1/BCL-1 proto-oncogene as the cyclin D1 gene, and the finding that cyclin E is overexpressed in breast and other solid tumors.[6,7] Most persuasively, the discovery of the CDK inhibitory proteins in the mid-1990s lent further support to the concept that CDKs were viable cancer targets. For example, the CDKN2 tumor suppressor gene was found to code for the CDK4 inhibitor p16, whereas the pan-CDK inhibitors p21 and p27 were discovered and shown to be able to mediate growth arrest when overexpressed.[8–12]

11.2 EARLY CDK INHIBITOR LEAD STRUCTURES

Around the time that the seminal discoveries of the involvement of CDKs in the cell cycle were being made, there were limited opportunities for designing small molecule inhibitors of this kinase family. The only known inhibitors that had been reported, as shown in Figure 11.1, were the flavonoid L86-8275 (**11.1**, later named flavopiridol), the *Aspergillus terreus*–derived natural product butyrolactone I (**11.2**), and a series of purine based inhibitors typified by olomoucine (**11.3**).[13–15]

Surveying these potential starting points for inhibitor design, we had some concerns with the potentially reactive functionality of **11.2** and also with an inhibitor design program based on a purine such as **11.3**. We therefore focused our initial attention on the design of a novel template derived from flavopiridol, with the aim of studying the structure–activity relationship (SAR) for CDK inhibition in a proprietary chemical series. A promising class of inhibitors that resulted from this effort consisted of the flavopiridol nucleus with a heteroatom inserted between the chromanone ring and the 2-aryl substituent (Figure 11.2).[16] Whereas flavopiridol is a relatively nonselective inhibitor within the CDK family (IC$_{50}$: CDK1/cycB1 = 30 nM; CDK2/cycE = 170 nM; CDK4/cycD1 = 100 nM), analogs **11.4** with an inserted

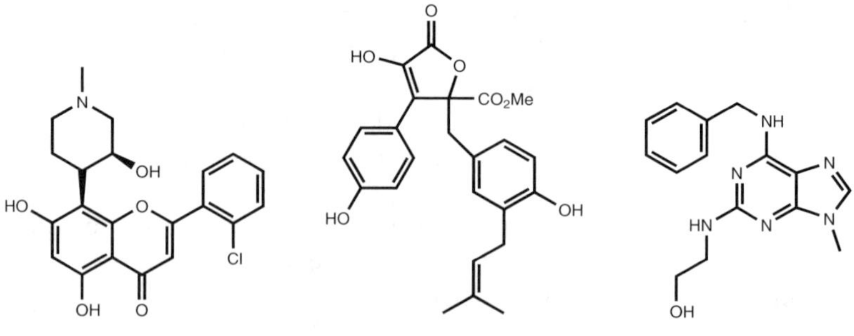

Flavopiridol (**11.1**)   Butyrolactone I (**11.2**)   Olomoucine (**11.3**)

FIGURE 11.1 Known CDK inhibitor lead structures.

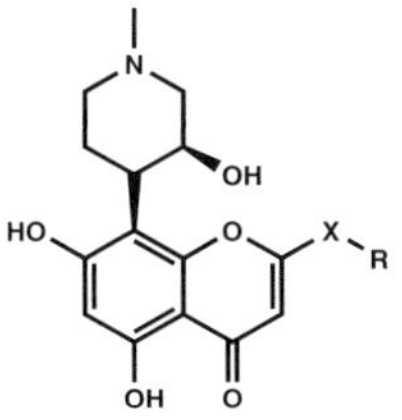

11.4a (X = S, R = 2-chlorophenyl (-))
11.4b (X = O, R = 2-chlorophenyl (-))
11.4c (X = S, R = t-butyl (+))

FIGURE 11.2 Novel flavonoid CDK inhibitor chemotype.

heteroatom were more selective for CDK1, by virtue of decreased inhibition of CDK2 and CDK4. For example, the IC_{50} values for CDK1/B1:CDK2/E:CDK4/D1 for the 2-chlorophenyl analog in which the linker is sulfur (**11.4a**) are 0.110:2.10:16.2 μM, respectively, whereas the corresponding values for the O-linked analog **11.4b** are 0.130:2.11:6.15 μM. Interestingly, an analog with a branched alkyl replacing the phenyl group (**11.4c**) also provided a potent inhibitor, with the corresponding IC_{50} values being 0.08:1.07:2.07 μM. We are unaware of similar structure–activity studies with flavopiridol that assess whether analogs with 2-alkyl substituents retain CDK inhibitory potency. In cellular assays such as inhibition of colony growth on plastic matrices, these new CDK inhibitors displayed potent activity, with some selectivity differences when compared to flavopiridol. For example, although analog **11.4a** was equipotent to flavopiridol in certain cell lines (e.g., the prostate cell line PC-3 and the pancreatic cell line Mia PaCa-2), in others it was > tenfold weaker as an inhibitor of colony growth (e.g., the colon cell line HCT116 and the ovarian cell line A2780).

11.3 DISCOVERY OF NOVEL CDK INHIBITOR LEAD STRUCTURES

Despite our success at obtaining proprietary flavopiridol analogs with interesting CDK inhibitory profiles, we remained interested in identifying novel chemotypes with CDK inhibitory activity. Toward this end, a high-throughput screen for inhibitors of CDK2/Cyclin E was run against the Bristol-Myers Squibb compound deck. From this effort, a number of interesting hits were identified, with two of the most attractive, the pyrazolopyridine SQ-67563 (**11.5**) and the aminothiazole BC-2626 (**11.6**), shown in Figure 11.3. Preliminary SAR studies were undertaken to determine if either or both of these hits could serve as viable starting points for lead optimization.

For the pyrazolopyridine lead SQ-67563, the pyrazolyl NH was determined to be a critical pharmacophore, as shown by the high IC_{50} value for the *N*-ethyl analog **11.8** (>25 μM) (Figure 11.4).[17] In agreement with this finding, the solid-state structure of **11.5** bound to CDK2 shows that the pyridine nitrogen and the pyrazolyl NH serve as a hydrogen bond acceptor and donor, respectively, to the backbone amide

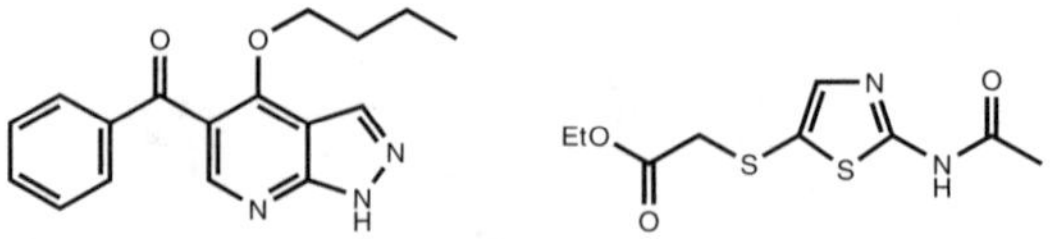

SQ-67563 **(11.5)**   BC-2626 **(11.6)**

FIGURE 11.3 CDK2 high-throughput screening hits.

of Leu83. The ketone carbonyl was initially thought to be an important pharmacophore, as evidenced by the poor CDK2 inhibitory activity of the analogous carbinol **11.9**, the methylene analog **11.10** or the olefin **11.11** (all with $IC_{50} > 25$ μM). However, the structure of the CDK2/compound **11.5** complex did not provide a ready explanation for the importance of the carbonyl group in CDK2 binding. The 4-side chain, which extends into the ribose-binding pocket, was tolerant of a variety of hydrophobic ethers. Small neutral heterocycles were the optimal substituents attached to the 5-carbonyl group, and the most potent inhibitors were afforded by phenyl groups substituted with fluorines at the 2,6-positions.[18] For example, the trisubstituted phenacyl analog **11.12** displayed IC_{50} values of 6 nM, 9 nM, and 230 nM against CDK1/cycB, CDK2/cycE, and CDK4/cycD, respectively. Despite this very high biochemical potency across the CDK family, the cytotoxicity IC_{50} value for **11.12** in the A2780 human ovarian tumor cell line was only 0.76 μM. Interestingly, the solid-state structure of CDK2 with a bound unsubstituted phenacyl analog or with the 2,6-difluoro-4-bromophenacyl analog showed that the carbonyl group assumed an opposite orientation, suggesting that its role is to position the attached aryl ring rather than to provide a polar binding interaction.

In spite of the fact that we were able to obtain pyrazolopyridines with remarkably potent biochemical inhibitory activity against the CDK family, there were several factors that led us to de-emphasize efforts on this chemotype in favor of the aminothiazoles. In particular, the pyrazolopyridines demonstrated low aqueous solubility, and their potency for tumor cell cytotoxicity was poor compared to their high biochemical potency.

11.7 (X = O, R_1 = H, R_2 = Me)   **11.12**
11.8 (X = O, R_1 = Et, R_2 = Me)
11.9 (X = H, OH; R_1 = R_2 =H)
11.10 (X = H, H; R_1 = R_2 = H)
11.11 (X = CH_2, OH; R_1 = R_2 = H)

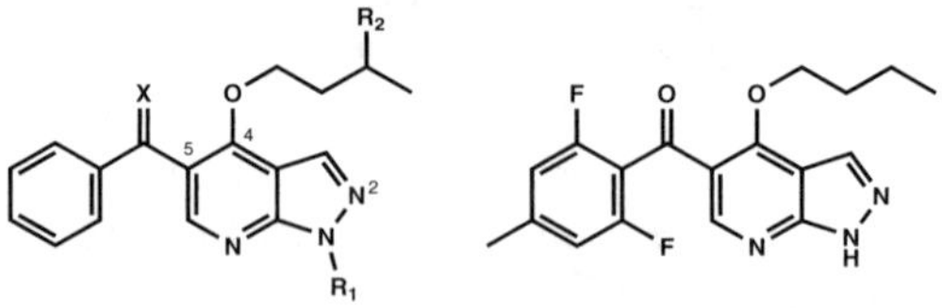

FIGURE 11.4 Pyrazolopyridine analogs.

11.4 STRUCTURE–ACTIVITY RELATIONSHIPS OF AMINOTHIAZOLE CDK INHIBITORS

In the aminothiazole series, SAR studies of the ester substituent of the screening hit **11.6** indicated requirements for a hydrogen bond acceptor and a hydrophobic group of moderate size.[19] Other key features of the lead series were identified through solid-state structural studies of inhibitors bound to CDK2 (the solid-state structure of **11.17** bound to CDK2 is representative of the class and is depicted in Figure 11.5). The solid-state structure of the t-butyl ester analog of **11.6** bound to CDK2 showed the presence of a signature hydrogen bond donor/acceptor pair between the 2-amino hydrogen and the carbonyl of Leu83 and between the aminothiazole nitrogen and the NH of Leu83. Interestingly, the inhibitor adopted a folded conformation in the CDK active site. Nevertheless, cellular studies indicated that the ester of this lead series was a liability for cellular activity, presumably because of hydrolysis. The search for an ester surrogate then led to the identification of the key 5-ethyloxazole analog **11.13** (Figure 11.6). This compound was a relatively selective and potent inhibitor of CDK2 (IC$_{50}$ values: CDK1/cycB = 300 nM; CDK2/cycE = 20 nM; CDK4/cycD = 1400 nM). In addition, **11.13** showed cytotoxicity toward the A2780 cell line with an IC$_{50}$value of 1.4 μM. A key SAR finding was that the positional analog with the ethyl group at position 4′ was a poorer CDK inhibitor (CDK2/cycE IC$_{50}$ = 200 nM). SAR studies of the 5′-position indicated that a t-butyl group provided optimal inhibition (IC$_{50}$ values for **11.14**: CDK1/cycB = 40 nM; CDK2/cycE = 5 nM; CDK4/cycD = 690 nM) and very potent cytotoxicity (A2780 IC$_{50}$ = 50 nM). Therefore, the 5′-t-butyloxazole was maintained during SAR studies of the remainder of the molecule.

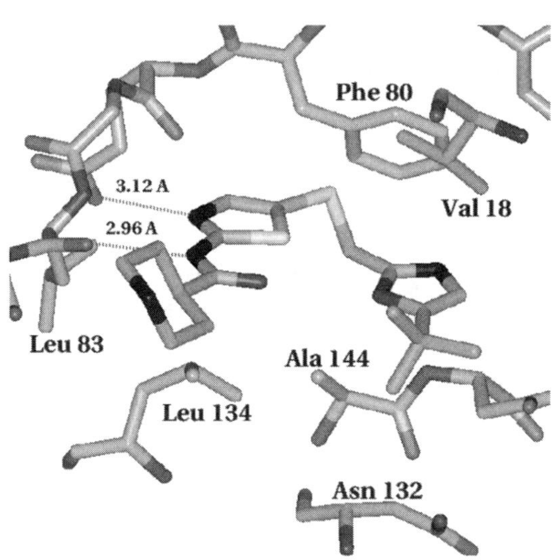

FIGURE 11.5 Solid-state structure of BMS-387032 (**11.17**) bound to CDK2.

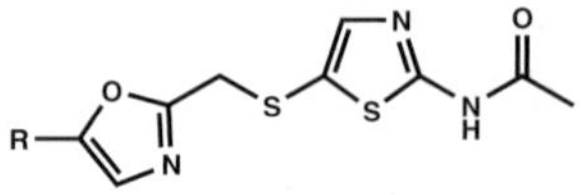

11.13 (R = Et)
11.14 (R = tBu)

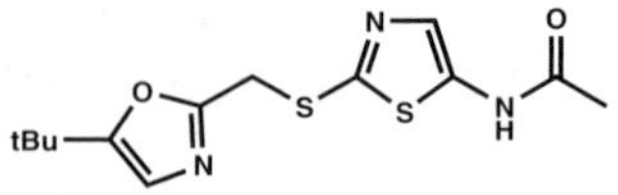

11.15

FIGURE 11.6 Initial oxazole analogs.

Studies indicated relatively stringent SAR requirements for the thiomethylene linker. For example, the sulfoxide and sulfonyl analogs of the compound with an isobutyryl group on the aminothiazole amine were 28- and 50-fold less potent than the parent compound against CDK2. As expected from the crystallographic data, the requirements for the aminothiazole group were also stringent. For example, **11.15**, the positional isomer of **11.14** in which the acylamino group and thiomethylene linker are attached at positions 5′ and 2′, respectively, was essentially inactive against CDK2 (IC$_{50}$ > 25 µM).

Crystallographic studies indicated that the acylamino side chain projected toward the solvent-exposed region of CDK2. This information, which was supported by SAR studies, suggested that this region of the aminothiazole CDK inhibitors could be extensively modified to optimize pharmacokinetic and pharmaceutics properties. Extensive exploration of acylamino substituents led us to focus on two classes of inhibitors (Figure 11.7). One class, phenylacetyl analogs with solublizing substituents on the phenyl ring, provided extremely potent CDK2 inhibitors such as **11.16** (IC$_{50}$ values: CDK1/cycB = 30 nM; CDK2/cycE = 3 nM; CDK4/cycD > 1000 nM). An alternative class contained saturated rings attached to the acyl group, with solubilizing substituents either contained within the ring or attached to it.[20] Pursuit of this latter class ultimately led to the identification of a clinical candidate BMS-387032 (**11.17**), and the preclinical profile of this compound will be described in some detail.

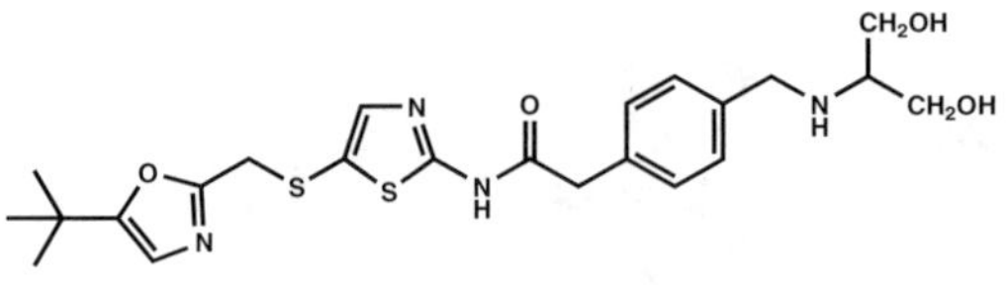

11.16

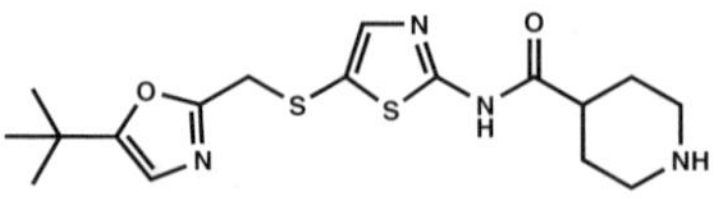

11.17
(BMS-387032)

FIGURE 11.7 Aminothiazole inhibitor lead series.

11.5 PROPERTIES OF BMS-387032

11.5.1 BIOCHEMICAL AND CELLULAR PROPERTIES

BMS-387032 is a potent inhibitor of CDK2, and it also inhibits CDK1 and CDK4 at higher concentrations (IC_{50} values: CDK1/cycB = 480 nM; CDK2/cycE = 48 nM; CDK4/cycD = 925 nM). It is highly selective vs. a panel of kinases, with IC_{50} values > 25 μM for PKCα, PKCβ, PKCδ, PKCε, PKCξ, IKK, EMT, Lck, FAK, ZAP70, EGFR, HER2, and IGF-1R. BMS-387032 inhibits the proliferation of many human tumor cell lines of a variety of histological types, with potency generally in the 100–200 nM range following exposure to drug for 72 h (Table 11.1).[21] Sensitivity to drug is not dependent on cellular p53 status, as demonstrated by the equivalent sensitivity of the A2780S and A2780R cell lines that express wild-type and mutant

TABLE 11.1
Tumor Cell Antiproliferative Effects of BMS-387032 (11.17)

Tumor Cell Line (Tissue Source)		IC_{50} (nM)
A2780S (ovarian)		95
A2780R (ovarian)		202
SKBR3 (breast)		125
PC-3 (prostate)		108
LX-1 (non-small-cell lung)		879
K562 (B-cell lymphoma)		185
A431 (squamous cell)		152
Colorectal Carcinoma	**Mdr 1 mRNA Expression**	
HCT116		215
HCT116-VM46	++++	4700
HT-29		125
CCD-18Co		266
CCD-33Co	++	314
Colo201		96
Colo205	+	91
Colo320DM	+++	2600
CX-1		415
HCT-8	+++	1490
LoVo	+	309
SK-CO-1		203
SW403	+	107
SW1417		752
SW480	+++	2010
SW620	++	319
SW837	+++	1830
T84	++	5470
Caco-2	+++	354
LS174T	++	427
MIP	+++	8180

TABLE 11.2
Time Dependence of Cytotoxicity of BMS-387032 (11.17)

	IC$_{50}$ (nM)	IC$_{90}$ (nM)
A2780 Clonogenic assay, 4 h exposure	302	> 1000
A2780 Clonogenic assay, 8 h exposure	154	303
A2780 Clonogenic assay, 24 h exposure	166	208

p53, respectively. Resistance does appear to be mediated by overexpression of mdr1. This is especially evident by the higher IC$_{50}$ value against the HCT116/VM46 cell line, a derivative of HCT116 that was selected for resistance to etoposide and which overexpresses mdr1.

A key piece of information for developing cancer drug therapies is the length of drug exposure needed to produce an antiproliferative effect. As shown in Table 11.2, the inhibition of colony formation by BMS-387032 is time dependent, and a minimum of 8 h exposure to drug is necessary to achieve the maximal antiproliferative effect. These data were important for defining the optimal pharmacokinetics for anti-tumor efficacy.

11.5.2 *In Vitro* Pharmacokinetic Properties

During the lead optimization process, *in vitro* studies of cytochrome P450 inhibition, microsomal stability, and serum protein binding were routinely conducted to identify analogs that could achieve desired exposures of free drug and that ultimately would be devoid of drug–drug interactions in humans. BMS-387032 showed minimal inhibition of isolated human cytochrome P450 enzymes, with IC$_{50}$ values greater than 100 μM for isozymes 1A2, 2C9, and 2D6, and IC$_{50}$ values of 48 μM for 2C19 and 80 μM for 3A4.[22] BMS-387032 showed good stability in mouse liver microsomes (0.05 nmol/min/mg protein) and a very high free fraction in mouse serum as measured by equilibrium dialysis (69% serum protein binding).[20] In fact, *in vitro* serum protein binding was similar and moderate in all species examined, including human (rat = 66%; dog = 62%; human = 63%).

11.5.3 *In Vivo* Pharmacokinetic and Anti-Tumor Efficacy Studies

These promising properties led to the progression of the compound into our pharmacokinetic screening model, namely intraperitoneal (ip) dosing in mice, in which BMS-387032 was shown to have good exposure and an acceptable half-life (23 μmol/kg ip dose, AUC = 4.0 μmol x h, T$_{1/2}$ = 2.5 h). The acceptable screening pharmacokinetic parameters for BMS-387032 justified its advancement into *in vivo* anti-tumor efficacy studies. Our initial screening model involved ip dosing of compound daily for 7 d to immunocompetent mice bearing P388 murine leukemia tumors. This model served both as an initial indicator of efficacy, as well as a model that could quickly determine a range of tolerated doses. In this model, BMS-387032

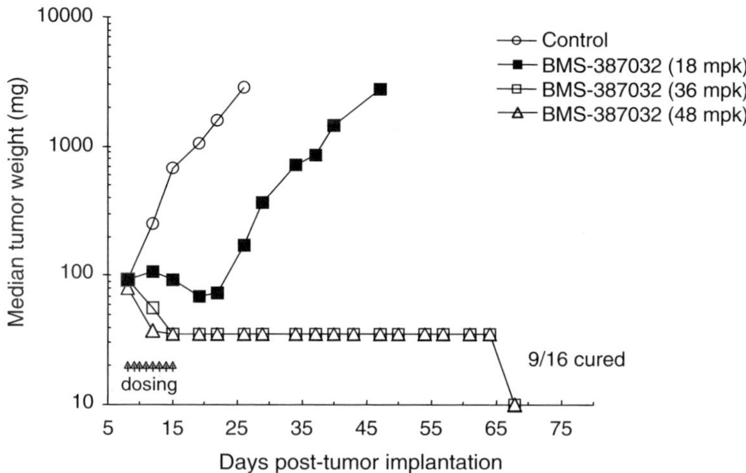

FIGURE 11.8 Anti-tumor efficacy of BMS-387032 in the A2780 human ovarian carcinoma xenograft model.

produced an increase in life span of 40%, with a life span prolongation of $\geq 25\%$ considered an active result. Our primary human tumor xenograft model involved ip dosing of compound daily for 8 d to immunocompromised mice bearing staged A2780 human ovarian tumors. As shown in Figure 11.8, BMS-387032 produced an active result at its MTD of 48 mg/kg, and also at doses of 36 mg/kg and 18 mg/kg. At both of the 48 mg/kg and 36 mg/kg doses, BMS-387032 produced not only dramatic tumor regressions, but also cures in a large percentage of the treated mice (9 cures out of 16 mice treated at these doses). Even at the 18 mg/kg dose, approximately one third of the MTD, the compound produced some tumor regression, as well a robust tumor growth delay equal to 2.1 log cell kill (where 1 log cell kill equals tumor growth delay equivalent to 3.3 times the tumor volume doubling time; see Reference 23 for descriptions of *in vivo* protocols).

The pharmacokinetic parameters for BMS-387032 were more extensively evaluated in rodent and nonrodent species (Table 11.3).[20] The compound exhibited a similar half-life in mouse, rat, and dog, and had a high volume of distribution in the two rodent species. BMS-387032 showed nearly complete oral bioavailability in the mouse and acceptable oral bioavailability in the rat and dog, suggesting the possibility that the compound could be effectively dosed using an oral route in humans.

The reason for the modest oral bioavailability in rats was further explored.[24] The exposure in rats after intraportal administration was shown to be similar to that after intraarterial administration, suggesting that the bioavailability is limited by absorption rather than first-pass metabolism. The permeability of BMS-387032 in Caco-2 cell monolayers, a model of intestinal epithelium, was shown to be low when examined in the apical to basolateral direction (Pc = 14 nm/sec at pH 6.5). In addition, the permeability in the opposite direction was much higher (basolateral to apical

TABLE 11.3
Preclinical Pharmacokinetic Parameters of BMS-387032 (11.17)

	Mouse[a]	Rat[a]	Dog[b]
Dose (μmol/kg)	24	24	1.2
$T_{1/2}$ (h), iv	4.8	5.3	7.0
MRT (h), iv	2.8	3.7	6.3
AUC_{tot} (μM*h), iv	4.8	6.3	2.1
Clearance (ml/min/kg), iv	81	64	9.4
Vss (l/kg), iv	14	15	3.6
C_{max}, (μM), po	1.4	0.52	0.33
T_{max}, (h), po	2.0	4.7	0.4
% Oral bioavailability	100	31	28

[a] Dosed in 1:9 E+OH/water vehicle. [b]Dosed in 1:19 dextrose/water, podose 2.4 μmols/kg.

Pc = 161 nm/sec), suggesting that the compound is a substrate for P-glycoprotein. Finally, the brain uptake in rats following an intravenous (iv) dose was shown to be higher in P-glycoprotein knockout mice. Together, these data suggest that the lower oral bioavailability of BMS-387032 in rats is likely a result of efflux mediated by P-glycoprotein in the intestine.

11.5.4 PHARMACEUTICS PROPERTIES AND CHEMICAL SYNTHESIS ROUTE

In addition to its good biological, pharmacological, and pharmacokinetic properties, there were additional aspects of BMS-387032 that were conducive to successful development as a clinical agent. The compound has reasonable solubility as a free base (0.28 mg/ml).[20] In addition, salt formation using the basic piperidine nitrogen produced highly soluble salts. In particular, the half-tartate salt (2:1 ratio of BMS-387032 to l-tartaric acid) displayed properties well suited to clinical development, including high aqueous solubility (5.5 mg/ml) and low hygroscopicity. In addition, the synthetic route to BMS-387032 used during our discovery program phase, as shown in Figure 11.9, consisted of only seven linear steps and proceeded in over 50% yield.

11.5.5 PRECLINICAL SUMMARY AND EARLY CLINICAL STUDIES

In summary, BMS-387032 demonstrated an overall excellent preclinical profile. It is a potent CDK inhibitor that shows some selectivity within the CDK family and good selectivity with respect to kinases outside the CDK family. It has good tumor cell antiproliferative activity, producing G1 block and apoptosis. It has good ADME (absorption/distribution/metabolism/excretion) properties, including low protein binding, moderate half-life, and good oral bioavailability. It has excellent pharmaceutical properties, including high solubility. It has an *in vitro* safety profile suggestive of a low probability for drug–drug interactions or cardiac liabilities. And it has broad-spectrum preclinical anti-tumor activity, observed with either

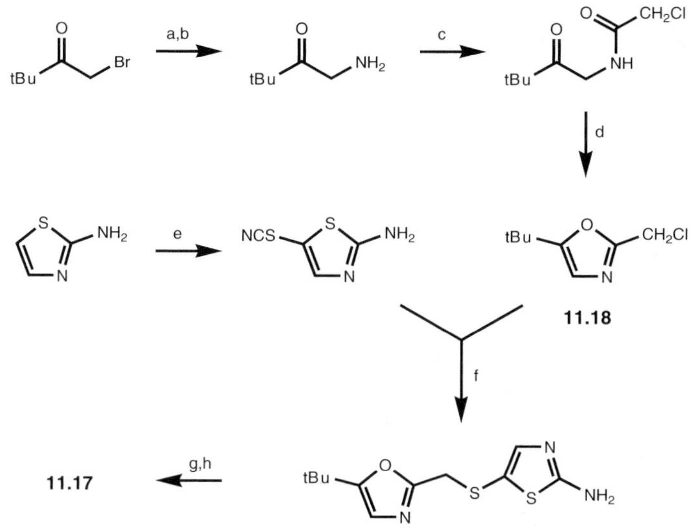

FIGURE 11.9 Reagents and conditions: (a) NaN$_3$, acetone, 25°C, 100%; (b) Pd/C, H$_2$, aq HCl, MeOH, 25°C, 91%; (c) ClCH$_2$COCl, Et$_3$N/CH$_2$Cl$_2$, −10°C, 98%; (d) POCl$_3$, 105°C, 80%; (e) Br$_2$, KSCN, MeOH, 26%; (f) NaBH$_4$ (2 eq), EtOH, 25°C, then **11.18** (1.1 eq), reflux, 94%; (g) N-t-butoxycarbonyl-piperidinyl-4-carboxylic acid, EDAC (2 eq), DMAP (0.5 eq), DMF, CH$_2$Cl$_2$, 25°C, 97%; (h) 4N HCl in dioxane, CHCl$_3$, 45–55°C, 83%.

parenteral or oral drug delivery. On the basis of this profile, BMS-387032 was advanced into clinical trials.

Reports have appeared on several of the initial Phase I clinical trials of BMS-387032. These reports disclosed the preliminary pharmacokinetics of the compound given as a single ascending dose to cancer patients using three different regimens, namely a 1 h infusion given every 3 weeks, a 1 h infusion given weekly, and a 24 h infusion given every 3 weeks.[25–27] All of the studies reported that BMS-387032 was well tolerated, that its pharmacokinetics were roughly dose proportional, and none of these preliminary reports indicated that a maximum tolerated dose had been reached.

11.6 CDK2 AS A TARGET FOR CANCER THERAPY

Finally, several recent reports have appeared that raise questions about the importance of CDK2 in cancer cell proliferation. In one report, continued proliferation of colon cancer cells was observed following downregulation of CDK2 using dominant negative constructs, siRNA, or antisense oligonucleotides.[28] These authors suggest that compensatory phosphorylation of CDK2 substrates by CDK4 underlies a portion of this lack of requirement for CDK2 activity. A second report showed that CDK2$^{-/-}$ mice are viable, and that CDK2$^{-/-}$ mouse embryo fibroblasts proliferate similarly to wild-type fibroblasts.[29] These authors also postulate compensatory activity by other kinases, and suggest that CDK1, CDK4, and/or CDK6 may be playing key roles.

It is important to note that a functional knockout of the CDK2 protein may cause different biological effects on the cell cycle as compared to the direct inhibition of the kinase. It can be hypothesized, for example, that in the absence of its preferred kinase partner, cyclins can partner with alternative CDKs to form an active kinase, resulting in continued cell cycle progression. Conversely, inhibition of the kinase activity may result in the sequestration of the cyclin in inactive complexes, preventing activation through binding of the cyclin to alternative CDKs and thereby causing aberrant progression or arrest of the cell cycle. The preceding studies may therefore not conclusively support the proposition that CDK2 is not essential for cell cycle progression. Furthermore, a number of reports provide contrasting data using dominant negative CDK2 as well as the biological inhibition of CDK2 with selective small molecule or peptide inhibitors of CDK2, suggesting that CDK2 plays a key role in the entry into and progression through S phase.[30–34] Taken together, the data support a role for the chemical inhibition of CDK2 as a viable approach to the treatment of at least some tumor types.

Several additional explanations are also possible for the substantial *in vitro* and *in vivo* antiproliferative effects demonstrated with BMS-387032. Although the compound demonstrates selectivity for CDK2, it is also capable of inhibiting CDK1 and CDK4 at approximately 10- and 20-fold higher concentrations, respectively. Such concentrations are achieved for at least some time following dosing regimens that lead to preclinical anti-tumor activity. It is also possible that inhibition of other CDK family members contributes to the antiproliferative effects of BMS-387032. For example, CDK7, CDK8, and CDK9 have been shown to phosphorylate the C-terminal domain (CTD) of RNA polymerase II and to affect various steps in mRNA formation.[35] BMS-387032 (SNS-032) has been shown to potently inhibit CDK7 and CDK9 *in vitro* (unpublished data), and therefore effects on substrates of these kinases such as RNA polymerase II may additionally contribute to the preclinical anti-tumor effects of this compound *in vivo*.

ACKNOWLEDGMENTS

I would like to acknowledge Tai Wong of Bristol-Myers Squibb and Duncan Walker of Sunesis for helpful suggestions during the writing of this chapter.

REFERENCES

1. Dowdy, S.F., Hinds, P.W., Louie, K., Reed, S.I., Arnold, A. et al. Physical interaction of the retinoblastoma protein with human D cyclins. *Cell* **1993**, *73*, 499–511.
2. Kato, J.-Y., Matsushime, H., Hiebert, S.W., Ewen, M.E., Sherr, C.J. Direct binding of cyclin D to the retinoblastoma gene product (pRb) and pRb phosphorylation by the cyclin D-dependent kinase CDK4. *Genes and Development* **1993**, *7*, 331–342.
3. Koff, A., Cross, F., Fisher, A., Schumacher, J., Leguellec, K. et al. Human cyclin E, a new cyclin that interacts with two members of the *CDC2* gene family. *Cell* **1991**, *66*, 1217–1228.
4. Dulic, V., Lees, E., Reed, S.I. Association of human cyclin E with a periodic G_1-S phase protein kinase. *Science* **1992**, *257*, 1958–1961.

5. Girard, F., Strausfeld, U., Fernandez, A., Lamb, N.J.C. Cyclin A is required for the onset of DNA replication in mammalian fibroblasts. *Cell* **1991**, *67*, 1169–1179.

6. Arnold, A. The cyclin D1/PRAD1 oncogene in human neoplasia. *Journal of Investigative Medicine* **1995**, *43*, 543–549.

7. Keyomarsi, K., O'Leary, N., Molnar, G., Lees, E., Fingert, H.J. et al. Cyclin E, a potential prognostic marker for breast cancer. *Cancer Research* **1994**, *54*, 380–385.

8. Serrano, M., Hannon, G.J., Beach, D. A new regulatory motif in cell-cycle control causing specific inhibition of cyclin D/CDK4. *Nature* **1993**, *366*, 704–707.

9. Xiong, Y., Hannon, G.J., Zhang, H., Casso, D., Kobayashi, R. et al. p21 is a universal inhibitor of cyclin kinases. *Nature* **1993**, *366*, 701–704.

10. Gu, Y., Turck, C.W., Morgan, D.O. Inhibition of CDK2 activity *in vivo* by an associated 20K regulatory subunit. *Nature* **1993**, *366*, 707–710.

11. Polyak, K., Lee, M.-H., Erdjument-Bromage, H., Koff, A., Roberts, J.M. et al. Cloning of p27^{Kip1}, a cyclin-dependent kinase inhibitor and a potential mediator of extracellular antimitogenic signals. *Cell* **1994**, *78*, 59–66.

12. Toyoshima, H. and Hunter, T. p27, a Novel inhibitor of G1 cyclin-cdk protein kinase activity, is related to p21. *Cell* **1994**, *78*, 67–74.

13. Losiewicz, M.D., Carlson, B.A., Kaur, G., Sausville, E.A., Worland, P.J. Potent inhibition of CDC2 kinase activity by the flavonoid L86-8275. *Biochemical and Biophysical Research Communications* **1994**, *201*, 589–595.

14. Kitagawa, M., Okabe, T., Ogino, H., Matsumoto, H., Suzuki-Takahashi, I. et al. Butyrolactone I, a selective inhibitor of cdk2 and cdc2 kinase. *Oncogene* **1993**, *8*, 2425–2432.

15. Vesely, J., Havlicek, L., Strnad, M., Blow, J.J., Donella-Deana, A. et al. Inhibition of cyclin-dependent kinases by purine analogs. *European Journal of Biochemistry* **1994**, *224*, 771–786.

16. Kim, K.S., Sack, J.S., Tokarski, J.S., Qian, L., Chao, S.T. et al. Thio- and oxoflavopiridols, cyclin-dependent kinase 1-selective inhibitors: synthesis and biological effects. *The Journal of Medicinal Chemistry* **2000**, *43*, 4126–4134.

17. Misra, R.N., Rawlins, D.B., Xiao, H.-Y., Shan, W., Bursuker, I. et al. 1*H*-Pyrazolo [3,4-*b*]pyridine inhibitors of cyclin-dependent kinases. *Bioorganic and Medicinal Chemistry Letters* **2003**, *13*, 1133–1136.

18. Misra, R.N.; Xiao, H.-Y., Rawlins, D.B., Shan, W., Kellar, K.A. et al. 1*H*-Pyrazolo [3,4-*b*]pyridine inhibitors of cyclin-dependent kinases: highly potent 2,6-difluorophenacyl analogues. *Bioorganic and Medicinal Chemistry Letters* **2003**, *13*, 2405–2408.

19. Kim, K.S., Kimball, S.D., Misra, R.N., Rawlins, D.B., Hunt, J.T. et al. Discovery of aminothiazole inhibitors of cyclin-dependent kinase 2. Synthesis, x-ray crystallographic analysis, and biological activities. *The Journal of Medicinal Chemistry* **2002**, *45*, 3905–3927.

20. Misra, R.N., Xiao, H.-Y., Kim, K., Lu, S., Han, W.-C. et al. N-(Cycloalkylamino) acyl-2-aminothiazole inhibitors of Cyclin-Dependent Kinase 2. N-[5[[[5-(1,1-Dimethylethyl)-2-oxazolyl]methyl]thio]-2-thiazolyl]-4-piperidinecarboxamide (BMS-387032), a highly efficacious and selective antitumor agent. *The Journal of Medicinal Chemistry* **2004**, *47*, 1719–1728.

21. Wong, T.W., Kimball, D., Misra, R.N., Kim, K., Lee, F. et al. BMS-387032, A potent and selective inhibitor of cyclin-dependent kinase 2 (CDK2), induces cell cycle arrest and apoptosis in human tumor cells. *American Association of Cancer Research 94th Annual Meeting*: Washington, D.C., 2003, Abstract #3125.

22. Kim, K.S., Misra, R.N., Xiao, H., Lu, S., Han, W. et al. BMS-387032: a selective cdk2 inhibitor with potent anti-tumor activity. *225th American Chemical Society National Meeting*: New Orleans, LA, 2003.

23. Lee, F.Y.F., Borzilleri, R., Fairchild, C., Kim, S.-H., Long, B.H. et al. BMS-247550: A novel epothilone analog with a mode of action similar to paclitaxel but possessing superior antitumor efficacy. *Clinical Cancer Research* **2001**, *7*, 1429–1437.

24. Kamath, A.V., Chong, S., Chang, M., Marathe, P.H. P-glycoprotein plays a role in the oral absorption of BMS-387032, a potent cyclin-dependent kinase 2 inhibitor, in rats. *Cancer Chemotherapy and Pharmacology* **2005**, *55*, 110–116.

25. Jones, S., III, Burris, H.A., Kies, M., Wilcutt, N., Degen, P. et al. A phase I study to determine the safety and pharmacokinetics (PK) of BMS-387032 given intravenously every three weeks in patients with metastatic refractory solid tumors. *Proceedings of the American Society of Clinical Oncology* **2003**, *22*, 199 (abstract 798).

26. Shapiro, G., Lewis, N., Leeuwen, B.V., Letrent, S., Woo, M. et al. A phase I study to determine the safety and pharmacokinetics (PK) of BMS-387032 with a 24-h infusion given every three weeks in patients with metastatic refractory solid tumors. *Proceedings of the American Society of Clinical Oncology* **2003**, *22*, 199 (abstract 799).

27. McCormick, J., Gadgeel, S.M., Helmke, W., Chaplen, R.A., Leuwen, B.V. et al. Phase I study of BMS-387032, a cyclin dependent kinase (CDK) 2 inhibitor. *Proceedings of the American Society of Clinical Oncology* **2003**, *22*, 208 (abstract 835).

28. Tetsu, O. and McCormick, F. Proliferation of cancer cells despite CDK2 inhibition. *Cancer Cell* **2003**, *3*, 233–245.

29. Ortega, S., Prieto, I., Odajima, J., Martin, A., Dubus, P. et al. Cyclin-dependent kinase 2 is essential for meiosis but not for mitotic cell division in mice. *Nature Genetics* **2003**, *35*, 25–31.

30. van den Heuvel, S. and Harlow, E. Distinct roles for cyclin-dependent kinases in cell cycle control. *Science* **1993**, *262*, 2050–2054.

31. Hu, B., Mitra, J., van den Heuvel, S., Enders, G.H. S and G_2 phase roles for Cdk2 revealed by inducible expression of a dominant-negative mutant in human cells. *Molecular and Cellular Biology* **2001**, *21*, 2755–2766.

32. Zhu, Y., Alvarez, C., Doll, R., Kurata, H., Schebye, X.M. et al. Intra-S-phase checkpoint activation by direct CDK2 inhibition. *Molecular and Cellular Biology* **2004**, *24*, 6268–6277.

33. Lane, M.E., Yu, B., Rice, A., Lipson, K.E., Liang, C. et al. A novel cdk2-selective inhibitor, SU9516, induces apoptosis in colon carcinoma cells. *Cancer Research* **2001**, *61*, 6170–6177.

34. Chen, Y.-N.P., Sharma, S.K., Ramsey, T.M., Jiang, L., Martin, M.S. et al. Selective killing of transformed cells by cyclin/cyclin-dependent kinase 2 antagonists. *Proceedings of the National Academy of Science of the United States of America* **1999**, *96*, 4325–4329.

35. Prelich, G. RNA Polymerase II carboxy-terminal domain kinases: emerging clues to their function. *Eukaryotic Cell* **2002**, *1*, 153–162.

12 Oxindole Inhibitors of Cyclin-Dependent Kinases as Anti-Tumor Agents

Philip A. Harris

CONTENTS

12.1 INTRODUCTION

The initial reports on the use of oxindoles (2H-indol-2-ones) as kinase inhibitors were published by Japanese researchers at Kanegafuchi in the late 1980s as a small subset within a series of patent applications on 4-hydroxycinnamamide-based derivatives.[1,2] The compounds were claimed to be tyrosine kinase inhibitors; for example, 3′,5′-diisopropyl-4′-hydroxybenzylidene-2-oxindole (**12.1**) was reported as having moderate activity against the epidermal growth factor receptor (EGFR), but possessing no activity against serine or threonine kinases. These derivatives were prepared by simple condensation of the appropriate benzaldehyde with oxindole in the presence of base. A few years later, researchers at Farmitalia, later Pharmacia, disclosed the use of 3-arylidene-oxindoles, such as compound **12.2**, as inhibitors of tyrosine kinases involved in signal transduction pathways modulating cellular growth.[3–6]

12.1

12.2

In the mid-1990s Sugen, later part of Pharmacia and Upjohn and then Pfizer, began an extensive research program developing oxindoles as kinase inhibitors with a primary focus on the inhibition of tyrosine kinases involved in angiogenesis, such as vascular endothelial growth factor receptor-2 (VEGFR-2), platelet-derived growth factor receptor (PDGFR), and fibroblast growth factor receptor (FGFR). The outcome of this work was the development of drug candidates SU5416 (**12.3**),[7] SU6668 (**12.4**),[8] and SU11248 (**12.5**),[9] all possessing a common 2,4-dimethylpyrrole moiety linked at C-3 of an oxindole. These inhibitors are under evaluation in a number of different clinical trials as potential anticancer agents. The most advanced inhibitor from this series, oxindole **12.5**, has recently been granted orphan drug status for the treatment of gastrointestinal stromal cell tumors (GIST) in patients who have stopped responding to Gleevec® (imatinib mesylate). The reported efficacy in this tumor type is proposed to be due to inhibition of the tyrosine kinases c-Kit (hematopoietic cell growth factor receptor kinase) and PDGFR. The clinical efficacy of oxindole **12.5** supports the utility of oxindoles as therapeutically effective kinase inhibitors in anticancer drug therapy.

12.3

12.4

12.5

12.6

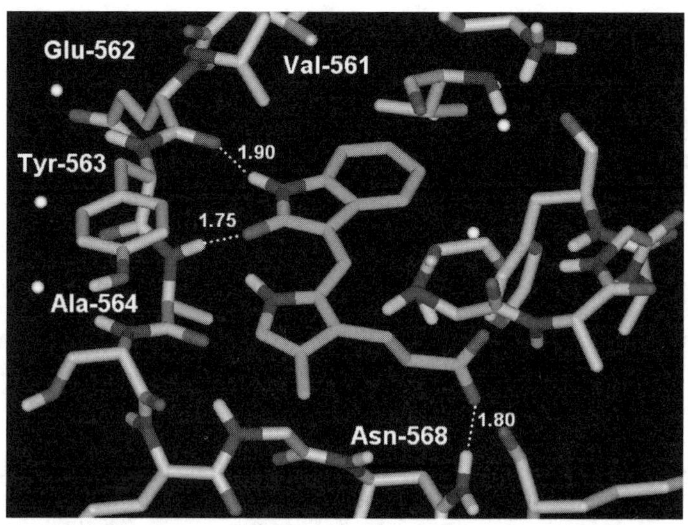

FIGURE 12.1 Crystal structure of SU5402 (**12.6**) in FGFR.

The first reported crystal structure of a member of the oxindole class of kinase inhibitors, SU5402 (**12.6**), bound in the kinase domain of FGFR, was published in 1997 (Figure 12.1).[10] This compound is a competitive inhibitor with respect to ATP and binds in the ATP-binding site of the kinase. In the crystal structure of compound **12.6** bound to the kinase domain of FGFR, the oxindole heterocyclic ring was shown to occupy the same site as the adenosine of ATP, although in a different orientation, with its C-3 pyrrole substituent emerging from the ATP cleft at approximate right angles to the direction of the ATP ribose and triphosphate (Figure 12.2). The oxindole

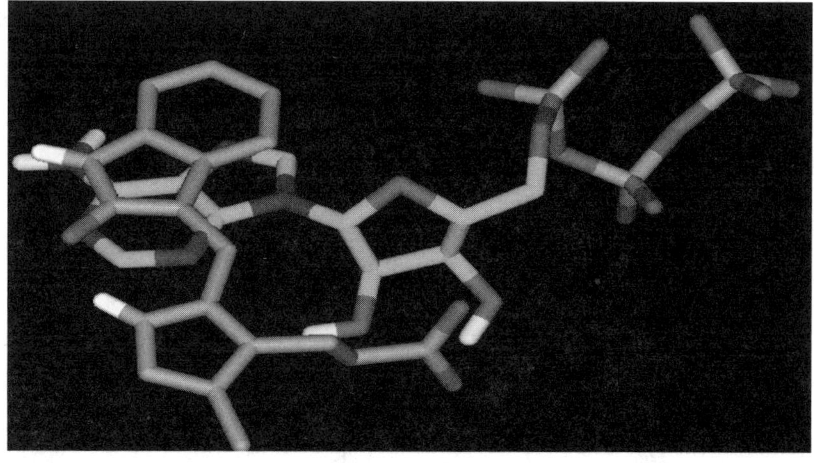

FIGURE 12.2 Overlay of SU5402 (**12.6**) (colored gray) and ATP in FGFR.

makes two hydrogen bonds to the backbone of the protein, one between the oxindole nitrogen (N-1) and backbone carbonyl oxygen of Glu562, and another between the oxindole carbonyl (O-2) and the amide nitrogen of Ala564.

12.2 IMIDAZOLYLMETHYLIDENE-OXINDOLES

Although most of the work reported by Sugen was focused on tyrosine kinase inhibitors, an imidazolylmethylidene-oxindole, SU9516 (**12.7**), was described in 2001 as a selective CDK2 inhibitor identified through high-throughput screening.[11] The compound was reported to inhibit CDK2 (IC_{50} = 0.022 μM), CDK1 (IC_{50} = 0.04 μM), and CDK4 (IC_{50} = 0.3 μM) and not to inhibit PDGFR, EGFR, p38, and protein kinase C (PKC). In subsequent biochemical studies, oxindole **12.7** was confirmed to bind competitively to CDK2/cyclin A with respect to ATP with a K_i of 0.031 μM.[12] However, it surprisingly exhibited binding to CDK4/cyclin D in an ATP-noncompetitive manner with a K_i of 1.1 μM. The authors speculate that the weaker binding affinity of **12.7** for CDK4 reflects the ability of both ATP and inhibitor to bind within the active site with **12.7** interfering with catalytic activity, although the possibility of allosteric inhibition cannot be excluded. In additional studies carried out in RKO and SW480 tumor cell lines, oxindole **12.7** caused a decrease in the phosphorylation of the retinoblastoma protein (pRb), an increase in apoptosis (as measured by caspase-3 activation), and cell cycle arrest in both G_0-G_1 and G_2-M stages.[13] The crystal structure of oxindole **12.7** bound to CDK2 showed a binding mode similar to that of **12.6** bound to FGFR. In the CDK2 structure, the oxindole nitrogen (N-1) and carbonyl (O-2) bind to the CDK2 backbone carbonyl of Glu81 and amide nitrogen of Leu83, respectively. An additional third backbone interaction was noted between the N-1 of the imidazole of inhibitor **12.7** and the carbonyl oxygen of Leu83.

In a 2002 patent application, researchers at Aventis also claimed imidazolidene-5-substituted oxindoles as specific inhibitors of CDK1.[14] For example, the 5-(3-pyridyl)-oxindole **12.8** was reported to inhibit CDK1 with an IC_{50} of 0.3 nM. The majority of these compounds contained an imidazolylmethylidene at the oxindole C-3 position, although 2-pyrrole and 2-furan heterocycles were also employed.

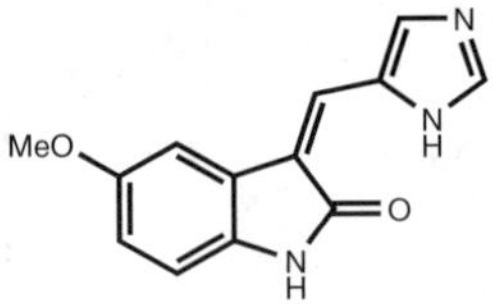

12.7 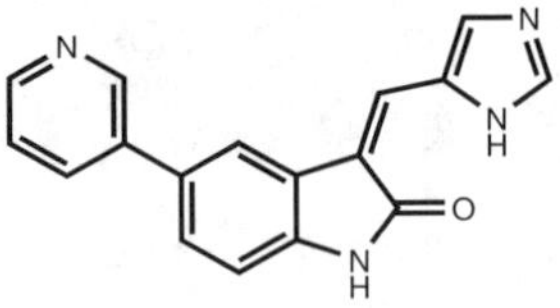   12.8

12.3 PYRROLYLLACTONE AND PYRROLYLLACTAM OXINDOLES

The screening hit **12.7** was subjected to a lead optimization effort by Sugen.[15,16] The imidazole heterocycle was replaced with either a pyrrolyllactone or pyrrolyllactam, with the additional lactone or lactam rings predicted to make further hydrogen

TABLE 12.1
Representative Examples of Pyrrolyllactone and Pyrrolyllactam CDK2 Inhibition

Compound	Class	R4	R5	CDK2 IC$_{50}$ (µM)
12.9	Pyrrolyllactone	H	SO$_2$NHMe	0.004
12.10	Pyrrolyllactone	H	SO$_2$NH$_2$	0.012
12.11	Pyrrolyllactone	H	CO$_2$H	0.004
12.12	Pyrrolyllactone	4-Piperidine	H	0.009
12.13	Pyrrolyllactam	H	SO$_2$NHMe	0.009
12.14	Pyrrolyllactam	H	CO$_2$H	0.068
12.15	Pyrrolyllactam	4-Piperidine	H	0.049

bonding interactions with Lys89 in CDK2. At the oxindole ring, the 5-position was optimized by the incorporation of a series of hydrogen bond-accepting sulfonamide or carboxylic acid functionalities, which are proposed to interact with Lys33. Furthermore, a piperidinyl ring was incorporated at the oxindole 4-position to interact with the Gln131 and Asn132 CDK2 residues. These modification resulted in a series of potent, low-nanomolar inhibitors of CDK2 (Table 12.1) with the pyrrolyllactone oxindoles (**12.9–12.12**) being generally more potent than their corresponding pyrrolyllactam analogs (**12.13–12.15**).

12.9-12.12   12.13-12.15

12.4   4-AZACYCLIC AND 4-ALKYNYL PYRROLYLMETHYLIDENE-OXINDOLES

Researchers at Roche have recently published two lead optimization efforts for CDK2 inhibition starting from the pyrrolylmethylidene-oxindole **12.16**, which was originally identified through screening. A series of saturated cyclic amines were investigated based on the analysis of x-ray crystal structures in which the 4-position of the oxindole ring points towards the open space that the ribose ring of ATP would occupy (Figure 12.2).[17] The desired inhibitors were synthesized via nucleophilic aromatic substitution with the appropriate cyclic amine on the corresponding 4-bromo-5-nitro-oxindoles. Such 4-substituents (compounds **12.17–12.20**) led to significant increases in both inhibition of CDK2 enzyme and inhibition of the proliferation of RKO tumor cell-lines (Table 12.2).

TABLE 12.2
Representative Examples of 4-Cyclicamino-Oxindole CDK2 Inhibition

Compound	R4	CDK2/Cyclin E IC$_{50}$ (μM)	RKO IC$_{50}$ (μM)
12.16	H	0.039	0.70
12.17	3-(aminocarbonyl)pyrrolidin-1-yl (H$_2$NOC)	0.006	0.19
12.18	3-hydroxypyrrolidin-1-yl (HO)	0.003	0.032
12.19	4-aminopiperidin-1-yl (NH$_2$)	0.005	<0.01
12.20	3-aminopyrrolidin-1-yl (H$_2$N)	0.003	<0.01

In a second lead optimization effort, Roche scientists took a slightly different approach beginning with screening hit pyrrolylmethylidene-oxindole **12.21**, which is structurally similar to their earlier lead oxindole **12.16**.[18,19] In this new series, alkynyl-linked heteroatoms were introduced at the 4-position of the oxindole via a palladium-catalyzed coupling of the corresponding 4-iodo, 5-fluoro-oxindole with a terminal alkyne. Incorporation of these substituents (compounds **12.22–12.25**) led to significant increases in potency against inhibition of CDK2 compared to the unsubstituted analog **12.21** (Table 12.3). An x-ray crystal structure of inhibitor **12.24** bound to CDK2 showed the expected three hydrogen bonding interactions involving the oxindole and pyrrole

TABLE 12.3

Representative Examples of 4-Alkynyl-Oxindole CDK2 Inhibition

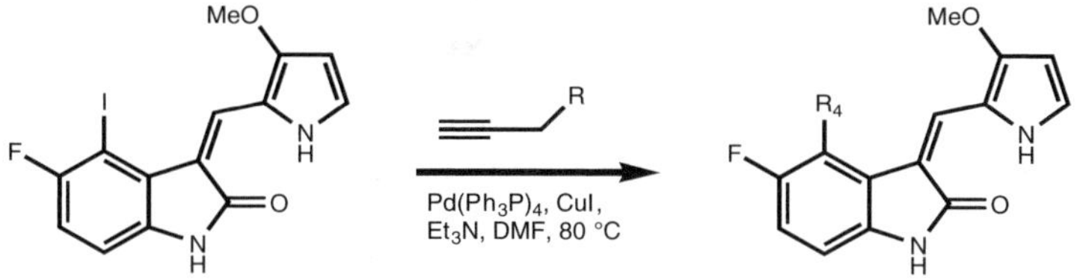

Compound	R4	CDK2/Cyclin E IC$_{50}$ (μM)	SW480 IC$_{50}$ (μM)
12.21	H	0.85	—
12.22	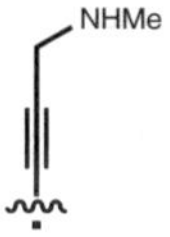	0.004	0.15
12.23	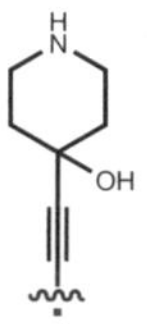	0.006	0.11
12.24	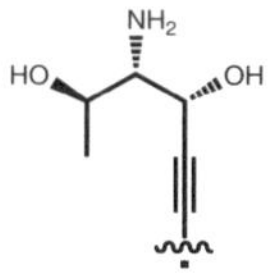	0.003	0.40
12.25	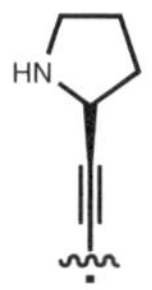	0.002	0.031

heteroatoms with the protein backbone of Glu81 and Leu83, analogous to those observed in SU9516. Two additional hydrogen bond interactions were also observed, one between the propargylic oxygen and Asp145, and another between the terminal hydroxyl and the Glu12 backbone. The interaction with Asp145 could explain the requirement for a propargyl heteroatom in this series for high levels of CDK2 inhibition.

A representative compound, oxindole **12.23**, causes a concentration-dependent decrease in the phosphorylation of the retinoblastoma protein (pRb) in SW480 tumor cell lines, a cell cycle block in the G_1 and G_2 stages, and also some apoptosis as evidenced by sub-G_0 cell fractions in cell cycle analysis.

12.5 ISATIN-3-PHENYLHYDRAZONES AND 3-(ANILINOMETHYLENE)-OXINDOLES

Two closely related classes of oxindole-based compounds, the isatin 3-arylhydrazones (**12.26**) and the 3-(anilinomethylene)-oxindoles (**12.27**), were shown to potently inhibit CDK2 by researchers at GlaxoSmithKline (GSK) in 2001.[20,21] Inhibitors in this series can be regarded as homologues of the 3-benzylidene-oxindole class as the aryl ring is linked to the C-3 oxindole by two atoms instead of one. The isatin hydrazones **12.26** were prepared by condensation of the substituted isatin with an arylhydrazine in ethanol. The 3-(anilinomethylene)-oxindoles **12.27** were obtained by initial conversion of a substituted oxindole to the corresponding dimethylamino-methinyloxindole or ethoxymethinyloxindole, followed by condensation with an aniline. In an initial screening set, the 5-bromo-isatin 3-arylhydrazones **12.28–12.30** were inactive (IC_{50} >10 μM) against all kinases tested, but compound **12.31** selectively inhibited CDK2 (IC_{50} = 0.06 μM) and served as the lead compound.

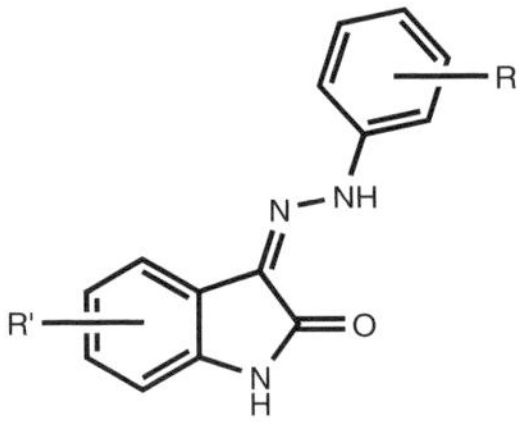

12.26 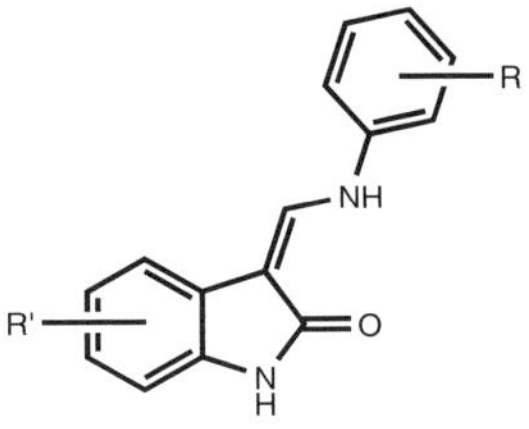   12.27 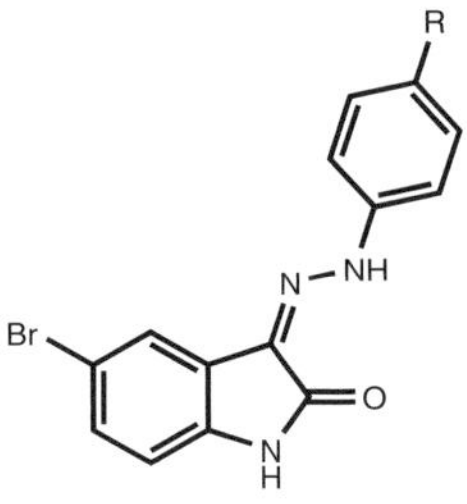   12.28 (R = Br)
                                                                      12.29 (R = H)
                                                                      12.30 (R = OMe)
                                                                      12.31 (R = SO$_2$NH$_2$)

To aid in structural design, the crystal structure of the lead **12.31** bound to CDK2 was solved, as shown in Figure 12.3. The lactam ring of oxindole **12.31** interacted with CDK2 in a manner analogous to that observed for oxindole **12.6** bound to FGFR kinase, with an additional, albeit weaker, third hydrogen bond interaction between the NH of the hydrazone linker and the carbonyl of Leu83. Furthermore, the sulfonamide moiety forms hydrogen bonds with the backbone NH and the side-chain carboxylate functionality of Asp86 toward the opening to the binding cleft of CDK2, thus accounting for the importance of this functionality for CDK2 potency as seen in the initial screening set.

Inspection of the CDK2/inhibitor **12.31** structure shown in Figure 12.3 suggests a number of opportunities for the design of improved inhibitors. The incorporation of lipophilic 4-substituents targets a hydrophobic region of the kinase (Val18 and Leu134), as exemplified by the increased potency of the 4-isopropyl oxindole **12.32** (Table 12.4). Supportive of this analysis is the fact that incorporation of hydrophilic groups at the 4-position is detrimental to activity; for example, the 4-amido oxindole

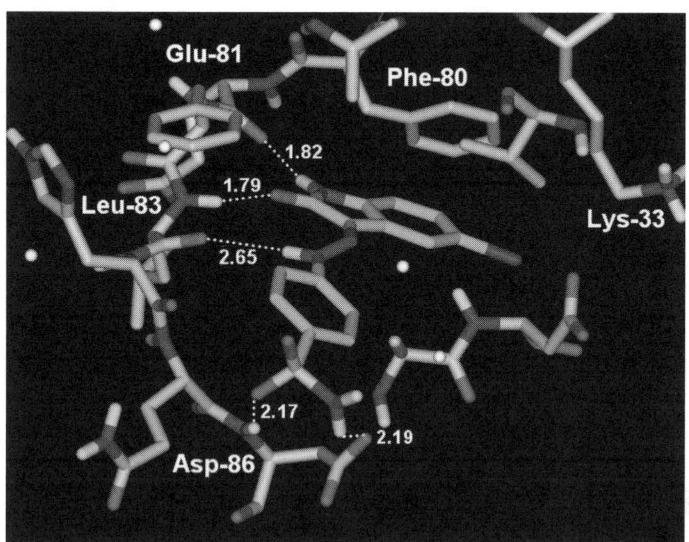

FIGURE 12.3 Crystal structure of 5-bromo-isatin 3-arylhydrazones **12.31** in CDK2.

12.33 shows very weak inhibition of CDK2. The propinquity of the 5-position and Lys33 allows for the incorporation of hydrogen bond acceptors at that position, such as oxindoles (**12.34-12.36**), with the expected enhancement of inhibitor affinity activity. Carboxylic esters such as **12.36** are particularly active inhibitors, an observation consistent with the ability of those substituents to form a hydrogen bond with Lys33 and position a hydrophobic group in the lipophilic region near the 4-position. The 6-position projects toward a small cavity at the back of the binding cleft and into the region affected by cyclin A association. This cavity is smaller in size relative to many other protein kinases because of the bulkiness of the Phe80 side chain, and is compatible with small substituents such as bromo (**12.37**) and ethyl (**12.38**), which yielded moderate contributions to affinity. In contrast, no measurable inhibition is observed for compounds containing larger substituents such as *tert*-butyl (**12.39**). The 7-position of the oxindole ring system is positioned close to the side chain of Phe80, and the lack of activity found for the 7-methyl oxindole (**12.40**) shows it to be too sterically crowded to allow for such substitution. Also consistent with structural and modeling considerations is the observation that the carbon atom of the enamine linkage can be substituted with groups such as methyl without negatively affecting inhibition of CDK2, as shown by oxindole **12.41**. Oxindoles with 4,5-fused heterocycles that contain a hydrogen bond acceptor at the 5-position and hydrophobic character at the 4-position, such as the thiazolo-oxindoles **12.43** and **12.44**, also display high affinity for CDK2.

The 4-sulfonamide moiety common to inhibitors **12.31–12.44** interacts with Asp86 at the opening to the binding cleft and points away from the protein into solution. This allows for the introduction of sulfonamide substituents with polar functionality, such as the pyridinyl of **12.45**, with the potential to improve aqueous solubility without

TABLE 12.4
Representative CDK2 Inhibitory Activities of Isatin-3-Phenylhydrazones and 3-(Anilinomethylene)-Oxindoles 12.31–12.44

12.31-12.44

Compound	R4	R5	R6	R7	X	CDK2 IC$_{50}$ (μM)
12.31	H	Br	H	H	N	0.06
12.32	—i-Pr	H	H	H	N	0.0025
12.33	CONH$_2$	H	H	H	N	>1
12.34	H	—OH	H	H	N	0.01
12.35	H	—SO$_2$Me	H	H	N	0.016
12.36	H	—CO$_2$Me	H	H	CH	0.002
12.37	H	H	Br	H	N	0.043
12.38	H	H	—Et	H	N	0.021
12.39	H	H	—t-Bu	H	N	>10
12.40	H	H	H	Me	CH	>10
12.41	H	5-Oxazolyl	H	H	C-Me	0.002
12.42	—N=N-NH—		H	H	N	0.01
12.43	—S-CH=N—		H	H	N	0.007
12.44	—S-CH=N—		H	H	CH	0.003

detrimentally affecting enzyme affinity. The structure of compound **12.45** bound to the CDK2/cyclin A complex shows the pyridyl sulfonamide at the opening of the binding site, as illustrated in Figure 12.4 with the protein shown as a surface representation. The thiazole nitrogen of the oxindole interacts with the side chain of Lys33, with a weaker additional hydrogen bond to the backbone NH of Asp145 (Figure 12.5).

12.45   **12.46**

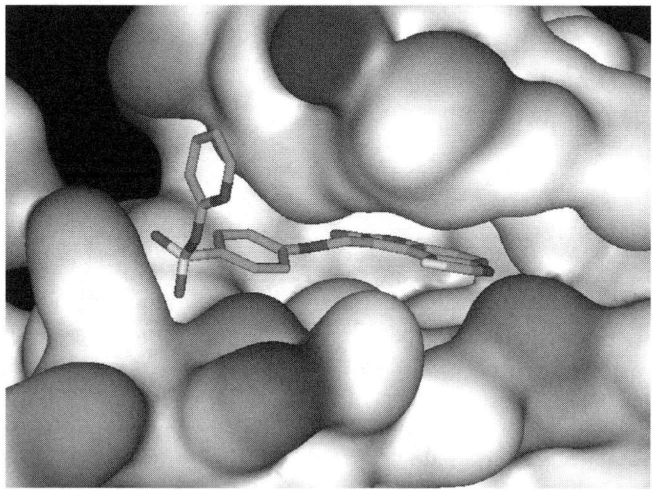

FIGURE 12.4 Surface representation of thiazolo-oxindole **12.45** bound to CDK2/cyclin A.

The most potent and selective CDK2 inhibitors were evaluated in two cellular assays, a mechanistic assay that measures progression from G_1 phase to S phase and a number of antiproliferation assays (Table 12.5). The antiproliferative activity of the inhibitors is up to 8-fold selective for the tumor cell lines relative to the non-tumor HFF cell line. The compounds were also active in blocking progression of cells into S phase, consistent with the role of CDK2 activity in the regulation of G_1/S phase progression.

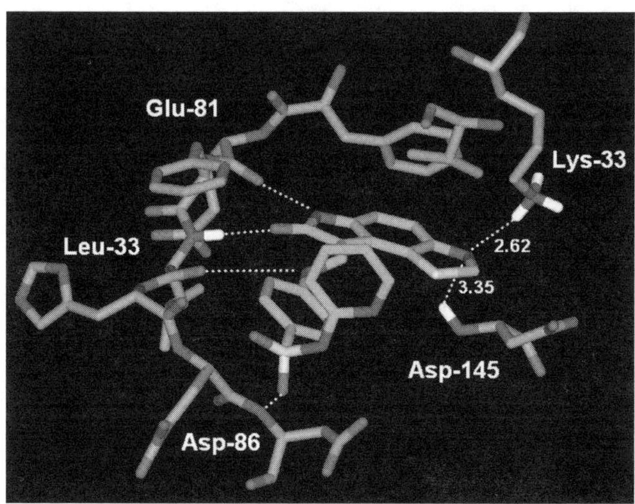

FIGURE 12.5 Crystal structure of thiazolo-oxindole **12.45** in CDK2/cyclin A surface representation.

TABLE 12.5
Cell-Based Data for Oxindoles 12.32 and 12.45 (IC$_{50}$, μM)

Compound	G$_1$/S	HFF	HT29	MDAMB468	RKO	SW620
12.32	1.8	5.9	14	4.4	1.4	4.8
12.45	2.6	13	5.1	6.8	1.7	10

Compounds in which the para-sulfonamide of the 3-(anilinomethylene)-oxindoles was replaced with a methylene-spaced tertiary amine moiety, as shown for compound **12.46**, were published as a patent application as CDK4 inhibitors.[22] The tertiary amine functionality, protonated at physiological pH, is thought to form a hydrogen bond with the hydroxyl of Thr102 in CDK4, thereby imparting selectivity for this kinase in a fashion analogous to that of the sulfonamide interaction with Asp86 in the similar location of CDK2.

12.6 3-(α-ANILINOBENZYLIDENE)-OXINDOLES

From 1999 onward, scientists at Boehringer Ingelheim have published a series of patent applications on 3-(α–anilinobenzylidene)-oxindoles as CDK inhibitors.[23–25] The distinctive feature of these inhibitors is the 3-phenyl 3-anilinomethylidene substitution at the C3 of the oxindole ring. The oxindoles thus exemplified typically contain an aniline 4-substituted by a methylene-piperidine (**12.47**) or an amino heterocycle such as 5-aminoindole (**12.48**). The oxindole is generally substituted at the 5–position with a carboxamide or another hydrogen bond acceptor. Compounds were generally prepared via initial treatment of an oxindole with triethyl orthobenzoate and acetic anhydride, followed by amination with an aniline or amino heterocycle.

12.47

12.48

12.7 INDIRUBIN AND DERIVATIVES

Indirubin (**12.49**), a 3′,2 bisindole isomer of indigo, is the active ingredient of a traditional Chinese medicine, Danggui Longhui Wan, and is used in the treatment of chronic myelocytic leukemia (CML). In clinical trials on patients with CML in the late 1970s, indirubin was dosed orally at 150–450 mg per day and showed promising complete (26%) and partial (33%) remission rates with relatively low toxicity.[26] Subsequently, in the late 1990s

TABLE 12.6
CDK Inhibitions (IC$_{50}$, µM) for Indirubin 12.49 and Derivatives 12.50–12.52

Compound	CDK2/Cyclin A	CDK2/Cyclin E	CDK1/Cyclin B	CDK4/Cyclin D1
12.49	2.2	7.5	10	12
12.50	0.44	0.25	0.18	3.3
12.51	0.75	0.55	0.4	6.5
12.52	0.035	0.15	0.055	0.3

it was discovered that indirubin and derivatives **12.50–12.52** selectively inhibit the CDKs (Table 12.6).[26,27] The oxime derivative **12.50** was also found to cause a cell cycle arrest in Jurkat cells in G$_1$ and G$_2$/M and reduce phosphorylation of the Rb protein.

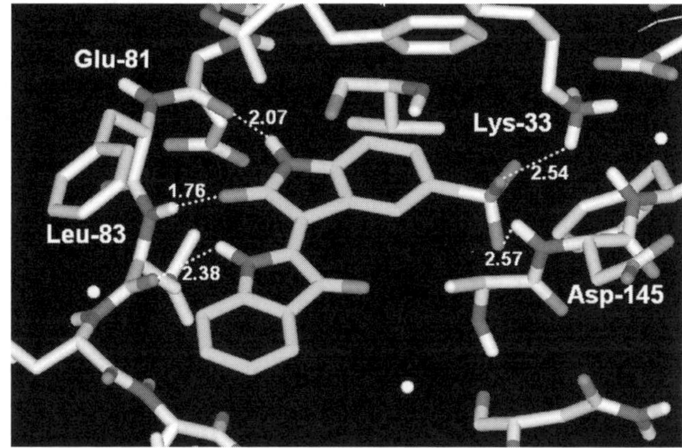

12.49   12.50   12.51   12.52

Indirubin-5-sulfonic acid (**12.52**) shares a similar binding mode (Figure 12.6) to the oxindoles described earlier, with the lactam nitrogen and carbonyl interacting with the Glu81 backbone carbonyl and Leu83 backbone NH, respectively, and the cyclic nitrogen forming a hydrogen bond with the Leu83 backbone carbonyl.[28] The

FIGURE 12.6 Crystal structure of indirubin-5-sulfonic acid (**12.52**) bound to CDK2/cyclin A.

high potency of compound **12.52** can be accounted for by the additional interaction of the 5-sulfonic acid with the amine of Lys33 and the backbone nitrogen of Asp145.

12.8 INDOLYMETHYLENE-OXINDOLES AND IMIDAZOTHIAZOLYLMETHYLENE-OXINDOLES

Researchers in Italy have published a series of indolymethylene-oxindoles and imidazothiazolylmethylene-oxindoles as CDK1 inhibitors.[29,30] The compounds have moderate enzymatic potencies; for example, **12.53** and **12.54** have IC_{50} values of 2.2 and 20 μM against CDK1, respectively, and show weak antiproliferative effects against HeLa cells.

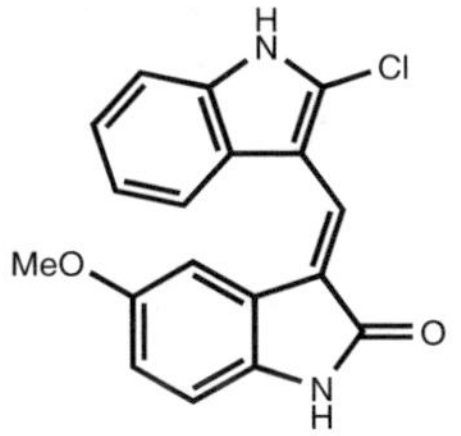

12.53 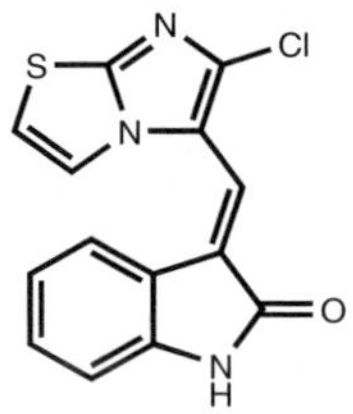   12.54

12.9 NAPHTHOSTYRILS

Following up on their earlier work in this area, Roche scientists have synthesized a series of naphthostyrils as conformationally constrained analogs of the pyrrolyl-methylidene-oxindole CDK2 inhibitor possessing a two-carbon bridge between the vinyl and C-4 oxindole positions.[31–33] The compounds were accessed via a one-pot

TABLE 12.7
Representative Examples of Naphthostyril CDK2 Inhibitors

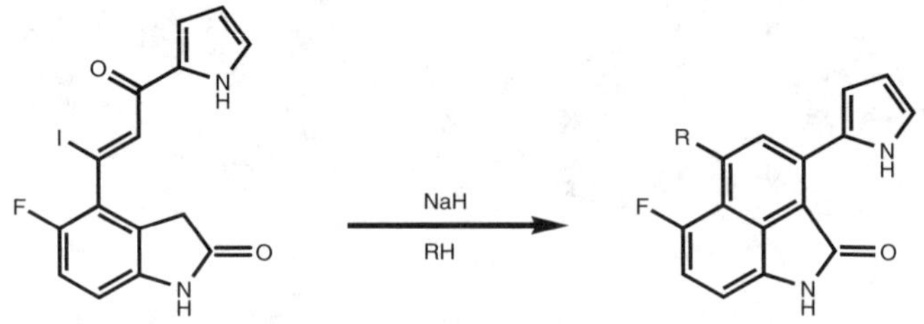

12.55   12.56–12.59

Compound	R	CDK2 IC_{50} (μM)
12.56	MeO—	0.018
12.57	$NH_2(CH_2)_2O$—	0.012
12.58	$NH_2(CH_2)_2NH$—	0.012
12.59	$NH_2(CH_2)_2S$—	0.003

cyclization of the vinyl iodide oxindole **12.55** with a variety of alcohols, amines, and thiols in the presence of NaH. The ethylamino side chain analogs (**12.57–12.59**) provided the most potent analogs (Table 12.7). The crystal structure of **12.58** in CDK2 showed that the lactam portion of the naphthostyril bonded in similar fashion to the backbone protein of CDK2 as seen with the oxindoles discussed earlier, and the terminal amine of the ethylamino side chain provides an additional hydrogen bond interaction with Asn132.

12.10 CONCLUSIONS

Oxindole kinase inhibitors targeting anti-tumor activity have been studied by a number of pharmaceutical companies and academic laboratories in recent years. Structural design, aided by extensive crystallographic work, has led to the optimization of oxindole binding interactions in the ATP-binding site, thereby achieving very potent inhibitory activities against CDK 1, 2, and 4. To date, only limited *in vivo* data have been reported, with indirubin being the only oxindole inhibitor of the CDKs reported to have been evaluated in clinical trials. The recent demonstration of clinical efficacy for the tyrosine kinase oxindole SU11248 (**12.5**) should encourage continued investigation of this chemotype as a CDK inhibitor for oncology indications.

REFERENCES

1. Shiraishi, T., Domoto, T., Imai, N., Shimada, Y., Watanabe, K. Specific inhibitors of tyrosine-specific protein kinase, synthetic 4-hydroxycinnamamide derivatives. *Biochem. Biophys. Res. Commun.* 1987, **147**, 322–328.
2. Shiraishi, T., Kameyama, K., Domoto, T., Imai, N., Shimada, Y., Ariki, Y., Hosoe, K., Kawatsu, M., Katsumi, I. et al. Preparation of hydroxystyrene derivatives as pharmaceuticals. *PCT Int. Appl.* 1988, WO8807035.
3. Buzzetti, F., Brasca, M.G., Crugnola, A., Fustinoni, S., Longo, A., Penco, S., Dalla Z.P., Comoglio, P.M. Cinnamamide analogs as inhibitors of protein tyrosine kinases. *Farmaco* 1993, **48**, 615–636.
4. Buzzetti, F., Longo, A., Colombo, M. Preparation of (hetero)arylacrylates useful as tyrosine kinase inhibitors. *PCT Int. Appl.* 1991, WO9113055.
5. Buzzetti, F., Longo, A., Colombo, M. Methyleneoxindole derivatives and process for their preparation. *PCT Int. Appl.* 1993, WO9301182.
6. Buzzetti, F., Longo, A., Brasca, M.G., Orzi, F., Crugnola, A., Ballinari, D., Mariani, M. Arylidene and heteroarylidene oxindole derivatives as tyrosine kinase inhibitors. *PCT Int. Appl.* 1995, WO9501349.
7. Mendel, D.B., Laird, A.D., Smolich, B.D., Blake, R.A., Liang, C., Hannah, A.L., Shaheen, R.M., Ellis, L.M., Weitman, S., Shawver, L.K., Cherrington, J.M. Development of SU5416, a selective small molecule inhibitor of VEGF receptor tyrosine kinase activity, as an anti-angiogenesis agent. *Anti-Cancer Drug Des.* 2000, **15**, 29–41.
8. Fabbro, D. and Manley, P.W. SU-6668, SUGEN. *Curr. Opin. Invest. Drugs* 2001, **2**, 1142–1148.
9. Sakamoto, K.M. SU-11248 SUGEN. *Curr. Opin. Invest. Drugs* 2004, **5**, 1329–1339.

10. Mohammadi, M., McMahon, G., Sun, L., Tang, C., Hirth, P., Yeh, B.K., Hubbard, S.R., Schlessinger, J. Structure of the tyrosine kinase domain of fibroblast growth factor receptor in complex with inhibitors. *Science* 1997, **276**, 955–960.

11. Lane, M.E., Yu, B., Rice, A., Lipson, K.E., Liang, C., Sun, L., Tang, C., McMahon, G., Pestell, R.G., Wadler, S. A novel cdk2-selective inhibitor, SU9516, induces apoptosis in colon carcinoma cells. *Cancer Res.* 2001, **61**, 6170–6177.

12. Moshinsky, D.J., Bellamacina, C.R., Boisvert, D.C., Huang, P., Hui, T., Jancarik, J., Kim, S.-H., Rice, A.G. SU 9516: biochemical analysis of cdk inhibition and crystal structure in complex with cdk2. *Biochem. Biophys. Res. Commun.* 2003, **310**, 1026–1031.

13. Yu, B., Lane, M.E., Wadler, S. SU9516, a cyclin-dependent kinase 2 inhibitor, promotes accumulation of high molecular weight E2F complexes in human colon carcinoma cells. *Biochem. Pharmacol.* 2002, **64**, 1091–1100.

14. Riou, J.-F., Maratrat, M., Grondard, L., Thompson, F., Petitgenet, O., Mailliet, P., Lavayre, J. CDK-1 inhibitor arylideneoxindoles. *PCT Int. Appl.* 2002, WO2002068411.

15. Li, X., Huang, P., Cui, J.J., Zhang, J., Tang, C. Novel pyrrolyllactone and pyrrolyllactam indolinones as potent cyclin-dependent kinase 2 inhibitors. *Bioorg. Med. Chem. Lett.* 2003, **13**, 1939–1942.

16. Tang, P.C., Miller, T.A., Li, X., Zhang, R., Cui, J., Huang, P., Wei, C.C. Synthesis of Pyrolyllactone-indolinone derivatives as kinase inhibitors. *PCT Int. Appl.* 2001, WO2001064681.

17. Dermatakis, A., Luk, K.-C., DePinto, W. Synthesis of potent oxindole CDK2 inhibitors. *Bioorg. Med. Chem.* 2003, **11**, 1873–1881.

18. Luk, K.-C., Simcox, M.E., Schutt, A., Rowan, K., Thompson, T., Chen, Y., Kammlott, U., DePinto, W., Dunten, P., Dermatakis, A. A new series of potent oxindole inhibitors of CDK2. *Bioorg. Med. Chem. Lett.* 2004, **14**, 913–917.

19. Chen, Y., Corbett, W.L., Dermatakis, A., Liu, J.-J., Luk, K.-C., Mahaney, P.E., Mischke, S.G. Preparation of 4-alkynyl-3-(pyrrolylmethylene)-2-oxoindoles as inhibitors of cyclin-dependent kinases, in particular CDK2. *PCT Int. Appl.* 2000, WO2000035908.

20. Bramson, H.N., Corona, J., Davis, S.T., Dickerson, S.H., Edelstein, M., Frye, S.V., Gampe, R.T., Jr., Harris, P.A., Hassell, A., Holmes, W.D., Hunter, R.N., Lackey, K.E., Lovejoy, B., Luzzio, M.J., Montana, V., Rocque, W.J., Rusnak, D., Shewchuk, L., Veal, J.M., Walker, D.H., and Kuyper, L.F. Oxindole-based inhibitors of cyclin-dependent kinase 2 (CDK2): design, synthesis, enzymatic activities, and x-ray crystallographic analysis. *J. Med. Chem.* 2001, **44**, 4339–4358.

21. Davis, S.T., Dickerson, S.H., Frye, S.V., Harris, P.A., Hunter, R.N., III, Kuyper, L.F., Lackey, K.E., Luzzio, M.J., Veal, J.M., Walker, D.H. Preparation of oxindoles as protein tyrosine kinase and protein serine/threonine kinase inhibitors. *PCT Int. Appl.* 1999, WO9915500.

22. Dickerson, S.H. and Drewry, D.H. Preparation of pyrrolo[3,2-f]quinolin-2-ones as CDK4 inhibitors. *PCT Int. Appl.* 2002, WO2002020524.

23. Heckel, A., Walter, R., Grell, W., Van, M., Jacobus C.A., Redemann, R. Substituted indolinones having an inhibiting effect on kinases and cyclin/CDK complexes. *PCT Int. Appl.* 1999, WO9952869.

24. Walter, R., Grell, W., Heckel, A., Himmelsbach, F., Eberlein, W., Roth, G., Van, M., Jacobus C.A., Redemann, N., Spevak, W., Tontsch-Grunt, U., Von Ruden, T. Preparation of 3-aminobenzylideneindolinones as cyclin dependent kinase inhibitors. *PCT Int. Appl.* 2000, WO2000018734.

25. Roth, G.J., Heckel, A., Walter, R., Meel Van, J., Redemann, N., Schnapp, G., Tontsch-Grunt, U., Spevak, W. Preparation of 3-aminomethylene-2-indolinones as kinase and cyclin/CDK complex inhibitors. *PCT Int. Appl.* 2001, WO2001027080.

26. Hoessel, R., Leclerc, S., Endicott, J.A., Nobel, M.E.M., Lawrie, A., Tunnah, P., Leost, M., Damiens, E., Marie, D., Marko, D., Niederberger, E., Tang, W., Eisenbrand, G., Meijer, L. Indirubin, the active constituent of a Chinese antileukemia medicine, inhibits cyclin-dependent kinases. *Nat. Cell Biol.* 1999, **1**, 60–67.
27. Eisenbrand, G., Hippe, F., Jakobs, S., Muehlbeyer, S. Molecular mechanisms of indirubin and its derivatives: novel anticancer molecules with their origin in traditional Chinese phytomedicine. *J. Cancer Res. Clin. Oncol.* 2004, **130**, 627–635.
28. Davies, T.G., Tunnah, P., Meijer, L., Marko, D., Eisenbrand, G., Endicott, J.A., Noble, M.E.M. Inhibitor Binding to Active and Inactive CDK2. The Crystal Structure of CDK2-Cyclin A/Indirubin-5-Sulphonate. *Structure*, 2001, **9**, 389–397.
29. Andreani, A., Granaiola, M., Leoni, A., Locatelli, A., Morigi, R., Rambaldi, M., Giorgi, G., Salvini, L., Garaliene, V. Synthesis and antitumor activity of substituted 3-(5-imidazo[2,1-b]thiazolylmethylene)-2-indolinones. *Anti-Cancer Drug Des.* 2001, **16**, 167–174.
30. Andreani, A., Cavalli, A., Granaiola, M., Leoni, A., Locatelli, A., Morigi, R., Rambaldi, M., Recanatini, M., Garnier, M., Meijer, L. Imidazo[2,1-b]thiazolyl-methylene- and indolylmethylene-2-indolinones: a new class of cyclin-dependent kinase inhibitors. Design, synthesis, and CDK1/cyclin B inhibition. *Anti-Cancer Drug Des.* 2001, **15**, 447–452.
31. Liu, J.-J., Dermatakis, A., Lukacs, C., Konzelmann, F., Chen, Y., Kammlott, U., Depinto, W., Yang, H., Yin, X., Chen, Y., Schutt, A., Simcox, M.E., Luk, K.-C. 3,5,6-Trisubstituted naphthostyrils as CDK2 inhibitors. *Bioorg. Med. Chem. Lett.* 2003, **13**, 2465–2468.
32. Liu, J.-J., Konzelmann, F., Luk, K.-C. A novel and convenient method for the synthesis of substituted naphthostyrils. *Tetrahedron Lett.* 2003, **44**, 2545–2548.
33. Liu, J.-J., Konzelmann, F., Luk, K.-C. Organometallic reagent-mediated one-pot synthesis of 3,5,6-trisubstituted naphthostyrils. *Tetrahedron Lett.* 2003, **44**, 3901–3904.

13 Indenopyrazoles as Cyclin-Dependent Kinase Inhibitors

David J. Carini, Catherine R. Burton, and Steven P. Seitz

CONTENTS

13.1 DEVELOPMENT OF THE INDENOPYRAZOLES

In 1995 the Dupont Merck Pharmaceutical Co., later the Dupont Pharmaceutical Co., entered into a collaboration with Mitotix to research a series of cyclin-dependent kinase (CDK) inhibitors for use as anticancer agents. Initially we were interested in selective inhibitors of CDK4/cyclinD1 (CDK4/D1), but we later expanded our focus to include selective inhibitors of CDK2/cyclinE (CDK2/E) as well as balanced inhibitors of both of these CDKs. High-throughput screening of our compound collection against CDK4/D1 afforded a series of hits. One of the most interesting of these leads proved to be the indenopyrazole **13.1** (Table 13.1). As described in the following text, the development of this lead has produced a series of highly potent and selective CDK inhibitors with *in vitro* and *in vivo* efficacy as anticancer agents.[1–4]

Our first breakthrough in this program came with the preparation of the 5-acetamido indenopyrazole **13.2**.[1] This analog is ~100-fold more potent than the original lead against both CDK4/D1 and CDK2/E. The acetamide **13.2** was originally prepared as a protected form of the aniline **13.3**; however, in a serendipitous result, **13.2** turned out to be much more potent than **13.3**. The increased acidity of the NH group in the amide is not responsible for the superior potency of **13.2**, as demonstrated by the fact that the corresponding phenol **13.4** does not show better activity than **13.3**. The importance of the position of the 5-acetamido group, adjacent to the

TABLE 13.1
Activity of Early Indenopyrazole Analogs

Compound	X	CDK4/D1 IC$_{50}$ (µM)	CDK2/E IC$_{50}$ (µM)
13.1	H	45	26
13.2	5-CH$_3$CONH	0.46	0.51
13.3	5-NH$_2$	23	8.4
13.4	5-OH	20	5.2
13.5	6-CH$_3$CONH	>150	> 300

indenone carbonyl, is demonstrated by the fact that the 6-acetamido analog **13.5** is inactive. These results led our group to explore a series of 5-amidoindenopyrazole analogs.[2]

The structure–activity relationships (SAR) of the 5-amido substituents on the indenopyrazole core are demonstrated by the compounds of Table 13.2. In the amides, branching at the α-carbon results in a loss of activity, especially against CDK4/D1, with the benzamide **13.7** showing a greater loss than the isobutyramide **13.6**. However, if a methylene spacer is incorporated between the phenyl ring and the carbonyl, the resulting phenylacetamides, e.g., **13.8**, are much more potent. Compound **13.8** is ~10-fold more active against CDK2/E than against CDK4/D1. This is not unusual. With few exceptions, the indenopyrazoles are either balanced in their activity against both kinases or are selective for CDK2/E over CDK4/D1.

Another series of amides to be explored were the glycinamides, e.g., **13.9–13.12**. These analogs are comparable in potencies to the phenylacetamide **13.8**, especially when the amino acid nitrogen is incorporated into a ring, such as in **13.10–13.12**. The glycinamides are relatively easy to prepare (see following discussion) and have the advantage of moderate solubility in acidic aqueous solutions.[2]

Further improvements in inhibitory potencies were observed when the amides were modified to a series of ureas and semicarbazides. For example, the urea **13.13** is ~10-100-fold more potent than the closely related acetamide **13.2**, depending on the kinase under consideration. A further comparison of the glycinamides **13.9–13.11** with the corresponding semicarbazides **13.14–13.16** shows modest increases in CDK2/E potency but more significant increases in CDK4/D1 potency. These semi-carbazides were some of the very first compounds prepared in this program to show low-nanomolar inhibitory activities for both kinases. Remarkably, these compounds represent a greater than 1000-fold improvement in potency compared to the original lead compound **13.1**.

TABLE 13.2
Activity of 5-Amidoindenopyrazole Analogs

Compound	R	CDK4/D1 IC$_{50}$ (μM)	CDK2/E IC$_{50}$ (μM)
13.2	CH$_3$	0.46	0.51
13.6	i-Pr	35	1.4
13.7	Ph	>250	41
13.8		0.48	0.038
13.9	Me$_2$NCH$_2$	0.90	0.044
13.10		0.195	0.021
13.11		0.125	0.012
13.12		0.020	0.012
13.13	NH$_2$	0.066	0.007
13.14	Me$_2$NNH	0.021	0.005
13.15		0.012	0.018
13.16		0.009	0.012

Another area of the SAR that was extensively investigated was the nature of the substituents at the 3-position on the indenopyrazole core.[3,4] Historically, as in all the derivatives shown so far, this position is occupied by a 4-methoxyphenyl group. One of the first alternatives investigated was to replace this aromatic substituent with a variety of simple alkyl groups. In the original 5-acetamidoindenopyrazole series, this resulted in a total loss of CDK4/D1 inhibitory activity, whereas against CDK2/E

TABLE 13.3
Activity of 5-Alkylindenopyrazole Analogs

Compound	R	CDK4/D1 IC$_{50}$ (µM)	CDK2/E IC$_{50}$ (µM)
13.17	i-propyl	0.340	0.048
13.18	c-propyl	0.430	0.055
13.19	c-hexyl	0.360	0.096

substitution by an isopropyl, cyclopropyl, or cyclohexyl group was moderately well tolerated.[3] In the much more potent series of glycinamide analogs shown in Table 13.3, these alkyl groups are responsible for a 10–20-fold loss of CDK4/D1 potency and a 5–10-fold loss against CDK2/E compared to compound **13.12** in Table 13.2.

In the original 5-acetamidoindenopyrazole series, analogs were prepared in which the 4-methoxy group on the phenyl at the 3-position of the indenopyrazole core was replaced with a variety of neutral, electron-donating substituents.[3] Generally, this had very little effect on the potencies of these derivatives relative to compound **13.2**, but it demonstrated the flexibility of the SAR at this position. Later in our program this flexibility proved useful when the para-position of the phenyl was substituted with basic groups. As shown in Table 13.4, the methoxy group in **13.15** can be replaced with a variety of piperazine, homopiperazine, and aminopiperidine rings with no loss of potency against CDK4/D1 or CDK2/E.[5] However, basic analogs such as **13.20-13.26** have one critical advantage over neutral analogs such as **13.15**. When formulated as acid salts, these compounds possess good to excellent solubilities in deionized water, typically in the range of 1–20 mg/ml. However, salts of the piperazines often show poor solubilities in a pH 7.4 aqueous phosphate buffer. Due to the relatively low basicity of the piperazines (pKa ~7.8), solutions of these salts at pH 7.4 are unstable and tend to precipitate the parent compounds. Salts of the homopiperazines (pKa ~8.8) and aminopiperidines (pKa ~9.1) are more reliably soluble in the pH 7.4 phosphate buffer.

Another series of piperidine derivatives that are interesting are the four compounds shown in Table 13.5. These chiral semicarbazides are selective for CDK4/D1

TABLE 13.4
Activity of 5-Arylindenopyrazole Analogs

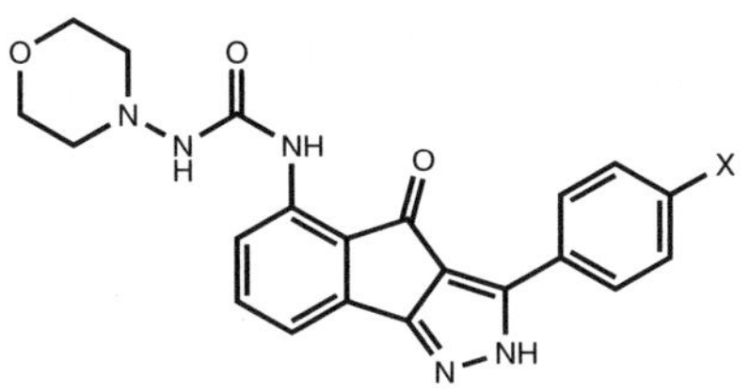

Compound	X	CDK4/D1 IC$_{50}$ (µM)	CDK2/E IC$_{50}$ (µM)
13.15	OMe	0.012	0.018
13.20	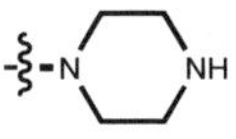 piperazin-1-yl (NH)	0.004	0.016
13.21	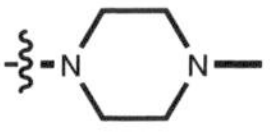 4-methylpiperazin-1-yl	0.007	0.015
13.22	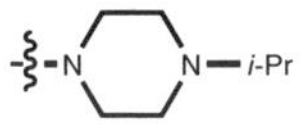 4-(i-Pr)piperazin-1-yl	0.006	0.022
13.23	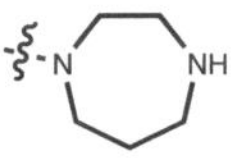 1,4-diazepan-1-yl (NH)	0.003	0.007
13.24	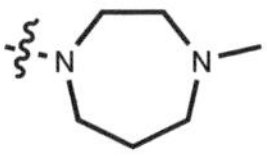 4-methyl-1,4-diazepan-1-yl	0.003	0.004
13.25	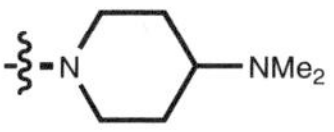 4-(NMe$_2$)piperidin-1-yl	0.003	0.012
13.26	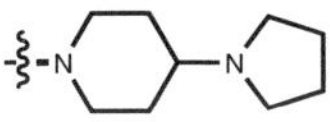 4-(pyrrolidin-1-yl)piperidin-1-yl	0.003	0.010

over CDK2/E. With selectivities of 20–25-fold, **13.27** and **13.30** are the most CDK4/D1-selective indenopyrazole CDK inhibitors that were prepared.

A final noteworthy series of inhibitors are the 3-heteroarylindenopyrazoles represented by the compounds of Table 13.6.[3,4] In contrast to the compounds in Table 13.5, this series includes some of our most CDK2/E-selective inhibitors, e.g., **13.35**.

TABLE 13.5
Activity of CDK4/D1-Selective Indenopyrazole Analogs

Compound	X	Z	CDK4/D1 IC_{50} (µM)	CDK2/E IC_{50} (µM)
13.27	CH_2OMe	H	0.008	0.165
13.28	H	CH_2OMe	0.008	0.039
13.29	$C(CH_3)_2OMe$	H	0.006	0.072
13.30	H	$C(CH_3)_2OMe$	0.017	0.405

13.2 BINDING OF THE INDENOPYRAZOLE CDK INHIBITORS

Kinetic studies with our indenopyrazoles show our compounds to be competitive with ATP in their inhibition of CDK4/D1 and CDK2/E, and our modeling was done on the assumption that the indenopyrazoles bind into the ATP-binding pocket of these kinases.[2] For a detailed discussion of the published x-ray structures of CDK2 and CDK2/A, with and without ATP bound, see Reference 2 and references cited therein. These published x-ray structures show that the shape of the ATP-binding pocket is not altered significantly by the binding of the CDK to the cyclin. Our assumption that the indenopyrazoles bind to the ATP-binding pocket was later confirmed by obtaining an x-ray structure of compound **13.19** bound to CDK2 (see Figure 13.1).[3] In this structure the pyrazole binds deep into the ATP-binding pocket and forms hydrogen bonds with amino acid residues in the back of the pocket. The inhibitor itself when bound is roughly U-shaped, with the amide group at the 5-position wrapping around toward the 3-cyclohexyl substituent. The mode of binding of the indenopyrazoles later was reconfirmed by a second x-ray structure of compound **13.36** bound to CDK2.[4]

13.3 SYNTHESIS OF THE INDENOPYRAZOLES

Our original route for the synthesis of the indenopyrazoles is demonstrated in Figure 13.2 for the preparation of **13.2**.[1,2] The condensation of dimethyl 3-acetamidophthalate and 4-methoxyacetophenone in DMSO with sodium hydride affords the trione **13.37**, and the treatment of **13.37** with hydrazine in refluxing ethanol in the presence

TABLE 13.6
Activity of 3-Heteroarylindenopyrazole Analogs

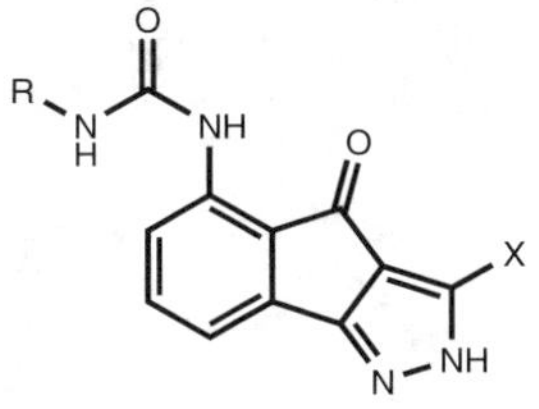

Compound	R	X	CDK4/D1 IC$_{50}$ (µM)	CDK2/E IC$_{50}$ (µM)
13.31	H	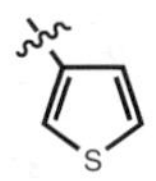 (3-thienyl)	0.067	0.009
13.32	H	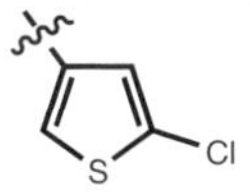 (5-chloro-2-thienyl)	0.057	0.013
13.33	H	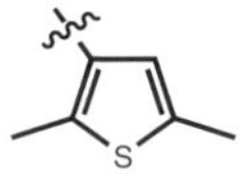 (2,5-dimethyl-3-thienyl)	>1.50	0.061
13.34	Me$_2$N	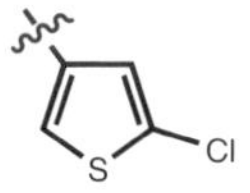 (5-chloro-2-thienyl)	0.021	0.007
13.35	Me$_2$N	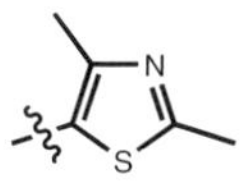 (2-methylthiazol-5-yl)	0.150	0.004
13.36	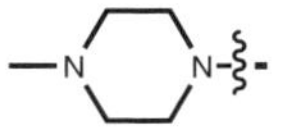 (4-methylpiperazin-1-yl)	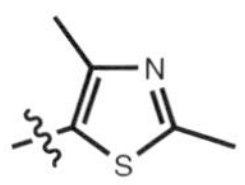 (2-methylthiazol-5-yl)	0.046	0.008

of catalytic acid furnishes **13.2**. The acetamido group was selected because it represents a protected form of the aniline that is stable to the highly basic conditions of the sodium hydride reaction; however, this group affords another important benefit. The presence of this amide directs the hydrazine reaction to produce the pyrazole **13.2** with very high regioselection.[1] As will be shown in the following text, any amido group at this position has the same directing effect.[2-4] Generally, the

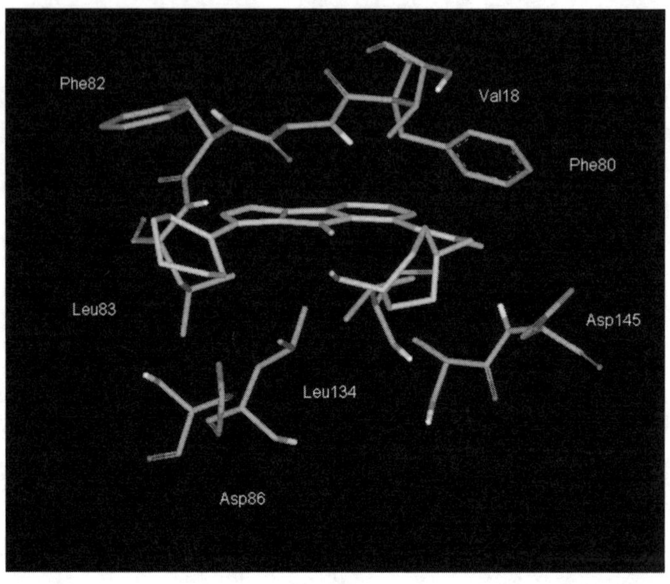

FIGURE 13.1 X-ray structure of compound **13.19** bound to CDK2.

alternative 8-amidoindenopyrazole regioisomers are not detected. Conversely, if the hydrazine reaction is run on a trione with a nitro group at this position, the only product observed is the 8-nitroindenopyrazole.[3]

Deprotection of **13.2** in refluxing hydrochloric acid and methanol provides the aniline **13.38**. Acylation of this aniline with acid chlorides in the presence of base affords the 5-amidoindenopyrazoles **13.39**. An efficient route to the glycinamide analogs was developed and is demonstrated in Figure 13.3.[2] Acylation of the aniline **13.38** with 2-chloroacetyl chloride provides the intermediate **13.40**. The reaction of this 2-chloroacetamide with a variety of amines produces the glycinamides **13.41**.

It is occasionally preferable to alter the order of the steps described in Figure 13.2. This is most commonly the case for the production of the urea and semicarbazide analogs. As demonstrated in Figure 13.4, the acetamidotrione **13.37** can be deprotected in high yield to furnish the aniline **13.42**. This intermediate is then converted to a urea or semicarbazide in one of two ways. The reaction of **13.42** with phenyl chloroformate, followed by the treatment of the resulting phenylcarbamate with an amine, affords the desired ureas **13.43** (Method A). This method sometimes can be

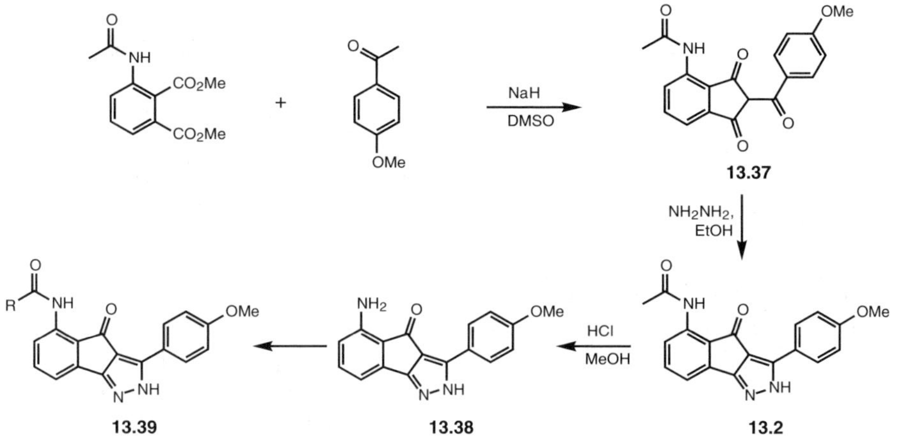

FIGURE 13.2 Original synthetic route to the indenopyrazoles.

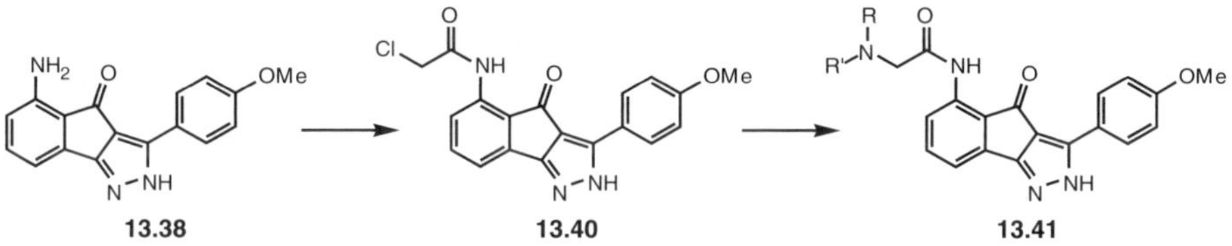

FIGURE 13.3 Synthesis of glycinamide analogs.

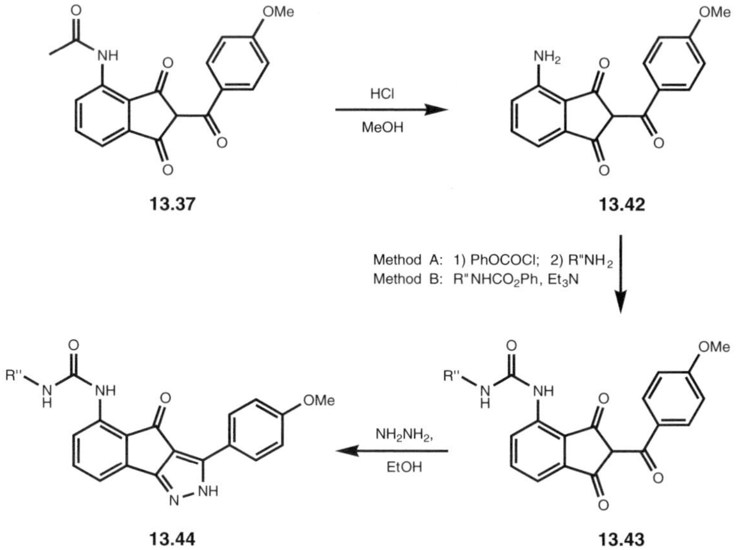

FIGURE 13.4 Synthesis of urea and semicarbazide analogs.

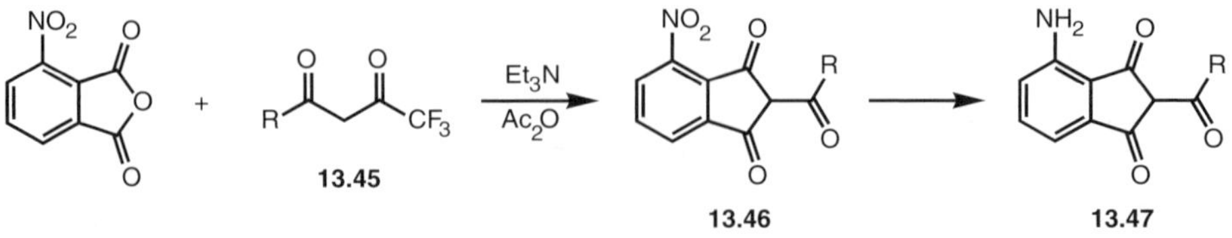

FIGURE 13.5 Alternative synthesis of anilinotriones **13.47**.

employed to produce semicarbazides from 1,1-disubstituted hydrazines, but the reaction of the carbamate with hydrazines can be unreliable. Fortunately, an alternate route to the semicarbazides has been developed. Treatment of the aniline **13.42** with reagents prepared from 1,1-disubstituted hydrazines and phenyl chloroformate furnishes the semicarbazide intermediates reliably and in high yields (Method B).[3] The amidotrione intermediates **13.43** all undergo the same regioselective pyrazole formation discussed previously to provide **13.44**.

A major problem with the chemistry described in Figure 13.2 was that it proved to be very limited in its scope. Although the sodium hydride condensation worked consistently with 4-methoxyacetophenone, it failed for many other acetophenones ($ArCOCH_3$) and heteroarylmethyl ketones ($HetCOCH_3$). This limitation significantly restricted our ability to explore this series of CDK inhibitors. For this reason a second and more versatile route to the key trione intermediates was developed. As shown in Figure 13.5, the reaction of 3-nitrophthalic anhydride with 1,1,1-trifluoro-2, 4-dicarbonyl compounds **13.45**, readily prepared from the corresponding methyl ketones, affords the nitrotriones **13.46**.[3,4] This method is a modification of chemistry first reported by Rotberg and Oshkaya[6] in which symmetrical 1,3-diketones were employed. Their original method was useful for the production of the requisite triones in which a small alkyl group was desired at the 3-position of the final indenopyrazole; however, it was impractical for more elaborate substituents. Finally, the reduction of the nitrotriones **13.46** to the anilines **13.47** provides intermediates that can be converted to the indenopyrazoles as described previously.

13.4 PROPERTIES OF THE INDENOPYRAZOLES IN CELLULAR ASSAYS

The CDKs are key regulators of the cell cycle. The frequent disruption of the orderly regulation of the cell cycle in transformed cells made the CDKs attractive targets for the treatment of neoplasia. At the onset of the project, the prevailing view was that CDK4/6 became active early in the cell cycle and phosphorylated the tumor suppressor gene product Rb. As Rb is a transcriptional co-repressor, phosphorylation was thought to modify gene expression to facilitate progression through G_1. Subsequently, CDK2 complexed with cyclins E and A became active and drove progression through S phase. Finally, CDK1 bound to cyclin B drove the completion of the cycle. Recently, some aspects of this sequential model have been questioned. The key experiments have been knockout mice in which cell cycle regulators are disrupted individually or in combination.[7] Mouse embryonic fibroblasts (MEFs) from

TABLE 13.7
Cellular Properties of Selected Indenopyrazoles (IC$_{50}$, µM)

Compound	HCT116	MIAPaCa-2	PC-3	HT1080	NCI-H460	Arrested AG1523
13.32	0.12	0.26	—	0.49	0.24	>29
13.36	0.019	0.022	0.071	0.046	0.078	>21
13.20	0.015	0.009	0.052	0.032	0.20	0.051
13.21	0.008	0.010	0.042	0.012	0.018	>5.5
13.22	0.007	0.007	0.031	0.012	0.011	—
13.26	0.013	0.013	0.046	0.02	0.03	—

CDK4$^{-/-}$ mice grow normally, but are slower to enter S phase from G$_0$. MEFs from the CDK6 knockout also grow normally. Remarkably, the double CDK4/6 knockout is not embryonically lethal, with the mice surviving until shortly after birth. A portion of the MEFs from these animals are able to enter the cell cycle. In addition, the required phosphorylation of Rb early in the cell cycle has been assumed by another kinase, presumably CDK2. These results suggest that there is considerable redundancy in the regulation of the cell cycle by the CDKs and that in some circumstances the system can compensate for the absence of one or more of the enzymes.

A critical component of our compound evaluation scheme was activity in cellular proliferation assays. As shown in Table 13.7, the inhibitors have broad-spectrum activity toward a range of human tumor cell lines, including colon (HCT116, HT1080), pancreas (MiaPaCa2), prostate (PC3), and lung (NCI-H460). There is no obvious dependence on p53 status, as cells with wild-type (HCT116, NCI-H460) or mutant/deleted p53 (MiaPaCa2, HT1080, PC3) behave comparably. Although not examined in detail, it appears that Rb status is also not a critical variable, with compound **13.36** having an IC$_{50}$ of 24 nM against the leiomyosarcoma Skut1A in which Rb is deleted.[5] The last column of Table 13.7 details the performance of the compounds against arrested AG1523 fibroblasts. Quiescent cells have very little CDK-associated kinase activity. We postulated that a relatively normal arrested cell such as AG1523 would be little affected by inhibition of CDKs. As the data in Table 13.7 show, some compounds have very little activity against the arrested fibroblasts. The first two entries show substantial gaps between their activity against the transformed cell lines and the arrested AG1523. Interestingly, this property is not shared among all analogs. For example, the third entry has an IC$_{50}$ in the AG1523 in the range observed for the transformed cell lines. We posit that this compound has significant off-target effects that are responsible for the toxicity to the normal fibroblasts. These data hint that there may be differences between how transformed cells and normal cells respond to CDK inhibition.

We examined in more detail the response of normal and transformed cells to treatment with compound **13.32**. Exposure of HCT116 cells to **13.32** results in loss of adherence cell death.[3] FACS analysis showed the formation of a significant sub-G$_1$ population. Analysis of the floating population of cells revealed that Rb was

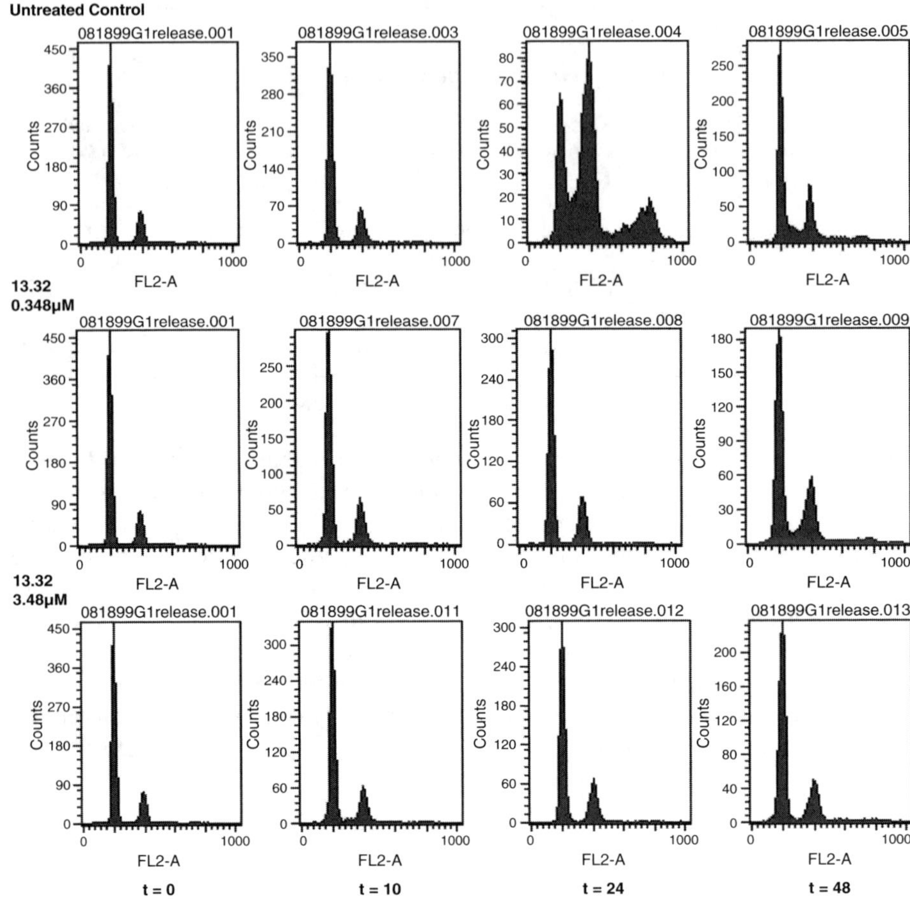

FIGURE 13.6 Effects of compound **13.32** on G1 arrested fibroblasts AG1523.

dephosphorylated. In addition, the floating population also showed evidence of PARP and caspase 3 cleavage. These data suggest that treatment of HCT116 with **13.32** results in inhibition of CDKs and initiation of an apoptotic response.

The response of AG1523 cells to treatment with **13.32** is markedly different. To simplify analysis, we first synchronized cells in G_1 by contact inhibition. The cells were induced to re-enter the cell cycle by subculture at a lower density. As shown in Figure 13.6, the untreated fibroblasts re-enter the cell cycle, and by 24 h many are in G_2/M. In contrast, the fibroblasts treated with **13.32** (348 nM) largely remain arrested in G_1. The G_1 arrest is complete when the concentration of compound is increased tenfold. This behavior is consistent with the potency of **13.32** as an inhibitor of CDK2/E (13 nM).

Release from a G_2/M blockade follows a different course. Compound **13.32** is also a moderate inhibitor of CDK1/B (IC_{50} = 44 nM). The compound is therefore expected to have effects late in the cell cycle as well. As shown in Figure 13.7, the

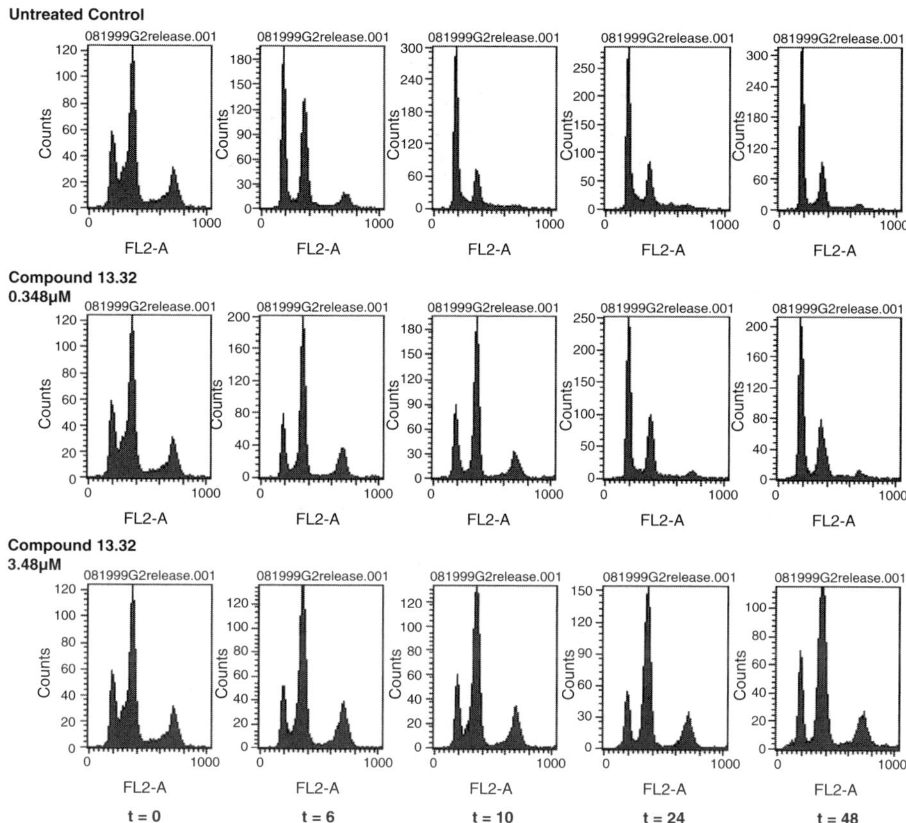

FIGURE 13.7 Effects of compound **13.32** on G2 arrested fibroblasts AG1523.

untreated fibroblasts re-enter the cell cycle and progress to G_1 within 10 h. The fibroblasts exposed to **13.32** (348 nM) took significantly longer to reach G_1 (24–48 h). Cells treated with tenfold higher concentration of **13.32** remain blocked in G_2/M. The observation that the maintenance of the G_2/M arrest requires a higher concentration of **13.32** than the G_1 arrest may be explained by the lower potency of inhibition of CDK1/B compared to CDK2/E. Finally, the activation of apoptotic markers seen in HCT116 is not observed in the normal fibroblasts. These results suggest that **13.32** affects normal and transformed cells differently, with cell cycle arrest being seen in the former and induction of the apoptotic program in the latter.

In summary, the indenopyrazoles are an interesting class of ATP-competitive inhibitors of the CDKs. The SAR of the series has been elaborated to permit significant modulation of the intra-CDK inhibitory selectivity. Members of the series display potent antiproliferative activity against a variety of transformed cells. Normal cells as represented by AG1523 fibroblasts respond to compound treatment with cell cycle arrest. The differential effect on transformed cells suggests that the compounds may have significant potential as anticancer agents.

ACKNOWLEDGMENTS

Over the course of this project, a great many individuals contributed to the design, preparation, and evaluation of the indenopyrazole series of CDK inhibitors. The authors would like to acknowledge the efforts of David A Nugiel, Eddy W. Yue, C. Anne Higley, Anup Vidwans, Chunhong He, Susan V. DiMeo, Carrie Benware, George Tora, and Anna-Marie Etzkorn for compound preparation, Chong-Hwan Chang, Jodi Muckelbauer, and Angela Smallwood for crystallography studies, Karen A. Rossi and Pieter Stouten for molecular modeling studies, and Marc R. Arnone, Pamela A. Benfield, John F. Boylan, Sarah, S. Cox, Philip M. Czerniak, Charity L. Dean, Deborah Doleniak, J. Gerry Everlof, Diane M. Sharp, Robert Grafstrom, and Lisa M. Sisk for *in vitro* and *in vivo* compound evaluation.

REFERENCES

1. Nugiel, D.A., Etzkorn, A.-M., Vidwans, A., Benfield, P.A., Boisclair, M., Burton, C.R., Cox, S., Czerniak, P.M., Doleniak, D., Seitz, S.P. Indenopyrazoles as novel cyclin dependent kinase (CDK) inhibitors. *J. Med. Chem.* **2001**, *44*, 1334–1336.
2. Nugiel, D.A., Vidwans, A., Etzkorn, A.-M., Rossi, K., Benfield, P.A., Burton, C.R., Cox, S., Doleniak, D., Seitz, S. Synthesis and evaluation of indenopyrazoles as cyclin-dependent kinase inhibitors. 2. probing the indeno ring substituent pattern. *J. Med. Chem.* **2002**, *45*, 5224–5232.
3. Yue, E.W., Higley, A., DiMeo, S.V., Carini, D.J., Benware, C., Benfield, P.A., Burton, C.R., Cox, S., Grafstrom, R.H., Sharp, D.M., Sisk, L.M., Boylan, J.F., Muckelbauer, J.K., Smallwood, A.M., Chen, H., Chang, C.-H., Seitz, S.P., Trainor, G.L. Synthesis and evaluation of indenopyrazoles as cyclin-dependent kinase inhibitors. 3. structure activity relationships at C3. *J. Med. Chem.* **2002**, *45*, 5233–5248.
4. Yue, E.W., DiMeo, S.V., Higley, C.A., Markwalder, J.A., Burton, C.R., Benfield, P.A., Grafstrom, R.H., Cox, S., Muckelbauer, J.K., Smallwood, A.M., Chen, H., Chang, C.-H., Trainor, G.L., Seitz, S.P. Synthesis and evaluation of indenopyrazoles as cyclin-dependent kinase inhibitors. Part 4. heterocycles at C3. *Bioorg. Med. Chem. Lett.* **2004**, *14*, 343–346.
5. Unpublished results.
6. Rotberg, Y.T. and Oshkaya, V.P. Synthesis of 2-acetyl-1,3-indanediones substituted in the benzene ring. *J. Org. Chem. USSR (Engl. Transl.)* **1972**, *8*, 86–88.
7. Sanchez, I. and Dynlacht, B.D. New insights into cyclins, CDKs, and cell cycle control. *Semin. Cell Dev. Biol.* **2005**, *16*, 311–321.

14 Development of Indolocarbazoles as Cyclin D1/CDK4 Inhibitors

Guoxin Zhu

CONTENTS

14.1 INTRODUCTION

The cell cycle is a series of highly regulated processes that results in the duplication of a cell.[1] Cell cycle progression is strictly controlled by the sequential activation of various cyclins that associate with specific cyclin-dependent kinases (CDKs).[2] For example, the serine/threonine-specific kinases, CDK2 and CDK4, play an essential role in the passage through G_1 to S phase of the cell cycle. These kinases regulate the release of the E2F family of transcription factors through hyperphosphorylation of the retinoblastoma tumor suppressor protein (Rb). This Rb signaling pathway is altered in many human cancers.[3] The importance of CDKs in cell cycle regulation coupled with the frequent documented deregulation of CDKs and their modulators in cancer have stimulated significant research in this area.[4,5] Thus, inhibitors of these CDKs that stop uncontrolled tumor cell growth are expected to be promising new therapeutic agents for the treatment of cancer.

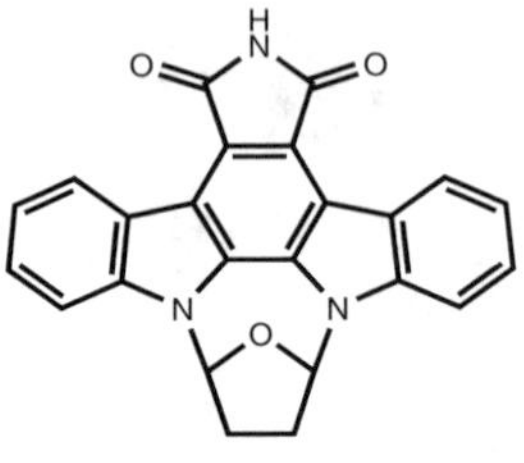

SB-218078 **(14.1)**

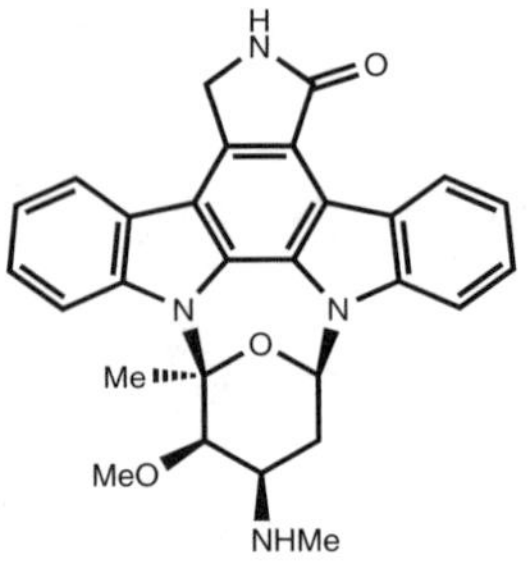

Staurosporine **(14.2)**

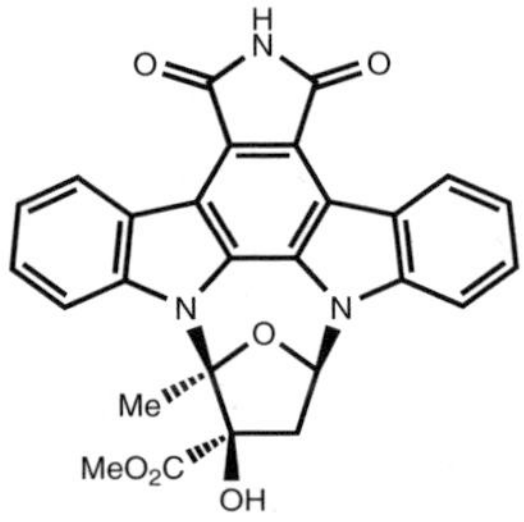

K-252a **(14.3)**

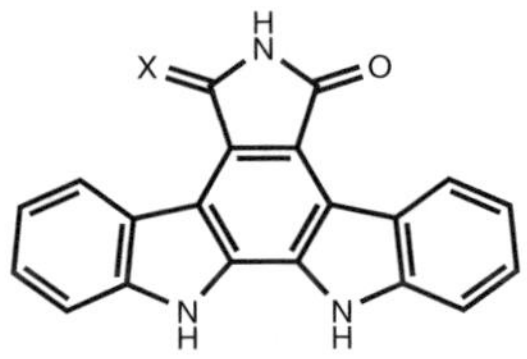

K-252c (X = H_2) **(14.4)**
Arcyriaflavin A (X = O) **(14.5)**

FIGURE 14.1 Structure of CDK4 inhibitors.

High-throughput screening (HTS) of the Lilly compound collection against the cyclin D1–CDK4 enzyme complex yielded several interesting scaffolds, including SB-218078 (**14.1**, Figure 14.1), with an IC_{50} of 0.54 μM in the D1/CDK4 kinase assay. Structurally related known PKC inhibitors **14.2–14.5** also showed moderate inhibition of D1/CDK4 (Table 14.1). Because of our interest in discovering selective and potent CDK4 inhibitors, we implemented a medicinal chemistry strategy to develop CDK4 inhibitors within this scaffold. Figure 14.2 summarizes the structure–activity relationship studies from our laboratories.

TABLE 14.1
Enzymatic Activities of Indolocarbazoles 14.1–14.5

Compound	D1/CDK4 IC_{50} (μM) (Rb^{ING})	E/CDK2 IC_{50} (μM) (Rb^{ING})
14.1	0.54	0.68
14.2	0.08	0.01
14.3	0.03	0.26
14.4	0.83	> 1.0
14.5	0.14	0.90

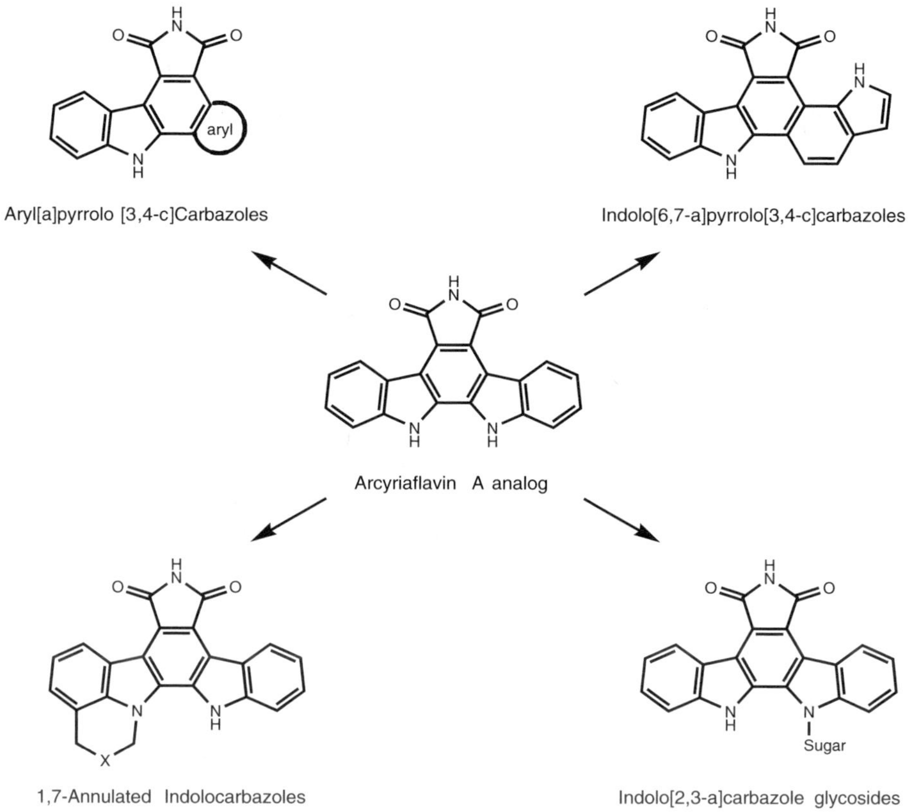

Aryl[a]pyrrolo [3,4-c]Carbazoles

Indolo[6,7-a]pyrrolo[3,4-c]carbazoles

Arcyriaflavin A analog

1,7-Annulated Indolocarbazoles

Indolo[2,3-a]carbazole glycosides

FIGURE 14.2 SAR studies of arcyriaflavin A.

14.2 ARCYRIAFLAVIN A ANALOGS

Bisindolomaleimides **14.6** are potent PKC inhibitors, but their inhibitory activity toward other kinases were less known.[6] During our SAR studies, a synthetic route involving oxidative cyclization of bisindolomaleimide **14.6** was developed for the synthesis of indolocarbazole **14.7** to assess the kinase inhibitory activities of both platforms (Figure 14.3).[7]

The inhibition of cyclin D1/CDK4 and cyclin E/CDK2 by **14.6** and **14.7** were evaluated in an enzymatic assay by measuring phosphorylation of the Rb[ING] substrate following standard protocols.[8] Staurosporine **14.2** (Figure 14.1), a known kinase inhibitor, was used as a standard compound for the assays. Table 14.2 summarizes the enzyme inhibitory activity. Maleimide (**14.6a**) was a moderate inhibitor of cyclin D1/CDK4; however, a methyl substitution on the indole nitrogen increased the binding affinity of compound **14.6b** by fourfold.[7] A variety of polar groups can be tolerated on the indole nitrogen (**14.6c–14.6g**), which provides an opportunity to

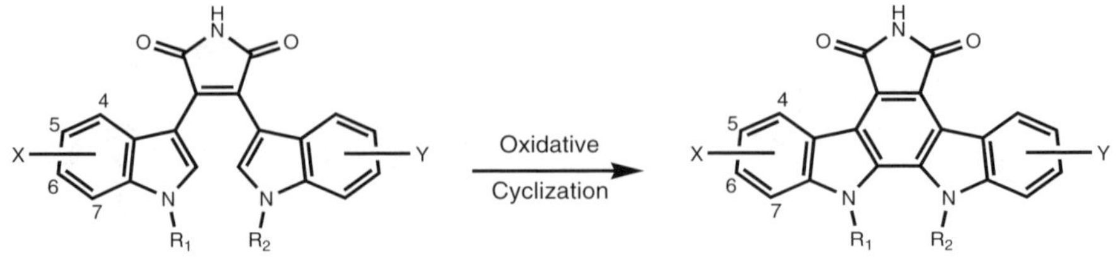

FIGURE 14.3 Synthesis of indolocarbazoles (**14.7**).

introduce ionizable functional groups for the improvement of aqueous solubility. Kinase activities were affected by substitution at different positions of the indole. Bromo substitution on the 4-, 5-, and 6-positions improved the binding affinity in D1/CDK4 (**14.6h**, **14.6i**, and **14.6j**). The 4-bromo analog **14.6h** is 27-fold more potent in D1/CDK4 compared to unsubstituted **14.6a**. An additive effect from 6-bromo indole ring substitution and N-methyl substitution on the indole was observed with compound **14.6l**. However, 6,6-bissubstituted analogs **14.6m** and **14.6n** led to

TABLE 14.2
SAR of Bisindolylmaleimides 14.6 and Indolocarbazoles 14.7

| | | | | Maleimide 14.6 | | | Carbazole 14.7 | | |
| | | | | | CDK4 IC_{50} (μM) (Rb[ING]) | CDK2 IC_{50} (μM) (Rb[ING]) | | CDK4 IC_{50} (μM) (Rb[ING]) | CDK2 IC_{50} (μM) (Rb[ING]) |
X	Y	R¹	R²	14.6			14.7		
H	H	H	H	**14.6a**	5.81	11.69	**14.7a**	0.14	0.90
H	H	H	Me	**14.6b**	1.25	1.95	**14.7b**	0.08	>1.00
H	H	H	$(CH_2)_3OH$	**14.6c**	1.17	—	**14.7c**	0.07	0.18
H	H	H	$(CH_2)_3NHMe$	**14.6d**	0.96	>1.00	**14.7d**[a]	0.05	0.16
H	H	H	$(CH_2)_3NMe_2$	**14.6e**	0.99	10.18	**14.7e**	0.13	>1.00
H	H	Me	$CH_2CHOH\text{-}CH_2OH$	**14.6f**	2.14	1.61	**14.7f**	0.26	<0.06
H	H	H	$(CH_2)_4OH$	**14.6g**	0.89	—	**14.7g**	0.04	0.09
4-Br	H	H	H	**14.6h**	0.21	—	**14.7h**	0.08	0.22
5-Br	H	H	H	**14.6i**	1.08	—	**14.7i**	4.40	4.30
6-Br	H	H	H	**14.6j**	1.59	9.12	**14.7j**	0.08	0.52
7-Br	H	H	H	**14.6k**	4.76	—	**14.7k**	16.2	19.4
6-Br	H	H	Me	**14.6l**	0.73	1.88	**14.7l**	0.11	>0.2
6-Br	6-OMe	H	H	**14.6m**	2.39	—	**14.7m**	0.52	—
6-OMe	6-OMe	H	H	**14.6n**	4.64	—	**14.7n**	0.18	>1.00

Note: — = not tested.

[a] Hydroiodic salt.

decreased D1/CDK4 activity compared with mono-substituted analog **14.6j**. In general, these maleimide compounds **14.6** were more potent in D1/CDK4 compared to E/CDK2.

A substantial improvement in D1/CDK4 activity was observed when the maleimide **14.6** was converted to the corresponding indolocarbazole **14.7**. For example, indolocarbazoles **14.7a** and **14.7d** were 42- and 20-fold more active in D1/CDK4, respectively, than their corresponding maleimides **14.6a** and **14.6d**. A similar phenomenon was observed for N-substituted analogs, **14.7b–14.7g**, which were 8- to 22-fold more potent in D1/CDK4 than their corresponding precursors **14.6b–14.6g**. Compounds **14.7i** and **14.7k** were the exceptions as they were relatively less potent than their maleimide precursors **14.6i** and **14.6k**; however, all of them were weak inhibitors of D1/CDK4. A substitution effect similar to the maleimides was observed for the N-methylation of the indolocarbazole. For example, N-methyl analog **14.7b** is more potent than the non-substituted analog **14.7a** in D1/CDK4 and various polar groups on the indole nitrogen were well tolerated (**14.7c–14.7g**). Indolocarbazole activity in D1/CDK4 was more sensitive to the substitution on the indole ring compared to the corresponding maleimides. Substitution on the 5- and 7-positions of the indolocarbazole decreases D1/CDK4 inhibition, whereas the indolocarbazole analog with substitution at the 4- or 6-position provided improvement. However, 6-substituted indolocarbazoles gave better kinase selectivity than the corresponding 4-substituted analogs.

With the 6-bromo indolocarbazole **14.7j** being one of the most potent D1/CDK4 inhibitors and having the desired selectivity profile within the kinase panel tested, a variety of substituents were studied at the 6-position of the indole.[9] The enzymatic activities of these 6-substituted indolocarbazoles are summarized in Table 14.3. A variety of substituent groups at the 6-position of the indole were well tolerated for maintaining potent D1/CDK4 activity. In general, these compounds were more potent in D1/CDK4 compared to E/CDK2 and PKA. However, the selectivity against CamKII was more sensitive to the nature of the substituent. Substituting the 6-position of one indole in the carbazoles with fluorine (**14.7o**), methyl (**14.7r**), or methoxyl (**14.7t**) led to potent CamK2 activity. By introducing a methyl group at the indole nitrogen position R^2, the potency of D1/CDK4 and selectivity against CamKII improved (**14.7o** vs. **14.7p**). A combination of the 6-bromo and 6-methoxy substitutions provided analog **14.7m**, which had diminished D1/CDK4 activity, but better selectivity against other panel kinases. Kinetic studies showed that selected compounds **14.7j**, **14.7o**, and **14.7q** exhibited pure competitive inhibition with respect to ATP and therefore certain 6-substituted indolocarbazoles are potent and selective ATP-competitive inhibitors of D1/CDK4.

The indolocarbazole **14.7g** was co-crystallized with CDK2. The x-ray structure of the complex confirmed that the indolocarbazole occupies the ATP-binding pocket located in the cleft between the N- and C-terminal domains (Figure 14.4). The proton from the carbazole NH was hydrogen-bonded to the carbonyl group of Glu81. One of the carbonyl groups from the carbazole also hydrogen bonds with the backbone amide NH of Leu83. The other carbonyl group of the carbazole is hydrogen bonded

TABLE 14.3
SAR of 6-Substituted Indolocarbazoles 14.7

14.7

					IC$_{50}$ (μM)			
Compound					D1-CDK4	E-CDK2	CamKII	PKA
14.7	**X**	**R^1**	**R^2**	**Y**	**RB21**	**RB21**	**Autocamtide**	**Histone**
14.7l	Br	H	Me	H	0.074	—	0.174	>2.0
14.7o	F	H	H	H	0.103	0.141	0.092	>2.0
14.7p	F	H	Me	H	0.042	0.144	>2.0	>2.0
14.7q	Cl	H	H	H	0.072	0.278	0.914	>2.0
14.7r	Me	H	H	H	0.226	>1.0[a]	0.082	>2.0
14.7s	CF$_3$	H	H	H	1.135	>2.0[a]	>2.0	>2.0
14.7t	OMe	H	H	H	0.134	2.287[a]	<0.125	>2.0
14.7m	Br	H	H	OMe	0.63	>2.0[a]	>2.0	>2.0

[a] Rb^{ING} as substrate.

Note: — = not tested.

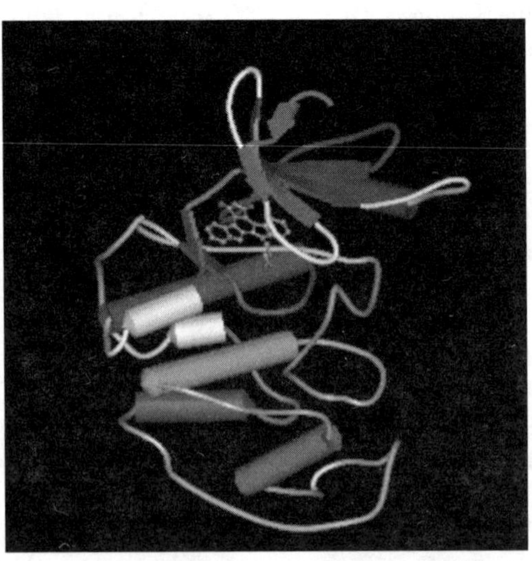

FIGURE 14.4 X-ray crystal structure of the human CDK2/**14.7g** complex.

to a water molecule underneath Phe80. The x-ray results revealed the importance of the two carbonyl groups and the NH on the indolocarbazole, which with the help of a molecular modeling study using a CDK4 homology model assisted us in developing new CDK4 inhibitors.

14.3 1,7-ANNULATED INDOLOCARBAZOLES

Based on the SAR information from the simple substituted arcyriaflavin A analogs and the x-ray co-crystal structural information, substitution of the indole nitrogen was well tolerated. To further elaborate the SAR and find novel indolocarbazoles as D1/CDK4 inhibitors, 1,7-annulated indolocarbazole analogs (**14.8**) were designed and evaluated.[10]

The ring-size effect was first studied, and similar potencies in D1/CDK4 were found for analogs of various ring sizes (5-, 6-, 7-, and 8 -membered, **14.8a–14.8d**) as shown in Table 14.4. Thus the ring size did not affect the D1/CDK4 activity or selectivity against cyclin E/CDK2. Studies of the nonannulated indole ring substitution revealed that the 6-position is tolerant of a variety of groups and, with the exception of the methyl ester (**14.8k**), compounds **14.8g-14.8j** were consistent with

---

TABLE 14.4
SAR of 1,7-Annulated Indolocarbazoles 14.8

14.8

	X	n	D1/CDK4 IC_{50} (µM) (Rb^{ING})	E/CDK2 IC_{50} (µM) (Rb^{ING})	HCT-116 IC_{50} (µM)	NCI-H460 IC_{50} (µM)
14.8a	H	0	0.24	>2.0	1.02	1.18
14.8b	H	1	0.11	>1.0	1.60	1.97
14.8c	H	2	0.09	1.0	5.16	1.88
14.8d	H-	3	0.12	>2.0	1.25	1.37
14.8e	4-F	1	>2.0	—	—	—
14.8f	5-F	1	>2.0	—	—	—
14.8g	6-F	1	0.11	>0.2	2.37	1.30
14.8h	6-Cl	1	0.11	>0.2	0.78	0.53
14.8i	6-Br	1	0.09	1.0	2.3	1.69
14.8j	6-CF$_3$	1	0.29	>2.0	1.48	1.53
14.8k	6-CO$_2$CH$_3$	1	>2.0	—	—	—

Note: — = not tested.

the arcyriaflavin A analogs. In contrast, the 4- and 5-positions did not tolerate small substitutions such as the fluoro group (**14.8e** and **14.8f**). In addition to the kinase activity discussed earlier, these potent D1/CDK4 inhibitors were also studied for their antiproliferative activity in a variety of tumor cell lines.[11] As shown in Table 14.4, most of these compounds were potent in inhibiting cell growth in the human colon carcinoma (HCT-116) and non-small-cell lung carcinoma (NCI-460) cell lines with activities ranging from 0.53 to 5.16 μM (IC$_{50}$).

The poor aqueous solubility of this class of compounds may have been due to the flat and hydrophobic nature of the indolocarbazoles. To optimize the biopharmaceutical properties of these indolocarbazole CDK4 inhibitors, the SAR strategy was focused on introducing ionizable groups into the molecule. Because substitution at the 6-position of the nonannulated indole ring is tolerated, a variety of basic groups were explored in this region (Table 14.5). The pyridyl groups were well tolerated in the 6-position, with the 3-pyridyl analog **14.9a** being slightly more potent than the corresponding 4-pyridyl analog **14.9b** in both the enzymatic and cellular assays. To further explore other amino groups in this position, a hydroxymethyl analog **14.9c** was synthesized, which was potent in both kinase and cellular assays. With this key intermediate in hand, displacement of the activated hydroxyl group with a variety of amines resulted in indolocarbazole analogs **14.9d–14.9k** with improved aqueous solubility. These amino-substituted indolocarbazoles **14.9d–14.9k** were potent D1/CDK4 inhibitors with IC$_{50}$ values ranging from 50 to 90 nM. The kinase activity of monoamino-substituted indolocarbazoles **14.9d–14.9i** correlated very well with their antiproliferative cellular activity in human colon carcinoma (HCT-116) and non-small-cell lung carcinoma (NCI-460) cell lines. However, poor cellular activities were observed for the diamino-substituted analogs **14.9j** and **14.9k** in spite of their potent D1/CDK4 activities. This is probably due to the poor cellular permeability of the diamino analogs.

The fact that the annulated ring size (**14.8a–14.8d**, Table 14.4) has little effect on D1/CDK4 activity indicates that this pocket might be tolerant of substitution. We also explored the possibility of introducing ionizable groups on the annulated ring. It was hypothesized that the substitution on the annulated ring could decrease the planarity of the overall molecule, thus improving compound solubility. A similar strategy of incorporating a hydroxymethyl group as a chemical handle for further elaboration was used for rapid SAR expansion.[12] Indolocarbazoles **14.9l**, **14.9m**, and **14.9n** exemplified the effect of hydroxymethyl substitution at varying positions of the annulated ring. All three compounds were potent inhibitors of cyclin D1/CDK4 and effective antiproliferative agents in the HCT-116 and NCI-460 cell lines. The evaluation of the substitution on the non-annulated indole ring was focused on the 6-position, as based on our early finding that substitution at this position was most favorable for D1/CDK4 activity. The fluoro (**14.9p**), bromo (**14.9q**), trifluoro methyl (**14.9r**) and methoxyl (**14.9s**) substituted compounds were all very potent inhibitors of cyclin D1/CDK4 with selectivity against cyclin E/CDK2. In addition, they also demonstrated antiproliferative effects in the HCT-116 colon carcinoma cell line and NCI-460 lung carcinoma cell line, with activities ranging between 0.36–1.98 μM. As expected, the ring size did not appear to affect the activity. The 7-membered analog **14.9t** was equally potent in cyclin D1/CDK4

TABLE 14.5
SAR of 1,7-Annulated Indolocarbazoles 14.9 with an Ionizable Group

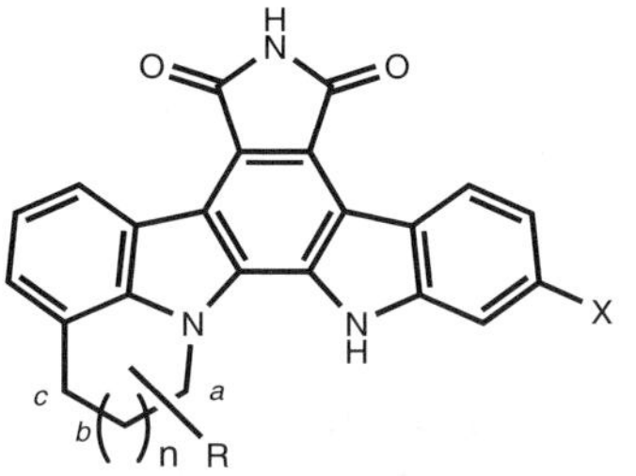

14.9

	X	n	R	D1/CDK4 IC$_{50}$ (μM)	E/CDK2 IC$_{50}$ (μM) (Rb^{ING})	HCT-116 IC$_{50}$ (μM)	NCI-H460 IC$_{50}$ (μM)
14.9a	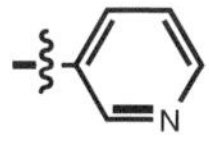 3-pyridyl	1	H	0.16	—	1.1	1.45
14.9b	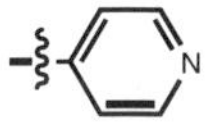 4-pyridyl	1	H	0.59	—	3.29	—
14.9c	CH_2OH	1	H	0.052[a]	—	0.34	0.465
14.9d	$CH_2NH_2 \cdot HCl$	1	H	0.170[a]	—	2.26	8.69
14.9e	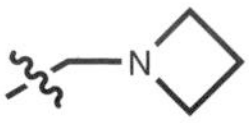 azetidinyl	1	H	0.075[a]	—	0.23	0.37
14.9f	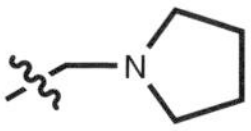 pyrrolidinyl	1	H	0.050[a]	—	0.17	0.14
14.9g	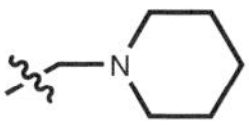 piperidinyl	1	H	0.087[a]	—	0.32	0.94
14.9h	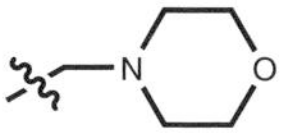 morpholinyl	1	H	0.075[a]	—	0.37	1.35
14.9i	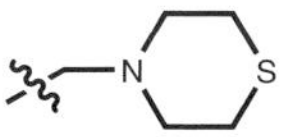 thiomorpholinyl	1	H	0.090[a]	—	0.60	0.72
14.9j	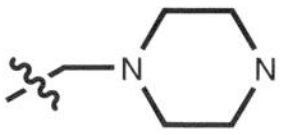 piperazinyl	1	H	0.065[a]	—	>10.0	8.80
14.9k	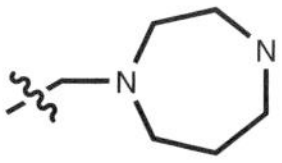 homopiperazinyl	1	H	0.079[a]	—	7.6	>10.0

(continued)

TABLE 14.5 (CONTINUED)
SAR of 1,7-Annulated Indolocarbazoles 14.9 with an Ionizable Group

	X	n	R	D1/CDK4 IC_{50} (μM)	E/CDK2 IC_{50} (μM) (Rb[ING])	HCT-116 IC_{50} (μM)	NCI-H460 IC_{50} (μM)
14.9l	H	1	a-CH$_2$OH-	0.018 [b]	0.22	0.92	0.82
14.9m	H	1	b-CH$_2$OH	0.071 [b]	0.22	0.91	0.96
14.9n	H	1	c-CH$_2$OH	0.031 [b]	—	0.37	0.54
14.9o	F	1	a-CH$_2$OH	0.002 [b]	0.079	0.82	0.26
14.9p	F	1	b-CH$_2$OH	0.022 [b]	0.141	0.36	0.43
14.9q	Br	1	b-CH$_2$OH	0.010 [b]	>2.0	0.50	0.36
14.9r	CF$_3$	1	b-CH$_2$OH	0.023 [b]	—	1.25	0.69
14.9s	OCH$_3$	1	b-CH$_2$OH	0.015 [b]	0.284	1.83	1.96
14.9t	H	2	b-CH$_2$OH	0.126 [a]	—	0.54	0.33
14.9u	F	1	b-CH$_2$Cl	0.064	—	3.27	5.65
14.9v	F	1	b-CH$_2$NHCH$_3$.HCl	0.008	—	1.73	5.15
14.9w	H	1	b-CH$_2$N(CH$_3$)$_2$	0.898 [a]	—	0.77	4.75
14.9x	F	1		0.671 [a]	—	>10.0	>10.0
14.9y	F	1		>2.0 [a]	—	>10.0	>10.0

Note: — = not tested.

[a] Rb[ING] as substrate.

[b] Rb[21] as substrate.

compared to the 6-membered compound **14.9m**. When the hydroxyl group was converted to the methyl amino analog **14.9v**, the D1/CDK4 kinase activity was well maintained. Further increase in the size of the amino group led to decreased D1/CDK4 activity (**14.9w–14.9y**).

With the success of modifying of the 1,7-all carbon-annulated indolocarbazole, the aza-annulated indolocarbazoles **14.10** were also hypothesized and synthesized.[13] The goal was to introduce additional binding interactions with the protein backbone in the ATP-binding site and to improve the biopharmaceutical properties. Using a similar synthetic strategy for the carbazole formation from the corresponding maleimide, a series of aza-annulated indolocarbazoles were prepared and evaluated in the D1/CDK4 assay. As summarized in Table 14.6, indolocarbazoles **14.10a–14.10k** were very potent inhibitors of cyclin D1/CDK4, with activity ranging from 0.027 to

TABLE 14.6
SAR of 1,7-Aza-Annulated Indolocarbazoles 14.10

	m	n	X	R	D1/CDK4 IC$_{50}$ (μM) (Rb^{ING})	E/CDK2 IC$_{50}$ (μM) (Rb^{ING})	HCT-116 IC$_{50}$ (μM)	NCI-H460 IC$_{50}$ (μM)
14.10a	1	1	H	H	0.027	0.038	0.031	—
14.10b	2	1	H	H	0.160	—	0.04	0.18
14.10c	1	3	H	H	0.084	0.200	0.10	—
14.10d	1	1	H	^{i}Pr	0.078	0.780	0.64	0.52
14.10e	1	1	H	SO$_2$Me	0.053	>0.20	1.60	1.74
14.10f	1	1	H	COCH$_2$NH$_2$	0.100	—	>10.0	>10.0
14.10g	1	1	5—F	H	0.034	>0.2	0.26	1.54
14.10h	1	1	6—F	H	0.031	0.016	0.20	0.67
14.10i	1	1	5—F, 6—F	H	0.110	>0.2	0.40	2.91
14.10j	1	1	6—CF$_3$	H	0.069	0.210	0.17	0.73
14.10k	1	1	7—CH$_2$OH	H	0.044	—	3.50	3.53

Note: — = not tested

0.26 μM. As expected from previous findings, the ring size of the 1,7-annulated ring had little effect on D1/CDK4 activity. Indolocarbazoles **14.10a–14.10c** demonstrated similar activity against D1/CDK4, which also indicated the nitrogen in the ring did not affect the D1/CDK4 activity nor the antiproliferative cell-based activity. To probe the importance of the NH on the 1,7-annulated ring, an isopropyl group was installed on the nitrogen of the 1,7-annulated ring (**14.10d**). The fact that compound **14.10d** had similar activity as **14.10a** suggested the proton on the 1,7-aza-annulated ring is not critical for binding in the D1/CDK4 ATP pocket. Further studies also concluded that the basicity of the nitrogen on the 1,7-annulated ring had little impact on D1/CDK4 activity, as exemplified in compounds **14.10d–14.10f**. The poor cellular activity of compound **14.10f** may be due to reduced cell penetration of the molecule. The substitution effect of the nonannulated ring was also examined. Compounds **14.10g–14.10k** were very potent inhibitors of D1/CDK4 with excellent activity in HCT-116 and NCI-460 cell lines. However, the aza-1,7-annulated indolocarbazoles were also potent cyclin E/CDK2 inhibitors.

14.4   ARYL[a]PYRROLO[3,4-c]CARBAZOLES

Based on the x-ray co-crystal structure of the complex of human CDK2 with indolo-carbazole **14.7g** and early SAR information, the carbonyl group and the acidic proton of the maleimide moiety play critical roles by acting as a hydrogen bond acceptor and donor in the ATP-binding pocket of CDK2. It has been proposed that the hydrogen bonding sites of the ATP pocket are conserved among many serine/ threonine kinases.[14] Novel kinase inhibitors were studied within this chemical platform, which resulted from the replacement of one indole moiety in arcyriaflavin with other aryl/heteroaryl groups (**14.11**, Figure 14.5), while maintaining the key interaction points of the maleimide moiety with the protein. Some of the examples are summarized in Table 14.7.[15]

7-Aza-indolyl analog **14.12** and benzothienyl analog **14.13** were hypothesized to maintain the molecular symmetry of the carbazole region. Moderate inhibitory activity of 0.25 μM and 0.22 μM (IC$_{50}$), against D1/CDK4 respectively, were observed with consistent cellular activity in the HCT-116 cell line. Naphthyl and quinolinyl modifications were also selected to replace one of the indole rings in arcyriaflavin A. Three possible orientations of the fused rings were explored (**14.14–14.16**). The up-angular derivatives **14.14a** and **14.14b** were most favorable, with the naphthyl analog **14.14a** being highly active in D1/CDK4 with moderate cellular activity in the HCT-116 cell line. However, the quinolinyl analog **14.14b** was fairly potent in the HCT-116 cellular assay despite the moderate binding activity. The linear orientated derivatives **14.15a** and **14.15b** were basically inactive in the binding assay (IC$_{50}$ > 10 μM). However, the down-angular regioisomers (**14.16a** and **14.16b**) exhibited a different activity profile. Quinolinyl analog **14.16a** was inactive, and naphthyl analog **14.16b** showed moderate activity in D1/CDK4 with an IC$_{50}$ of 0.43 μM. The discovery of the preferred up-angular orientation led to other 6,6-fused ring systems. The tetrahydronaphthyl analog **14.17** had moderate binding activity, but was tenfold less potent compared to the corresponding naphthyl analog **14.14b.** This finding further confirmed that the flat structure was preferred in the ATP-binding pocket of the D1/CDK4 enzyme. For the quinolinyl/isoquinolinyl fused analogs **14.18–14.21**, the nitrogen position in the 6,6-fused rings had a big impact on the D1/CDK4 inhibitory activity. When positions 5 and 6 of the isoquinolinyl group were fused to the carbazole, the resulting compound **14.20** demonstrated potent D1/CDK4 activity, as well as cellular activity in the HCT-116 cell line. Another fused isoquinolinyl analog **14.18** and fused quinolinyl analogs **14.19** and **14.21** were much less active in the

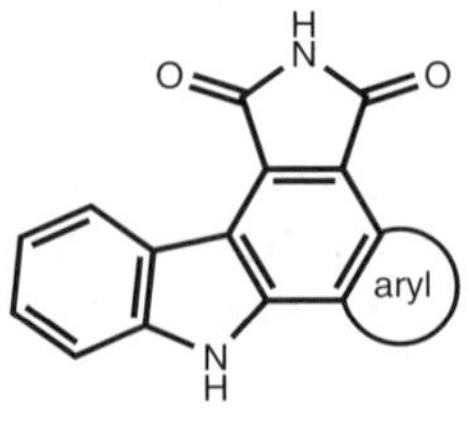

14.11

FIGURE 14.5 Aryl analogs of arcyriaflavin A.

TABLE 14.7
SAR of Aryl[a]pyrazole[3,4-c]carbazoles 14.12–14.26

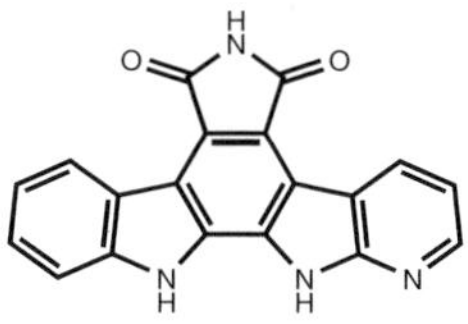

14.12

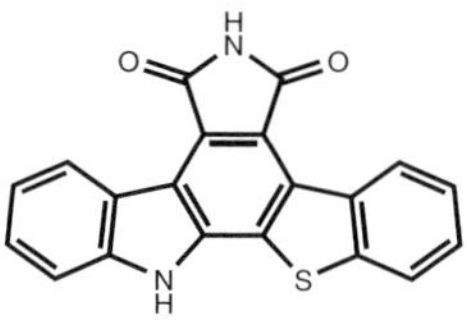

14.13

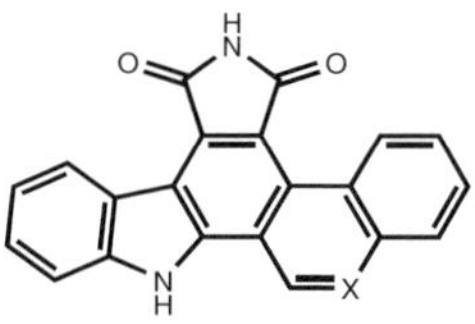

14.14a–b

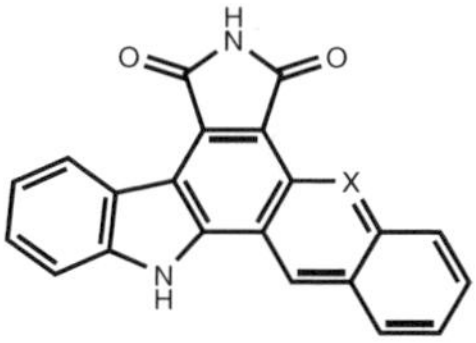

14.15a–b

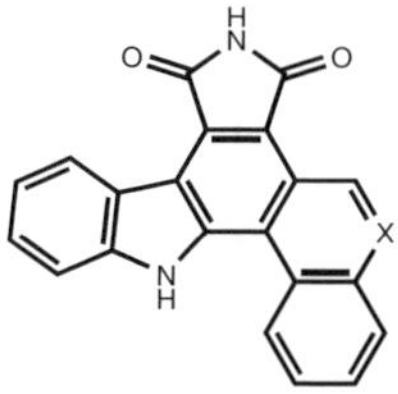

14.16a–b

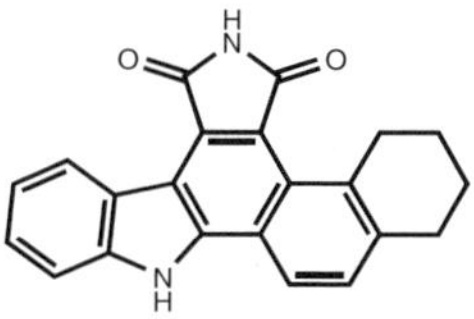

14.17

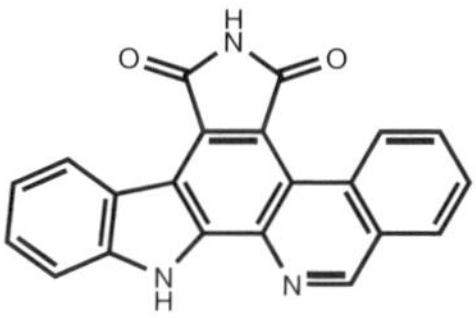

14.17

14.18

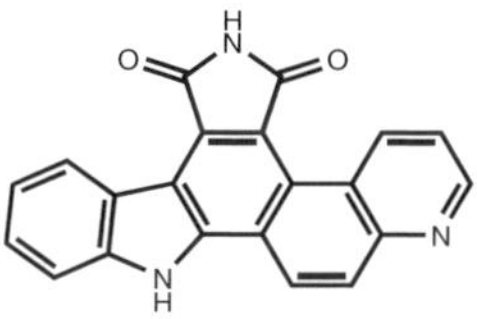

14.19

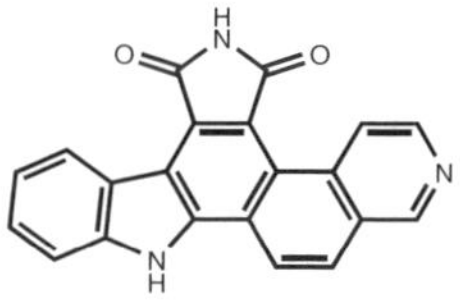

14.20

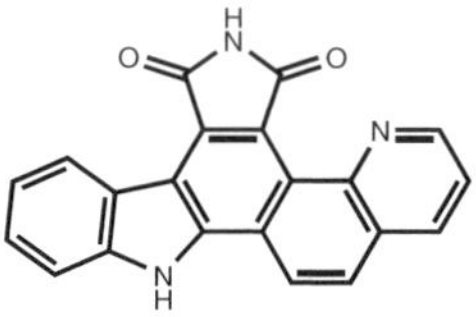

14.21

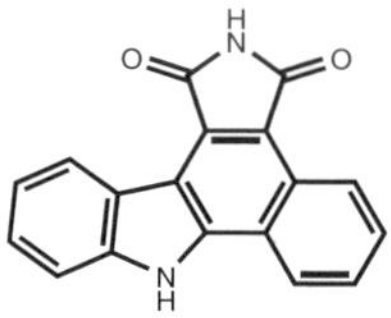

14.22

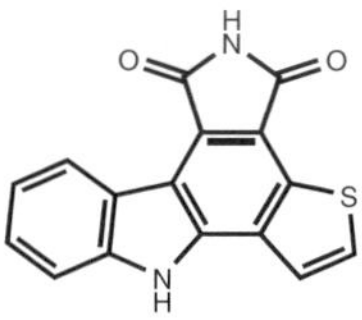

14.23

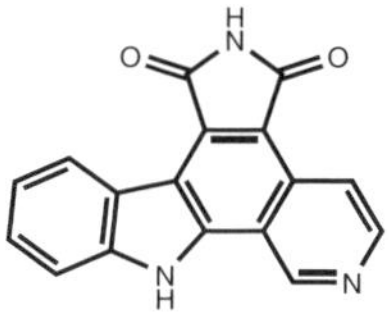

14.24

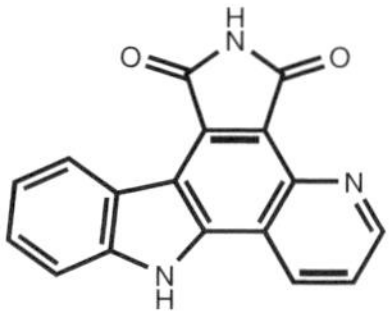

14.25

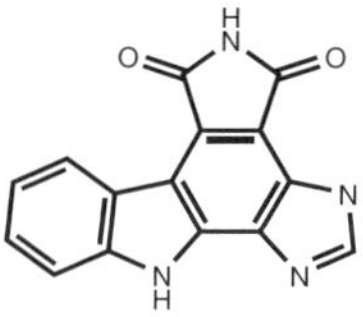

14.26

Compound	D1/CDK4 IC$_{50}$ (µM)	HCT-116 IC$_{50}$ (µM)
14.12	0.25	1.05
14.13	0.22	3.02
14.14a (X = N)	0.683	0.39

(*continued*)

TABLE 14.7 (CONTINUED)
SAR of Aryl[a]pyrazole[3,4-c]carbazoles 14.12–14.26

Compound	D1/CDK4 IC$_{50}$ (μM)	HCT-116 IC$_{50}$ (μM)
14.14b (X = CH)	0.045	5.05
14.15a (X = N)	>10.0	—
14.15b (X = CH)	>10.0	—
14.16a (X = N)	>10.0	—
14.16b (X = CH)	0.43	—
14.17	0.41	—
14.18	>2.0	—
14.19	>0.2	—
14.20	0.069	0.13
14.21	>2.0	—
14.22	0.26	5.83
14.23	0.15	—
14.24	1.81	—
14.25	0.18	—
14.26	0.74	—

Note: — = not tested.

D1/CDK4 assays. This result indicated the electronic density of the fused rings may play a critical role in the binding mode of the carbazole to the D1/CDK4 protein backbone.

Replacement of the indole moiety with monoheteroaromatic rings (**14.22–14.26**) was also studied. Small monoheterocyclic rings, such as phenyl (**14.22**), thienyl (**14.23**), pyridyl (**14.24–14.25**), and imidazolyl (**14.26**) led to decreased D1/CDK4 activity compared to either arcyriaflavin A or naphthyl analog **14.14b**. Presumably, the decrease of binding affinity of these monoaromatic fused ring carbazoles, **14.22–14.26**, was due to the loss of protein backbone interactions.

14.5 INDOLO[6,7-a]PYRROLO[3,4-c]CARBAZOLES

As discussed in the previous section, replacement of the indole moiety in arcyriaflavin A (**14.5**) with an up-angular naphthyl group (**14.14b**) is well tolerated for D1/CDK4 activity. This result indicated that the carbazole fused with a [6,6]-member fused ring instead of a [5,6]-member fused ring can preserve the D1/CDK4 activity, but when a naphthyl group (**14.14b**, IC$_{50}$ = 0.045 μM) was replaced by a phenyl ring (**14.22**, IC$_{50}$ = 0.26 μM), the D1/CDK4 activity decreased, presumably because of a lack of binding interaction. To occupy the interaction space, a novel indolocarbazole derived from a carbazole fused with an indole ring at the 6,7-position, indolo [6,7-a]pyrrolo[3,4-c]carbazole (**14.27a**), was designed and synthesized (Table 14.8).[16] Compound **14.27a** inhibited both D1/CDK4 and E/CDK2 potently and displayed antiproliferative activity in two human carcinoma cell lines, HCT-116 and NCI-H460 (lung). Methyl and ethyl substitutions (**14.27b** and **14.27c**) were tolerated in R^1 position, but other bulkier groups at R^1 led to significant activity loss in both D1/CDK4 and the cellular assays. The 4-

TABLE 14.8
SAR of Indolocarbazoles 14.27

14.27

Compound	R¹	R²	X	R³	R⁴	CDK4 IC$_{50}$ (μM) (Rb^{ING})	CDK2 IC$_{50}$ (μM) (Rb^{ING})	HCT–116 IC$_{50}$ (μM)	NCI–H460 IC$_{50}$ (μM)
14.27a	H	H	H	H	H	0.036	0.064	1.44	1.43
14.27b	CH$_3$	H	H	H	H	0.047	0.075	2.11	1.17
14.27c	CH$_2$CH$_3$	H	H	H	H	0.059	—	4.52	>10
14.27d	CH$_3$	H	4-F	H	H	0.071	>2	>2	>10
14.27e	CH$_3$	H	4-Br	H	H	0.171	2.86	4.71	3.52
14.27f	CH$_3$	H	4-OCH$_3$	H	H	>2	—	>10	>10
14.27g	CH$_3$	H	4-CH$_2$OH	H	H	0.155	>2	>10	>10
14.27h	CH$_3$	H	5-F	H	H	0.084	>0.2	>10	>10
14.27i	CH$_3$	H	5-Br	H	H	>2	—	—	—
14.27j	CH$_3$	H	5-OCH$_3$	H	H	0.519	>2	>10	>10
14.27k	CH$_3$	H	5-CF$_3$	H	H	>2	—	7.04	7.85
14.27l	CH$_3$	H	5-CH$_3$	H	H	0.393	1.07	2.87	3.35
14.27m	CH$_3$	H	6-F	H	H	0.038	0.009	>10	>10
14.27n	CH$_3$	H	6-Br	H	H	0.038	0.075	1.08	0.72
14.27o	CH$_3$	H	6-OCH$_3$	H	H	0.053	0.112	1.80	1.09
14.27p	H	H	6-OCH$_3$	H	H	0.067	0.125	2.16	1.92
14.27q	CH$_3$	H	6-CF$_3$	H	H	0.076	0.225	>10	3.51
14.27r	CH$_3$	H	6-CH$_3$	H	H	0.048	0.320	3.84	3.41
14.27s	CH$_3$	H	6-CH$_2$CH$_3$	H	H	0.101	0.360	3.51	7.49
14.27t	CH$_3$	H	6-CN	H	H	0.029	0.010	>10	>10
14.27u	CH$_3$	H	6-OH	H	H	0.033	—	0.67	1.08
14.27v	H	H	6-OH	H	H	0.037	0.006	0.17	0.23
14.27w	CH$_3$	H	6-OCH$_2$CH$_3$	H	H	0.044	0.324	2.35	3.20
14.27x	CH$_3$	H	7-Br	H	H	0.093	—	17.1	13.8
14.27y	CH$_3$	H	7-OCH$_3$	H	H	0.062	>2	1.00	2.71
14.27z	CH$_3$	H	7-OBn	H	H	0.123	0.328	6.52	>10
14.27aa	CH$_3$	H	H	CH$_3$	H	0.028	0.070	1.98	2.26
14.27bb	CH$_3$	H	H	F	H	0.067	0.086	3.49	2.90
14.27cc	CH$_3$	H	H	OCH$_3$	H	0.032	0.067	1.90	0.69
14.27dd	CH$_3$	H	H	H	OCH$_3$	0.054	0.035	1.13	0.71
14.27ee	CH$_3$	H	6-F	CH$_3$	H	0.050	0.023	0.58	1.02
14.27ff	CH$_3$	H	6-OCH$_3$	CH$_3$	H	0.042	0.057	1.43	1.30
14.27gg	CH$_3$	H	6-CF$_3$	CH$_3$	H	0.092	>2	8.78	2.42
14.27hh	CH$_3$	CH$_3$	H	CH$_3$	H	0.092	0.267	4.70	>10

Note: — = not tested.

and 5-positions did not tolerate substitution. Although fluoro analogs **14.27d** and **14.27h** exhibited good enzymatic activity, no measurable cellular activity was observed for both compounds. Other substitutions larger than fluorine at the 4- and 5-positions (**14.27e–14.27g**, **14.27i–14.27l**, respectively) exhibited substantially lower activity compared to the non-substituted parent analog **14.27b**. Consistent with previous findings, the 6-position can tolerate a wide range of substitution with respect to D1/CDK4. Potent D1/CDK4 activities were observed for all the 6-substituted analogs **14.27m-14.27w**. A slight selectivity for D1/CDK4 vs. E/CDK2 was found for most cases with the exception of compounds **14.27m**, **14.27t**, and **14.27v**. The 7-substituted analogs **14.27x-14.27z** also exhibited good binding activity in D1/CDK4 with moderate cellular activity. Substitutions were also incorporated at the R^3 and R^4 positions of carbazole **14.27**. Compounds bearing substitution at position R^3 gave good enzymatic and cellular activity (**14.27aa–14.27cc**, **14.27ee–14.27gg**); however, a significant decrease in cellular activity was found for compound **14.27hh**, which has substitutions at both R^2 and R^3 positions.

The challenge of discovering a pharmaceutical in this class of compounds is because of their poor aqueous solubility. One major area of focus was to introduce ionizable groups to improve the aqueous solubility of this chemical platform. Hydroxy-alkyl groups were selected as a chemical handle for further rapid SAR elaboration. Compounds **14.28a**, **14.29a**, **14.30a,** and **14.31a** were potent inhibitors of D1/CDK4 with good cellular activity in HCT-116 and NCI-H460 (Table 14.9). This result indicated that large substitutions such as hydroxyethyl and hydroxypropyl were well tolerated in positions 7 and R^1 of carbazoles (**14.28–14.31**). Further SAR was focused on incorporating amino groups at these two positions with the aim of improving aqueous solubility. Table 14.9 summarizes the results for the amino-alkyl-substituted carbazoles **14.28–14.31**. Carbazoles bearing an aminoalkyl group in positions 7 or R^1 were all potent inhibitors of D1/CDK4, with activity ranging from 0.033 μM to 0.19 μM (IC_{50}). Solubility studies using the *in situ* salts of the amino-alkyl substituted carbazoles confirmed the improved solubility in D5W compared to non-basic analogs. In addition to the improved solubility profile of these aminoalkyl carbazole analogs, significantly higher selectivities (with many having >20 to >40-fold selectivity and some >40-fold with respect to E/CDK2) were observed for carbazoles **14.28–14.31** when compared to the indolo[6,7-*a*]pyrrolo[3,4-*c*]carbazoles lacking aminoalkyl side chains. Kinetic analysis of selected examples indicated that these compounds are reversible ATP-competitive inhibitors. Most of the indolo[6,7-*a*] pyrrolo[3,4-*c*]carbazoles bearing aminoalkyl substitution were also potent antiproliferative agents in HCT-116 and NCI-H460 cell lines, with compound **14.28b** being an exception.

14.6 INDOLO[2,3-A]CARBAZOLE GLYCOSIDE ANALOGS

To continue improving the aqueous solubility of the indolo[2,3-a]carbazole scaffold, indolo[2,3-a]carbazole glycosides **14.32** bearing naturally occurring sugars with one gly-cosidic linkage were synthesized and evaluated.[17] As shown in Table 14.10, indolo[2,3-a]carbazoles glycosidically linked with a variety of carbohydrates, for example, glucose (**14.32a**), galactose (**14.32b**), fucose (**14.32c**), rhamnose (**14.32d**), xylose (**14.32e**), and maltose (**14.32f**), and exhibited good to moderate inhibitory effect against D1/CDK4; however, with diminished selectivity against E/CDK2 and PKCβ II. Compound **14.32d**,

TABLE 14.9
SAR of Indolocarbazoles 14.28–14.31

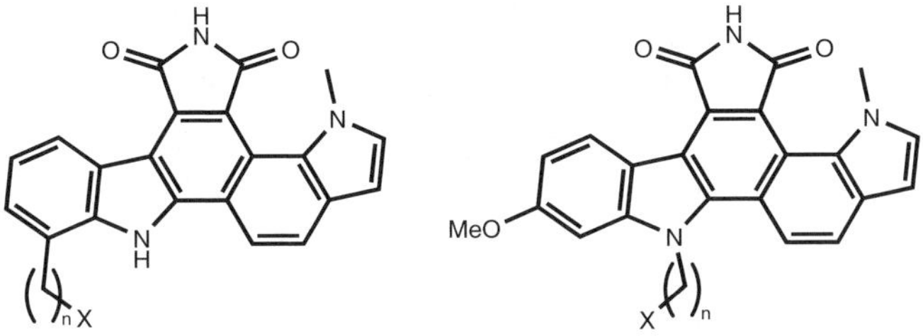

14.28 (n = 2)
14.29 (n = 3)

14.30 (n = 2)
14.31 (n = 3)

X

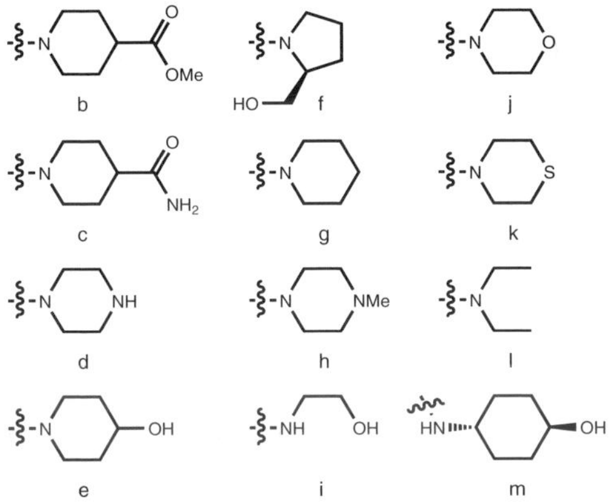

	CDK4 IC$_{50}$ (μM) (Rb^{ING})	CDK2 IC$_{50}$ (μM) (Rb^{ING})	HCT-116 IC$_{50}$ (μM)	NCI-H460 IC$_{50}$ (μM)
Compound / X				
14.28a OH	0.082	0.210	1.13	—
14.28b b	0.047	0.972	>10	>10
14.28c c	0.051	0.796	1.00	1.35
14.28d d	0.049	2.03	1.04	1.11
14.28e e	0.045	0.978	0.20	0.14
14.28f f	0.039	0.685	0.42	0.21
14.28g g	0.045	1.06	2.12	2.47
14.28i i	0.033	0.265	0.17	0.33
14.28j j	0.054	1.28	1.63	1.91
14.28k k	0.082	1.80	3.41	3.41
14.28m m	0.053	0.450	1.23	2.82
14.29a OH	0.168	0.558	3.02	2.35
14.29h h	0.052	1.67	1.38	1.48
14.29i i	0.040	0.567	0.77	1.10

(continued)

TABLE 14.9 (CONTINUED)
SAR of Indolocarbazoles 14.28–14.31

Compound	X	CDK4 IC$_{50}$ (μM) (Rb^{ING})	CDK2 IC$_{50}$ (μM) (Rb^{ING})	HCT-116 IC$_{50}$ (μM)	NCI-H460 IC$_{50}$ (μM)
14.29j	j	0.056	1.33	2.82	1.68
14.29l	l	0.034	1.50	0.95	0.37
14.29m	m	0.049	1.03	1.05	1.67
14.30a	OH	0.047	0.101	4.07	1.32
14.30f	f	0.084	0.942	2.69	3.80
14.30g	g	0.135	0.928	3.74	2.47
14.30i	i	0.035	0.089	1.47	1.31
14.30m	m	0.063	0.902	0.89	1.05
14.31a	OH	0.054	0.113	1.62	0.50
14.31c	c	0.077	1.00	1.96	2.44
14.31e	e	0.048	2.10	1.26	1.13
14.31h	h	0.095	1.23	0.45	0.90
14.31i	i	0.072	0.432	1.55	1.28
14.31j	j	0.069	0.716	1.69	1.63
14.31k	k	0.190	1.26	2.02	2.11

Note: — = not tested.

TABLE 14.10
SAR of Indolo[2,3-a]carbazole Glycoside Analogs 14.32

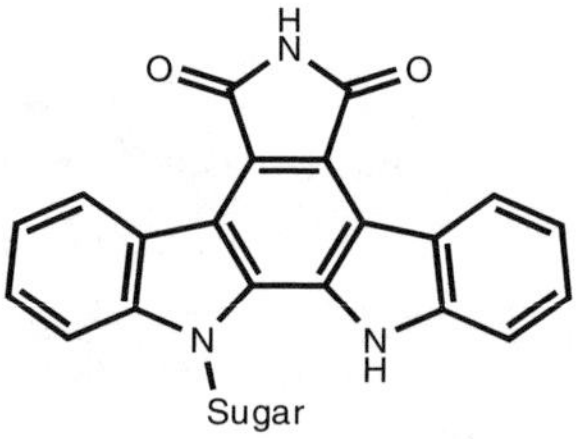

14.32

Compound	Sugar	CDK4 IC$_{50}$ (μM) (Rb^{ING})	CDK2 IC$_{50}$ (μM) (Rb^{ING})	PKCβ II IC$_{50}$ (μM)
14.32a	D-glucose	0.785	6.880	0.594
14.32b	D-galactose	1.762	2.063	5.268
14.32c	L-Fucose	1.600	1.830	2.040
14.32d	L-Rhamnose	0.076	0.296	1.145
14.32e	D-Xylose	0.130	0.404	0.318
14.32f	D-Maltose	0.803	16.18	0.748

derived from L-rhamnose, was found to be the most potent inhibitor in D1/CDK4 with an IC_{50} of 0.076 μM, which also demonstrated selectivity against E/CDK2 and PKCβ II.

14.7 EFFECT ON CELL CYCLE AND pRb PHOSPHORYLATION BY D1/CDK4 INHIBITORS

CDK4 plays a critical role in the G_1-S transition of the cell cycle by specifically phosphorylating the Ser780 residue on the Rb protein.[18] As discussed in previous sections, the carbazole D1/CDK4 inhibitors demonstrated antiproliferative activity in a variety of tumor cell lines. Thus, inhibition of cellular CDK4 activity will result in the inhibition of Rb phosphorylation on Ser-780 and cell cycle arrest in the G_1 phase.

Cell cycle effects in HCT-116 cells were examined by flow cytometry.[19] A cell cycle inhibition study was conducted by treating HCT-116 cells with different concentrations of the CDK4 inhibitors for 24 h and then analyzing them by flow cytometry. Treatment of the HCT-116 cells with these indolocarbazole D1/CDK4 inhibitors resulted in a significant accumulation of cells in the G_1 population and decreased the S and G_2/M populations (Table 14.11). In most cases, G_1 arrest occurred at the same concentration range as that for the corresponding cell growth inhibition (cellular inhibition IC_{50}).

TABLE 14.11
G_1 Arrest and Inhibition of Rb (Ser780) Phosphorylation in HCT 116 Cells of Carbazoles

Compound	IC_{50} HCT–116 (μM)	G_1 Arrest (Fold Increase over Control)		Percentage Inhibition of Rb Phosphorylation (Ser780)	
		@ 1 × IC_{50}	@ 3 × IC_{50}	@ 1 × IC_{50}	@ 3 × IC_{50}
14.7q	2.47	2.6	3.6	72	98
14.7o	3.53	2.3	1.8	66	97
14.7p	0.76	2.3	2.8	82	89
14.8b	1.60	2.0	2.9	60	91
14.9m	0.91	2.0	2.0	21	70
14.10a	0.31	1.5	1.8	82	79
14.12	1.05	—	—	24	82
14.14a	0.66	1.0	1.2	3	82
14.14b	5.05	—	—	77	75
14.20	0.11	1.9	1.2	73	86
14.27b	2.11	1.4	2.5	63	97
14.27o	1.80	2.6	2.6	89	97
14.27ff	1.43	2.0	2.2	83	95
14.28c	1.00	1.9	2.2	52	55
14.29m	1.05	2.9	3.4	45	83
14.30m	0.89	1.8	2.0	67	95
14.31i	1.55	1.8	2.7	57	78
14.32c	1.70	3.0	6.7	33	94

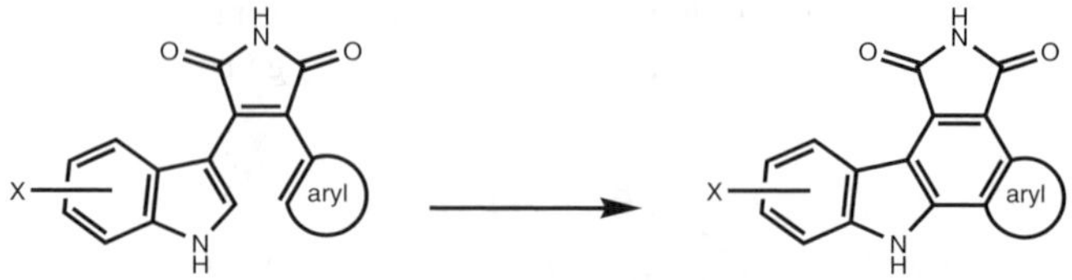

FIGURE 14.6 Synthesis of aryl indolocarbazoles (**14.33**).

Inhibition of Rb phosphorylation at the CDK4-specific Ser780 residue was investigated by Western blot analysis by measuring phospho-Rb^{S780} levels using a phospho-specific antibody.[20] Colon carcinoma cells (HCT-116) were treated with compounds at 1× and 3× antiproliferation IC_{50} concentrations for 24 h, followed by Western blot analysis. As shown in Table 14.11, these compounds potently inhibited phosphorylation of Rb^{S780}, demonstrating that they inhibit cellular CDK4 activity. These specific observations on the G_1 cell cycle arrest and strong inhibition of phosphorylation of Ser780 on pRb were consistent with inhibition of cyclin D1/CDK4.

14.8 CHEMISTRY

The general synthetic approach to the indolocarbazole compounds **14.33** is outlined in Figure 14.6. The key step in the synthesis of these compounds involved the cyclization of indolo-aryl-maleimides **14.32** to the corresponding indolocarbazole **14.33**. This was accomplished by oxidative cyclization mediated by a variety of oxidation systems, for example, DDQ in the presence or absence of a catalytic amount of p-toluene sulfonic acid using benzene, toluene, or dioxane as solvent,[21] Pd(OAc)$_2$,[22] or PIFA.[23] Alternatively, the cyclization of bisindolylmaleimide **14.32** could be carried out under photochemical conditions with or without iodine in a benzene/methanol solvent combination.[24]

A Heck cyclization for 2-bromoheteroaryl indolylmaleimides **14.34** was also successful for the synthesis of certain carbazole analogs (Figure 14.7).[25] Aryl- and heteroarylindolyl maleimide **14.32** was prepared in good yield via base-promoted condensation of an appropriately substituted indolyl 3-acetamide **14.35** with an aryl/heteroaryl glyoxylate **14.36** in the presence of potassium tert-butoxide in THF based on Faul's procedure. Modifications such as substituted indolyl 3-glyoxylate **14.37** with an aryl 3-acetamide **14.38** (Figure 14.8) are well tolerated.[26] In some

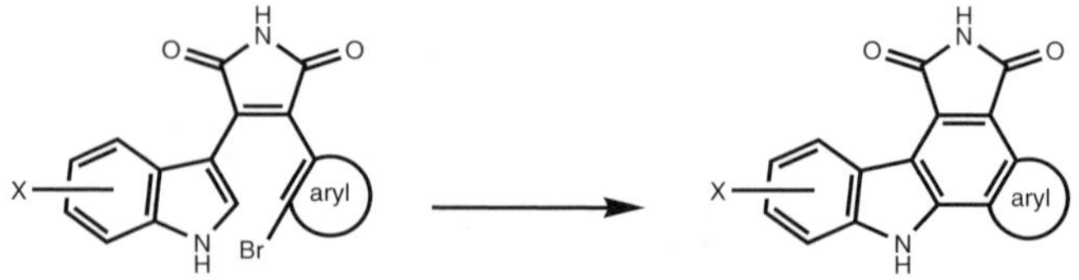

FIGURE 14.7 Alternative synthesis of aryl indolocarbazoles (**14.33**).

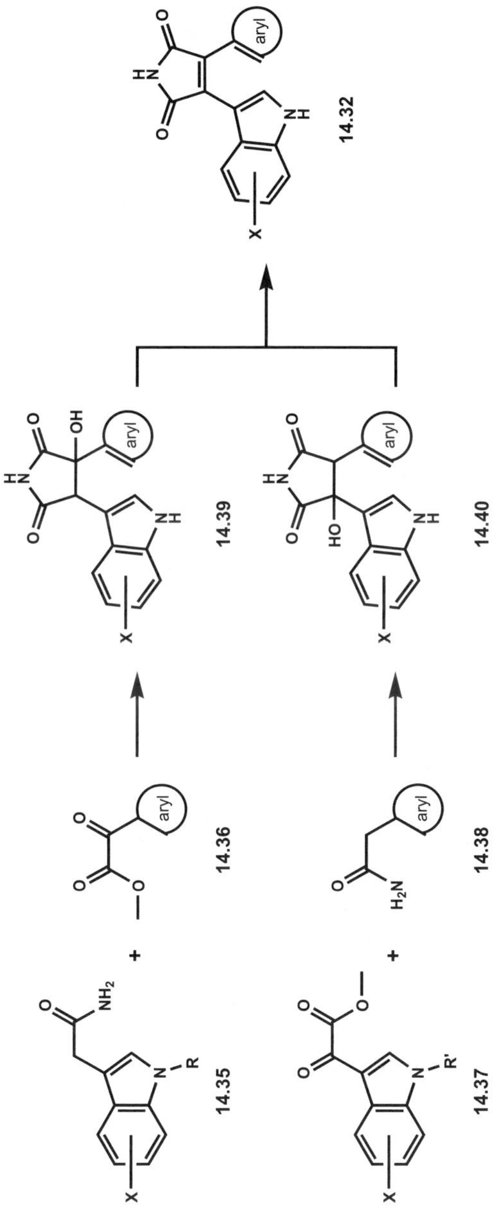

FIGURE 14.8 Synthesis of maleimides (**14.32**).

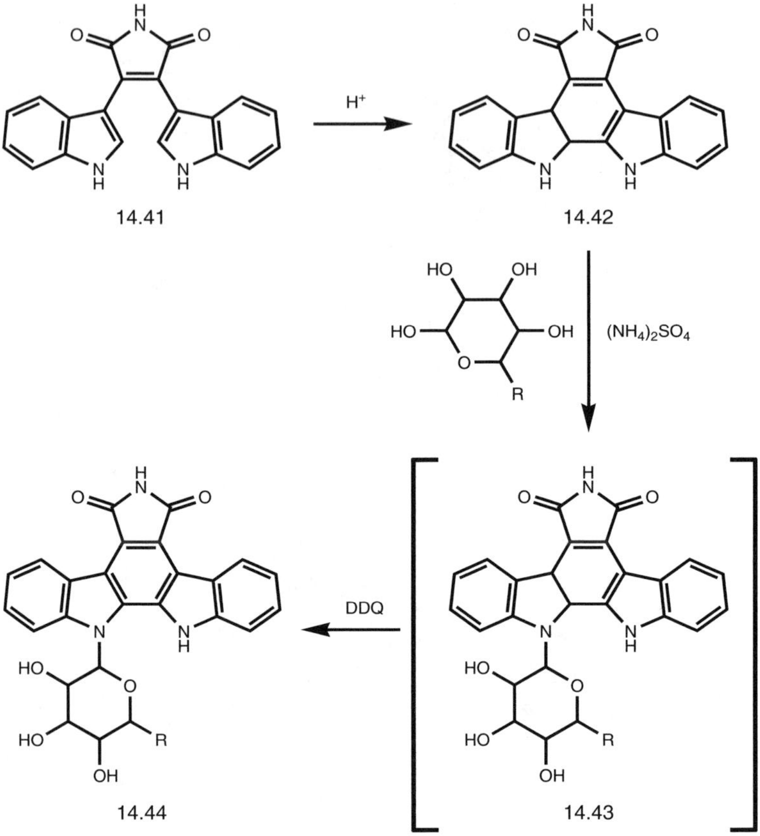

FIGURE 14.9 Synthesis of indolocarbazole glycosidic analogs (**14.44**).

cases, dehydration of the hydroxyimide intermediates (**14.39** and **14.40**) was slow under basic conditions, so concentrated hydrochloric acid was added to accelerate the dehydration of this intermediate. Both the starting aryl acetamides (**14.35** and **14.38**) and the aryl glyoxylates (**14.36** and **14.37**) are readily available by simple transformations from easily accessible starting materials.[27]

Indolo[2,3-a]carbazole glycoside analogs were prepared from direct glycosidation of an indole-indoline intermediate **14.42** using conditions similar to those developed by Chisholm and Van Vranken (Figure 14.9).[28] Subsequent oxidation of intermediate **14.43** in the presence of DDQ affords the desired glycosidated indolocarbazole **14.44**.[29]

14.9 SUMMARY

A novel class of potent CDK inhibitors based on carbazole is described. Structure–activity studies as well as x-ray co-crystal analysis of the inhibitor/CDK2 complex revealed the binding mode of the compound in the ATP pocket and was successfully utilized in the discovery of new potent CDK4 inhibitors. The selectivity for CDK4 vs. other kinases can be achieved by tuning different substituents on the carbazole. Moreover, these carbazole-based CDK4 inhibitors are ATP competitive and capable of inhibiting cell growth in the human tumor cell lines HCT-116 and NCI-H460. In addition, the specific G_1 cell cycle arrest and selective inhibition of phosphorylation of Ser780 on pRb are consistent with the *in vitro* D1/CDK4 inhibitory activity.

ACKNOWLEDGMENT

The author thanks the Lilly CDK4 program team members for their contributions and collaborations on this research. The author is also grateful to Mark Chappell and Scott Conner for reading the manuscript.

REFERENCES

1. (a) Dynlacht, B.D. *Nature* **1997**, *389*, 149. (b) Elledge, S.J. *Science*, **1996**, *274*, 1664.
2. (a) Pines, J. *Nat. Cell Biol.* **1999**, *1*, E73. (b) Pines, J. *Trends Biochem. Sci.* **1993**, *18*, 195. (c) Morgan, D.O. *Annu. Rev. Cell Dev. Biol.*, **1997**, *13*, 261.
3. (a) Owa, T., Yoshino, H., Yoshimatsu, K., Nagasu, T. *Current Med. Chem.*, **2001**, *8*, 1487–1503. (b) Walker, D.H. *Cur. Top. Microbiol. Immunol.* **1998**, *227*, 149–165. (c) Hunter, T., Pines, J. *Cell* **1991**, *66*, 1071–1074. (d) Kamb, A., Gruis, N.A., Weaver-Feldhaus, J., Liu, Q., Harshman, K., Tavtigian, S.V., Stockert, E., Day, R.S. III, Johnson, B.E., Skolnick, M.H. *Science* **1994**, *264*, 436–440. (e) Palmero, I., Peters, G. *Cancer Surv.*, **1996**, *27*, 351–367. (f) Kim, H., Ham, E.K., Kim, Y.I., Chi, J.G., Lee, H.S., Park, S.H., Jung, Y.M., Myung, N.K., Lee, M.J., Jang, J.-J. *Cancer Lett.*, **1998**, *131*, 177–183.
4. (a) Hunter, T., Pines. J. *Cell* **1991**, *66*, 1071. (b) Hartwell, L.H., Kastan, M.B. *Science*, **1994**, *266*, 1821. (c) Kamb, A, Gruis, N.A., Weaver-Feldhaus, J., Liu, Q., Harshman, K., Tavtigian, S.V., Stockert, E., Day, R.S., Johnson, B.E., III, Skolnick, M.H. *Science* **1994**, *264*, 436. (d) Palmero, I., Peters, G. *Cancer Surv.*, **1996**, *27*, 351. (e) Kim, H., Ham, E. K., Kim, Y. I., Chi, J. G., Lee, H. S., Park, S. H., Jung, Y. M., Myung, N. K., Lee, M. J., Jang, J.J. *Cancer Lett.*, **1998**, *131*, 77.
5. (a) Senderowicz, A.M. *Oncogene*, **2003**, *22*, 6609. (b) Sielecki, T.M., Boylan, J.F., Benfield, P.A., Trainor, G.L. *J. Med. Chem.* **2000**, *43*, 1. (c) Shapiro, G.I. *Clin. Cancer Res.* **2004**, *10*, 4270 and references cited therein. (d) Meijer, L., Raymond, E. *Acc. Chem. Res.* **2003**, *36*, 417. (e) McLaughlin, F., Finn, P., Thangue, N.B.L. *Drug Discovery Today*, **2003**, *8*, 793. (f) Knockaert M., Greengard, P., Meijer, L. *Trends Pharmacol. Sci.*, **2002**, *23*, 417.
6. Davis, P.D., Hill, C.H., Lawton, G., Nixon, J.S., Wilkinson, S.E., Hurst, S.A., Keech, E., Turner, S.E. *J. Med. Chem.*, **1992**, *35*, 177.
7. Sanchez-Martinez, C., Shih, C., Zhu, G., Li, T., Brooks, H.B., Patel, B., Schultz, R.M., DeHahn, T., Spencer, C.D., Watkins, S.A., Ogg, C., Considine, E., Dempsey, J.A., Zhang, F. *Bioorg. Med. Chem. Lett.*, **2003**, *13*, 3841.
8. The Rb[ING] substrate is a 12 amino acid peptide taken from the primary sequence of Rb protein. Kinase inhibitory assays were done as described in Konstantinidis, A.K., Radhakrishnan, R., Gu, F., Rao, R.N., Yeh, W.K. *J. Biol. Chem.*, **1998**, *273*, 28508.

9. Zhu, G., Conner, S.E., Zhou, X., Shih, C., Li, T., Anderson, B.D., Brooks, H.B., Campbell, R.M., Considine, E., Dempsey, J.A., Faul, M.M., Ogg, C., Patel, B., Schultz, R.M., Spencer, C.D., Teicher, B., Watkins, S.A. *J. Med. Chem.* **2003**, *46*, 2027.

10. Al-awar, R.S., Ray, J.E., Kecker, K.A, Huang, J., Waid, P.P, Shih, C., Brooks, H.B., Spencer, C.D., Watkins, S.A., Patel, B.R., Stamm, N.B., Ogg, C.A., Schultz, R.M., Considine, E.L., Faul, M.M., Sullivan, K.A., Kolis, S.P., Grutsch, J.L., Joseph, S. *Bioorg. Chem. Lett.* **2004**, *14*, 3217.

11. Schultz, R.M., Merriman, R.L., Toth, J.E., Zimmermann, J.E., Hertel, L.W., Andis, S.L., Dudley, D.E., Rutherford, P.G., Tanzer, L.R., Grindey, G.B. *Oncology Res.,* **1993**, *5*, 223.

12. Zhu, G., Conner, S.E., Zhou, X., Chan, H., Shih, C., Engler, T.A., Al-awar, R.S., Brooks, H.B., Watkins, S.A., Spencer C.D., Schultz, R.M., Dempsey, J.A., Considine, E.L., Patel, B.R., Ogg, C.A., Vasudevan, V., Lytle, M.L. *Bioorg. Chem. Lett.* **2004**, *14*, 3057.

13. Al-awar, R.S., Ray, J.E., Kecker, K.A, Joseph, S. Huang, J., Shih, C., Brooks, H.B., Spencer, C.D., Watkins, S.A., Schultz, R.M., Considine, E.L., Faul, M.M., Sullivan, K.A., Kolis, S.P., Carr, A.A., Zhang, F. *Bioorg. Chem. Lett.* **2004**, *14*, 3925.

14. (a) Honma, T., Hayashi, K., Aoyama, T., Hashimoto, N., Machida, T., Fukasawa, K., Iwama, T., Ikeura, C., Ikuta, M., Suzuki-Takahashi, I., Iwasawa, Y., Hayama, T., Nishimura, S., Morishima, H. *J. Med. Chem.,* **2001**, *44*, 4615. (b) Honma, T., Yoshizumi, T., Hayashi, K., Kawanishi, N., Hashimoto, N., Fukasawa, K., Takaki, T., Ikeura, C., Ikuta, M., Suzuki-Takahashi, I., Hayama, T., Nishimura, S., Morishima, H. *J. Med. Chem.,* **2001**, *44*, 4628.

15. (a) Zhu, G., Conner, S., Zhou, X., Shih, C., Brooks, H.B., Considine, E., Dempsey, J.A., Ogg, C., Patel, B., Schultz, R.M., Spencer, C.D., Teicher, B., Watkins, S.A. *Bioorg. Med. Chem. Lett.* **2003**, *13*, 1231–1235. (b) Sanchez-Martinez, C., Shih, C., Faul, M.M., Zhu, G., Paal, M., Somoza, C., Li, T., Krumrich, C.A., Winneroski, L.L., Zhou, X., Brooks, H.B., Patel, B., Schultz, R.M., DeHahn, T., Kirmani, K., Spencer, C.D., Watkins, S.A., Considine, E., Dempsey, J.A., Ogg, C. *Bioorg. Med. Chem. Lett.* **2003**, *13*, 3835.

16. Engler, T.A., Furness, K., Malhotra, S., Sanchez-Martinez, C., Shih, C., Xie, W.,Zhu, G., Zhou, X., Conner, S., Faul, M.M., Sullivan, K.A., Kolis, S.P., Brooks, H.B., Patel, B., Schultz, R.M., DeHahn, T.B., Kirmani, K., Spencer, C.D., Watkins, S.A., Considine, E.L., Dempsey, J.A., Ogg, C.A., Stamm, N.B., Anderson, B.D., Campbell, R.M., Vasudevan V., Lytle, M.L. *Bioorg. Med. Chem. Lett.* **2003**, *13*, 2261.

17. Faul, M.M., Sullivan, K.A., Grutsch, J.L, Winneroski, L.L, Shih, C., Sanchez-Martinez, C., Cooper, J.T. *Tetrahedron Lett.* **2004**, *45*, 1095.

18. Brugarolas, J., Moberg, K., Boyd, S.D., Taya, Y., Jacks, T., Lees, J.A. *Proc. Natl. Acad. Sci. USA,* **1999**, *96*, 1002.

19. Darzynkiewicz, Z., Juan, G. In *Current Protocols in Cytometry,* Robinson, J. P., Ed., John Wiley & Sons: New York, **1997**, pp. 7.5.1–7.5.24.

20. (a) Kitagawa, M., Higashi, H., Jung, H. K., Suzuki-Takahashi, I., Ikeda, M., Tamai, K., Kato, J. Y., Segawa, K., Yoshida, E., Nishimura, S., Tava, Y. *EMBO J.* **1996**, *15*, 7060–7069. (b) Brugarolas, J., Moberg, K., Boyd, S. D., Taya, Y., Jacks, T., Lees, J.A. *Proc. Natl. Acad. Sci. USA* **1999**, *96*, 1002–1007.

21. (a) Joyce, R.P., Gainir, J.A., Weinreb, S.M. *J. Org. Chem.* **1987**, *52*, 1177, (b) Kleinschroth, J., Hartenstein, J., Rudolph, C., Schaechtele, C. *Bioorg. Chem. Med. Lett.* **1995**, *5*, 55. (c) Ohkubo, M., Nishimura, T., Jona, H., Nonma, T., Morishima, H. *Tetrahedron,* **1996**, *52*, 8099.

22. Harris, W., Hill, C.H., Keech, E., Malsher, P. *Tetrahedron Lett.* **1993**, *34*, 8361.

23. Faul, M.M., Sullivan, K.A. *Tetrahedron Lett.* **2001**, *42*, 3271.
24. Gallant, M., Link, J.T., Danishefsky, S.J., *J. Org. Chem.* **1993**, *58*, 343.
25. Sanchez-Martinez, C., Faul, M.M., Shih, C., Sullivan, K.A., Grutsch, J.L., Cooper, J.T., Kolis, S.P. *J. Org. Chem.* **2003**, *68*, 8008.
26. (a) Faul, M.M., Winneroski, L.L., Kumrich, C. *J. Org. Chem.,* **1998**, *63*, 6053, (b) Faul, M.M., Winneroski, L.L., Krumrich, C.A. *Tetrahedron Lett.,* **1999**, *40*, 110.
27. (a) Kolis, S.P., Clayton, M.T., Grutsch, J.L., Faul, M.M. *Tetrahedron Lett.* **2003**, *44*, 5707, (b) Faul, M.M., Grutsch, J.L., Kobierski, M.E., Kopach, M.E., Krumrich, C.A., Staszak, M.A., Udodong, U., Vicenzi, J.T., Sullivan, K.A. *Tetrahedron* **2003**, *59*, 7215.
28. Chisholm, J.D., Van Vranken, D.L. *J. Org. Chem.* **2000**, *65*, 7541–7553.
29. Faul, M.M., Sullivan, K.A., Grutsh, J.L., Winneroski, L.L., Shih, C., Sanchez-Martinez, C., Cooper, J.T. *Tetrahedron Lett.* **2004**, *45*, 1095–1098.

15 Pyrazoles as Efficient Adenine-Mimetic Heterocycles for the Discovery of CDK Inhibitors

Paolo Pevarello and Anna Vulpetti

CONTENTS

15.1 INTRODUCTION

Pyrazoles are very common heterocyclic nuclei in medicinal chemistry. Examples can be easily retrieved for several therapeutic areas in drug discovery, and some marketed compounds contain an embedded 1H-pyrazole ring (MDL, Comprehensive Medicinal Chemistry Database).

In recent years there has been an intensive application of isolated and fused 1H-pyrazole heterocycles in the field of kinase inhibition. Cyclin-dependent kinases (CDKs) constitute one of the most popular families of kinase targets in drug discovery. In particular, CDK2 has been chosen by many companies as a suitable target for blocking excessive proliferative activity, a hallmark of all cancers.

With regard to ATP-competitive CDK2 inhibitors, two interactions are consistently observed in the ATP-binding pocket: (1) at least one hydrogen bond with the hinge region of CDK2 and (2) lipophilic (mostly aromatic) interactions with two hydrophobic regions. The two regions are the adenine region occupied by the adenine moiety and the buried region in the back of the ATP pocket, in front of Phe80 (Figure 15.1) (Vulpetti and Bosotti, 2004). The pyrazole nucleus is very much suited to providing,

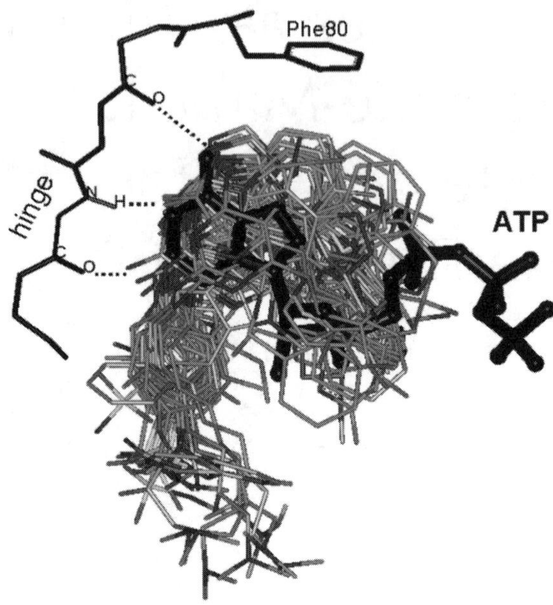

FIGURE 15.1 Overlay of different CDK2 inhibitors (gray carbon atoms) on the structure of CDK2/cyclin A complexed with ATP (1QMZ PDB code, black). With the exception of the gatekeeper Phe80 residue, only the backbone of the residues forming the hinge region of CDK2 is shown.

through the adjacent nitrogens, the donor–acceptor interactions required to harbor the hinge region of kinases (Figure 15.2). In addition, the chemistry of the pyrazole ring is usually straightforward and allows rapid parallel expansion using a variety of substituents to reach other regions of the ATP-binding pocket of kinases.

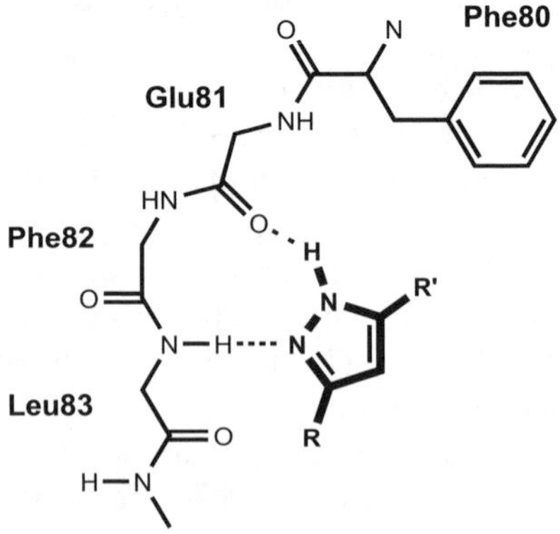

FIGURE 15.2 Schematic representation of the pyrazole scaffold interacting with the hinge region of CDK2.

In this chapter we will discuss the discovery and some preclinical development of pyrazole-based CDK2 inhibitors. In particular, we will show how, starting from simple and low molecular weight 3-aminopyrazole hits, we rapidly achieved orally bioavailable CDK2 inhibitors for preclinical development. This early class served also as a foundation for the search of related, yet different, pyrazole-based scaffolds able to probe previously unexplored regions of the CDK2 ATP pocket. Further examples of pyrazole compounds in the recent literature of CDK inhibition will also be highlighted.

15.2 3-AMINOPYRAZOLES

The 3-aminopyrazole class emerged from an effort aimed at finding potent CDK inhibitors (Pevarello et al. 2004). High-throughput screening (HTS) was performed against CDK2/cyclin E, and the hits obtained were ranked according to several criteria, including the following: the accessibility of the basic scaffold of the hit compounds (e.g., the 3-aminopyrazole nucleus), the existence of several hits with a common substructure and, potentially, some preliminary understanding of a possible SAR, the identification of low molecular weight basic scaffolds that are easily functionalized with both commercially available and proprietary building blocks, the amenability to rapid chemical expansion (e.g., suitability to both solution and solid phase parallel synthesis), and patentability.

The 3-aminopyrazole class satisfied these criteria and was thus selected for further expansion. The crystal structure of CDK2 with **15.1** was determined (Figure 15.3), and a basic pattern of protein–inhibitor interactions soon became apparent. The three nitrogens of the 3-aminopyrazole scaffold were involved in a donor–acceptor–donor hydrogen bond triad, making contact with the hinge region (Glu81-Leu83) of the kinase ATP pocket. The 5-cyclopropyl ring occupied the narrow lipophilic pocket formed by Val18, Ala31, Val64, and Phe80. The plane of the cyclopropyl ring was

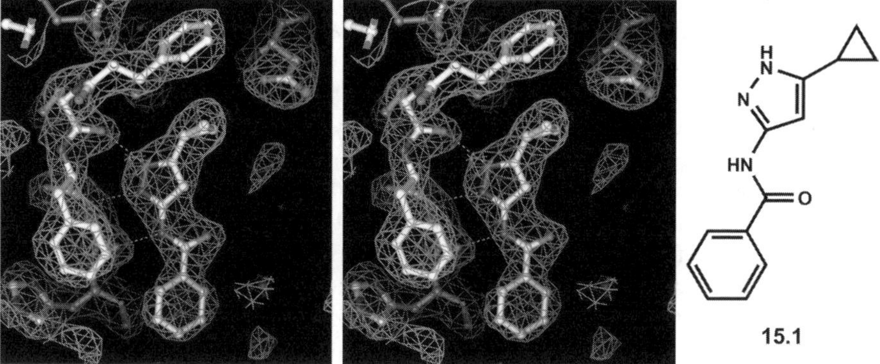

15.1

FIGURE 15.3 (See color insert following page 142.) ATP-binding pocket residues of the x-ray structure of CDK2 in complex with **15.3**. The three hydrogen bonds between the CDK2 hinge region and the inhibitor are drawn as dashed lines.

almost perpendicular to the plane of the pyrazole ring and was packed against the phenyl ring of Phe80. Finally, the benzene ring of **15.1** was directed toward the solvent-accessible region of the enzyme pocket, contributing to the overall binding through lipophilic interactions with the side chains of Ile10 and Phe82 and the main chain of His84 and Asp86.

A classical medicinal chemistry approach to hit expansion consists of fixing one position of the hit molecule (e.g., the aryl pointing toward the solvent-accessible region) while varying the other position (e.g., the 5-substituent tucking into the narrow lipophilic pocket). This approach was applied to the 3-aminopyrazole series, and according to structural evidence the cyclopropyl group (or a small cycloalkyl group) was found to be best suited to fill the buried region pocket (Table 15.1).

A strong synthetic methodology suitable for rapid parallel expansion is always desirable. Synthetic schemes should ideally entail simple and inexpensive reagents and the presence of a key intermediate that can be scaled up easily to be functionalized with the widest variety of commercially available and proprietary building blocks. These criteria were fulfilled by the syntheses devised for the 3-aminopyrazole CDK2 inhibitors. Compound **15.2**, for example, is a key intermediate in these syntheses (Orsini et al., 2005) (Figure 15.4). The 3-aminopyrazole system exists in several tautomeric forms, and its reactivity toward acylating agents is deceptively simple. Double acylation of the 3-aminopyrazole scaffold (Brinkmeyer and Terando, 1989) and subsequent selective hydrolysis of the more labile endocyclic residue is not an optimal choice in terms of atom economy and results in unsatisfactory yields. To get a suitably reactive 3-aminopyrazole substrate (e.g., **15.2**), we masked the exocyclic amino group via an oxidation–reduction sequence, finding that the 3-nitro group (Figure 15.4) can act simultaneously as a protecting and directing group, both for steric and electronic reasons. Expansion at the 3-position could then take advantage of this straightforward synthetic scheme and was accomplished in an exhaustive manner. Some results obtained with different 3-arylacetamido groups are reported in Table 15.2. Some compounds were single-digit nanomolar inhibitors of CDK2/cyclin A. Among them, compound **15.3** was not the most active in the biochemical assay but displayed potent antiproliferative activity against different human tumor cell lines (Table 15.3) and showed an overall druglike profile compatible with an *in vivo* evaluation. The suitability of the aminopyrazole scaffold as an adenine-mimetic is demonstrated by the superposition of **15.3** and *(R)*-roscovitine complexed to CDK2/cyclin A. The two compounds can be placed in identical positions with respect to the protein (Figure 15.5). In particular, the cyclopropyl of **15.3** and the isopropyl of *(R)*-roscovitine overlap perfectly and make similar interactions within the buried region. Furthermore, the naphthyl moiety of **15.1** lies in the same plane as the phenyl moiety of the *(R)*-roscovitine N-6-benzyl group, allowing hydrophobic interactions with the side chains of Ile10 and Phe82 (Figure 15.5b).

Importantly, **15.3** showed remarkable selectivity when tested in a panel of 33 kinases, with CDK2/cyclin E, CDK1/cyclin B, and CDK5/p25 being the only kinases inhibited at a comparable concentration (Table 15.4), and it displayed potent *in vitro* activity in tumor cell proliferation assays. Furthermore, it exhibited significant *in vivo* activity in A2780 xenografted *nu/nu* mice (Pevarello et al., 2004).

TABLE 15.1
SAR of 3-Propylamido- and 3-Benzamidoaminopyrazoles

Entry	R	R_1	CDK2/Cyclin A IC$_{50}$ (μM)
	Methyl	Phenyl	1.50
	Methyl	Propyl	>10
	Cyclopropyl	Propyl	0.22
15.1	Cyclopropyl	Phenyl-	0.29
	Cyclopropyl	$4\text{-Br–C}_6\text{H}_4-$	0.034
	Cyclopropyl	$4\text{-Cl–C}_6\text{H}_4-$	0.085
	Cyclopropyl	$4\text{-OMe–C}_6\text{H}_4-$	0.70
	Cyclopropyl	$4\text{-COOH–C}_6\text{H}_4-$	0.16
	Cyclopropyl	$4\text{-CONH}_2\text{–C}_6\text{H}_4-$	0.068
	Cyclopropyl	$3\text{-Br–C}_6\text{H}_4-$	0.99
	Cyclopropyl	$3\text{-Cl–C}_6\text{H}_4-$	1.00
	Cyclopropyl	$3\text{-OMe–C}_6\text{H}_4-$	0.60
	Cyclopropyl	$3\text{-CF}_3\text{–C}_6\text{H}_4-$	0.50
	Cyclopropyl	$2\text{-Br–C}_6\text{H}_4-$	1.50
	Cyclopropyl	$2\text{-Cl–C}_6\text{H}_4-$	1.00
	Cyclopropyl	$2\text{-OMe–C}_6\text{H}_4-$	>10
	Cyclopropyl	$2,6\text{-diCl–C}_6\text{H}_4-$	>10
	Cyclopropyl	$3,4\text{-diCl–C}_6\text{H}_4-$	0.083
	Cyclobutyl	$4\text{-Br–C}_6\text{H}_4-$	0.090
	Cyclopentyl	$4\text{-Br–C}_6\text{H}_4-$	0.050
	Cyclohexyl	$4\text{-Br–C}_6\text{H}_4-$	1.00
	Methyl	$4\text{-Br–C}_6\text{H}_4-$	6.0
	Ethyl-	$4\text{-Br–C}_6\text{H}_4-$	0.63
	Propyl	$4\text{-Br–C}_6\text{H}_4-$	0.65
	Isopropyl	$4\text{-Br–C}_6\text{H}_4-$	0.20
	Sec-butyl	$4\text{-Br–C}_6\text{H}_4-$	>10
	Tert-butyl	$4\text{-Br–C}_6\text{H}_4-$	>10
	Phenyl	$4\text{-Br–C}_6\text{H}_4-$	>10
	Benzyl	$4\text{-Br–C}_6\text{H}_4-$	>10

Further optimization of the lead compound **15.3** continued (Pevarello et al., 2005). Compound **15.3** displayed low solubility (below 50 μM) in aqueous buffer. The extent of dissolution was sufficiently high to evaluate the compound *in vivo* but precluded its possible use in an intravenous administration setting in the clinic. Furthermore, **15.3** was a strong human serum albumin (HSA) binder in a preliminary assay. A lead optimization program was therefore launched with the aim of finding a compound retaining the good pharmacological properties of **15.3** while improving

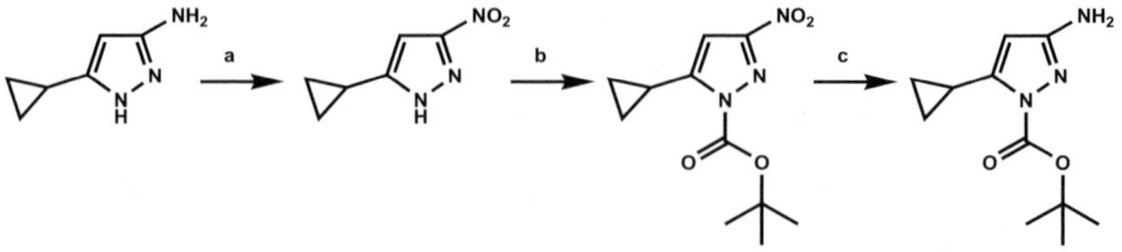

FIGURE 15.4 Synthesis of aminopyrazoles. Reagents and conditions: (a) Oxone®, NaOH, NaHCO$_3$, water, acetone, 0°C, 56%; (b) (Boc)$_2$O, CH$_2$Cl$_2$/aq. NaHCO$_3$, 95%; (c) H$_2$/Pd-C 10% 50 psi, EtOH, 95%.

physicochemical properties such as solubility and plasma protein binding (PPB). The cyclopropyl ring was maintained in the new series of molecules, and variations were made to the part of the molecule pointing toward the solvent accessible region, i.e., the arylacetamido portion. The presence of a small group at the benzylic position (a methyl or a fluorine) was tolerated (Table 15.5), and according to modeling indications the *S*-enantiomer proved to be the active isomer. Adding a methyl at this position also allowed an increase of solubility in the administration vehicles commonly

TABLE 15.2
SAR of 3-Arylacetamido-aminopyrazoles

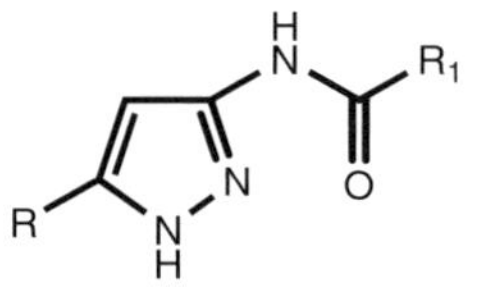

Entry	R	R$_1$	CDK2/Cyclin A IC$_{50}$ (μM)	A2780 Cells IC$_{50}$ (μM) $\pm$ SD
	Cyclopropyl	Phenylacetyl	0.048	16.40 $\pm$ 1.12
	Cyclopropyl	4-CONH$_2$-phenylacetyl	0.074	2.83 $\pm$ 0.23
	Cyclopropyl	3-OMe-phenylacetyl	0.029	>20
	Cyclopropyl	4-(2-Pyrrolidin-1-yl)-ethoxy	0.095	6.98 $\pm$ 0.81
	Cyclopropyl	4-OCF$_3$-phenylacetyl	0.067	4.99 $\pm$ 0.57
	Cyclopropyl	4-Biphenylacetyl	0.056	0.111 $\pm$ 0.039
	Cyclopropyl	4-(3-Fluorobiphenyl)acetyl	0.004	0.182 $\pm$ 0.003
	Cyclopropyl	4-(3-Methylbiphenyl)acetyl	0.009	0.205 $\pm$ 0.007
	Cyclopropyl	4-(4-Carboxybiphenyl)acetyl	0.011	>20
	Cyclopropyl	4-(4-Carboxamidobiphenyl)acetyl	0.004	1.46 $\pm$ 0.16
	Cyclopropyl	4-(2-Thenyl)-phenylacetyl	0.003	0.121 $\pm$ 0.003
15.3	Cyclopropyl	4-(2-Naphthyl)acetyl	0.037	0.29 $\pm$ 0.049
	Cyclopropyl	4-(1-Naphthyl)acetyl	0.038	0.90 $\pm$ 0.11

TABLE 15.3
Antiproliferative Activity of 15.3 on Different Human Tumor Cell Lines

	A2780 Cells IC_{50} (μM) ± SD	HT-29 Cells IC_{50} (μM) ± SD	HCT116 Cells IC_{50} (μM) ± SD	DU145 Cells IC_{50} (μM) ± SD
15.3	0.29 ± 0.049	0.170 ± 0.001	0.073 ± 0.030	0.651 ± 0.089

used for *in vivo* testing, such as aqueous Tween20 or PEG50. This modification was therefore retained throughout the optimization path. Compound **15.3** contained a β-naphthyl moiety, a very lipophilic entity that most likely caused its high PPB. A heterocycle replacing the second phenyl ring of **15.3** would point outside the ATP pocket and maintain the lipophilic contacts with the amino acids lining the pocket, and thus the activity, but would also likely decrease the affinity to plasma proteins and possibly increase solubility. A first set of compounds replaced the β-naphthyl moiety of **15.3** with different 4-lactam-1-yl-phenyl moieties. The *para* position was chosen for the heterocyclic substitution because it allowed the substituent to directly point toward the solvent-accessible region. A simple unsubstituted 5-membered lactam (**15.4**) was as active as the parent lead against CDK2/cyclin A (Table 15.6). Improvements in activity were seen with the benzo-fused analogs of **15.4** and by substituting the 5-membered ring lactam, though the latter compounds turned out to be less active in the cellular assay. Compound **15.4** was also active in cells in the submicromolar range and scored a >tenfold increase in buffer solubility over **15.3**. The complex of **15.4** with CDK2/cyclin A was solved (data not shown), but it was difficult to unambiguously define the orientation of the pyrrolidone ring.

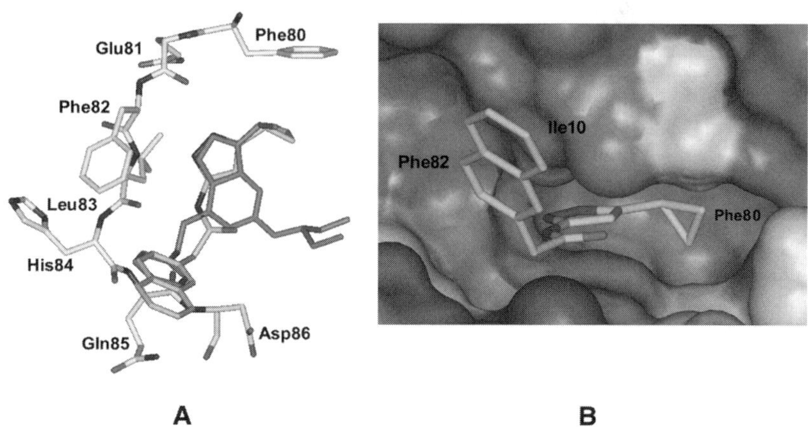

FIGURE 15.5 Crystal structure of **15.3** in complex with CDK2/cyclin A. (a) Overlap of **15.3** (atom-type colored) and roscovitine (magenta) complexed to CDK2/cyclin A. (b) Surface representation of ATP-binding cleft, colored by residue hydrophobicity, in complex with **15.3**. Red indicates most hydrophobic residues and blue indicates the most polar ones.

TABLE 15.4
Selectivity Profile of 15.3 in a Panel of 33 Kinases

Kinase	IC_{50} (μM)	Enzyme Concentration (nM)	Substrate	Buffer	Assay Format
CDK2/cyclin A	0.037	1.1	HISTONE H1	A	SPA
CDK2/cyclin E	0.092	1.1	HISTONE H1	A	SPA
CDK5/p25	0.114	4	HISTONE H1	A	SPA
CDK1/cyclin B	0.270	4	HISTONE H1	A	SPA
CDK4/cyclin D1	>10	60	RB	A	Multiscreen
ABL	>10	1.2	MBP	B	SPA
AKT-1	>10	5	AKTtide	A	Dowex
AUR2	>10	2.5	CHOCKtide 4X	C	SPA
CDC7/DBF4	>10	18	MCM2	C	SPA
CHK-1	>10	12	CHKtide	B	SPA
CK-2	>10	0.7	MCM2	C	SPA
EGFR-1	>10	9	MBP	B	SPA
ERK-2	4.60	3	MBP	A	Dowex
FGFR-1	>10	4	MBP	B	SPA
GSK3β	1.90	20	MBP	B	SPA
IGFR-1	>10	6	IRStide	D	SPA
IKK-2	>10	15	IKBalpha	C	SPA
IKKi	>10	5	IKBalpha	B	Dowex
IR	>10	5	IRStide	C	SPA
KIT	>10	4	KITtide	B	SPA
LCK	>10	1.3	MBP	B	SPA
LYN	>10	2	HISTONEH1	B	SPA
MET	>10	10	MBP	B	Dowex
p38α	>10	5	MBP	B	Dowex
PAK4	>10	2	PAKtide	B	SPA
PDGFR-1	>10	9	MBP	B	SPA
PDK1	>10	15	PDKtide	C	SPA
PKCbeta	>10	2	HISTONE H1	B	SPA
PLK1	>10	50	αCASEIN	E	Dowex
RET	>10	5	MBP	B	SPA
STLK2	>10	1	MBP	B	SPA
TRKA	>10	1.5	MBP	B	Dowex
VEGFR-2	>10	4.7	MBP	D	SPA
VEGFR-3	>10	10	MBP	B	SPA
ZAP70	>10	8	GASTRIN	D	SPA

Table 15.7 reports different 5-membered heterocycles. Among them, the imidazolinone **15.4b** was similar in potency and other properties to **15.4**. Contrary to **15.4**, compound **15.4b** was very well defined in the x-ray structure with the protein (Figure 15.6). The carbonyl of the imidazolinone ring interacted with a water molecule that in turn bridged to Lys20. The NH group of the imidazolinone was within hydrogen bond distance of Glu8, although, from the electron density this latter residue was obviously

TABLE 15.5

SAR of α-Substituted 3-Phenylacetamido-5-Cyclopropyl-1H-Pyrazoles

R	Configuration	CDK2/Cyclin A IC$_{50}$ (μM)
Methoxy	R,S	5.0
Methoxy	S	2.0
Methoxy	R	>10
Fluoro	R,S	0.099
Amino	S	0.96
Phenyl	R,S	>10
Oxo		2.60
Methyl	R,S	0.17
Methyl	S	0.084
Methyl	R	6.0

somewhat mobile. Bulkier groups such as benzyl were detrimental for activity, solubility, and PPB. Introducing unsaturation into the 5-membered ring of the imidazolinone did not hamper CDK2/cyclin A inhibitory activity but considerably worsened cellular activity. Adding a second carbonyl moiety on the lactamyl ring

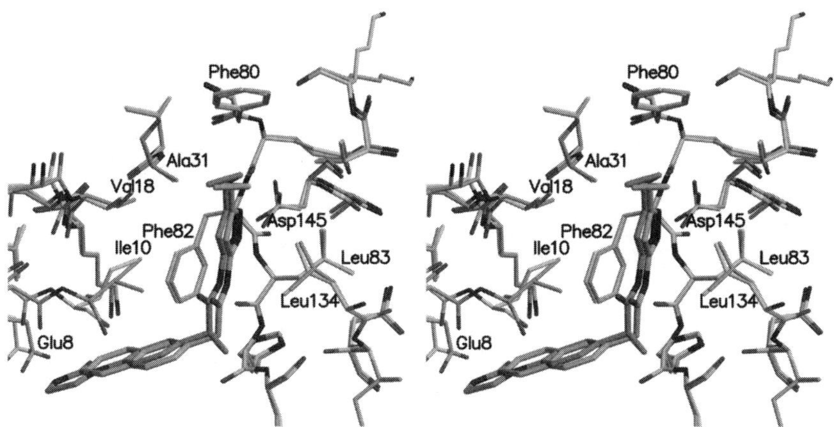

FIGURE 15.6 Superposition of the CDK2/cyclin A complex with compound **15.4b** (yellow/brown carbon atoms) on the complex with **15.3** (green/cyan carbon atoms). It can be seen that the two inhibitors have very similar binding modes. Ile10 appears to adopt a different conformation in the two complexes. The electron density for this residue is better defined in the complex with **15.3** than for that with compound **15.4b**.

TABLE 15.6
SAR of 3-(4-Lactam-1-yl)Phenylacetamido-5-Cyclopropyl-1H-Pyrazoles

Entry	Het	α-Methyl Configuration	CDK2/ Cyclin A IC$_{50}$ (μM)[a]	A2780 IC$_{50}$ (μM)[a]	Solubility (μM; Buffer pH 7)	Plasma Protein Binding (%)
		R,S	0.17	0.63	>250	72
		R	2.50	12.2	>250	70
		S	0.037	0.80	>250	74
15.4		R,S	0.24	>10	240	59
		R,S	0.020	5.71	120	57
		R,S	0.15	6.40	222	67
		R,S	0.20	15.3	50	97
		R,S	0.013	4.75	12	ND

TABLE 15.6 (CONTINUED)
SAR of 3-(4-Lactam-1-yl)Phenylacetamido-5-Cyclopropyl-1H-Pyrazoles

Entry	Het	α-Methyl Configuration	CDK2/ Cyclin A IC$_{50}$ (μM)[a]	A2780 IC$_{50}$ (μM)[a]	Solubility (μM; Buffer pH 7)	Plasma Protein Binding (%)
		R	0.17	ND	51	ND
		S	0.002	1.02	5	ND
15.3			0.037	0.29	22	99

[a] At least two independent experiments were performed for each compound in order to determine IC$_{50}$ values. Potency is expressed as the mean of IC$_{50}$ values obtained by nonlinear least-squares regression fitting of the data. The coefficient of the variation of the mean ranges from 10% to 24%.

of **15.4** gave a moderately active compound devoid of cellular activity. This latter finding, although surprising, is rather common in the practice of the medicinal chemistry in oncology; even minor modifications can impact cellular activity positively or negatively, as discussed in Pevarello et al., 2004. Other heterocyclic moieties were tried as a replacement of the naphthyl group of **15.3**, but none of them worked better than the imidazolinone group of **15.4** (Pevarello et al., 2005). Attempts were made to fine-tune the 5-membered heterocycle, but substituted and unsubstituted imidazolinones, imidazolindiones, imidazolones, triazolones, pyrrolidinones, and sultams (Table 15.7 and Table 15.8), although producing in some cases potent CDK2 inhibitors in the bioassay, yielded only the oxazolidinone **15.5** with an activity/preliminary druglike profile comparable to that of the imidazolinone **15.4**. In general, the SAR of this portion of the molecule shows that although a variety of substituents pointing toward the solvent-accessible region are allowed, the prediction of the potency of the compounds remains difficult because the region is characterized by a high degree of conformational motion for both the protein and the inhibitor. Two compounds, **15.4** and **15.5**, were advanced into further cellular and selectivity studies and an *in vivo* preliminary assessment to assess them in an *in vivo* pharmacological tumor model. Results showing a G1 block in synchronized HT-29 cells with a concomitant block of BrdU incorporation in treated cells, coupled with a phosphorylation block of the

TABLE 15.7
SAR of 3-(4-Heterocycl-1-yl)Phenylacetamido-5-Cyclopropyl-1H-Pyrazoles

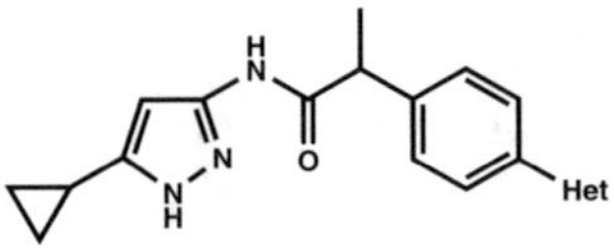

Entry	Het	α-Methyl Configuration	CDK2/Cyclin A IC$_{50}$ (μM)[a]	A2780 IC$_{50}$ (μM)[a]	Caco-2 Permeability	Solubility (μM; Buffer pH 7)	Plasma Protein Binding (%)
	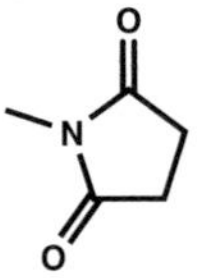	R,S	0.077	>10	Moderate	220	48
	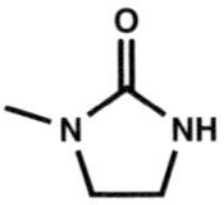	R,S	0.012	2.25	Moderate	224	74
	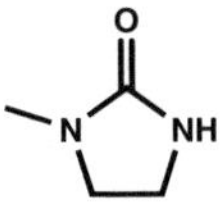	R	0.45	13.2	Moderate	220	74
15.4b	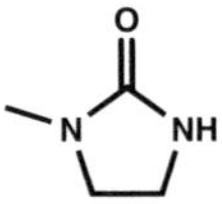	S	0.002	1.27	Moderate	>225	74
	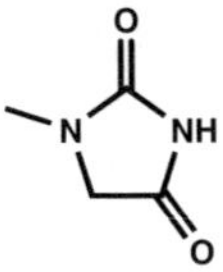	R,S	0.017	4.54	Low	201	67
	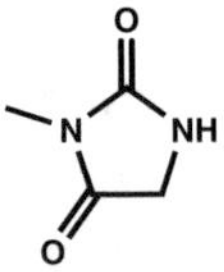	R,S	0.12	>10	Low	218	50
	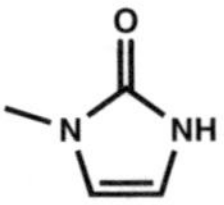	R,S	0.011	6.98	Low	233	74
	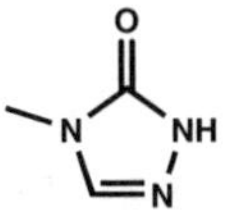	R,S	0.073	8.39	Low	109	58

TABLE 15.7 (CONTINUED)
SAR of 3-(4-Heterocycl-1-yl)Phenylacetamido-5-Cyclopropyl-1H-Pyrazoles

Entry	Het	α-Methyl Configuration	CDK2/ Cyclin A IC_{50} (μM)[a]	A2780 IC_{50} (μM)[a]	Caco-2 Permeability	Solubility (μM; Buffer pH 7)	Plasma Protein Binding (%)
	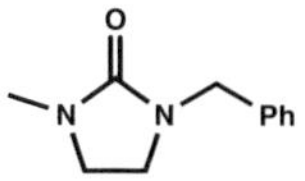 (imidazolidinone–CH$_2$Ph)	R,S	1.45	ND	ND	2	100
	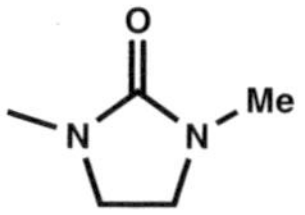 (imidazolidinone–Me)	R,S	0.029	1.68	Moderate	221	83
	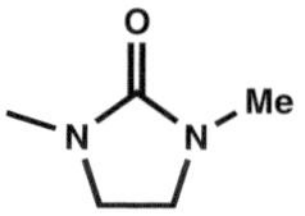 (imidazolidinone–Me)	R	8.50	ND	Moderate	218	83
	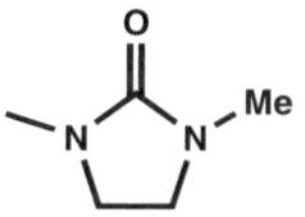 (imidazolidinone–Me)	S	0.017	1.40	Moderate	210	83
15.3	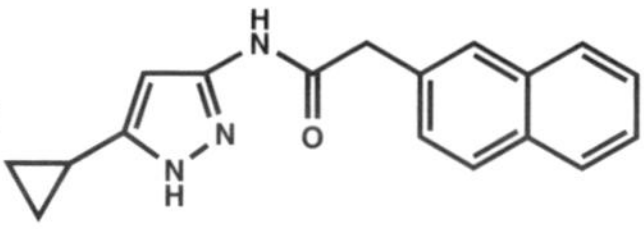 (cyclopropyl-pyrazole naphthylacetamide)		0.037	0.29	High	22	99

[a] At least two independent experiments were performed for each compound in order to determine IC_{50} values. Potency is expressed as the mean of IC_{50} values obtained by nonlinear least-squares regression fitting of the data. The coefficient of the variation of the mean ranges from 10% to 24%.

substrate pRb are shown in Figure 15.7 and Table 15.9. This data collectively show that both compounds are able to block the cell cycle of tumor cells in the G1/S phase and that they do so by blocking phosphorylation of one of the major substrates of CDK2. Table 15.10 reports kinase selectivity data for **15.4** and **15.5** together with the IC_{50} against three tumor cell lines. The compounds are similar in profile in biochemical and cellular potency and in selectivity toward other kinases.

To prioritize a compound for *in vivo* efficacy studies, it is very important to gain knowledge about its PK/ADME properties both *in vitro* and *in vivo*. As shown in Table 15.11, such an assessment was made for both **15.4** and **15.5** and the overall profile turned out to be more favorable for **15.4**. Solubility was deemed acceptable for an oral compound; Caco-2 cell penetration was, respectively, high and moderate; the compounds were not subjected to CYP4503A4 metabolism in a preliminary assay; and PPB was moderate. A low to moderate clearance and a low volume of distribution were common features of both compounds. Oral bioavailability was

TABLE 15.8
SAR of 3-(4-Heterocycl-1-yl)Phenylacetamido-5-Cyclopropyl-1H-Pyrazoles

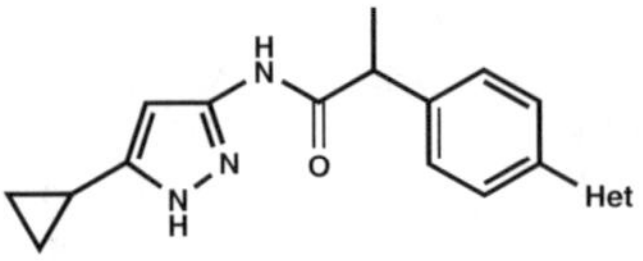

Entry	Het	α-Methyl Configuration	CDK2/ cyclin A IC$_{50}$ (μM)[a]	A2780 IC$_{50}$ (μM)[a]	Caco-2 Permeability	Solubility (μM; Buffer pH 7)	Plasma Protein Binding (%)
	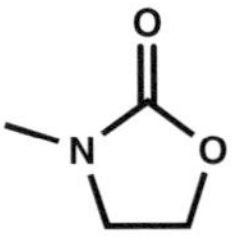 (oxazolidinone)	R,S	0.025	1.04	Moderate	222	73
	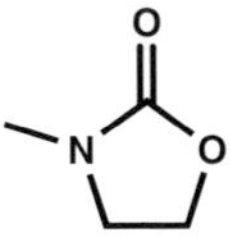 (oxazolidinone)	R	2.5	ND	Moderate	220	72
15.5	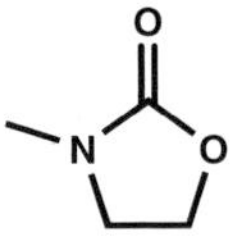 (oxazolidinone)	S	0.020	0.79	Moderate	232	74
	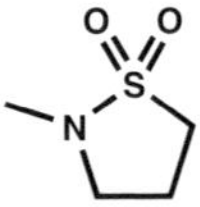 (isothiazolidine 1,1-dioxide)	R,S	0.013	5.24	Moderate	135	78
	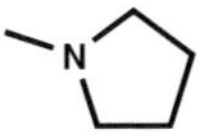 (pyrrolidine)	R,S	0.20	0.37	Low	3	96
	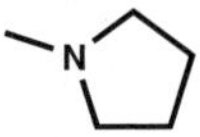 (pyrrolidine)	S	0.080	0.22	Low	24	98
15.3	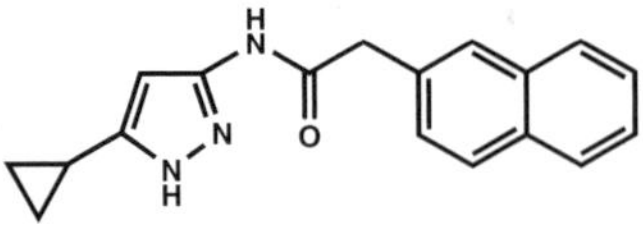 (naphthyl)		0.037	0.29	High	22	99

[a] At least two independent experiments were performed for each compound in order to determine IC$_{50}$ values. Potency is expressed as the mean of IC$_{50}$ values obtained by nonlinear least-squares regression fitting of the data. The coefficient of the variation of the mean ranges from 10% to 24%.

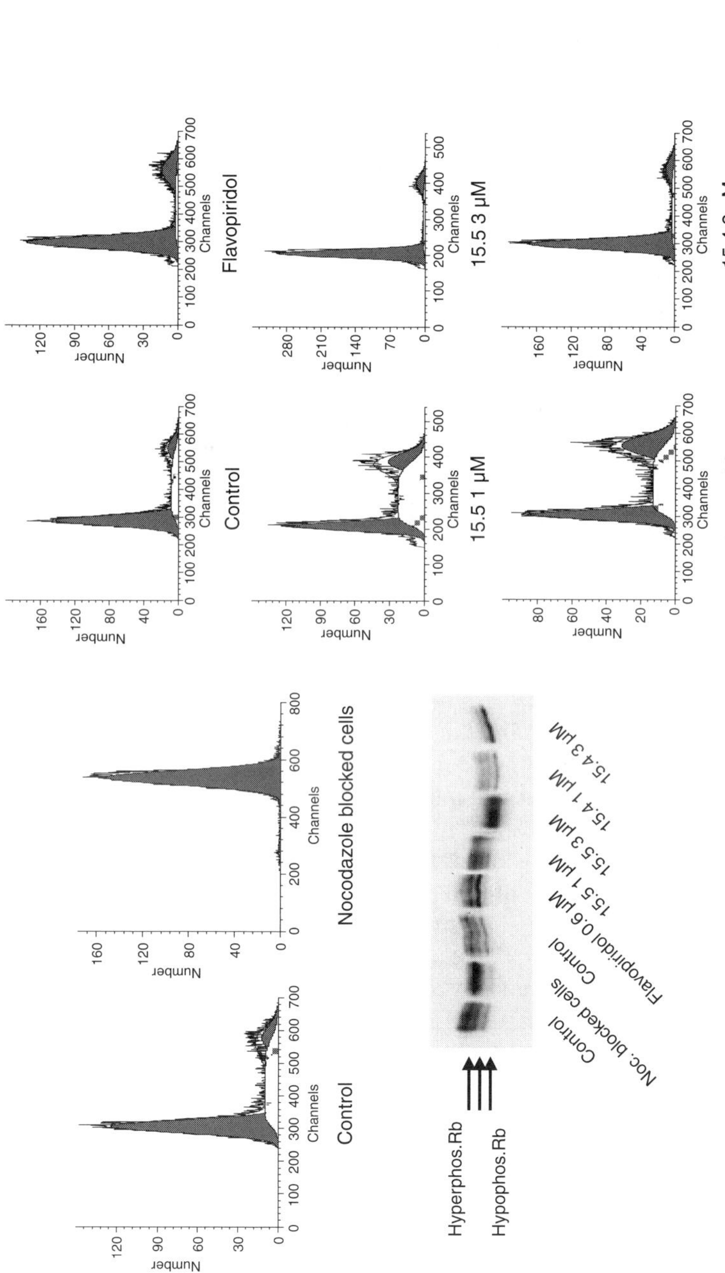

FIGURE 15.7 HT-29 cells were synchronized in G$_2$/M by a nocodazole treatment of 12 h at the concentration of 75 ng/ml. G$_2$/M-blocked HT-29 cells were recovered by shake-off and reseeded in the presence or absence of compounds at the concentration of 1 and 3 μM. Flavopiridol at the concentration of 0.6 μM was chosen as positive reference. HT-29 cells were collected 24 h after the release from the nocodazole block and analyzed by FACS for the cell cycle analysis. Cell lysates obtained from the same cells were used to follow the status of pRb phosphorylation. The data reported in the picture indicate the ability of both **15.4** and **15.5** (as well as flavopiridol) to induce an accumulation of the hyposphosphorylated forms of pRb as compared to the control and nocodazole blocked cells.

TABLE 15.9

Percentage of BrdU Incorporation in Treated and Untreated HT-29 Cells

	Control	Flavopiridol (1 μM)	15.5 (1 μM)	15.5 (3 μM)	15.4 (1 μM)	15.4 (3 μM)
Brdu positive cells (%)	40	1	65	0.7	25	4

TABLE 15.10

Kinase Selectivity and Tumor Cell Antiproliferative Activity of Compounds 15.4 and 15.5

	CDK2/A IC_{50} (μM)	CDK2/E IC_{50} (μM)	CDK1/B IC_{50} (μM)	CDK5/p25 IC_{50} (μM)	CDK4/D1 IC_{50} (μM)	Other Kinases[a] IC_{50} (μM)	A2780 IC_{50} (μM)	HT-29 IC_{50} (μM)	HCT116 IC_{50} (μM)
15.4	0.037 (n = 90)	0.055 (n = 6)	0.21 (n = 6)	0.065 (n = 5)	>10 (n = 4)	GSK-3β: 0.73 (n = 13)	0.74 ± 0.22	0.64 ± 0.18	1.35 ± 0.33
15.5	0.020 (n = 11)	0.037 (n = 4)	0.24 (n = 4)	0.057 (n = 4)	>10 (n = 2)	GSK-3β: 0.41 (n = 6)	0.72 ± 0.13	0.75 ± 0.23	1.13 ± 0.67

[a] From a kinase panel of 30 serine-threonine and tyrosine kinases as previously described. (Pevarello, P. et al. (2004) *Journal of Medicinal Chemistry* **47(13)**, 3367–3380.)

definitely higher for **15.4** vs. **15.5**, and thus the former was selected for an *in vivo* efficacy trial in the A2780 xenograft nu/nu mice. As shown in Figure 15.8, compound **15.4** displayed a consistent tumor growth inhibition (TGI) in the A2780 mice (>70%) after 20 consecutive days of administration at a dose of 7.5 mg/kg twice a day. This result, compared to the 50% TGI obtained with the *in vitro* more potent **15.3**, shows the power of balancing activity vs. druglikeness as opposed to trusting only in the pharmacological potency of a suboptimal druglike compound such as **15.3**. Compound **15.4** also allowed the establishment of a relationship between the anti-tumor activity observed *in vivo* in the xenograft model and markers of CDK2 inhibition in tumor

TABLE 15.11

***In Vitro* and *In Vivo* ADME Parameters for Compounds 15.4 and 15.5**

	In Vitro ADME Parameters				*In Vivo* ADME Parameters (CD-1 Mice)			
	Solubility (μM)	Caco-2 Cell Permeability	CYP4503A4 (Percentage of Remaining)	PPB (%)	Clearance (ml/min/kg)	Half-Life $t_{1/2}$ (min)	Volume of Distribution V_{ss} (ml/kg)	Oral Bioavailability; F_{os} (%)
15.4	239	High	96	74	5.8	0.8 (iv) 3.7 (os)	280	100
15.5	225	Moderate	92	73	4.7	2.8 (iv) 3.1 (os)	573	63

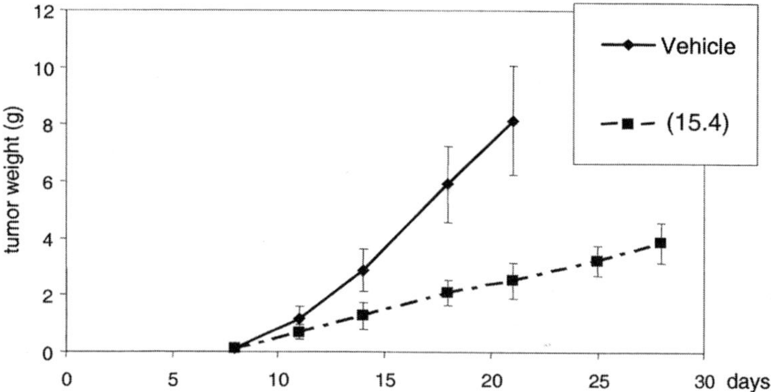

FIGURE 15.8 Compound **15.4** shows 73% tumor growth inhibition (TGI) against a human ovarian cancer model (A2780) transplanted into nude mice, when administered at 7.5 mg/kg twice a day for 20 consecutive days.

sections taken from treated animals. BrdU incorporation and phosphorylation of pRb on Ser795 (this site being phospho-specific for CDK2 and CDK4) were chosen as markers of CDK2 inhibition. Sections of tumors were thus processed as described in Figure 15.9. The IHC analysis of BrdU staining showed a significant reduction in the percentage of positive cells in the tumor sections obtained from the treated

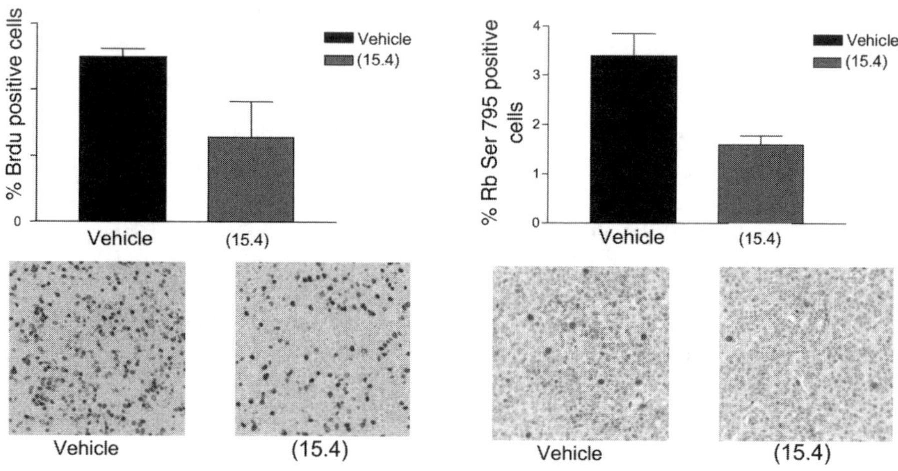

FIGURE 15.9 (See color insert.) Analysis by IHC of the BrdU incorporation (A) and pRb staining (B) of the sections of A2780 xenograft tumors at the end of treatment. Tumor sections were fixed in formalin, paraffin embedded, and stained for BrdU incorporation and phospho-pRb expression. The analysis shows a significant reduction of both BrdU and phospo-pRb positive cells in the treated (**15.4**) vs. control tumors. Representative examples of the stainings are also reported.

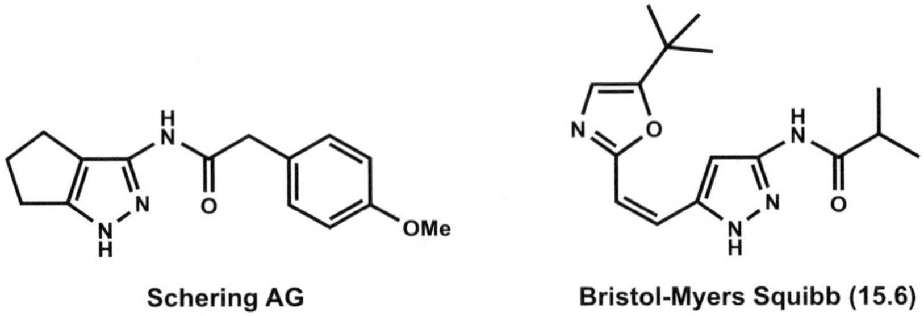

Schering AG   **Bristol-Myers Squibb (15.6)**

FIGURE 15.10 Other 3-aminopyrazole CDK inhibitors.

group compared to the section obtained from the vehicle group, thus demonstrating a clear effect on DNA replication. A comparative analysis of the phospho-Rb positive cells between the tumor sections obtained from the vehicle and the treated group indicated a significant reduction of 50% (B), confirming that the tumor growth inhibition obtained was related to the inhibition of the target CDK2. The 3-aminopyrazole class clearly shows that it is suitable for identifying potent *in vivo* inhibitors of the target enzyme and that the class expansion and refinement are greatly facilitated by the chemical versatility of the scaffold.

Schering AG scientists reported compounds with a different substitution pattern at positions 4- and 5- on the 3-aminopyrazole nucleus (Krueger et al., 2003). In this case, a tetrahydrocyclopentane ring was condensed onto the 3-aminopyrazole ring (Figure 15.10), thus providing a platform for a potentially longer-range exploration of the pocket as compared to simple 4H, 5-substituted-3-amino-1H-pyrazoles. Also, for this scaffold, protocols for rapid parallel synthesis were devised and published (Seelen et al., 2003).

15.3 BENZODIPYRAZOLES

In the isolated 3-aminopyrazoles discussed earlier, the ATP pocket is only partially occupied. Prolonging the chain that exits at position 5 is a strategy for placing suitable groups in the phosphate or sugar region of the pocket. This approach has been reported in a patent (Salvati and Kimball, 2001) (Figure 15.10). Compounds such as **15.6** project, by all evidence, an oxazole ring into the phosphate-binding region similar to the aminothiazoles published from the same company (Kim et al., 2002).

Another way of establishing a different pattern of interactions within the ATP-binding site while retaining the adenine-mimetic anchoring of a pyrazole entails a condensation of the 3-aminopyrazole ring onto another ring. If the latter ring can be easily functionalized, then substituents can be directed toward the sugar- and phosphate-binding regions. The 1,3,4,6-tetrahydropyrrole[3,4-c]pyrazole scaffold recently disclosed by Fancelli (Fancelli et al., 2005) exemplifies this approach well (Figure 15.11).

This concept can be further extended to a tricyclic benzodipyrazole, as recently published (D'Alessio et al., 2005). A prototype of this chemical class is shown in complex with CDK2/cyclin A (Figure 15.12). In this scaffold the two nitrogens of the pyrazole ring A bind to the hinge region of CDK2. Groups R1 and R2 are allowed to point toward, respectively, the sugar region/solvent-accessible areas and the phosphate-binding region.

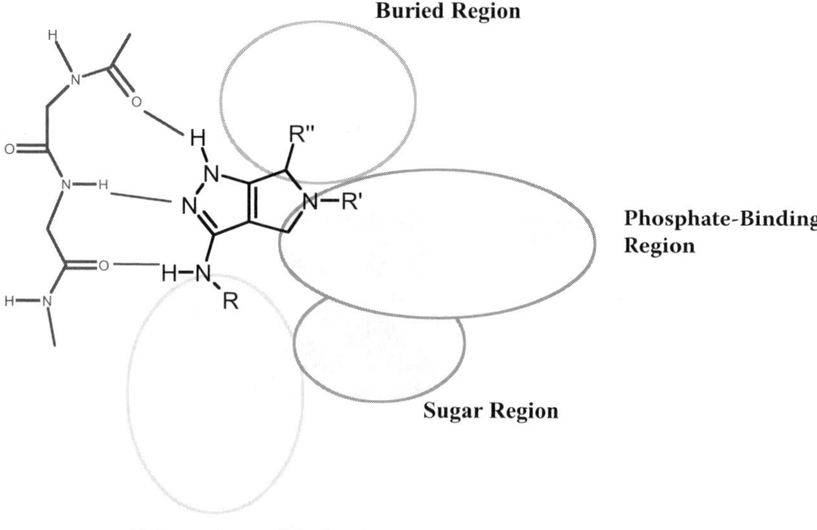

FIGURE 15.11 Schematic model of a bicyclic pyrazole placed into the ATP pocket of a generic kinase.

The R3 bridge, in turn, can fit into the buried region. This compound was very potent in the biochemical assay, whereas activity in cells was very weak. It was speculated that the presence of both the sulfamoyl and carbamoyl moieties may prevent this compound from penetrating cells efficiently. In particular, low lipophilicity (clogP: 0.49) and the presence of several NH bonds were deemed to contribute to its poor cell permeability. To improve these features, some modifications were designed, trying to retain those functionalities directly involved in the binding to the protein as much as possible.

As for the R_1 group pointing toward the sugar region of the ATP pocket, the replacement of the primary sulfonamido group with other moieties (Table 15.12) led

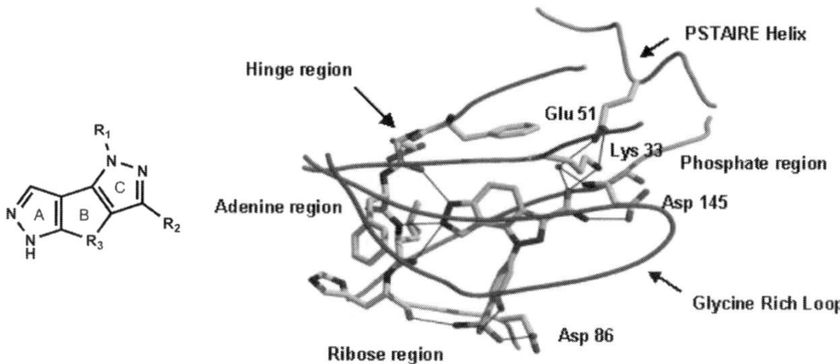

FIGURE 15.12 (See color insert.) General formula of tricyclic benzodipyrazoles (BDP) and binding mode of a derivative into the ATP-binding pocket.

TABLE 15.12
Variations at R1[a]

| | | CDK2/Cyclin A IC$_{50}$ (μM) | | A2780 Cells IC$_{50}$ (μM) | |
| | R2 = –CONH$_2$ | | R3 | | |
Compound	R1	–CH$_2$–CH$_2$–	–CH=CH–	–CH$_2$–CH$_2$–	–CH=CH–
	–Ph-4–SO$_2$NH$_2$	0.004	0.0003	>20	8.74
	–Ph-4–SO$_2$NHMe	0.04	0.002	>20	>20
	–Ph-4–SO$_2$NMe$_2$	0.54	0.010	13.92	1.89
	–Ph-4–SO$_2$NHBu	0.21	0.002	9.76	1.25
	–Ph-4–SO$_2$N(CH$_2$–CH$_2$)$_2$N–Me	>10	0.150		>20
	–Ph-4–SO$_2$Me	0.03	0.001	>20	8.81
	–H	0.46	0.078	18.19	3.54
	–Ph	0.15	0.014	5.27	3.86
	–Ph-4–Me	0.14	0.015	8.05	0.88
	–Ph-4–OMe	0.14	0.009	6.05	1.46
	–Ph-4–Cl	0.16	0.008	>20	2.73
	–Ph-4–F	0.27	0.036	8.50	1.65
	–Ph-4–CF$_3$	0.08	0.013	>20	1.77
	–Ph-4–OCF$_3$	0.51	0.037	>20	2.53
	–Ph-4–CN	0.19	0.005	>20	1.94
	–Ph-4–N(CH$_2$–CH$_2$)$_2$O	0.56	0.018	8.75	4.07
	–Ph-4-(2-imidazolo)	0.25	0.075	>20	>20
	–Ph-3-Me	0.16	0.009	1.88	0.40
	–Ph-3–Cl	0.36	0.029	4.50	2.10
	–Ph-3–F	0.42	0.028	4.10	1.50
	–2-pyridyl	0.26	0.008	4.06	0.65
	–3-pyridyl	1.60	0.062		4.48
	–Bn	2.20	0.140		2.98
	–Me	0.29	0.018	13.45	0.58
	–CH$_2$–CF$_3$	0.02	0.002	2.67	0.26
	–CH$_2$–CH$_2$–OH	1.15	0.028		0.92

Note: Only compounds with IC$_{50}$(CDK2/cyclinA) < 1 μM have been tested in A2780 cell line.

[a] Refer to the general formula in Figure 15.12.

to an increase of the antiproliferative activity. The slight loss in biochemical activity was thus compensated by increased cell penetration, providing compounds with sub-micromolar activity in an A2780 antiproliferation assay. Some variations were also performed at R$_2$ in an attempt to conveniently replace the carboxamido group (Table 15.13). Primary carboxamides proved to be the most active derivatives with a reduction of the activity upon N-alkylation. Esters were considerably less active on CDK2 than the corresponding amides, whereas acids and a methylketone retained some activity, according to the binding mode of the parent compound. Among other functionalities explored, the hydroxamic acid was the most promising one, showing activity

TABLE 15.13
Variations at R2[a]

| | | CDK2/Cyclin A IC$_{50}$ (μM) | | A2780 Cells IC$_{50}$ (μM) | |
| | | R3 | | | |
R1	R2	$-CH_2-CH_2-$	$-CH=CH-$	$-CH_2-CH_2-$	$-CH=CH-$
–Ph-4–OMe	–COOEt	8.1	7.5		
	–CONH$_2$	0.14	0.009	6.05	1.46
	–COOH	2.1	0.21		>20
	–CONHNH$_2$	2.1	1.3		
	–CONHOH	0.69	0.029	7.5	2.82
	–CONHOCH$_2$CH=CH$_2$	4.5			
	–CN		1.75		
	–C=N(OH)NH$_2$		0.31		4.6
	–NH$_2$		>10		
–Ph-4–SO$_2$NHBu	–COOEt	10			
	–CONH$_2$	0.21	0.002	9.76	1.25
	–COOH	1.2			
	–CONHMe	0.6	0.16	10.2	2.9
	–CONHOCH$_2$CH=CH$_2$	2.85	0.2		8.7
–Ph-4–Me	–COOEt	8.5			
	–CONH$_2$	0.14	0.015	8.05	0.88
	–COMe	0.9		>20	

Note: Only compounds with IC$_{50}$(CDK2/cyclin A) < 1 μM have been tested in A2780 cell line.

[a] Refer to the general formula in Figure 15.12.

comparable to the parent amide and conferring higher solubility on the scaffold. A considerable improvement in activity and selectivity was achieved through modifications of the central ring B (i.e., R$_3$, Figure 15.12). Dihydrobenzodipyrazoles (DBP) were always more active than their tetrahydrobenzodipyrazole (TBP) counterparts, both in terms of potency on CDK2/cyclin A and antiproliferative activity on the A2780 tumor cell lines. According to the information obtained in the isolated 3-aminopyrazole series (in which a small-to-medium-size alkyl group such as cyclopropyl or isopropyl gives best results; see Table 15.1), we added a 4,4-gemdimethyl moiety to the central ring of benzodipyrazoles and found a compound that was similar in activity compared to the unsaturated or saturated/unsubstituted ring B, but had a considerable advantage in terms of kinase selectivity. GSK3β is a well-known cross-reacting kinase in many

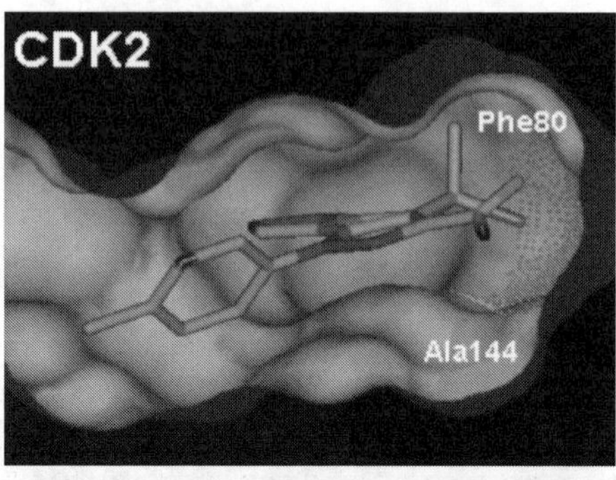

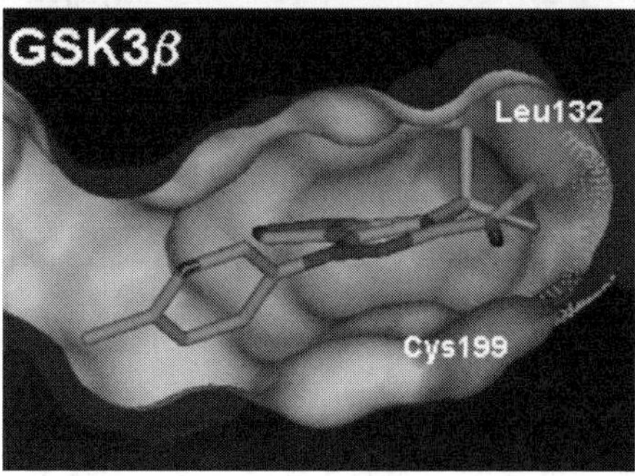

FIGURE 15.13 Compound **15.7** docked in CDK2 and GSK3β crystal structure.

CDK inhibitor programs (Meijer et al., 2004). Although there is no proven relationship between GSK3β inhibition and any undesirable side effect, it is important to isolate the two kinase activities. Figure 15.13 shows a 4,4-gem-dimethyl derivative docked in CDK2 and GSK3β crystal structures solved in-house (Bertrand et al., 2003). The larger size of GSK3β/Cys199 vs. CDK2/Ala144 and the non-planar nature of GSK3β/Leu132 vs. CDK2/Phe80 would be responsible for the selectivity due to gem-dimethyl clashing with the GSK3 residues (Vulpetti et al., 2005). As shown in Table 15.14, compound

TABLE 15.14
4,4-Gem-Dimethyl Series[a]

Compound	R1	R2	R3	CDK2/ Cyclin A IC$_{50}$ (μM)	GSK3β IC$_{50}$ (μM)	A2780 Cells IC$_{50}$ (μM)
	–Ph-4–Me	–CONH$_2$	–CH$_2$–CH$_2$–	0.14	0.46	8.05
			–CH=CH–	0.015	0.023	0.88
15.7			–CH$_2$–C(CH$_3$)$_2$–	0.087	>10	1.65

[a] Refer to the general formula in Figure 15.12.

15.7 was able to inhibit the CDK2/cyclin A complex at concentrations comparable to those of the corresponding dihydrobenzodipyrazole without affecting GSK3β kinase activity.

15.4 OTHER PYRAZOLE-BASED CDK INHIBITORS

In parallel to our efforts on pyrazole-based scaffolds, other groups have worked on and published CDK2 inhibitors that contain an embedded pyrazole acting as an adenine-mimetic ring. Structural analysis of 3-aryl-1H pyrazole in complex with CDK2 (Figure 15.14) performed at Novartis revealed that both of the tautomeric forms of the pyrazole can bind to the hinge region of CDK2 (Furet et al., 2002).

In binding mode (I) the pyrazole ring makes two hydrogen bond interactions with the carbonyl oxygen of Glu81 and the amide nitrogen of Leu83. In binding mode (II) both the pyrazole nitrogens, N1 and N2-H, interact respectively with the carbonyl oxygen and the amide nitrogen of Leu83.

Binding mode (I) resembles the binding mode of the 3-amino pyrazoles and benzodipyrazoles described in the previous paragraphs, whereas binding mode (II) is similar to that of the indenopyrazole class developed at DuPont (Nugiel et al., 2001).

At variance with the binding mode displayed by the pyrazole moieties discussed previously, a pyrazole containing scaffold showed a different placement into the ATP pocket. Bristol-Myers Squibb's 1H-pyrazolo[3,4-b]pyridine inhibitors of CDK1/cyclin B and CDK2/cyclin E (Misra et al., 2003) bound into the ATP pocket of CDK2 alone showed involvement of both the pyrazolo and the pyridine nitrogens to form hydrogen bonds with Leu83 (Figure 15.15).

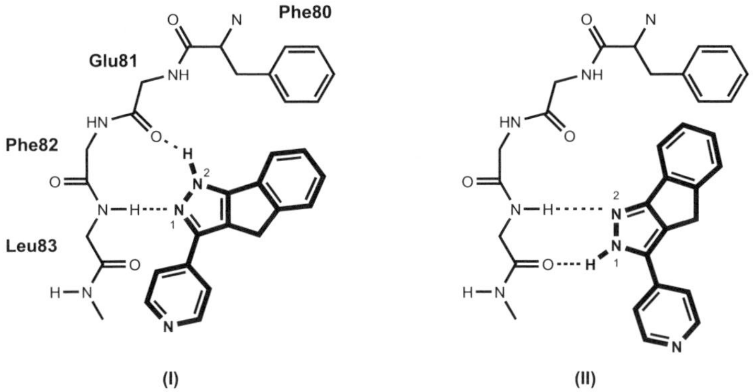

FIGURE 15.14 Schematic representation of the x-ray structure of PKF049-365 in complex with CDK2 (1JVP PDB code, CDK2 IC$_{50}$ = 1.6 μM). With the exception of the gatekeeper Phe80 residue, only the backbone of the residues forming the hinge region of CDK2 is shown.

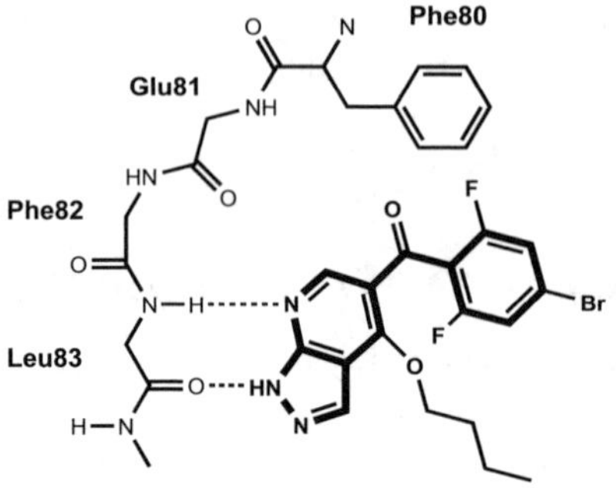

FIGURE 15.15 Schematic representation of the x-ray structure of Bristol-Myers Squibb's 1H-pyrazolo[3,4-b]pyridine inhibitor in complex with CDK2 (CDK2/cyclin E $IC_{50} = 0.020$ μM). With the exception of the gatekeeper Phe80 residue, only the backbone of the residues forming the hinge region of CDK2 is shown.

15.5 CONCLUSION

We and others have shown the potential of using a pyrazole moiety as a mimic of the adenine ring of ATP to the hinge region of its pocket in the CDK2/cyclin A enzyme. By exploiting the lipophilic regions not occupied by ATP, it is possible to both improve affinity to the enzyme and increase selectivity over other enzymes of the kinase superfamily. It was shown that a full occupancy of the ATP pocket is not a necessary feature for very potent CDK2 inhibition. Furthermore, the rapid production of compounds suitable for *in* vivo testing with excellent oral bioavailability confirms the druglikeness of the scaffold. Evolution of the simple, isolated aminopyrazole scaffold can give rise to a variety of condensed heterocycles that allow substituents to be placed in other regions of the ATP pocket (e.g., the sugar- and the phosphate-binding regions), generating different chemotypes for the inhibition of CDK2/cyclin E-A.

REFERENCES

Betrand, J.A., Thieffine, S., Vulpetti, A., Christiani, C., Valsasina, B., Knapp, S., Kalisz, H.M., Flocco, M. (2003). Structural characterization of the GSK3β active site using selective and non-selective ATP-mimetic inhibitors. *Journal of Molecular Biology*, **333**, 393–407.

Brinkmeyer, N.S. and Terando, N.H. (1989). *Journal of Heterocyclic Chemistry* **26**, 1713–1717.

D'Alessio, R., Bargiotti, A., Metz, S., Brasca, M.G., Cameron, A., Ermoli, A., Marsiglio, A., Polucci, P., Roletto, F., Tibolla, M., Vazquez, M.L., Vulpetti, A., Pevarello, P. (2005). Benzodipyrazoles: a new class of potent CDK2 inhibitors. *Bioorganic and Medicinal Chemistry Letters* **15**, 1315–1319.

Fancelli, D., Berta, D., Bindi, S., Cameron, A., Cappella, P., Carpinelli, P., catanna, C., Forte, b., Giordano, P., Giorgini, M.L., Mantegani, S., Marsiglio, A., Meron, M., Moll, J., Pittalà, V., Roletto, F., Severino, D., Soncini, C., Storici, P., Tonani, R., Varasi, M., Vulpetti, A., Vianello, P. (2005). Potent and selective Aurora inhibitors identified by the expansion of a novel scaffold for protein kinase inhibition. *Journal of Medicinal Chemistry* **48(8)**, 3080–3084.

Furet, P., Meyer, T., Strauss, A., Raccuglia, S., and Rondeau, J.M. (2002). Structure-based design and protein x-ray analysis of a protein kinase inhibitor. *Bioorganic Medicinal Chemistry Letters 12*, 221–224.

Kim, K.S., Kimball, S.D., Misra, R.N., Rawlins, D.B., Hunt, J.T., Xiao, H.-Y., Lu, S., Qian, L., Han, W.-C., Shan, W., Mitt, T., Cai, Z.-W., Poss, M.A., Zhu, H., Sack, J.S., Tokarski, J.S., Chang, C.Y., Pavletich, N., Kamath, A., Humphreys, W.G., Marathe, P., Bursuker, I., Kellar, O.K.A., Roongta, U., Batorsky, R., Mulheron, J.G., Bol, D., Fairchild, C.R., Lee, F.Y., Webster, K.R. (2002). Discovery of aminothiazole inhibitors of cyclin-dependent kinase 2: synthesis, x-ray crystallographic analysis, and biological activities. *Journal of Medicinal Chemistry,* **45(18),** 3905–3927.

Krueger, M., Prien, O., Steinmeyer, A., Kroll, J., Ernst, A., Siemeister, G., Haberey, M., Hoffmann, J. (2003). Preparation of Tetrahydrocyclopentapyrazolamides as Cyclin-Dependent Kinase Inhibitors. Ger. Offen. DE10219294 A1, 20031113, CAN 139:381509 AN 2003:891926.

MDL Information System Inc., 14600 Catalina Street, San Leandro, CA 14577, Comprehensive Medicinal Chemistry Database: Zyloprim™, Winstrol™, Antizol™, Histalog™, ThiopurinolTM.

Meijer, L., Flajolet, M., Greengard, P. (2004). Pharmacological inhibitors of glycogen synthase kinase 3. *Trends in Pharmacological Sciences,* **25(9),** 471–480.

Misra, R.N., Xiao, H.-Y., Rawlins, D.B., Shan, W., Kellar, K., Mulheron, J., Sack, J.S., Tokarski, J.S., Kimball, S.D., Webster, K.R. (2003). 1*H*-Pyrazolo[3,4-*b*]pyridine inhibitors of cyclin-dependent kinases: highly potent 2,6-difluorophenacyl analogues. *Bioorganic and Medicinal Chemistry Letters,* **13,** 2405–2408.

Nugiel, D.A., Etzkorn, A.-M., Vidwans, A., Benfield, P.A., Boisclair, M., Burton, C.R., Cox, S., Czerniak, P.M., Doleniak, D., and Seitz, S.P. (2001). Indenopyrazoles as novel cyclin dependent kinase (CDK) inhibitors. *Journal of Medicinal Chemistry* **44,** 1334–1336.

Orsini, P., Traquandi, G., Sansonna, P., Pevarello, P. (2005). 3-Acylaminopyrazole derivatives via a regioselectively N-protected-3-nitropyrazole. *Tetrahedron Letters* **46,** 933–935.

Pevarello, P., Brasca, M.G., Orsini, P., Traquandi, G., Longo, A., Nesi, M., Orzi, F., Piutti, C., Sansonna, P., Varasi, M., Cameron, A., Vulpetti, A., Roletto, F., Alzani, R., Ciomei, M., Pastori, W., Marsiglio, A., Pesenti, E., Fiorentini, F., Bischoff, J.R., Mercurio, C. (2005). 3-Aminopyrazole inhibitors of CDK2/cyclin A as antitumor agents. 2. Lead optimization. *Journal of Medicinal Chemistry* **48(8),** 2944–2956.

Pevarello, P., Brasca, M.G., Amici, R., Orsini, P., Traquandi, G., Corti, L., Piutti, C., Sansonna, P., Villa, M., Pierce, B.S., Pulici, M., Giordano, P., Martina, K., Fritzen, E.L., Nugent, R.A., Casale, E., Cameron, A., Ciomei, M., Roletto, F., Isacchi, A., Fogliatto, G., Pesenti, E., Pastori, W., Marsiglio, A., Leach, K.L., Clare, P.M., Fiorentini, F., Varasi, M., Vulpetti, A., Warpehoski, M.A. (2004). 3-Aminopyrazole inhibitors of CDK2/cyclin A as antitumor agents. 1. Lead finding. *Journal of Medicinal Chemistry* **47(13),** 3367–3380.

Salvati, M.E. and Kimball, S.D. (2001). Preparation of 3-aminopyrazole inhibitors of cyclin dependent kinases. *PCT Int. Appl.* WO2001057034 A1, 20010809, CAN 135:152803, AN 2001:581885.

Seelen, W., Schaefer, M., Ernst, A. (2003). Selective ring N-protection of aminopyrazoles. *Tetrahedron Letters* **44,** 4491–4493.

Vulpetti, A. and Bosotti, R. (2004). Sequence and structural analysis of kinase ATP pocket residues. *Il Farmaco* **59,** 759–765.

Vulpetti, A., Crivori, P., Cameron, A., Bertrand, J., Brasca, M.G., D'Alessio, R., Pevarello, P. (2005). Structure-based approaches to improve selectivity: CDK2-GSK3β binding site analysis. *Journal of Chemical Information and Modeling,* **45,** 1282–1290.

16 Pyrazolo[3,4-*d*]pyrimidin-4-ones: Exploring the Structural Determinants of Potency and Selectivity in Cyclin-Dependent Kinase Inhibition

Jay A. Markwalder and Steven P. Seitz

CONTENTS

16.1 INTRODUCTION

Misregulation of the cell cycle is one of the hallmarks of the transformed cell. The key role played by the cyclin-dependent kinases (CDKs) in regulating the cell cycle makes them interesting targets for pharmacological intervention in cancer.[1] Our initial interest focused on CDK4/cyclin D1 (K4/D). There were several reasons for our choice of target. CDK4 was thought to be important for cell cycle entry through its phosphorylation of the Rb protein. The tumor suppressor Rb is frequently inactivated in transformed cells. In addition, p16, a specific inhibitor of CDK4, has been identified as a product of a tumor suppressor gene, whereas the CDK4-activating cyclin D1 is a proto-oncogene product. Earlier work from our program showed that conditional expression of p16 in the setting of an *in vivo* xenograft model could lead to tumor regression.[2] Recent developments in cell cycle biology have cast doubt on the simple model of sequential activation of CDKs during cell cycle progression. For example, mouse embryonic fibroblasts (MEFs) from CDK4- and CDK6-null (the two G1 CDKs) mice proliferate normally.[3] Interestingly, MEFs from a CDK4/6-null mouse can also enter the cell cycle, at least in a fraction of the population. It has been suggested that in the absence of CDK4 and CDK6, another kinase, presumably CDK2, can partner with cyclin D and fulfill the early G1 Rb phosphorylation requirement. Although these results suggest that there is significant redundancy in the regulation of cell cycle progression by the CDKs, they do not diminish the potential of this family as chemotherapeutic targets.

16.2 SCREENING STRATEGY

The initial target of our cell cycle drug discovery effort was the CDK4/cyclin D1 complex. Screening a subset of the DuPont compound library generated numerous inhibitors of the CDK4/cyclin D1-mediated phosphorylation of pRb.[4–6] These hits were validated by retest and assay to obtain discrete IC_{50} values against CDK4/cyclin D1 and later, other CDK/cyclin complexes. Initial follow-up also involved similarity searching of the full library and testing of close analogs.

16.3 PRELIMINARY STRUCTURE–ACTIVITY RELATIONSHIPS

16.3.1 GENERAL SYNTHETIC ROUTES

Using this screening strategy, we identified several modestly potent ($IC_{50} < 10\ \mu M$) pyrazolo[3,4-*d*]pyrimidin-4-ones, of which **16.1** is representative (Figure 16.1).[7] A synthetic program aimed at improving the potency of this series was initiated. The two-step literature preparation starting from bis(methylthio)methylidenecyanoacetamide shown in Figure 16.2 was utilized in the first round of compound synthesis.[8]

A more general preparation of related analogs starts from hydrazones A of the various aldehydes R₃CHO and substituted phenylhydrazines (Figure 16.3).[9] The hydrazones react with *N*-bromosuccinimide in cold DMF to afford, after extractive workup, bromohydrazones B. Preparation of the penultimate intermediates, pyrazole aminocarboxamides D, is accomplished by heating the bromohydrazones with

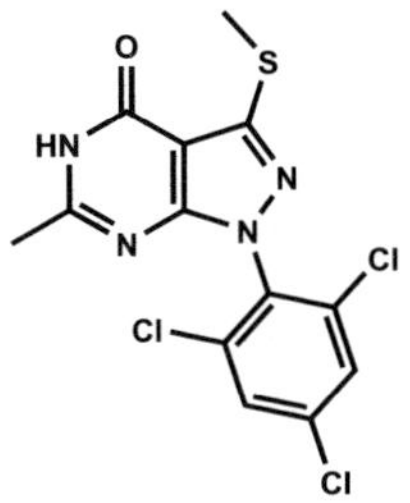

16.1

FIGURE 16.1 Screening hit.

the preformed sodium salt of cyanoacetamide. Intermediate B can also react with the sodium salt of malononitrile to give pyrazole aminonitrile C, which is converted to D by the action of concentrated sulfuric acid. The latter route is usually cleaner but more time consuming than the former. Conversion of D to pyrazolopyrimidinones E is accomplished by heating with ethanolic sodium ethoxide in an excess of ethyl acetate for Table 16.1 and Table 16.2 compounds. Preparation of other R_6 analogs is accomplished using the appropriate ester, anhydride, or lactone in place of ethyl acetate.

16.3.2 SAR of the Pyrazole N1

Preliminary examination of the SAR at pyrazole N1 revealed a strict requirement for an aryl or heteroaryl group having at least one ortho substituent. Substitution at the ortho position should increase the dihedral angle between the two aromatic rings. The presence of a second ortho substituent further enhanced binding to CDK4/cyclin D1 (Table 16.1, entries **16.2-16.4**). Entries **16.5**, **16.6**, and **16.7** highlight less potent analogs in which methyl, trifluoromethyl, and halogen groups appear to be inter-changeable. Attempts to apply these changes to the 2,4,6-trisubstituted phenyl motif, however, yielded compounds (**16.8–16.11**) with greatly diminished potency and suggested that at least one ortho-chloro group was required for good binding. Consequently, the 2,4,6-trichlorophenyl group was initially considered to be optimal, at least in the *in vitro* assays we were conducting at the outset of our program. Our preliminary studies also failed to reveal substituents that were favored over methyl at the pyrimidine C6 position. Accordingly, these two groups were held constant when we probed the SAR at the pyrazole C3 position.

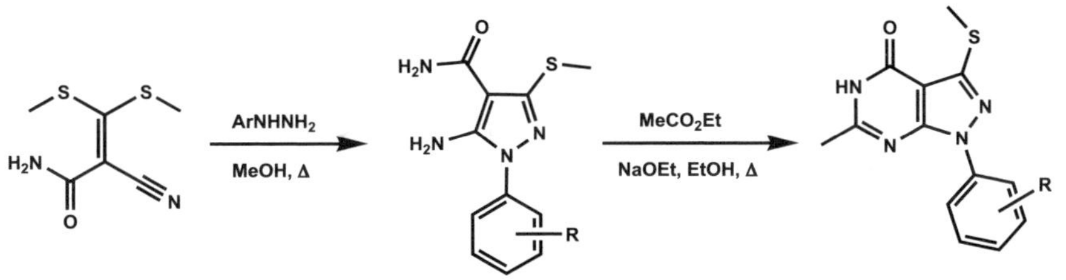

FIGURE 16.2 Initial synthesis of 1-aryl-4,5-dihydro-1*H*-pyrazolo[3,4-*d*]pyrimidin-4-ones.

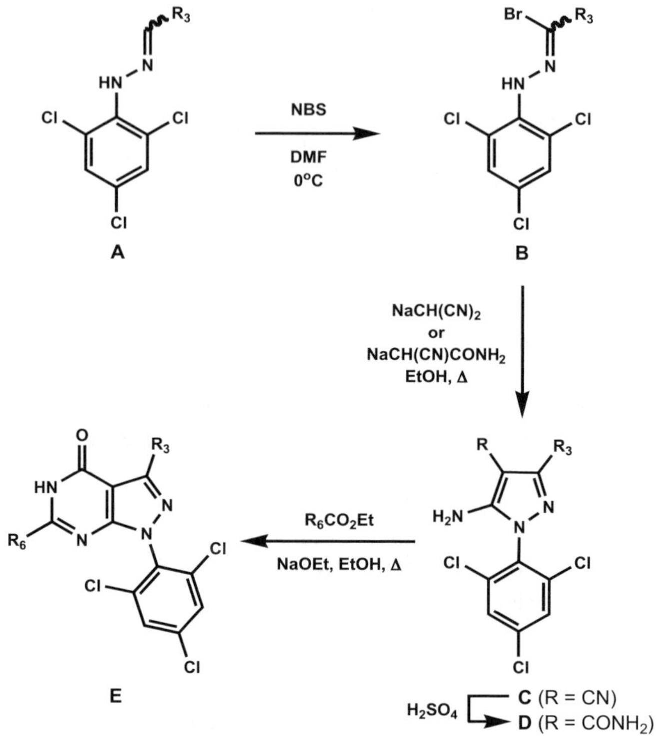

FIGURE 16.3 Synthesis of 1-aryl-4,5-dihydro-1H-pyrazolo[3,4-d]pyrimidin-4-ones.

16.3.3 SAR of the Pyrazole C3 Position

A series of early analogs (Table 16.2) that surveyed small, polar R_3 substituents demonstrated generally poor inhibition of CDK4/cyclin D1. Extreme perturbations in the NMR spectrum of **16.12** suggested that this might result from distortion of the ring electronics or unfavorable side-chain conformation due to intramolecular hydrogen bonding. Examples having aprotic electron-withdrawing groups at this position (**16.14** and **16.15**), however, are also weak CDK4/cyclin D1 inhibitors. This suggested that substitution of a small alkyl group for the thiomethyl in **16.1** might yield a superior enzyme inhibitor. With IC_{50} values of 1.1 µM and 0.14 µM in CDK4/cyclin D1 and CDK2/cyclin (K2/E), respectively, the ethyl analog **16.18** is slightly better than the parent compound. The steric requirements at this position were next probed by the synthesis of inhibitors with smaller, larger, branched, and cyclic alkyl groups at pyrazole C3. Significant, stepwise losses in CDK4/cyclin D1 activity were observed with linear addition (**16.19** and **16.20**) or truncation (**16.16** and **16.17**) of one, then two methylene units from ethyl. Three- and four-carbon units at C3 are quite well tolerated in CDK4/cyclin D1 when constrained as rings (**16.21** and **16.22**). Furthermore, these analogs exhibit slightly higher levels of CDK2/ cyclin E inhibition than **16.18**. Potency against both enzymes drops off dramatically

TABLE 16.1
SAR for Pyrazole N1 Substituents

Compound	R_1	CDK4/D IC_{50} (μM)
16.1	2,4,6-Trichloro	2.1
16.2	2-Chloro	19
16.3	2,6-Dichloro	5.3
16.4	2-Chloro-6-fluoro	6.8
16.5	2,5-Dichloro	23
16.6	2,5-Dimethyl	25
16.7	2-Chloro-5-(trifluoromethyl)	15
16.8	2,4-Dichloro-6-methyl	11
16.9	2,4-Dichloro-6-(trifluoromethyl)	29
16.10	2,4,6-Trimethyl	37
16.11	2,6-Dichloro-4-(trifluoromethyl)	8.0

with the cyclohexyl analog **16.23**. Examples **16.24** and **16.25** demonstrate that placement of a phenyl or benzyl substituent at this position essentially abolishes CDK inhibition. The improved potency seen with small cycloalkyl derivatives compared to their linear counterparts prompted the preparation of isopropyl analog **16.26**. The introduction of branching in this acyclic derivative resulted in a significant boost in potency over cyclopropyl derivative **16.21** against both CDKs in our panel.

The antiproliferative effects of selected compounds were assayed by measurement of growth inhibition in the NCI HCT116 (HCT116) cell line in a sulforhodamine B colorimetric assay. IC_{50} values in this assay are presented in Table 16.2 for some of the more potent members of this class.

16.3.4　SAR of the Pyrimidine C6 Position

Concurrent with the investigation of pyrazole SAR, inhibitors were prepared that addressed substitution of the pyrimidine ring. As alluded to earlier, relatively inactive compounds resulted when changes were made at the C6 position of the screening hit. Entries in Table 16.3 show that reduction or abolition of CDK activity is seen with replacement of the methyl group by hydrogen or a charged or otherwise highly polar group. Steric bulk at this position adversely affects CDK4/cyclin D1 affinity but has a smaller effect on CDK2/cyclin E binding. Although not highly polar, the electron-withdrawing fluoromethyl, difluoromethyl, and trifluoromethyl groups are

TABLE 16.2
Preliminary SAR for Pyrazole C3 Substituents

| | | IC$_{50}$ (μM) | | |
Compound	R$_3$	CDK4/D	CDK2/E	HCT116
16.1	SMe	2.1	0.36	2.5
16.12	CH$_2$CH$_2$OH	>120		
16.13	CONH$_2$	12		
16.14	CN	>28		
16.15	SO$_2$Me	14		
16.16	H	12		
16.17	Me	3.5	>0.73	
16.18	Et	1.1	0.14	1.1
16.19	*n*-Pr	5.5	0.31	6.4
16.20	*n*-Bu	>6.5	>2.5	
16.21	*c*-Pr	0.91	0.053	0.49
16.22	*c*-Bu	1.3	0.075	
16.23	*c*-Hex	>6.5	6.5	
16.24	Ph	>120		
16.25	CH$_2$Ph	>48	>24	>1.2
16.26	*i*-Pr	0.24	0.013	0.25

also associated with greatly diminished activity (**16.36–16.38**). It is postulated that the lower pKa for the pyrimidinone NH in compounds having electron-withdrawing substituents might affect their ability to form hydrogen bonds to CDKs.

Although compounds having R$_6$ groups of two carbon length (**16.39** and **16.40**) are tolerated in both CDK2 and CDK4, a clear drop-off in CDK4/cyclin D1 inhibition occurs with *n*-propyl R$_6$.

16.4 FEATURES OF THE CDK4/CYCLIN D1 HOMOLOGY MODEL

Further insight into the design of more potent inhibitors was provided by docking experiments involving **16.18** and a CDK4/cyclin D homology model based on published crystal structures of the CDK2/ATP/cyclin A complex.[10] The salient

TABLE 16.3
SAR for Pyrimidinone C6 Substituents

| | | | IC$_{50}$ (μM) | | |
Compound	R$_3$	R$_6$	CDK4/D	CDK2/E	HCT116
16.18	Et	Me	1.1	0.14	2.5
16.27	SMe	H	95	2.7	
16.28	SMe	2-Furyl	36		12
16.29	Et	3,4-dimethoxyphenyl	>21	0.58	>21
16.30	SMe	i-Pr	10		
16.31	Et	i-Bu	6.3	0.11	1.1
16.32	SMe	BocN(Me)CH$_2$CH$_2$	>48	19	
16.33	SMe	MeNHCH$_2$CH$_2$·HCl	28	6.6	
16.34	SMe	CO$_2$H	>120		
16.35	SMe	CO$_2$Me	>200		
16.36	SMe	CH$_2$F	44	1.6	
16.37	Et	CHF$_2$	>25	8.5	
16.38	SMe	CF$_3$	116		
16.39	Et	Vinyl	2.9	0.31	0.61
16.40	Et	Et	0.94	0.096	0.55
16.41	SMe	n-Pr	9.1		
16.42	SMe	HO(CH$_2$)$_3$	13	0.76	
16.43	Et	HO(CH$_2$)$_4$	2.4	0.25	
16.44	SMe	HO(CH$_2$)$_5$	10	0.42	

feature of this model is a bidentate hydrogen bond having Val96 NH shared with the pyrimidinone CO and the pyrimidinone NH shared with Val96 CO. The ethyl group is projected into a small, hydrophobic pocket, consistent with the observed SAR for that position. The N1 aryl group is directed toward the triphosphate pocket, and one ortho-chloro substituent is in close proximity with the side chain NH$_2$ of the conserved residue Lys35 (Lys33 in CDK2). Additionally, the model suggests that a pyrimidinone R$_6$ substituent of sufficient length could access charged residues in the ATP-binding pocket of CDK4. For three of these, Asp99, Asp97, and His95, the corresponding residues in CDK2 are neutral. Lys22 (Lys20 in CDK2) is common to both enzymes but is not in close proximity to the ATP-binding pocket in CDK2.

16.5   DISCOVERY OF OPTIMIZED CDK4/CYCLIN D1 LIGANDS

In an attempt to pick up a hydrogen-bonding interaction with one of these residues, we prepared a series of tethered alcohols (**16.42–16.44**). The relatively weak binding of all three indicates that they cannot donate or accept hydrogen bonds to/from either enzyme or if they do, the interactions are not sufficiently favorable to overcome entropy losses associated with maintaining the required conformation.

A significant breakthrough occurred when we examined incorporation of hydrogen bond donors into more conformationally restricted R_6 substituents. The phenols **16.45–16.47** (Table 16.4) place highly preorganized hydrogen bond donor–acceptor pairs at various distances from the pyrimidine core. Surprisingly, all three inhibitors are well accommodated in CDK2/cyclin E, as is the parent benzyl analog **16.48**. This provides further evidence that in CDK2/cyclin E the R_6 substituent binds in a highly hydrophobic environment.

The large boost in potency against CDK4/cyclin D1 for **16.46** vs. nonconstrained aliphatic alcohols, regioisomeric phenols, and the unsubstituted benzyl analog **16.48** suggests that it donates and/or accepts a hydrogen bond to a relatively inflexible region of the enzyme. That the 3-methoxy analog **16.49** is no more potent than **16.48** argues that the primary characteristic responsible for the potency of **16.46** against CDK4/cyclin D1 is its ability to serve as a hydrogen bond donor. Rossi has suggested that pyridine *N*-oxide **16.50** is a closer analog of **16.46** being of similar size and differing in its lack of an acidic proton.[10] The *N*-oxide is somewhat more potent than **16.48**, suggesting that it may accept a hydrogen bond, possibly from Lys22. The relatively weak inhibition of CDK4/cyclin D1 by aniline **16.51** compared to its methanesulfonamide derivative **16.52** suggests that the acidity of the donated hydrogen is critical.

Compounds **16.53–16.67** were prepared to further investigate the R_6 SAR, and their enzymatic potency and selectivity profile mirrors that of other benzyl analogs lacking a meta phenol. These compounds almost universally exhibit low μM potency against CDK4/cyclin D1, are about tenfold more potent against CDK1/cyclin B (K1/B), and a further tenfold more potent against CDK2/cyclin E. Similarly, meta phenols **16.68–16.72** differ little from **16.46**, being highly potent against CDK2/cyclin E, and losing ~2-fold and an order of magnitude against CDK4/cyclin D1 and CDK1/cyclin B, respectively. Energy minimization of catechol **16.68** in the CDK4/cyclin D1 homology model (*vide supra*) suggests that the meta phenol OH serves as a donor to the His95 side chain and an acceptor from the Lys22 side chain. In CDK2 the residue that aligns with the CDK4 His95 residue is Phe82, possibly explaining the lack of an observable effect for incorporation of a meta phenol on CDK2/cyclin E binding. Most compounds in this R_6 benzyl series exhibit reasonably potent antiproliferative activity in the HCT116 and other cell lines, but no strong correlation with CDK inhibition is immediately obvious.

The most potent of these "balanced" CDK inhibitors is derived from ethyl isohomovanillate and possesses a 3-hydroxy-4-methoxybenzyl moiety at R_6. Reprobing the R_3 SAR with this side chain (Table 16.5) served to confirm the trends for CDK2/cyclin E and CDK4/cyclin D1 inhibition shown in Table 16.2 for the R_6 methyl series.

TABLE 16.4
SAR for Pyrimidinone C6 Arylmethyl Substituents

Compound	R_6	IC$_{50}$ (μM)			
		K4/D	K2/E	K1/B	HCT116
16.45	2-Hydroxybenzyl	>11	0.12		5.5
16.46	3-Hydroxybenzyl	0.068	0.014	0.19	0.40
16.47	4-Hydroxybenzyl	2.0	0.016	0.10	0.60
16.48	Benzyl	3.0	0.024	0.26	0.60
16.49	3-Methoxybenzyl	5.0	0.032	0.41	1.7
16.50	1-Oxo-pyridin-3-ylmethyl	0.98	0.079	1.5	0.74
16.51	3-Aminobenzyl	2.9	0.040	0.27	0.47
16.52	3-(Methanesulfonamido)benzyl	0.16	0.26		0.47
16.53	3-Methylbenzyl	>2.2	0.039	>0.56	2.1
16.54	3-Amino-2-methylbenzyl	>2.2	0.019	>0.27	0.19
16.55	Pyrid-2-ylmethyl	9.7	0.23		3.7
16.56	2-(Hydroxymethyl)benzyl	2.9	0.027	>0.27	0.49
16.57	2-Methoxybenzyl	>22	0.10		3.4
16.58	Pyrid-3-ylmethyl	1.3	0.017	0.22	0.30
16.59	Pyrid-4-ylmethyl	3.0	0.038	0.49	0.55
16.60	4-Aminobenzyl	2.7	0.015	0.15	0.42
16.61	4-Methoxybenzyl	2.2	0.016	0.15	0.60
16.62	4-(Dimethylamino)benzyl	>2.1	0.023	>0.26	0.34
16.63	4-Methoxy-3-methylbenzyl	>2.1	0.067	0.87	2.1
16.64	4-Hydroxy-3-methylbenzyl	>2.2	0.027	>0.27	0.50
16.65	3-Methoxy-4-methylbenzyl	2.1	0.11		9.5
16.66	4-Hydroxy-3-methoxybenzyl	6.6	0.076	0.86	0.32
16.67	3,4-Dimethoxybenzyl	4.5	0.11	>0.51	0.61
16.68	3,4-Dihydroxybenzyl	0.044	0.020	0.24	0.72
16.69	3,5-Dihydroxybenzyl	0.059	0.022		0.30
16.70	3-Hydroxy-4-methoxybenzyl	0.024	0.029	0.20	0.13
16.71	3-Hydroxy-4-methylbenzyl	0.088	0.023	0.52	0.67
16.72	3-Hydroxy-4-methoxy-5-nitrobenzyl	0.037	0.022	>0.23	0.33

TABLE 16.5
Further SAR for Pyrazole C3 Substituents

		IC$_{50}$ (μM)			
Compound	R$_3$	CDK4/D	CDK2/E	CDK1/B	HCT116
16.70	Et	0.024	0.029	0.20	0.13
16.73	Benzyl	>1.8	>9.2		0.80
16.74	*n*-Bu	0.70	0.20		3.8
16.75	*t*-Bu	0.063	0.035		2.9
16.76	CF$_3$	0.14			0.21
16.77	CHF$_2$	0.10	0.032		3.0
16.78	*c*-Bu	0.040	0.022	>0.12	0.16
16.79	*c*-Pr	0.017	0.0096	0.097	0.070
16.80	*i*-Pr	0.011	0.012	0.099	0.060

Consistent with the modeling and diffraction studies (*vide infra*), the optimal R$_3$ substituents for both CDK2 and CDK4 were found to be small alkyls. As in the R$_6$ methyl series, bulky (**16.73** and **16.75**), extended (**16.74**), or electron-withdrawing (**16.76** and **16.77**) substituents resulted in diminished enzymatic potency.

The HCT116 growth inhibition data, however, raise some questions as to the mechanisms by which members of this series are acting. Several compounds having small alkyl R$_3$ groups (**16.70**, **16.78–16.80**) are relatively potent in the HCT116 assay with IC$_{50}$ values generally in the range of four- to sixfold higher than their IC$_{50}$ values in the CDK4/cyclin D1 biochemical assay. This modest shift suggests that the antiproliferative effects for these compounds might be CDK4/cyclin D1-mediated as the relatively high K$_{M, ATP}$ for this complex (400 μM) should result in a small IC$_{50}$ shift under physiological (~3 mM) ATP concentrations. The K$_{M, ATP}$ values of 3.6 μM and 2.3 μM for CDK2/cyclin E and CDK1/cyclin B, respectively, would predict IC$_{50}$ shifts on the order of 50-fold for a cellular assay vs. a biochemical assay run in the presence of 50 μM of ATP. One limitation of this analysis arises from binding of inhibitors to inactive forms of kinases that may have different K$_{M, ATP}$ values from the active form.[11] The R$_3$ = *t*-Bu and R$_3$ = CHF$_2$ analogs, however, lose a few fold in CDK4/cyclin D1 and CDK2/cyclin E potency, yet they are 22 to 50-fold less potent in HCT116 growth inhibition than the small alkyl derivatives. The subtle structural differences involved make it difficult to believe that cell penetrance issues alone are responsible for this disconnect. Furthermore, although the benzyl homolog

TABLE 16.6
Effect of α-Substitution at Pyrimidinone C6

[chemical structure]

			IC$_{50}$ (μM)			
Compound	R$_3$	R	CDK4/D	CDK2/E	CDK1/B	HCT116
16.48	Et	H	3.0	0.024	0.26	0.60
16.81	Et	OH	>11	0.18		8.0
16.82	*i*-Pr	Me	44	0.13		

16.73 has no significant CDK inhibitory activity, it is a modestly potent growth inhibitor in the HCT116 line. This suggests that at least some members of the series have significant off-target effects.

Further conformational constraint can be placed on the R$_6$ sidechain by substitution at the benzylic position (Table 16.6). Unfortunately this bias appears to disfavor the biologically relevant conformation, as evidenced by the loss of CDK4/cyclin D1 and CDK2/cyclin E activity.

16.6 CRYSTAL STRUCTURE OF AN INHIBITOR/CDK2 COMPLEX

The validity of the modeling predictions was evaluated by obtaining an x-ray structure of an inhibitor complexed with a CDK. A structure was determined of **16.68** bound to CDK2. The 1.85 Å resolution structure clearly showed the inhibitor in the ATP-binding cleft between the N- and C-terminal domains of the protein. Key hydrogen-bonding interactions were observed with the backbone of Leu83, with the pyrimidinone oxygen being 3.04 Å distant from the amide nitrogen and the inhibitor N5 interacting with the amide carbonyl (2.67 Å). The pyrazole R$_3$ substituent interacts with the side chain of Val64 and projects toward the closed selectivity pocket of the enzyme. An interesting interaction was observed with pyrazole N2 forming an apparent hydrogen bond with Nε of Lys33 (2.90 Å). The trichlorophenyl N-1 substituent is inclined to the plane of the pyrazolopyrimidinone ring with a dihedral angle of about 65°. This group appears to engage in hydrophobic interactions with both the N- and C-terminal domains. The R$_6$ benzyl projects toward the linker region of the kinase, with the dihydroxyphenyl ring of

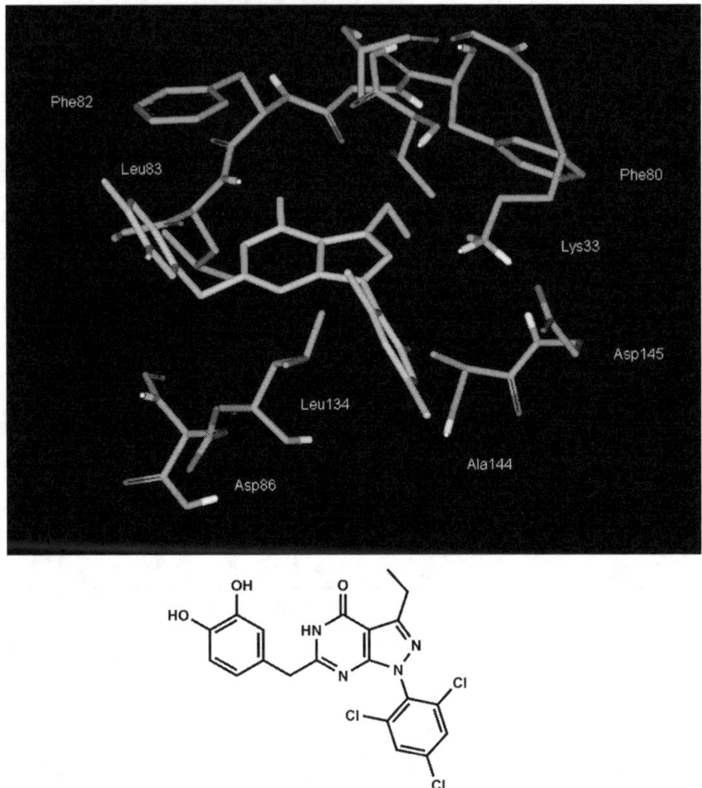

16.68

FIGURE 16.4 View of the active site of CDK2 complexed with compound **16.68**.

16.68 being adjacent to Phe82. Groups attached to the periphery of the R_6 substituent project out of the enzyme and are likely solvent exposed. Interestingly, the dihydroxyphenyl ring of **16.68** is nearly parallel to the R_1 trichlorophenyl group. Although pyrazolopyrimidines are similar to the purine ring system, the binding mode of **16.68** is substantially different from that reported for ATP[12,13] and olomoucine[14] in complex with CDK2. A view of the active site structure of **16.68** bound to CDK2 is shown in Figure 16.4.

16.7 PHENOL REPLACEMENTS

We were interested in examining other hydrogen bond donors as potential phenol replacements. Additional cycles of synthesis and screening resulted in the discovery of several bicyclic arylmethyl groups that gave potent CDK2/cyclin E inhibitors when incorporated into the R_6 position (Table 16.7). The uniformly low CDK2/cyclin E values for members of this series caused us to question whether our IC_{50} values might be approaching the enzyme concentration under the assay conditions. Although we

TABLE 16.7
SAR for Bicyclic Arylmethyl Pyrimidinone C6 Substituents

Compound	R_3	R_6	IC$_{50}$ (μM)			
			CDK4/D	CDK2/E	CDK1/B	HCT116
16.83	*c*-Pr	Indazol-4-ylmethyl	>1.0	0.010	>0.26	0.12
16.84	*c*-Pr	Indazol-5-ylmethyl	>1.0	0.010	0.047	0.033
16.85	*c*-Pr	Indazol-6-ylmethyl	0.44	0.010	0.17	0.048
16.86	*i*-Pr	3-Aminoindazol-5-ylmethyl	0.26	0.012		0.021
16.87	*c*-Pr	Benzoxazol-2-on-5-ylmethyl	0.14	0.012	>0.25	0.062
16.88	*i*-Pr	Benzoxazol-2-on-6-ylmethyl	0.073	0.011		0.007

had observed excellent reproducibility (($+/-$)-flavopiridol mean IC$_{50}$ = 1.50 μM, SD = 0.31 μM, [n = 10]), the use of insect cell lysates as a source of kinase/regulatory unit complexes complicated accurate measurements of enzyme concentrations. Although the indazole derivatives **16.83–16.85** were envisioned as phenol mimics, none appeared to achieve the favorable interaction with CDK4/cyclin D1 seen with meta phenols. Further, the cyclic carbamates **16.87** and **16.88**, which were reasonably good CDK4/cyclin D inhibitors, had a more relaxed requirement for placement of the hydrogen bond donor than the phenols. When adjusted for the slightly superior R_3 group in **16.88**, the two carbamate regioisomers are accommodated with similar affinity in CDK4/cyclin D1. This suggests that the acyloxy group in **16.88** accepts a favorable hydrogen bond from the enzyme or that an amide NH in the *para* position can engage in a similar, albeit weaker, hydrogen-bonding interaction to that of an OH at the meta position in a phenol. These bicyclic heterocycle derivatives are all quite potent antiproliferatives in the HCT116 cell line.

Additional derivatives incorporating bicyclic heterocycles at this position are shown in Table 16.8. Gains of greater than tenfold in potency against CDK4/cyclin D1 seen for **16.91** and **16.93** vs. unsubstituted analog **16.89** and methoxy derivatives **16.90** and **16.92** suggest that the indol-3-yl linker is a competent phenyl replacement for engaging this interaction in CDK4/cyclin D1. This change also results in significant enhancement of affinity for CDK2/cyclin E, hinting at subtle differences for the binding modes of these analogs and the meta phenols. The last eight compounds in the table exemplify a curious phenomenon in which an indole group is accommodated at R_6, but only when attached at the 2-position. The relatively weak inhibition by

TABLE 16.8
Further SAR for Bicyclic Aryl and Arylmethyl Pyrimidinone C6 Substituents

			IC$_{50}$ (μM)		
Compound	R$_3$	R$_6$	CDK4/D	CDK2/E	HCT116
16.89	Et	Indol-3-ylmethyl	>2.1	0.10	
16.90	Et	5-(Methoxy)indol-3-ylmethyl	>2.0	0.22	
16.91	Et	5-(Hydroxy)indol-3-ylmethyl	0.23	0.08	0.86
16.92	*i*-Pr	5-(Methoxy)indol-3-ylmethyl	>1.9	0.20	
16.93	*i*-Pr	5-(Hydroxy)indol-3-ylmethyl	0.13	0.031	0.58
16.94	*i*-Pr	Indol-4-yl	>2.2	>0.21	
16.95	*i*-Pr	Indol-5-yl	>1.1	>0.10	
16.96	*i*-Pr	Indol-6-yl	0.61	0.11	
16.97	*i*-Pr	Indol-2-yl	0.10	0.038	>2.0
16.98	*i*-Pr	5-(Methoxy)indol-2-yl	0.067	0.077	>2.0
16.99	*i*-Pr	5-(Hydroxy)indol-2-yl	0.041	0.018	1.2
16.100	*i*-Pr	5-Hydroxybenzofuran-2-yl	>1.0	0.20	
16.101	SMe	Phenyl	>100.0	1.4	

benzofuran and phenyl derivatives **16.100** and **16.101**, respectively, implies a favorable interaction with the indole NH in both enzyme systems.

16.8 DISCOVERY OF OPTIMIZED ANTIPROLIFERATIVES

16.8.1 GENERAL SYNTHETIC ROUTES

The recognition that most of these compounds have rather high logP values and limited solubility in aqueous media led us to prepare a series of *para*-substituted benzyl derivatives having tethered amine groups (Figure 16.5). Anilines F could be coupled to glycine, alanine, and β-alanine derivatives G using 1-(3-dimethylaminopropyl)-3-ethylcarbodiimide hydrochloride (EDCI) and triethylamine in dichloromethane. Treatment of F with chloroacetyl chloride and *N*-methylmorpholine in tetrahydrofuran provided chloromethylamides H. These furnished additional glycine analogs I upon treatment with excess amines in tetrahydrofuran.

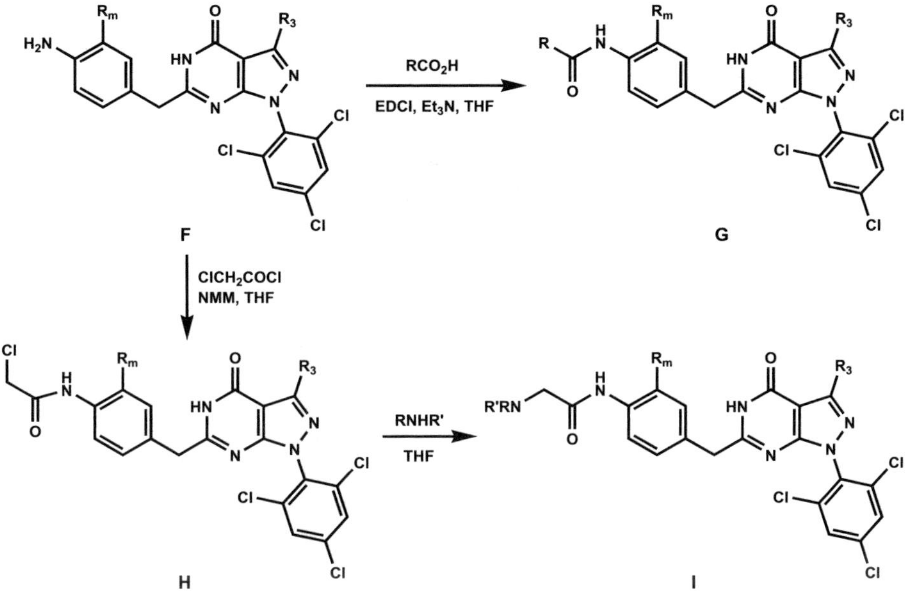

FIGURE 16.5 Synthesis of R_6 anilide derivatives.

16.8.2 α-Amino Acid Derivatives

Modeling studies predicted that the region into which these amines projected in CDK2 was solvent exposed, so it was not surprising that ionizable groups were well accommodated in this enzyme. CDK4/cyclin D1 inhibition in this series was generally about an order of magnitude weaker than that for CDK2/cyclin E. Two exceptions are the *N*-methyl derivative **16.109**, which is essentially inactive against CDK4/cyclin D1, and the meta phenol **16.108**, which shows a slight preference for CDK4/cyclin D1 over CDK2/cyclin E (Table 16.9).

Analogs in this series were potent antiproliferatives in the HCT116 cell line. The modest structural modification of **16.104** (*vide supra*) to give the improved CDK4/cyclin D1 inhibitor **16.108** resulted in no enhancement of this HCT116 activity, although *N*-methyl derivative **16.109** showed significant loss of antiproliferative potency and CDK4/cyclin D1 activity. The *N,N*-dimethylglycinamide derivative **16.105** showed uniformly high activity when tested in a large panel of cell lines with differing levels of pRb and p53 status including NCI HCT116, MDA MB468, NCI H460, A498, T47D, MCF7, DU145, and COLO205. When dosed at 10 mg/kg i.p. in nude mouse H460 and HCT116 xenograft models, **16.105** yielded tumor growth inhibition (tgi) values (H460: 41% tgi, HCT116: 46% tgi) comparable to those seen with racemic flavopiridol. Dehydration, weight loss, and death were observed in animals receiving higher doses of compound.

TABLE 16.9
SAR for Anilide C6 Substituents

Compound	R_p	R_m	R_3	K4/D	K2/E	K1/B	HCT116
					IC_{50} (μM)		
16.102	$H_2NCH_2CONH \cdot HCl$	H	i-Pr	0.31	0.023		0.031
16.103	$MeNHCH_2CONH$	H	i-Pr	0.26	0.013		0.018
16.104	$(Me)_2NCH_2CONH$	H	i-Pr	0.12	0.008	0.047	0.008
16.105	$(Me)_2NCH_2CONH \cdot HCl$	H	Et	0.59	0.018	0.10	0.034
16.106	$MeN(CH_2CH_2)_2NCH_2CONH$	H	i-Pr	0.11	0.017	0.021	0.007
16.107	$(Me)_2NCH_2CONH$	OMe	i-Pr	0.39	0.011	0.14	0.026
16.108	$(Me)_2NCH_2CONH$	OH	i-Pr	0.011	0.022	0.086	0.013
16.109	$(Me)_2NCH_2CONMe$	H	i-Pr	>1.8	0.084		0.091
16.110	$(+/-)$-$(Me)_2NCH(Me)CONH$	H	i-Pr	0.18	0.014		0.015

16.8.3 STABILIZED ANILIDE DERIVATIVES

Recognizing the metabolic liability posed by the relatively unhindered amide linkages in these molecules, we prepared a large series of aniline derivatives of which those in Table 16.10 are representative. Simple derivatives in this series were prepared by direct acylation of anilines F with isocyanates, chloroformates, or sulfonyl chlorides and N-methylmorpholine in tetrahydrofuran (Figure 16.6). A more versatile intermediate was isocyanate M, prepared from anilines F and triphosgene and triethylamine in 1,2-dichloroethane. This suffers addition of amines in tetrahydrofuran at ambient temperature to furnish ureas J. Treatment of M with alkoxides in tetrahydrofuran at reflux gives carbamates K.

The behavior of these ureas, carbamates, and semicarbazides in biochemical and cellular growth inhibition assays closely mirrors that of the Table 16.9 anilides. With no hydrogen-bond-donating group appended to the aromatic ring, **16.118** was a surprisingly good CDK4/cyclin D1 inhibitor with an IC_{50} of 0.22 μM. The dimethylaminomethyl moiety is likely protonated under physiological pH, and this group could potentially access His95. Interestingly, sulfonamide **16.117** binds poorly to CDK4/cyclin D1 even though it retains an acidic proton. The tetrahedral geometry of the sulfur atom adjacent to the NH may hinder the approach of the enzyme's

TABLE 16.10
SAR for C6 Ureides and Other Anilide Derivatives

Compound	R_p	R_m	R_3	IC$_{50}$ (μM)			
				K4/D	K2/E	K1/B	HCT116
16.111	MeNHCONH	H	i-Pr	0.17	0.013	0.030	0.006
16.112	MeNHCONMe	H	i-Pr	1.5	0.040	0.33	0.090
16.113	MeNHCONH	OMe	i-Pr	0.29	0.015	0.014	0.007
16.114	MeNHCONH	OH	i-Pr	0.020	0.018	0.082	0.012
16.115	MeOCONH	H	i-Pr	0.19	0.022	0.061	0.009
16.116	Me$_2$N(CH$_2$)$_3$OCONH	H	i-Pr	0.20	0.008		0.016
16.117	MeSO$_2$NH	H	Et	>1.9	0.023	>0.24	0.10
16.118	(Me)$_2$NCH$_2$	H	i-Pr	0.22	0.020	>0.23	0.031
16.119	(Me)$_2$NCH$_2$CH$_2$N(Me)CONH	H	i-Pr	0.25	0.027	0.073	0.020
16.120	MeN(CH$_2$CH$_2$)$_2$NNHCONH	H	i-Pr	0.078	0.024		0.008

hydrogen bond acceptor. Anilide **16.106** and its semicarbazide isostere **16.120** differ little in their biochemical and cellular activities, calculated logP values (**16.106**: 4.8, **16.120**: 4.9), and solubilities in pH 7.4 buffer (both 0.1 mg/ml), yet **16.120** would be expected to have greater plasma stability.

In summary, the pyrazolopyrimidin-4-ones represent a novel class of small molecule, ATP-competitive CDK inhibitors with a unique binding mode. Molecular modeling and single-crystal x-ray diffraction studies have identified key features affecting the potency and selectivity seen in this series and suggest where further improvements might be realized. Selected members of the series showed no significant affinity for other kinases when screened against a limited panel, although disconnects between CDK inhibition and cellular activity hint at off-target effects. Efficacy was exhibited in two mouse xenograft models. The wide variety of substituents that are accommodated at the R_6 position in the CDKs evaluated suggests many additional targets to further optimize pharmacokinetic and physical properties. Additionally, the reduced homology of the CDKs in this region implies that further modification could yield analogs with high intrafamily selectivity.

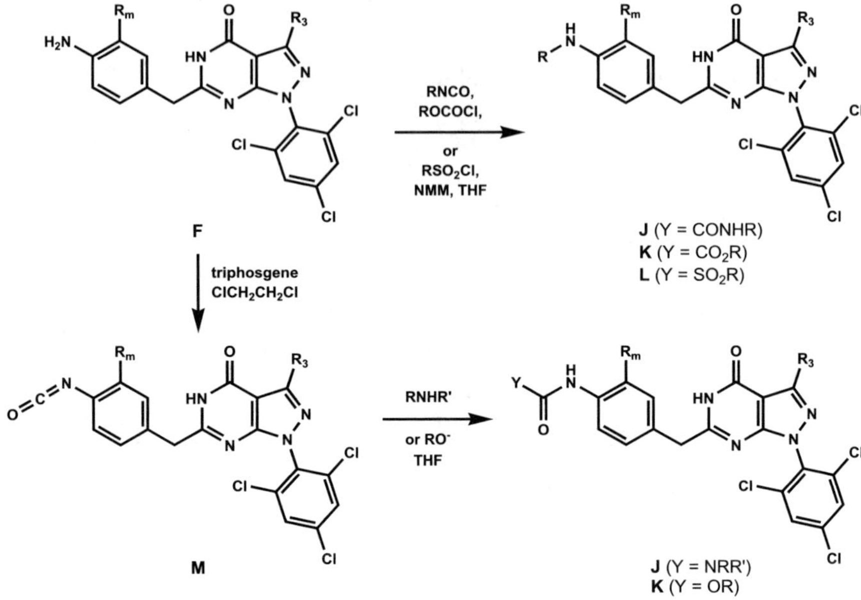

FIGURE 16.6 Synthesis of stabilized anilide derivatives.

ACKNOWLEDGMENTS

Over the course of this project, a great many individuals contributed to the design, preparation, and evaluation of the pyrazolo[3,4-*d*]pyrimidin-4-ones as CDK inhibitors. The authors would like to acknowledge the efforts of Anthony Cocuzza, Barbara A. Harrison, Robert F. Kaltenbach III, Vanessa A. Moore, David A. Nugiel, and Susan R. Sherk for compound preparation, Chong-Hwan Chang and Jodi Muckelbauer for crystallography studies, Jim Krywko, Karen A. Rossi, and Pieter Stouten for molecular modeling studies, and Marc R. Arnone, Pamela A. Benfield, John F. Boylan, Catherine R, Burton, Sarah S. Cox, Philip M. Czerniak, Charity L. Dean, Deborah Doleniak, J. Gerry Everlof, Robert Grafstrom, Diane M. Sharp, and Lisa M. Sisk for *in vitro* and *in vivo* compound evaluation.

REFERENCES

1. Sielecki, T.M., Boylan, J.F., Benfield, P.A., Trainor, G.L. Cyclin-dependent kinase inhibitors: useful targets in cell cycle regulation. *J Med Chem* 2000, **43**: 1–18.
2. Burton, C.A., Boylan, J., Robinson, C., Kerr, J., Benfield, P. Constitutive expression of a tumor suppressor leads to tumor regression in a xenograft model. *Inflammatory Processes: Molecular Mechanisms and Therapeutic Opportunities* 2000, pp. 67–76.
3. Sanchez, I., Dynlacht, B.D. New insights into cyclins, CDKs, and cell cycle control. *Semin Cell Dev Biol* 2005, **16**: 311–321.

4. Carini, D.J., Kaltenbach, R.F., Liu, J., Benfield, P.A., Boylan, J., Boisclair, M., Brizuela, L., Burton, C.R., Cox, S., Grafstrom, R. et al. Identification of selective inhibitors of cyclin dependent kinase 4. *Bioorg Med Chem Lett* 2001, **11**: 2209–2211.
5. Nugiel, D.A., Etzkorn, A.M., Vidwans, A., Benfield, P.A., Boisclair, M., Burton, C.R., Cox, S., Czerniak, P.M., Doleniak, D., Seitz, S.P. Indenopyrazoles as novel cyclin dependent kinase (CDK) inhibitors. *J Med Chem* 2001, **44**: 1334–1336.
6. Sielecki, T.M., Johnson, T.L., Liu, J., Muckelbauer, J.K., Grafstrom, R.H., Cox, S., Boylan, J., Burton, C.R., Chen, H., Smallwood, A. et al. Quinazolines as cyclin dependent kinase inhibitors. *Bioorg Med Chem Lett* 2001, **11**: 1157–1160.
7. Markwalder, J.A., Arnone, M.R., Benfield, P.A., Boisclair, M., Burton, C.R., Chang, C.H., Cox, S.S., Czerniak, P.M., Dean, C.L., Doleniak, D. et al. Synthesis and biological evaluation of 1-aryl-4,5-dihydro-1H-pyrazolo[3,4-d]pyrimidin-4-one inhibitors of cyclin-dependent kinases. *J Med Chem* 2004, **47**: 5894–5911.
8. Tominaga, Y., Honkawa, Y., Hara, M., Hosomi, A. Synthesis of 1,5-dihydro-1H-pyrazolo[3,4-d]pyrimidine derivatives using ketene dithioacetals. *J Heterocyc Chem* 1990, **27**: 775–783.
9. Miyashita, A., Iijima, C., Higashino, T. Studies on 1,5-dihydro-1H-pyrazolo[3,4-d]pyrimidine derivatives. XVIII. Facile preparation of 1H-1,5-dihydro-1H-pyrazolo[3,4-d]pyrimidin-4(5H)-ones. *Heterocycles* 1990, **31**: 1309–1314.
10. Rossi, K.A., Markwalder, J.A., Seitz, S.P., Chang, C.H., Cox, S., Boisclair, M.D., Brizuela, L., Brenner, S.L., Stouten, P.F. Understanding and modulating cyclin-dependent kinase inhibitor specificity: molecular modeling and biochemical evaluation of pyrazolopyrimidinones as CDK2/cyclin A and CDK4/cyclin D1 inhibitors. *J Comput Aided Mol Des* 2005, **19**: 111–122.
11. Knight, Z.A., Shokat, K.M. Features of selective kinase inhibitors. *Chem Biol* 2005, **12**: 621–637.
12. Kim, S.H., Schulze-Gahmen, U., Brandsen, J., de Azevedo Junior, W.F. Structural basis for chemical inhibition of CDK2. *Prog Cell Cycle Res* 1996, **2**: 137–145.
13. Schulze-Gahmen, U., De Bondt, H.L., Kim, S.H. High-resolution crystal structures of human cyclin-dependent kinase 2 with and without ATP: bound waters and natural ligand as guides for inhibitor design. *J Med Chem* 1996, **39**: 4540–4546.
14. Hardcastle, I.R., Arris, C.E., Bentley, J., Boyle, F.T., Chen, Y., Curtin, N.J., Endicott, J.A., Gibson, A.E., Golding, B.T., Griffin, R.J. et al. N2-substituted O6-cyclohexylmethylguanine derivatives: potent inhibitors of cyclin-dependent kinases 1 and 2. *J Med Chem* 2004, **47**: 3710–3722.

Section IV

Therapeutic Perspective

17 Cyclin-Dependent Kinase Inhibitors and Combination Therapy: Experimental and Clinical Status

Lloyd R. Kelland

CONTENTS

17.1 OVERVIEW

A number of cyclin-dependent kinase (CDK) inhibitors have now been evaluated in clinical trials, including flavopiridol, UCN-01 (7-hydroxystaurosporine), R-roscovitine (CYC202, seliciclib), BMS-387032, Ro31-7453, and E7070. A current focus, especially in the clinic, is on their use in combination with various cytotoxic agents. This is based on the dual facts that single-agent activity of CDK inhibitors in Phase II clinical trials has been generally disappointing, with low percentages of objective responses, and that synergy has been demonstrated with CDK inhibitors combined with several cytotoxics using cell lines *in vitro*. Most such combination studies performed *in vitro* show that cell kill is sequence dependent, with greater kill observed when the cytotoxic is added to cells before the CDK inhibitor (rather than after or concurrently). In some, but not all cases, these *in vitro* findings have been translated successfully to clinical trials. For the most clinically advanced CDK inhibitor, flavopiridol, single-agent activity is modest but appears greater when using a 1 h bolus infusion rather than a continuous infusion over 3 d. Phase II combination studies are taking place with irinotecan in patients with hepatocellular carcinoma and with cisplatin in ovarian cancer, and more such studies are planned. UCN-01 inhibits cell cycle checkpoint kinases as well as CDKs and shows synergy against cell lines, especially when combined with DNA-damaging agents. Phase II trials are taking place with UCN-01 in combination with 5 fluorouracil (5FU) in pancreatic cancer and with topotecan in ovarian cancer. In some cases, toxicity has precluded the robust clinical evaluation of CDK inhibitors combined with cytotoxics (e.g., flavopiridol with docetaxel or carboplatin, and UCN-01 with cisplatin). Roscovitine is currently undergoing Phase II monotherapy trials by the oral route. Phase I combination studies have started with Ro31-1453 plus gemcitabine or paclitaxel. The results of all of these combination Phase II trials are awaited with great interest and will define the role of the currently available CDK inhibitors in current cancer therapy.

17.2 INTRODUCTION

Over the past 30 years, first in yeast and then extending to human cells, the molecular players involved in the control and regulation of the cell cycle and how they interact have been largely elucidated. From these findings, pharmacological intervention or modulation of the cell cycle has emerged as an attractive anticancer strategy because aberrant cell cycling has been shown to occur in the majority of cancers. A master switch of the cell cycle is the retinoblastoma protein family. These are phosphorylated, resulting in cell proliferation by the family of CDKs. In turn, the CDKs themselves are activated by various cyclins and naturally inactivated by the Ink (e.g., $p16^{INK4a}$) and CIP/KIP (e.g., $p21^{CIP1}$) families. Therefore, because of the pivotal role of CDKs in cancer cell proliferation, the discovery and development of CDK inhibitors has received much attention in the past decade.

Consequently, a number of CDK inhibitors have now entered clinical trials including flavopiridol, 7-hydroxystaurosporine (UCN-01), R-roscovitine (CYC202, seliciclib), an aminothiazole BMS-387032, Ro 31-7453 and the chloroindolyl sulfonamide E7070 (see Figure 17.1 for structures). This chapter focuses on the concept and rationale of using these agents in combination for attaining maximum therapeutic benefit and describes the status of such combination studies at both the preclinical and clinical level.

17.3 COMBINATION CHEMOTHERAPY

The concept of using cancer drugs in combination goes back around 50 years. Impetus was originally provided by key clinical findings in the late 1950s in childhood leukemia and Hodgkin's disease (Frei et al., 1958). Since then, it has become apparent that

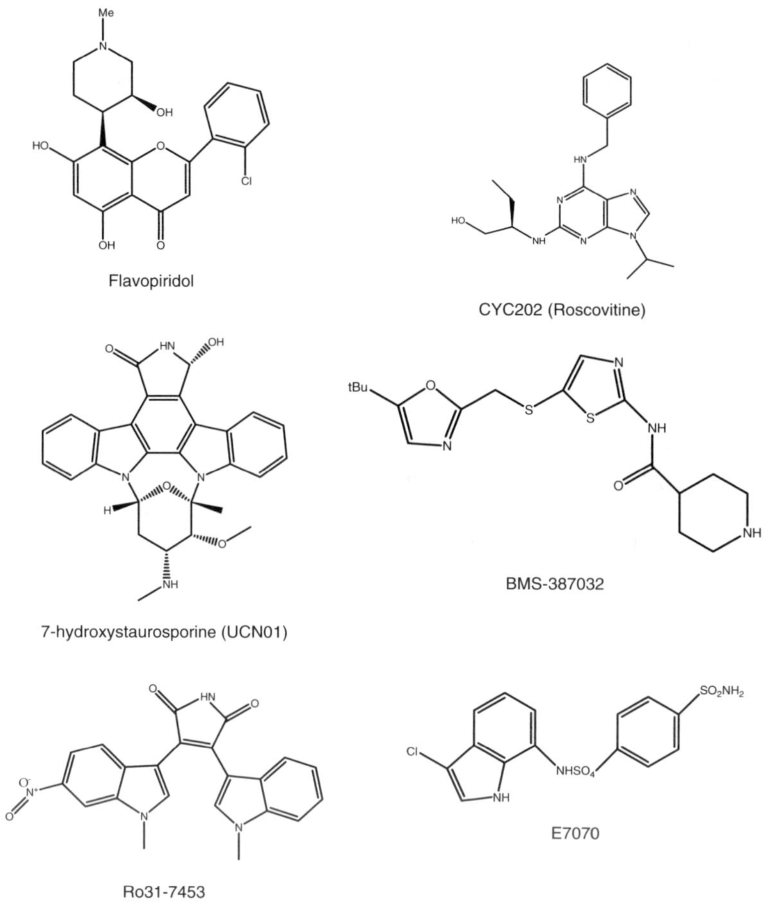

Flavopiridol

CYC202 (Roscovitine)

7-hydroxystaurosporine (UCN01)

BMS-387032

Ro31-7453

E7070

FIGURE 17.1 CDK inhibitors undergoing clinical trials.

except in rare cases standard single drugs are not curative. This is widely believed to be because of the inherent genetic instability of tumor cells, leading to somatic mutations and the emergence of resistant genotypes during repeated drug exposure. Therefore, a number of principles have emerged and are currently practiced to guide the selection and use of cancer drugs in combination (Table 17.1). These include using agents that each possess at least partial activity, have differing mechanisms of action and resistance, and drugs with nonoverlapping side effects. It should be noted, however, that these principles were established predominantly using combinations of cytotoxics. Modifications to these principles may be applicable where cytostatic drugs such as many of the recently approved molecularly targeted drugs (e.g., erlotinib, sorafenib) are to be included in combinations. In the particular case of CDK inhibitor-based combinations, special consideration needs to be given to "cytokinetic" interactions — that is where cell cycle inhibitors and cell-cycle-specific drugs may antagonize or synergize depending on their sequence of administration.

TABLE 17.1
Guiding Principles of Combining Anticancer Drugs in the Clinic

Serial Number	Principle
1.	Each drug should possess single-agent activity
2.	Use drugs, each with a different mechanism of action
3.	Use drugs, each with a different mechanism of resistance (e.g., membrane glycoprotein)
4.	Use drugs with nonoverlapping dose-limiting toxicities
5.	Consideration should be given to potential adverse drug–drug interactions (e.g., pharmacokinetic or cytokinetic)

Drug combinations may be studied at the preclinical level using either *in vitro* models (typically small panels of human cancer cell lines representative of the target tumor in humans) or *in vivo* using tumors grown in mice (generally as subcutaneous human tumor xenografts in immune-suppressed mice). Such studies are often used to guide or rank agents for combining in the clinic. A widely used method of evaluating the effect of combining drugs *in vitro* is median effect analysis, in which a combination index (CI) is calculated from pooled data from 3 to 5 individual experiments each consisting of full dose–response curves for each drug alone and for the combination (for an example, see Rogers et al., 2002). The CI represents the degree of synergy (value less than 0.8), additivity (values from 0.8 to 1.2), or antagonism (value greater than 1.2) occurring for any given drug combination. However, these *in vitro* studies are often of limited value in addressing the critical issue of whether such combinations *in vivo* could result in a gain in therapeutic index (i.e., achievement of greater anti-tumor effects without a similar increase in toxicity).

17.4 CLINICALLY USED CDK INHIBITORS

17.4.1 FLAVOPIRIDOL

The prototype CDK inhibitor and the first to enter clinical trial, flavopiridol (NSC649890, L86-8275), is a derivative of a natural product, the plant alkaloid rohitukine, a component of traditional Indian medicine. Flavopiridol is a relatively broad-spectrum CDK inhibitor inhibiting CDKs 1, 2, 4, 6, and 7 as well as being a potent inhibitor of the CDK9/cyclin T complex, thereby blocking transcription (Shapiro, 2004; Chao et al., 2000). Other "off-target" non-CDK-based effects such as direct binding to DNA may also contribute to its biological properties (Bible et al., 2000a).

17.4.1.1 Preclinical Studies

Flavopiridol possesses potent cell killing properties *in vitro* (typical IC_{50} values ranging from 25 to 150 n*M*) and moderate anti-tumor activity *in vivo* as a single agent against some human tumor xenografts (e.g., prostate cancer) (Drees et al., 1997). A variety of preclinical studies have highlighted the potential for combining flavopiridol with other drugs in the clinic. Using a lung cancer cell line, synergy was shown for the combination of flavopiridol and cisplatin regardless of the sequence of drug administration (Bible and Kaufmann, 1997). By contrast, for paclitaxel, ara C, topotecan, doxorubicin, and etoposide synergy was more pronounced when these drugs were administered before flavopiridol rather than together or following it (Bible and Kaufmann, 1997). The sequence dependence appeared to be due to flavopiridol causing arrest of cells in G1 and G2 phases of the cell cycle for up to 24 h after exposure. Notably, if ara C or 5FU (both S-phase-specific drugs) were added to flavopiridol exposed cells 2 to 3 d after flavopiridol removal, striking synergy was observed. Additional studies using gastric and breast cancer cell lines and mitomycin C or paclitaxel also showed a similar sequence-dependent-effect with mitomycin C or paclitaxel followed by flavopiridol, leading to a greater potentiation of apoptosis than concomitant exposure or the reverse sequence (Schwartz et al., 1997; Motwani et al., 1999).

Finally, a cancer-selective approach was shown using a human cancer cell line (from an osteosarcoma) that was deficient in functional retinoblastoma protein (a relatively common abnormality in human tumors) in that flavopiridol, when used at a relatively low concentration of 100 nM, selectively sensitized such cells to the widely used drug doxorubicin (Li et al., 2001).

As with most anticancer drugs, acquired tumor resistance to CDK inhibitors is likely to limit their clinical utility. As referred to earlier, the prevention of resistance provides a major underlying rationale for using CDK inhibitors in combination, aside from the fact that curative effects have not generally been seen in either preclinical or clinical studies with any CDK inhibitor to date. Clues as to how resistance may occur with flavopiridol are beginning to emerge from studies of acquired resistant cell lines. A human colon cancer cell line was repeatedly exposed *in vitro* to flavopiridol over 3 months, resulting in a stably resistant subline possessing around eightfold resistance (Smith et al., 2001). Thus, eight times more flavopiridol is required to cause the same level of cell kill as in the parent cells. Interestingly, several cancer drugs (etoposide, doxorubicin, cisplatin, paclitaxel, and topotecan) and other chemical classes of CDK inhibitors (roscovitine, purvalanol A, 9-nitropaullone, and hymenialdisine) all retained activity against the acquired resistant cells, similar to that observed in the parent line. Resistance was not related to any changes in drug accumulation but appeared to be at least partially due to increased cyclin-E- associated kinase activity. Further studies showed that acquired resistance to flavopiridol in this model was associated with upregulation of the telomerase catalytic subunit and telomere elongation; combination studies of flavopiridol with the telomere interactive G-quadruplex inhibitor BRACO-19 showed promising activity in these cells (Incles et al., 2003). By contrast, in a second cell line (ovarian) that spontaneously developed fivefold resistance to flavopiridol (and threefold to cisplatin), there was reduced drug accumulation in the resistant subline (Bible et al., 2000b).

17.4.1.2 Clinical Studies

The first Phase I trial of flavopiridol employed a 3 d continuous infusion given every 2 weeks; dose-limiting toxicities were diarrhea and hypotension (Senderowicz et al., 1998).

A second Phase I used a daily 1 h infusion for 1 to 5 d; the dose-limiting toxicity was neutropenia (Tan et al., 2002). Phase II monotherapy studies of flavopiridol have used these two schedules, initially 50 mg/m^2/d administered over 3 d by continuous infusion every 2 weeks (CIV regime) and, more recently, 50 mg/m^2 given as a 1 h bolus daily for 3 consecutive d every 3 weeks (bolus regime). Studies have been published for patients with renal cancer, CIV regime (Stadler et al., 2000), and bolus regime (van Veldhuizen et al., 2005); gastric carcinoma, CIV regime (Schwartz et al., 2001); non-small-cell lung cancer, CIV regime (Shapiro et al., 2001); colorectal cancer, CIV regime (Aklilu et al., 2003); mantle-cell lymphoma, bolus regime (Kouroukis et al., 2003); prostate cancer, CIV regime (Liu et al., 2004); endometrial carcinoma, bolus regime (Grendys et al., 2005); melanoma, bolus regime (Burdette-Radoux et al., 2004); chronic lymphocytic leukemia, 24 h continuous infusion (Flinn et al., 2005), and CIV and bolus regimes (Byrd et al., 2005). Overall, there have been very few objective responses; one complete response and three partial responses in renal carcinoma cancer patients receiving the bolus regime; four partial responses in chronic lymphocytic leukemia receiving the bolus regime; three partial responses in mantle-cell lymphoma patients, again receiving the bolus regime; one partial response in a renal carcinoma patient in the original Phase I trial. Hence, although the bolus regime appears to be more active than the continuous infusion regime, the overall low single-agent activity of flavopiridol does not warrant continued development as a monotherapy.

Therefore, in addition, and building on the previously described preclinical observations, a number of combination trials involving flavopiridol are taking place or Phase I trials have been completed: with paclitaxel (Schwartz et al., 2002), with docetaxel in breast cancer (Tan et al., 2004), with irinotecan (Shah et al., 2005), with cisplatin or carboplatin (Bible et al., 2005), and ara C and mitoxantrone in adult leukemias (Karp et al., 2005). With docetaxel followed 24 h later by flavopiridol, severe grade 3 or 4 toxicities prevented the feasible application of this combination in patients; this occurred either using flavopiridol in the CIV regime (dose-limiting toxicity of neutropenia) or with the bolus regime (dose-limiting toxicity of hypotension). Evidence of a biological effect with flavopiridol was demonstrated by increased p53 staining and decreased phosphorylated retinoblastoma in buccal mucosa biopsies.

By contrast to the findings with docetaxel, investigators recommended a Phase II dose for the combination of paclitaxel given on day 1 as a 3 h infusion (175 mg/m^2) followed on day 2 by a 24 h infusion of flavopiridol at 70 mg/m^2. At these doses, there were signs of anti-tumor activity, with neutropenia and pulmonary toxicity being dose limiting. Results of any completed Phase II combination trials of this combination have not been reported for date.

The Phase I combination trial using flavopiridol (24 h infusion) combined with platins showed contrasting effects. Although the combination with cisplatin was feasible, unexpectedly high toxicities occurred with carboplatin (thromboembolism, CNS hemorrhage). Mention is made of a planned Phase II study of the cisplatin (60 mg/m^2)/flavopiridol (100 mg/m^2/24 h) combination in patients with ovarian cancer.

Encouraging results have been published for the combination of flavopiridol and the DNA topoisomerase I inhibitor irinotecan, with the CDK inhibitor administered as a 1 h weekly infusion 7 h after the cytotoxic. The recommended Phase II doses of flavopiridol with irinotecan (100 mg/m^2 or 125 mg/m^2) are 60 mg/m^2, and 50 mg/m^2, respectively; three partial responses were reported in patients (2 colorectal, 1 gastric cancer), as well as several long-term (>6 months) disease stabilizations; dose-limiting toxicities were diarrhea, myelosuppression, hyperbilirubinemia, and fatigue. This combination is now being evaluated in a Phase II trial in hepatocellular carcinoma.

Finally, a recent report in patients with relapsed and refractory acute leukemias demonstrated the feasibility of combining flavopiridol using the bolus regime of 50 mg/m^2 daily for 3 d followed by ara C (2 g/m^2/3 d) beginning on day 6 and 40 mg/m^2 mitoxantrone given on day 9. Profound neutropenia was the dose-limiting toxicity; overall response rates were 31% in 26 acute myelogenous leukemia patients and 12.5% in acute lymphoblastic leukemia patients.

17.4.2 UCN-01 (7 HYDROXYSTAUROSPORINE)

UCN-01 is derived from the *Streptomyces* alkaloid staurosporine, a potent but non-specific protein and tyrosine kinase inhibitor possessing a low therapeutic index in mice. In addition to inhibiting CDK1 and CDK2, UCN-01 inhibits calcium-dependent protein kinase C, promotes p53-independent apoptosis through targeting the cell cycle checkpoint kinases Chk1 and Chk2 and inhibits the Akt/PI3 kinase signaling survival pathway by inhibiting pdk1.

17.4.2.1 Preclinical Studies

UCN-01 has potent antiproliferative activity against several human tumor cell lines *in vitro* (typical IC$_{50}$ of around 25 nM) as well as demonstrating anti-tumor activity *in vivo* against human tumor xenografts. It causes cell cycle arrest at the G1/S boundary (probably via effects on pdk1 and CDKs) and/or abrogation of arrest at the G2. Studies in cell lines have demonstrated synergy between UCN-01 and several anticancer drugs, in particular those that damage DNA (e.g., cisplatin and mitomycin C and with ionizing radiation) (Wang et al., 1996). It is thought that this synergy is due mainly to its ability to abrogate the G2/M checkpoint (via inhibition of Chk1 and Chk2), thereby allowing less time for the repair of DNA damage prior to mitosis. This may be particularly relevant and applicable to tumors that harbor inactivated p53 (over 50% of human cancers) (Wang et al., 1996), although other studies have shown no association between p53 status and sensitivity to combinations of UCN-01 and ionizing radiation (Yu et al., 2002). Potentiation of cytotoxicity in colon cancer cells with UCN-01 and the topoisomerase I poison SN38 (the active metabolite of irinotecan) was sequence dependent; SN38 followed by UCN-01 resulted in enhanced cytotoxicity via p53-dependent G2 checkpoint abrogation and mitotic catastrophe. However, the reverse sequence or concomitant exposure led to p53-independent S phase checkpoint override and did not increase apoptosis (Tse and Schwartz, 2004).

UCN-01 has also shown synergistic effects *in vitro* with antimetabolites such as gemcitabine, 5FU, fludarabine, and cytosine arabinoside (Ara C). In some cases, the interaction was sequence dependent; with 5FU an enhancement in apoptosis was observed to be greater when tumor cells were first exposed to FU for 24 h followed by UCN-01 and appeared to be related to UCN-01's suppression of thymidylate synthase gene expression (Hsueh et al., 1998).

Recently, synergy between UCN-01 and the heat shock protein 90 inhibitor 17AAG (currently undergoing clinical evaluation) has been reported in leukemia cell lines; this effect appeared to be due to interference with both the PI3K/akt and Raf-1/MEK/MAP kinase survival pathways (Jia et al., 2003). Finally, synergy in prostate or lung cancer cell lines has been observed using the combination of UCN-01 with the membrane-modifying alkylphospholipid perifosine at concentrations that are clinically achievable for both agents (Dasmahapatra et al., 2004).

17.4.2.2 Clinical Studies

Phase I studies of UCN-01 studied schedules of 72 h and 3 h intravenous infusions. A particular feature of the drug in man is tight binding to human α-1 acid glyco-protein, resulting in a long plasma half-life (several weeks), and consequent low volume of distribution and systemic clearance (Sausville et al., 2001). The recommended Phase II dose for the 72 h schedule was 42.5 mg/m^2/d every 4 weeks. Dose-limiting toxicities were hyperglycemia, pulmonary dysfunction, nausea, vomiting, and hypotension (Sausville et al., 2001). One partial response (melanoma) and a prolonged stable disease (anaplastic large cell lymphoma) were reported.

Following single-agent Phase I trials, the focus of additional trials has been combination with several different drugs in which synergy has been seen in preclinical studies, as described earlier. Full reports have recently been published for Phase I combinations of UCN-01 with 5FU (Kortmansky et al., 2005), cisplatin (Lara et al., 2005), the DNA topoisomerase I inhibitor topotecan (Hotte et al., 2005), and ara C in acute myelocytic leukemia (Sampath et al., 2005). The combination regime of UCN-01 and 5FU comprised weekly infusions of 5FU (24 h infusion) and a monthly infusion of UCN-01 (135 mg/m^2 over 72 h in cycle 1 and 67.5 mg/m^2 over 36 h in subsequent cycles). Dose-limiting toxicities included syncope, arrhythmia, hyperglycemia, and headache. A dose of 2600 mg/m^2 5FU was safely administered with UCN-01 and recommended for Phase II studies (planned for patients with pancreatic cancer).

In the case of cisplatin, where UCN-01 was used at a fixed dose of 45 mg/m^2/d as a 72 h continuous intravenous infusion, only dose-level two for cisplatin (30 mg/m^2) was achievable (note that it was planned to reach 75/m^2 cisplatin). In accordance with preclinical findings, cisplatin was administered 22 h before the start of the 3 d infusion of UCN-01. However, at the 30 mg/m^2 dose of cisplatin, although the plasma pharmacokinetics of UCN-01 were unaffected, severe dose-limiting toxicities were encountered (grade 5 sepsis with respiratory failure associated with grade 3 creatinine; grade 3 atrial fibrillation). The investigators advocated further studies using alternative dose schedules of the combination (shorter infusion times of UCN-01) or using alternative platinum-based drugs such as carboplatin or oxaliplatin.

The Phase I combination trial of UCN-01 and topotecan was relatively well tolerated (neutropenia and nausea and vomiting being dose limiting), showed some preliminary evidence of activity, and is now being studied in a Phase II trial in patients with ovarian cancer. The trial involved giving UCN-01 at 70 mg/m^2 on day 1 as a 3 h infusion followed immediately by topotecan at 1 mg/m^2; the topotecan dose was then repeated on days 2–5. Cycles were repeated every 21 d and with the UCN-01 dose reduced by 50% from cycle 2 to take account of its long plasma half-life (see earlier text).

Finally, the combination of UCN-01 (45 mg/m^2/d for 3 d) and ara C (1 g/m^2/d for 4 d) administered concurrently has been studied in a pilot trial in patients with relapsed acute myelocytic leukemia. Blasts from these patients showed decreased phosphorylated chk1 and akt as well as an activation of JNK.

17.4.3 R-ROSCOVITINE (CYC202, SELICICLIB)

Roscovitine is an aminopurine analog (Figure 17.1) being studied in clinical trials as the R enantiomer (CYC202).

17.4.3.1 Preclinical Studies

It is a broad-spectrum ATP-binding site competitive inhibitor of various CDKs including CDK2/cyclin E, CDK7/cyclin H, CDK9/cyclin T1 and, to a lesser extent, CDK4/cyclin D1 (McClue et al., 2002). It may also possess biological properties owing to effects on non-CDK- related proteins such as pyridoxal kinase (Bach et al., 2005). Roscovitine inhibits retinoblastoma protein phosphorylation, causes loss of cyclin D1, and activates the mitogen-activated protein kinase pathway in human colon cancer cells (Whittaker et al., 2004). It has been shown to induce cell cycle arrest and cell death from all compartments of the cell cycle, to require at least 16 h drug exposure to cause maximum cell kill, and to confer anti-tumor activity by the oral route in mice bearing colorectal or uterine cancer xenografts (McClue et al., 2002; Raynaud et al., 2005). In addition, inhibition of RNA polymerase II-dependent transcription and downregulation of the antiapoptotic protein Mcl-1 has been observed following exposure of myeloma cells to CYC202 (MacCallum et al., 2005).

With respect to combination studies, *in vitro* synergism has been reported using median effect analysis in myeloma cell lines for CYC202 combined with doxorubicin or the proteosome inhibitor bortezomib, which is approved for clinical use in this disease (Raje et al., 2005). Recently, roscovitine has been shown in some cell types to modulate DNA repair by inhibiting DNA double-strand break repair (Crescenzi et al., 2005). In combination with doxorubicin (or etoposide), roscovitine caused either protection or sensitization, dependent on cell cycle effects: in cells in which the dominant effect was G1 arrest (via p21^{CIP} and retinoblastoma proteins), protection was observed; whereas in cells in which the dominant cell cycle effect was accumulation in G2/M, sensitization was seen (by inhibition of DNA double-stand break repair). Finally, in human leukemia cells, roscovitine potentiated the cell killing and apoptotic effects of the histone deacetylase inhibitor LAQ824 (Rosato et al., 2005).

17.4.3.2 Clinical Studies

Clinical trials using oral dosing, either twice daily for 7 out of every 21 d or twice daily for 5 d every 21 d, have begun (Blagden and de Bono, 2005). In the 7-d schedule at

800 mg twice a day, dose-limiting toxicities of skin rash and hypokalemia were observed. In the 5 d trial a recommended dose for further study is reported as 1250 mg twice a day. Reported toxicities included emesis, asthenia, and skin rash. No objective responses were reported. Phase II trials have started in lung and B-cell malignancies, although it is unclear whether further studies in combination are taking place.

17.4.4   BMS-387032

BMS-387032 is an aminothiazole (Figure 17.1) and a potent (IC_{50} of 48 nM) CDK2-selective inhibitor (10-fold selective for CDK2/cyclin E relative to CDK1/cyclin B and 20-fold relative to CDK4/cyclin D) (Misra et al., 2004). It has demonstrated anti-tumor activity in preclinical leukemia, and ovarian and squamous cell carcinoma cancer models *in vivo*. Combination studies with cisplatin using a colon cancer cell line show sequence-dependent synergism; optimal (synergistic) cytotoxicity was seen when BMS-387032 was administered 24 h prior to cisplatin, but with antagonism reported when cisplatin was administered first. Phase I trials are ongoing using a 1 h intravenous infusion every 3 weeks at doses up to 85 mg/m² so far, or using a 24 h infusion every 3 weeks or once weekly dosing (Blagden and de Bono, 2005).

17.4.5   Ro 31-7453

Although its precise mechanism of action has not been identified, Ro 31-7453 (which bears some structural similarity to staurosporine) inhibits CDKs 1, 2, and 4 (and tubulin polymerization) in cell-free systems (Salazar et al., 2004). It shows additive or synergistic anti-tumor effects with the following: the antimitotic drugs paclitaxel and vinorelbine in breast cancer cell lines, gemcitabine in a non-small-cell lung cancer model, and capecitabine in breast and colorectal xenograft models. It also showed evidence of anti-tumor activity against a range of human tumor xenografts in mice (e.g., breast, colorectal, lung, and prostate). The results of two Phase I studies have been reported. Ro 31-7453 given as a 7 or 14 d oral twice daily dosing schedule every 4 weeks to patients with advanced solid tumors resulted in myelosuppression and mucositis as dose-limiting toxicities, with MTDs of 200 mg/m² and 125 mg/m² twice daily for the 7 and 14 d schedules, respectively (Salazar et al., 2004). One partial response was seen in a patient with non-small-cell lung cancer, and there was evidence of stable disease in some additional patients. In a second Phase I study patients received oral Ro 31-7453 twice daily for 4 consecutive days every 3 weeks (Dupont et al., 2004). There were no complete or partial responses; the recommended dose for further studies using this schedule was 1000 mg per day as a flat dose. Ro 31-7453 behaves as a prodrug and is activated by metabolism. Both studies showed large interpatient variability in pharmacokinetics. Phase II monotherapy studies have been initiated in patients with either breast, lung, or colorectal cancer using the 14 d schedule at a twice daily dose of 125 mg/m². Phase I combination studies with either gemcitabine or paclitaxel have also been initiated, using twice daily dosing for 4 d.

17.4.6 E7070

E7070 is a synthetic chloroindolyl sulfonamide. Although not a direct CDK inhibitor, it has been shown to deplete cyclin E with a reduction in CDK2 catalytic activity and cause transcriptional repression of cyclin H, thereby reducing CDK7. These effects result in cell cycle arrest at the G1 S boundary, accompanied by hypophosphorylation of the retinoblastoma protein family (Ozawa et al., 2001). Tumor regression was observed in human tumor xenografts representative of colon and lung cancers, with greater effects seen with daily (for 4 or especially 8 d) dosing (Ozawa et al., 2001). A Phase I clinical study using weekly dosing to patients with advanced solid tumors resulted in an MTD of 500 mg/m^2/week, with reversible neutropenia and thrombocytopenia as dose-limiting toxicities (Dittrich et al., 2003). There were hints of anti-tumor activity; a partial response in a patient with an endometrial adenocarcinoma who had had prior radiotherapy, and a prolonged stable disease in a patient with metastatic melanoma. However, Phase II monotherapy studies reported to date in patients with squamous cell carcinoma of the head and neck (Haddad et al., 2004) and melanoma (Smyth et al., 2005) (both using 700 mg/m^2 every 3 weeks) suggest that E7070 does not merit further development as single-agent therapy in these tumor types. Although E7070 has been reported (in an abstract) to be synergistic with the topoisomerase I inhibitor CPT11 in preclinical tumor models, it is unclear whether this combination strategy is currently being pursued in the clinic.

17.5 CONCLUSIONS AND FUTURE DIRECTIONS

Evidence to date from both preclinical xenograft and clinical studies all point to the optimum use of CDK inhibitors being as part of combination chemotherapy. An overall summary is shown in Table 17.2.

Although there are theoretical grounds to support the use of CDK inhibitors as monotherapy in tumors with certain genotypes (for example, where there is loss of retinoblastoma function [Li et al., 2001]); or for combining the checkpoint inhibitor UCN-01 with DNA-damaging agents such as cisplatin in tumors possessing mutant p53 (Wang et al., 1996), clinical evidence to support these uses is still lacking. A number of preclinical studies, mostly using cancer cell lines, demonstrate synergy when CDK inhibitors are combined with cytotoxics, especially when the cytotoxic precedes the CDK inhibitor (e.g., flavopiridol). The field is at an exciting juncture in that a number of trials of different CDK inhibitors (e.g., flavopiridol, UCN-01, and roscovitine) in combination with standard cytotoxics (e.g., platins and topoisomerase inhibitors) are taking place. In some cases (e.g., flavopiridol and docetaxel or carboplatin; UCN-01 and cisplatin), although preclinical studies showed promising effects, such combinations have been difficult to deliver in the clinic because of toxicity. Pivotal phase II trials are awaited and will provide answers as to whether there is a gain in efficacy over the use of standard-of-care drugs by the addition of a CDK inhibitor.

TABLE 17.2
Summary of Preclinical and Clinical Combination Studies with CDK Inhibitors

CDK Inhibitor	Model	Drugs	Effect/Comment
Flavopiridol	Lung cancer cells	Cisplatin	Synergy — sequence independent
			Synergy — cytotoxic before flavopiridol
	Lung cancer cells	Paclitaxel, ara C, topotecan, doxorubicin, etoposide, 5FU	
			Increased apoptosis
	Gastric and breast cancer cells	Mitomycin C, paclitaxel	Increased apoptosis
	Retinoblastoma-deficient sarcoma cells	Doxorubicin	Feasible — some signs of activity
			Encouraging activity (3 × partial responses)
	Phase I solid tumors	Paclitaxel	Cisplatin feasible (planned for ovarian)
	Phase I solid tumors	Irinotecan	Carboplatin too toxic
			Not feasible because of toxicity
	Phase I solid tumors	Cisplatin, carboplatin	Feasible — warrants further study
	Phase I breast cancer	Docetaxel	
	Phase I acute leukemia	Ara C, mitoxantrone	
UCN-01	Ovarian, breast cells	Cisplatin	Synergy
	Various cell lines	5 fluorouracil, camptothecin, mitomycin C, temozolomide, gemcitabine, ara C	Synergy
	Leukemia cell lines	17AAG	Synergy
	Prostate, lung cells	Perifosine	Synergy
	Phase I solid tumors	Cisplatin	72h UCN-01 infusion not feasible
	Phase I solid tumors	5 fluorouracil	Ongoing
	Phase I AML	ara C	Ongoing
	Phase I solid tumors	Topotecan	Ongoing
Roscovitine/ CYC202	Myeloma cells	Doxorubicin, bortezomib	Synergism
	Lung, breast cancer cells (retinoblastoma dysfunctional)	Low dose Doxorubicin, etoposide	Potentiation
	Leukemia cells	LAQ824 HDAC inhibitor	Synergism

TABLE 17.2 (CONTINUED)
Summary of Preclinical and Clinical Combination Studies with CDK Inhibitors

CDK Inhibitor	Model	Drugs	Effect/Comment
BMS 387032	Preclinical colon carcinoma cells	Cisplatin	Synergism
	Preclinical NSCLC, breast, colorectal xenografts	Gemcitabine Capecitabine	
Ro31-7453	Preclinical breast cancer cell lines	Paclitaxel, vinorelbine	Synergism
	Phase I oral	Gemcitabine, paclitaxel	
E7070	Preclinical	CPT11	Synergism

REFERENCES

Aklilu, M., Kindler, H.L., Donehower, R.C., Mani, S., and Vokes, E.E. (2003). Phase II study of flavopiridol in patients with advanced colorectal cancer. *Annals of Oncology*, **14:** 1270–1273.

Bach, S., Knockaert, M., Reinhardt, J., Lozach, O., Schmitt, S., Baratte, B., Koken, M., Coburn, S.P., Tang, L., Liang, D., Galons, H., Dierick, J.-F., Pinna, L.A., Meggio, F., Totzke, F., Schachtele, C., Lerman, A.S., Carnero, A., Wan, Y., Gray, N., and Meijer, L. (2005). Roscovitine targets: protein kinases and pyridoxal kinase. *Journal of Biological Chemistry*, **280:** 31208–31219.

Bible, K.C., Bible, R.H., Jr., Kottke, T.J., Svingen, P.A., Xu, K., Pang, Y.-P., Hajdu, E., and Kaufmann, S.H. (2000a). Flavopiridol binds to duplex DNA. *Cancer Research*, **60:** 2419–2428.

Bible, K.C., Boerner, S.A., Kirkland, K., Anderl, K.L., Bartelt, D., Jr., Svingen, P.A., Kottke, T.J., Lee, Y.K., Eckdahl, S., Stalboerger, P.G., Jenkins, R.B., and Kaufmann, S.H. (2000b). Characterization of an ovarian carcinoma cell line resistant to cisplatin and flavopiridol. *Clinical Cancer Research*, **6:** 661–670.

Bible, K.C. and Kaufmann, S.H. (1997). Cytotoxic synergy between flavopiridol (NSC 649890, L86-8275) and various antineoplastic agents: the importance of sequence of administration. *Cancer Research*, **57:** 3375–3380.

Bible, K.C., Lensing, J.L., Nelson, S.A., Lee, Y.K., Reid, J.M., Ames, M.M., Isham, C.R., Piens, J., Rubin, S.L., Rubin, J., Kaufmann, S.H., Atherton, P.J., Sloan, J.A., Daiss, M.K., Adjei, A.A., and Erlichman, C. (2005). Phase I trial of flavopiridol combined with cisplatin or carboplatin in patients with advanced malignancies with the assessment of pharmacokinetic and pharmacodynamic end points. *Clinical Cancer Research*, **11:** 5935–5941.

Blagden, S. and de Bono, J. (2005). Drugging cell cycle kinases in cancer therapy. *Current Drug Targets*, **6:** 325–335.

Burdette-Radoux, S., Tozer, R.G., Lohmann, R.C., Quirt, I., Ernst, D.S., Walsh, W., Wainman, N., Colevas, A.D., and Eisenhauer, E.A. (2004). Phase II trial of flavopiridol, a cyclin dependent kinase inhibitor, in untreated metastatic malignant melanoma. *Investigational New Drugs*, **22:** 315–322.

Byrd, J.C., Peterson, B.L., Gabrilove, J., Odenike, O.M., Grever, M.R., Rai, K., Larson, R.A., and the Cancer and Leukemia Group B (2005). Treatment of relapsed chronic lymphocytic leukemia by 72-hour continuous infusion of 1-h bolus infusion of flavopiridol: results from cancer and leukemia group B study 19805. *Clinical Cancer Research*, **11**: 4176–4181.

Chao, S.-H., Fujinaga, K., Marion, J.E., Taube, R., Sausville, E.A., Senderowicz, A.M., Peterlin, B.M., and Price, D.H. (2000). Flavopiridol inhibits P-TEFb and blocks HIV-1 replication. *Journal of Biological Chemistry*, **275**: 28345–28348.

Crescenzi, E., Palumbo, G., and Brady, H.J.M. (2005). Roscovitine modulates DNA repair and senescence: implications for combination chemotherapy. *Clinical Cancer Research*, **11**: 8158–8171.

Dasmahapatra, G.P., Didolkar, P., Alley, M.C., Ghosh, S., Sausville, E.A., and Roy, K.K. (2004). In vitro combination treatment with perifosine and UCN-01 demonstrates synergism against prostate (PC-3) and lung (A549) epithelial adenocarcinoma cell lins. *Clinical Cancer Research*, **10**: 5242–5252.

Dittrich, C., Dumez, H., Calvert, H., Hanauske, A., Faber, M., Wanders, J., Yule, M., Ravic, M., and Fumoleau, P. (2003). Phase I and pharmacokinetic study of E7070, a chloroindolyl-sulfonamide anticancer agent, administered on a weekly schedule to patients with solid tumors. *Clinical Cancer Research*, **9**: 5195–5204.

Drees, M., Dengler, W.A., Roth, T., Labonte, H., Mayo, J., Malspeis, L., Grever, M., Sausville, E.A., and Fiebig, H.H. (1997). Flavopiridol (L86-8275): selective antitumour activity *in vitro* and activity *in vivo* for prostate carcinoma cells. *Clinical Cancer Research*, **3**: 273–279.

Dupont, J., Bienvenu, B., Aghajanian, C., Pezzulli, S., Sabbatini, P., Vongphrachanh, P., Chang, C., Perkell, C., Ng, K., Passe, S., Breimer, L., Zhi, J., DeMario, M., Spriggs, D., and Soignet, S.L. (2004). Phase I and pharmacokinetic study of the novel oral cell-cycle inhibitor Ro 31-7453 in patients with advanced solid tumors. *Journal of Clinical Oncology*, **22**: 3366–3374.

Flinn, I.W., Byrd, J.C., Bartlett, N., Kipps, T., Gribben, J., Thomas, D., Larson, R.A., Rai, K., Petric, R., Ramon-Suerez, J., Gabrilove, J., and Grever, M.R. (2005). Flavopiridol administered as a 24-h continuous infusion in chronic lymphocytic leukemia lacks clinical activity. *Leukemia Research*, **29**: 1253–1257.

Frei, E. III, Holland, J.F., Schneiderman, M.A., Pinkel, D., Selkirk, G., Freireich, E.J., Silver, R.T., Gold, G.L., and Regelson, W. (1958). A comparative study of two regimens of combination chemotherapy in acute leukemia. *Blood*, **13**: 1126–1148.

Grendys, J.E.C., Blessing, J.A., Burger, R., and Hoffman, J. (2005). A phase II evaluation of flavopiridol as second-line chemotherapy of endometrial carcinoma: a Gynecologic Oncology Group study. *Gynecologic Oncology*, **98**: 249–253.

Haddad, R.I., Weinstein, L.J., Wieczorek, T.J., Bhattacharya, N., Raftopoulos, H., Oster, M.W., Zhang, X., Latham, V.M., Jr., Costello, R., Faucher, J., DeRosa, C., Yule, M., Miller, L.P., Loda, M., Posner, M.R., and Shapiro, G.I. (2004). A Phase II clinical and pharmacodynamic study of E7070 in patients with metastatic, recurrent, or refractory squamous cell carcinoma of the head and neck: modulation of retinoblstoma protein phosphorylation by a novel chloroindolyl sulfonamide cell cycle inhibitor. *Clinical Cancer Research*, **10**: 4680–4687.

Hotte, S.J., Oza, A., Winquist, E.W., Moore, M., Chen, E.X., Brown, S., Pond, G.R., Dancey, J.E., and Hirte, H.W. (2006). Phase I trial of UCN-01 in combination with topotecan in patients with advanced solid cancers: a Princess Margaret Hospital Phase II consortium study. *Annals of Oncology*, **17**: 334–340.

Hsueh, C.T., Kelsen, D., and Schwartz, G.K. (1998). UCN-01 suppresses thymidylate synthase gene expression and enhances 5-fluorouracil-induced apoptosis in a sequence-dependent manner. *Clinical Cancer Research*, **4:** 2201–2206.

Incles, C.M., Schultes, C.M., Kelland, L.R., and Neidle, S. (2003). Acquired cellular resistance to flavopiridol in a human colon carcinoma cell line involves up-regulation of the telomerase catalytic subunit and telomere elongation. Sensitivity of resistant cells to combination treatment with a telomerase inhibitor. *Molecular Pharmacology*, **64:** 1101–1108.

Jia, W., Yu, C., Rahmani, M., Krystal, G., Sausville, E.A., Dent, P., and Grant, S. (2003). Synergistic antileukemic interactions between 17-AAG and UCN-01 involve interruption of RAF/MEK- and AKT-related pathways. *Blood*, **102:** 1824–1832.

Karp, J.E., Passaniti, A., Gojo, I., Kaufmann, S., Bible, K., Garimella, T.S., Greer, J., Briel, J., Douglas Smith, B., Gore, S.D., Tidwell, M.L., Ross, D.D., Wright, J.J., Colevas, A.D., and Bauer, K.S. (2005). Phase I and pharmacokinetic study of flavopiridol followed by 1-beta-D-arabinofuranosylcytosine and mitoxantrone in relapsed and refractory adult acute leukemias. *Clinical Cancer Research*, **11:** 8403–8412.

Kortmansky, J., Shah, M.A., Kaubisch, A., Weyerbacher, A., Yi, S., Tong, W., Sowers, R., Gonen, M., O'Reilly, E., Kemeny, N., Ilson, D., Saltz, L.B., Maki, R.G., Kelsen, D.P., and Schwartz, G.K. (2005). Phase I trial of the cyclin-dependent kinase inhibitor and protein kinase C inhibitor 7-hydroxystaurosporine in combination with fluorouracil in patients with advanced solid tumors. *Journal of Clinical Oncology*, **23:** 1875–1884.

Kouroukis, C.T., Belch, A., Crump.M., Eisenhauer, E., Gascoyne, R.D., Meyer, R., Lohmann, R., Lopez, P., Powers, J., Turner, R., and Connors, J.M. (2003). Flavopiridol in untreated or relapsed mantle-cell lymphoma: results of a phase II study of the National Cancer Institute of Canada Clinical Trials Group. *Journal Clinical Oncology*, **21:** 1740–1745.

Lara, J.P.N., Mack, P.M., Synold, T., Frankel, P., Longmate, J., Gumerlock, P.H., Doroshow, J.H., and Gandara, D.R. (2005). The cyclin-dependent kinase inhibitor UCN-01 plus cisplatin in advanced solid tumors: a California Cancer Consortium Phase I pharmacokinetic and molecular correlative trial. *Clinical Cancer Research*, **11:** 4444–4450.

Li, W.W., Fan, J., and Bertino, J.R. (2001). Selective sensitization of retinoblastoma protein-deficient sarcoma cells to doxorubicin by flavopiridol-mediated inhibition of cyclin-dependent kinase 2 kinase activity. *Cancer Research*, **61:** 2579–2765.

Liu, G., Gandara, D.R., Lara, P.N., Jr., Raghavan, D., Doroshow, J.H., Twardowski, P., Kantoff, P., Oh, W., Kim, K.-M., and Wilding, G. (2004). A Phase II trial of flavopiridol (NSC #649890) in patients with previously untreated metastatic androgen-independent prostate cancer. *Clinical Cancer Research*, **10:** 924–928.

MacCallum, D.E., Melville, J., Frame, S., Watt, K., Anderson, S., Gianella-Boradori, A., Lane, D.P., and Green, S.R. (2005). Seliciclib (CYC202, R-Roscovitine) induces cell death in multiple myeloma cells by inhibition of RNA polymerase II-dependent transcription and down-regulation of Mcl-1. *Cancer Research*, **65:** 5399–5407.

McClue, S.J., Blake, D., Clarke, R., Cowan, A., Cummings, L., Fischer, P.M., MacKenzie, M., Melville, J., Stewart, K., Wang, S., Zhelev, N., Zheleva, D., and Lane, D.P. (2002). In vitro and in vivo antitumor properties of the cyclin dependent kinase inhibitor CYC202 (R-Roscovitine). *International Journal of Cancer*, **102:** 463–468.

Misra, R.N., Xiao, H., Kim, K.S., Lu, S., Han, W.-G., Barbosa, S.A., Hunt, J.T., Rawlins, D.B., Shan, W., Ahmed, S.Z., Qian, L., Chen, B.C., Zhao, R., Bednarz, M.S., Kellar, K.A., Mulheron, J.G., Batorsky, R., Roongta, U., Kamath, A., Marathe, P., Ranadive, S.A., Sack, J.S., Tokarski, J.S., Pavletich, N.P., Lee, F.Y.F., Webster, K.R., and

Kimball, S.D. (2004). N-(cycloalkylamino)acyl-2-aminothiazole inhibitors of cyclin-dependent kinase 2. N-[5-[[[5-(1, 1-dimethylethyl)-2-oxazolyl]methyl]thiol]-2-thiazolyl]-4-piperidinecarboxamine (BMS-387032), a highly efficacious and selective antitumor agent. *Journal of Medicinal Chemistry*, **47**: 1719-1728.

Motwani, M., Delohery, T.M., and Schwartz, G.K. (1999). Sequential dependent enhancement of caspase activation and apoptosis by flavopiridol on paclitaxel-treated human gastric and breast cancer cells. *Clinical Cancer Research*, **5**: 1876–1883.

Ozawa, Y., Sugi, N.H., Nagasu, T., Owa, T., Wantanabe, T., Koyanangi, N., Yoshino, H., Kitoh, K., and Yoshimatsu, K. (2001). E7070, a novel sulphonamide agent with potent antitumour activity in vitro and in vivo. *European Journal of Cancer*, **37**: 2275–2282.

Raje, N., Kumar, S., Hideshima, T., Roccaro, A., Ishitsuka, K., Yasui, H., Shiraishi, N., Chauhan, D., Munshi, N.C., Green, S.R., and Anderson, K.C. (2005). Seliciclib (CYC202 or R-roscovitine), a small-molecule cyclin-dependent kinase inhibitor, mediates activity via down-regulation of Mcl-1 in multiple myeloma. *Blood*, **106**: 1042–1047.

Raynaud, F.I., Whittaker, S.R., Fischer, P.M., McClue, S., Walton, M.I., Barrie, S.E., Garrett, M.D., Rogers, P., Clarke, S.J., Kelland, L.R., Valenti, M., Brunton, L., Eccles, S., Lane, D.P., and Workman, P. (2005). In vitro and in vivo pharmacokinetic-pharmacodynamic relationships for the trisubstituted aminopurine cyclin-dependent kinase inhibitors olomoucine, bohemine and CYC202. *Clinical Cancer Research*, **11**: 4875–4888.

Rogers, P., Boxall, F.E., Allott, C.P., Stephens, T.C., and Kelland, L.R. (2002). Sequence-dependent synergism between the new generation platinum agent ZD0473 and paclitaxel in cisplatin-sensitive and -resistant human ovarian carcinoma cell lines. *European Journal of Cancer*, **38**: 1653–1660.

Rosato, R.R., Almenara, J.A., Maggio, S.C., Atadja, P., Craig, R., Vrana, J., Dent, P., and Grant, S. (2005). Potentiation of the lethality of the histone deacetylase inhibitor LAQ824 by the cyclin-dependent kinase inhibitor roscovitine in human leukemia cells. *Molecular Cancer Therapeutics*, **4**: 1772–1785.

Salazar, R., Bissett, D., Twelves, C., Breimer, L., DeMario, M., Campbell, S., Zhi, J., Ritland, S., and Cassidy, J. (2004). A Phase I clinical and pharmacokinetic study of Ro 31-7453 given as a 7- or 14-d oral twice daily schedule every 4 weeks in patients with solid tumors. *Clinical Cancer Research*, **10**: 4374–4382.

Sampath, D., Cortes, J., Estrov, Z., Du, M., Shi, Z., Andreeff, M., Gandhi, V., and Plunkett, W. (2006). Pharmacodynamics of cytarabine alone and in combination with 7-hydroxystaurosporine (UCN-01) in AML blasts in vitro and during a clinical trial. *Blood*, **107**: 2517–2574.

Sausville, E.A., Arbuck, S.G., Messmann, R., Headlee, D., Bauer, K.S., Lush, R.M., Murgo, A.M., Figg.W.D., Lahusen, T., Jaken, S., Jing, X., Roberge, M., Fuse, E., Kuwabara, T., and Senderowicz, A.M. (2001). Phase I trial of 72-h continuous infusion UCN-01 in patients with refractory neoplasms. *Journal Clinical Oncology*, **19**: 2319–2333.

Schwartz, G.K., Farsi, K., Mastak, P., Kelsen, D.P., and Spriggs, D. (1997). Potentiation of apoptosis by flavopiridol in mitomycin-C treated gastric and breast cancer cells. *Clinical Cancer Research*, **3**: 1467–1472.

Schwartz, G.K., Ilson, D., Saltz, L., O'Reilly, E., Tong, W., Maslak, P., Werner, J., Perkins, P., Stoltz, M., and Kelsen, D. (2001). Phase II study of the cyclin-dependent kinase inhibitor flavopiridol administered to patients with advanced gastric cancer. *Journal Clinical Oncology*, **19**: 1985–1992.

Schwartz, G.K., O'Reilly, E., Ilson, D., Saltz, L., Sharma, S., Tong, W., Maslak, P., Stolz, M., Eden, L., Perkins, P., Endres, S., Barazzuol, J., Spriggs, D., and Kelsen, D. (2002).

Phase I study of the cyclin-dependent kinase inhibitor flavopiridol in combination with paclitaxel in patients with advanced solid tumors. *Journal Clinical Oncology*, **20:** 2157–2170.

Senderowicz, A.M., Headlee, D., Stinson, S.F., Lush, R.M., Kalil, N., Villalba, L., Hill, K., Steinberg, S.M., Figg, W.D., Tompkins, A., Arbuck, S.G., and Sausville, E.A. (1998). Phase I trial of continuous infusion flavopiridol, a novel cyclin-dependent kinase inhibitor, in patients with refractory neoplasms. *Journal of Clinical Oncology*, **16:** 2986–2999.

Shah, M.A., Kortmansky, J., Motwani, M., Drobnjak, M., Gonen, M., Yi, S., Weyerbacher, A., Cordon-Cardo, C., Lefkowitz, R., Brenner, B., O'Reilly, E., Saltz, L., Tony, W., Kelsen, D.P., and Schwartz, G.K. (2005). A Phase I clinical trial of the sequential combination of irinotecan followed by flavopiridol. *Clinical Cancer Research*, **11:** 3836–3845.

Shapiro, G.I. (2004). Preclinical and clinical development of the cyclin-dependent kinase inhibitor flavopiridol. *Clinical Cancer Research*, **10:** 4270s–4275s.

Shapiro, G.I., Supko, J.G., Patterson, A., Lynch, C., Lucca, J., Zacarola, P.F., Muzikansky, A., Wright, J.J., Lynch, J.T.J., and Rollins, B.J. (2001). A phase II trial of the cyclin-dependent kinase inhibitor flavopiridol in patients with previously untreated stage IV non-small cell lung cancer. *Clinical Cancer Research*, **7:** 1590–1599.

Smith, V., Raynaud, F.I., Workman, P., and Kelland, L.R. (2001). Characterization of a human colorectal carcinoma cell line with acquired resistance to flavopiridol. *Molecular Pharmacology*, **60:** 885–893.

Smyth, J.E., Aamdal, S., Awada, A., Dittrich, C., Caponigro, F., Schoffski, P., Gore, M., Lesimple, T., Djurasinovic, N., Baron, B., Ravic, M., Fumoleau, P., Punt, C.J., and EORTC New Drug Development and Melanoma Group (2005). Phase II study of E7070 in patients with metastatic melanoma. *Annals of Oncology*, **16:** 158–161.

Stadler, W.M., Vogelzang, N.J., Amato, R., Sosman, J., Taber, D., Liebowitz, D., and Vokes, E.E. (2000). Flavopiridol, a novel cyclin-dependent kinase inhibitor, in metastatic renal cancer: a University of Chicago Phase II Consortium Study. *Journal of Clinical Oncology*, **18:** 371–375.

Tan, A.R., Headlee, D., Messman, R., Sausville, E.A., Arbuck, S.G., Murgo, A.J., Melillo, G., Zhai, S., Figg, W.D., Swain, S.M., and Senderowicz, A.M. (2002). Phase I clinical and pharmacokinetic study of flavopiridol administered as a daily 1-hour infusion in patients with advanced neoplasms. *Journal Clinical Oncology*, **20:** 4074–4082.

Tan, A.R., Yang, X., Berman, A., Zhai, S.S.A., Parr, A.L., Chow, C., Brahim, J.S., Steinberg, S.M., Figg, W.D., and Swain, S.M. (2004). Phase I trial of the cyclin-dependent kinase inhibitor flavopiridol in combination with docetaxel in patients with metastatic breast cancer. *Clinical Cancer Research*, **10:** 5038–5047.

Tse, A.N. and Schwartz, G.K. (2004). Potentiation of cytotoxicity of topoisomerase I poison by concurrent and sequential treatment with the checkpoint inhibitor UCN-01 involves disparate mechanisms resulting in either p53-independent clonogenic suppression or p53-dependent mitotic catastrophe. *Cancer Research*, **64:** 6635–6644.

van Veldhuizen, P.J., Faulkner, J.R., Lara, P.N., Jr., Gumerlock, P.H., Goodwin, W.J., Dakhil, S.R., Gross, H.M., Flanigan, R.C., and Crawford, E.D. (2005). A phase II study of flavopiridol in patients with advanced renal cell carcinoma: results of Southwest Oncology Trial 0109. *Cancer Chemotherapy and Pharmacology*, **56:** 39–45.

Wang, Q., Fan. S., Eastman, A., Worland. P.J., Sausville, E.A., and O'Connor, P.M. (1996). UCN-01: a potent abrogator of G2 checkpoint function in cancer cells with disrupted p53. *Journal National Cancer Institute*, **17:** 956–965.

Whittaker, S.R., Walton, M.I., Garrett, M.D., and Workman, P. (2004). The cyclin-dependent kinase inhibitor CYC202 (R-Roscovitine) inhibits retinoblastoma protein phosphorylation, causes loss of cyclin D1, and activates the mitogen-activated protein kinase pathway. *Cancer Research*, **64:** 262–272.

Yu, Q., La Rose, J., Zhang, H., Takemura, H., Kohn, K.W., and Pommier, Y. (2002). UCN-01 inhibits p53 up-regulation and abrogates gamma-radiation-induced G(2)-M checkpoint independently of p53 by targeting both of the check-point kinases, Chk2 and Chk1. *Cancer Research*, **62:** 5743–5748.

18 CDK Inhibitors as Anticancer Agents: Perspectives for the Future

Jayalakshmi Sridhar, Nagarajan Pattabiraman, Eliot M. Rosen, and Richard G. Pestell

CONTENTS

18.1 INTRODUCTION

Orderly progression through the normal cell cycle involves a coordinated series of activation and inactivation steps monitored by cyclin and CDK enzyme complexes. The regulatory subunit (cyclin) heterodimerizes with its catalytic partner (CDK) to form holoenzymes that are activated by phosphorylation (Morgan, 1995) and, in turn, phosphorylate key target substrates. Phosphorylation and activation of these substrates lead to cell cycle progression (reviewed in Liu, Marshall et al., 2004). At least 13 CDKs and 25 proteins with homology in the cyclin box domain have been identified in the human genome. Transition through the mid and late G_1 phase of the cell cycle is promoted by the D-type cyclins and by cyclin E in late G_1 phase. Cyclin D1 was initially cloned by three different groups of investigators (Matsushime, Roussel et al., 1991; Motokura, Bloom et al., 1991; Xiong, Connolly et al., 1991; Geisen and Moroy, 2002). The human cyclin D1 gene was cloned during the analysis of the breakpoint rearrangement of the *PTH* gene in parathyroid adenomas and was shown to share homology with the yeast cyclins. Human cyclin E1 was identified as a G_1 cyclin on the basis of its ability to complement mutations of the CLN genes in budding yeast (Koff, Cross et al., 1991; Lew, Dulic et al., 1991). Cyclin E1 as well as a second E-type cyclin (cyclin E2, Lauper, Beck et al., 1998) function as activators of CDK2 (reviewed in Moroy and Geisen, 2004). Cyclin D1 abundance is labile and both growth factor- and oncogene-inducible; cyclin D1 kinase activity rises 6 h after serum stimulation, with cyclin D1 in complex with CDK4 or CDK6, to phosphorylate and inactivate the retinoblastoma (Rb) protein. Cyclin E levels rise sequentially, allowing the formation of cyclin E/CDK2 complexes, with levels subsequently declining upon degradation through the SCF (Fbw7) ubiquitin ligase (Carrano, Eytan et al., 1999; Sutterluty, Chatelain et al., 1999; Tomoda, Kubota et al., 1999; Tsvetkov, Yeh et al., 1999; Nakayama, Nagahama et al., 2000; Rodier, Montagnoli et al., 2001; Ishida, Hara et al., 2002) and KPC (Kamura, Hara et al., 2004). Cyclin E functions to promote the assembly of a pre-replication complex through loading CDC6 to the origins of replication and recruiting MCM2 to several proteins. CDK2/cyclin E kinase phosphorylates Rb and Rb-related proteins, together with several other substrates, including those involved in pre-mRNA splicing, histone biosynthesis, centrosome duplication, CDC25, and p27^{KIP1}, a cyclin-dependent kinase inhibitor (Moroy and Geisen, 2004).

18.2 CELL CYCLE REGULATORY COMPONENTS

Molecular genetic analysis through gene deletion in mice or fibroblasts has helped provide important insights into the normal function of cell cycle regulatory control proteins in development. Studies of mice deleted of cyclins E1 and E2 have demonstrated that cyclin E is dispensable for cell cycle progression but is essential for cells to reenter the cell cycle from the quiescent G_0 state (King, Lara et al., 2003; Parisi, Beck et al., 2003). Whereas CDK2 knockout mice survive, cyclin E1, E2 knockout mice die *in utero*. CDK2 was found to be dispensable for cell cycle progression and cell division (Ortega, Prieto et al., 2003; Tetsu and McCormick, 2003). *Cyclin D1*$^{-/-}$ mice are slightly smaller than littermate controls and show features of retinal degeneration, abnormalities in terminal alveolar breast bud development, impaired macrophage migration, and altered angiogenesis (reviewed in Fu, Wang et al., 2004).

The activity of the CDKs is in turn regulated by the cell cycle inhibitor proteins. INK4 protein family members bind CDK4 and CDK6 to inhibit their interaction with the D-type cyclins. The CIP/KIP family (p21^{CIP1}, p27^{KIP1}, and p57^{KIP2}) associate with CDK2/cyclin E and CDK2/cyclin A complexes. Although complexes of CDK4/6 and cyclin D1 associate with p21^{CIP1} and p27^{KIP1}, they do not interfere with kinase activity under stoichiometric conditions (Blain, Montalvo et al., 1997; LaBaer, Garrett et al., 1997). A growing body of evidence suggests that the p21^{CIP1} and p27^{KIP1} proteins associate with complexes of CDK4/6 and cyclin D during the early phase of the cell cycle to enhance the stability of the complexes (LaBaer, Garrett et al., 1997; Cheng, Olivier et al., 1999). p21^{CIP} and p27^{KIP1} physically interact with a number of other proteins, including RhoA, strathmin, ASK1, SAPK, cysteine proteases, caspase 3, and the 14-3-3 proteins (Shim, Lee et al., 1996; Suzuki, Tsutomi et al., 1998; Suzuki, Tsutomi et al., 1999; Suzuki, Kawano et al., 2000; Fujita, Sato et al., 2002; Tanaka, Matsumura et al., 2002; Huang, Shu et al., 2003). The accumulating evidence that endogenous CDK inhibitors physically associate with and regulate the activity of a broad array of substrates has led to *in vivo* analysis of the role of these proteins in tumor suppression.

18.3 RATIONALE FOR TARGETING CDK INHIBITORS

Cyclin D1 dysregulation is a common finding in human tumorigenesis (Arnold and Papanikolaou, 2005). Cyclin D1 is frequently overexpressed in various human cancers. Genomic rearrangement or amplification of the chromosomal region including cyclin D1 is commonly found as a clonal lesion in human cancers. Cyclin D1 expression is frequently increased when the gene is amplified. Targeted expression of cyclin D1 to selective tissue compartments in transgenic mice is sufficient for the induction of hyperplasia and tumorigenesis, including breast cancer (Wang, Cardiff et al., 1994). Cyclin D1$^{-/-}$ mice are resistant to mammary tumorigenesis induced by the ErbB2 or Ras oncogenes and are resistant to colonic tumorigenesis induced by mutation of the Apc (adenomatous polyposis coli) gene (Yu, Geng et al., 2001; Hulit, Wang et al., 2004). Importantly, mice heterozygous for cyclin D1 showed a 50% reduction in colonic tumorigenesis induced by the Apc^{min} mutation, suggesting that cyclin D1 may provide a logical target for the control of mammary cancer, colon cancer, and other tumor types in humans. Cyclin E dysregulation has also been implicated in cancer (Bortner and Rosenberg, 1997). High and persistent levels of cyclin E are observed in human tumors, especially aggressive cancers (Porter, Malone et al., 1997; Donnellan and Chetty, 1999; Malumbres, Sotillo et al., 2004). Moreover, cyclin E-deficient cells are resistant to transformation by Ras (King, Lara et al., 2003). The mechanism by which cyclin E promotes tumorigenesis may include the induction of chromosome instability (Spruck, Won et al., 1999). A CDK2-independent role for cyclin E in cell transformation is suggested by the finding that cyclin E1 mutants defective for stimulating kinase activity may still collaborate in cell transformation with Ras (Geisen and Moroy, 2002). An intriguing parallel exists with cyclin D1, in which kinase-deficient mutants of cyclin D1 can also collaborate in oncogenesis with Ras (Hinds, Dowdy et al., 1994; Zwicker, Brusselbach et al., 1999).

Based largely on *in vitro* experiments, the CDK inhibitors were shown to interfere with activities of cyclin D- and A-dependent kinases. Evidence supporting a role for CIP and KIP family proteins in tumorigenesis includes the finding that lower levels of p27^{KIP1} occur frequently in human tumors and may correlate with poor prognosis (Loda, Cukor et al., 1997). p27^{KIP1} heterozygous mice are susceptible to both radiation- and ENU-induced tumors (Kiyokawa, Kineman et al., 1996; Fero, Randel et al., 1998; Yu, Geng et al., 2001; Rowlands, Pechenkina et al., 2003; Kamura, Hara et al., 2004). Furthermore, the tumors maintain the wild-type allele, indicating p27 is haploinsufficient for tumor suppression (Kiyokawa, Kineman et al., 1996; Fero, Randel et al., 1998; Kamura, Hara et al., 2004). Furthermore, a p27^{KIP1} deficiency promotes tumorigenesis in transgenic mice (Fero, Randel et al., 1998) and enhances mammary tumorigenesis induced by ErbB2 (Ortega, Prieto et al., 2003). Targeted deletion of p21[CIP1] in mice results in the induction of a variety of tumors (Martin-Caballero, Flores et al., 2001). A deficiency of p21[CIP1] accelerates mammary tumorigenesis induced by a mammary-targeted Ras oncogene, suggesting that p21[CIP1] may function as a mammary tumor suppressor *in vivo* (Bearss, Lee et al., 2002). However, a p21[CIP1] deficiency did not enhance mammary tumorigenesis induced by myc (Bearss, Lee et al., 2002), suggesting that the tumor suppressor function of endogenous CDK inhibitors is oncogene selective. Thus, p27[KIP1] functions as a tumor suppressor with heterozygosity to inhibit mammary tumorigenesis in transgenic mice. Ablation of CDK2 does not suppress the p27[−/−] phenotype of pituitary tumors (Martin, Odajima et al., 2005), again suggesting that tumor-type specific interactions occur between CDK inhibitors and their *in vitro* kinase target.

18.4 APPROACHES TO TARGETING CDK INHIBITORS

Collectively, these studies have demonstrated both allele- and oncogene-specific interactions of CDK inhibitors in regulating tumorigenesis (Bearss, Lee et al., 2002). The rationale of targeting the activity of CDKs requires a deeper understanding of the components of these kinases *in vivo*. In this regard, CDKs form multiprotein, high molecular weight complexes in cells in response to growth factors and onco-genic signals. Nuclear/cytoplasmic shuffling is regulated by phosphorylation-dependent events that occur during cell cycle progression. Evidence that p27[KIP1] and p21[CIP1] block cellular proliferation in cultured cells has led to detailed structure–function analyses that have, in turn, provided important insights into the functionally and structurally conserved components of these proteins. The CDK inhibitors evolved to regulate a broad family of cell CDKs. It has been postulated that the unstructured nature of much of the p21[CIP1] and p27[KIP1] as disordered polypeptides (Lacy, Filippov et al., 2004) facilitates interaction with several protein partners (Kriwacki, Hengst et al., 1996). The structural properties of p27[KIP1] in solution are markedly different from those of p27[KIP1] bound to CDK2/cyclin A in crystals (Russo, Jeffrey et al., 1996; Lacy, Filippov et al., 2004). In general terms, it is hypothesized that p21[CIP1], p27[KIP1], and p57[KIP2] bind CDK/cyclin complexes through a sequential binding mechanism that regulates binding-induced protein folding. Models explaining the mechanism by which the CDK inhibitors associate with the CDK/cyclin complexes have been forced to account for the selectivity with which these interactions occur. Thus, p21[CIP1] and p27[KIP1] target the cell cycle CDKs (CDK1, CDK2, CDK3, CDK4, and CDK6) but do not bind or inhibit CDK5 and CDK7 (Harper, Elledge et al., 1995).

It has been proposed that p27[KIP1] recognizes residues within the cyclin partner of the holoenzyme complex. This motif within the cyclin partner typically contains a conserved hydrophobic surface patch created by an α-helix. In the case of cyclin A, this is referred to as the MRAIL α-helix. The CDK inhibitor is thought of as a molecular staple. The prongs of the staple (domains 1 and 2) are unstructured and flexible before binding the CDK/cyclin holoenzyme complex. The spacer between flexible domains 1 and 2 (or linker helices) form the bridge of the staple. This model is consistent with a finding that residues in the linker helix are partially structured in solution and are evolutionarily conserved. The secondary structure of the segment is conserved between the CDK inhibitors based on circular dichroism studies. The MRAIL helix of cyclin A interacts with a conserved motif within domain 1 of p27[KIP1] (RXLFG) (Schulman, Lindstrom et al., 1998; Brown, Noble et al., 1999). Once p27[KIP1] is bound to this substrate recognition site, the protein folds into an extended structure that extends to the kinase subunit. This sequential mechanism of binding-induced protein folding may convey key components of the selectivity involved in the tumor suppressor function of these proteins (Lacy, Filippov et al., 2004). Thus, although the outcome of *in vitro* studies of the addition of p27[KIP1] is simply the inhibition of CDK2/cyclin E kinase activity, substantial complexity remains embedded within the molecular interactions.

Structure–function analysis of the INK family of CDK inhibitors is also instructive for rational drug design. Discovered in 1993 by Beech and Scholnick (Serrano, Hannon et al., 1993; Kamb, Gruis et al., 1994), p16[INK4A] is a negative regulator of the G_1 to S transition. p16[INK4A] induces cell cycle arrest and inhibits tumor cell proliferation in cultured cells and in transgenic mice. Knockout of INK4A has provided compelling evidence that this gene functions as a *bona fide* tumor suppressor *in vivo*. Inactivation of p16[INK4A] by a variety of mechanisms, including point mutations in more than 70 different types of tumors to date, led to structure–function analyses with the goal of targeting CDK/cyclin complexes *in vivo*. p16[INK4A] comprises 4 ankyrin repeats, a structural motif approximately 33 amino acids long, believed to be involved in protein–protein interactions. Each ankyrin repeat exhibits a helix–turn–helix structure. It has been proposed the ankyrin repeats of p16[INK4A] physically associate with CDK4 within the N-terminus (Byeon, Li et al., 1998).

18.5 STRUCTURE–FUNCTION OF CDKS

More than 100 crystal structures of CDKs (with resolution of 1.3 to 3.1 Å) in apo form, complexed with natural or synthetic inhibitors with or without bound cyclins, have been reported (Protein Data Bank; Berman, Battistuz et al., 2002; http://www.rcsb.org). Two CDK2 structures complexed to substrate peptides — one with an ATP inhibitor analog and the other with ADP and NO_3 ions — have provided valuable insight into the CDK2 substrate-binding pocket (Brown, Noble et al., 1999; Cook, Matsui et al., 2002). In addition, the structures of six natural protein CDK inhibitors in the unbound form determined by NMR are available (Luh, Archer et al., 1997; Byeon, Li et al., 1998; Li, Byeon et al., 1999; Yuan, Li et al., 1999; Yuan, Selby et al., 2000). The structure of CDK2 has been studied extensively, and predicted structures of CDK5 (with ATP and natural and synthetic inhibitors; Tarricone, Dhavan et al., 2001; Mapelli, Massimilinao et al., 2005), CDK6 (with natural inhibitors bound to cyclin; Brotherton, Dhanaraj et al., 1998; Russo, Tong et al., 1998; Jeffrey, Tong et al., 2000; Schulze-Gahmen and Kim 2002; Lu, Chang et al., 2005) and CDK7 (with ATP; Lolli, Lowe et al., 2004) have also been reported. A wealth of knowledge exists about the ATP-binding pocket and the shape and size of the pocket induced by the binding of cyclin to CDK2 (Ikuta, Kamata et al., 2002). Thus, structure-based analysis and drug discovery and design have become feasible. Crystal and NMR structures of CDK1 and CDK4 are also available for structural analysis (Liu, Marshall et al., 2004). Three crystal structures of the chimeric structure of CDK2 containing seven critical residues of the CDK4 ATP-binding residues complexed with inhibitors have been reported (Liu, Marshall et al., 2004).

The human CDK2 protein contains 298 amino acids and has the classic bi-lobal fold (De Bondt, Rosenblatt et al., 1993; Morgan, 1997). The catalytic pocket is formed between the N-terminal and C-terminal lobes. The N-terminal lobe is formed predominantly by five anti-parallel β-strands and one α-helix (the C-helix also called the PSTAIRE helix in CDK for a conserved cyclin recognition motif within it), whereas the C-terminal lobe, which is larger than the N-terminal domain, is mainly α-helical. The two lobes are linked by a flexible hinge that is made up of five to six residues. The so called Gly-rich loop is located in the N-terminal lobe. The CDK family provides an example of allosteric regulation of catalytic activity via the C-helix. Activation of CDK2 is required for the rearrangement of resides that form the ATP and the substrate-binding pockets and the critical catalytic residues such as Lys33. CDK2 is activated by the binding of the cognate cyclin, which results in the C-helix (in particular, the residue Glu51) rotating toward the catalytic residues in the ATP-binding site. Cyclin binds directly to the C-helix and surrounding elements in the N lobe, inducing the reverse rotation that restores the Lys33-Glu51 ion pair (Jeffrey, Russo et al., 1995). Next, the residue Thr160 in the activation loop has to be phosphorylated to position the substrate in the proper orientation for the transfer of the γ-phosphate group from ATP to the hydroxyl group of serine in the substrate. CDK2 activation results in the rotation of the two domains by approximately 5° and widens the ATP-binding pocket, resulting in its deepening. One of the lysine residues in the substrate forms a salt bridge with the phosphorylated T160. Figure 18.1 shows a ribbon diagram of the crystal structure of CDK2/cyclin-A-substrate peptide and an ATP analog inhibitor. The N- and C-terminal lobes are shown in cyan and yellow, respectively, whereas the activation loop with phosphorylated T160 is colored green and cyclin A is colored cornflower blue.

A relatively high-resolution analysis of the structure of CDK2/cyclin E1 has been obtained (Honda, Lowe et al., 2005). The crystal structure of CDK2 complexed with a truncated cyclin E1 at a 2.25 Å resolution revealed that the N-terminal cyclin box fold of cyclin E1 is similar to that of cyclin A. However, the C-terminal box of cyclin E1 is significantly different from that of cyclin A, allowing additional inter-actions with CDK2, particularly in the region of the activation loop. These additional interactions may contribute to CDK2-independent binding sites of cyclin E. Thus, a typical kinase fold of CDK2 comprises an N-terminal lobe, which is mostly β-sheet with one helix and the C helix, containing the PSTAIRE sequence. The C-terminal lobe is mostly α-helical and contains the catalytic residues responsible for phospho-rylation. The ATP-binding site is located between the two lobes, and the ATP recognition motif involves residues from both lobes. The CDK2 C-terminal residues differ in the context of cyclin E1 *vs.* cyclin A complexes. In the CDK2/cyclin A complex, C-terminal residues wrap around the outside of the kinase. In the CDK2/cyclin E1 complex, the polypeptide chain from residue 289 onward is almost 180° from that in the CDK2/cyclin A complex. Amino acids 230–249 of human cyclin E are required for binding to centrosomes and promoting DNA synthesis in a CDK2-independent manner (Matsumoto and Maller, 2004), and there are relatively few contacts to CDK2 in this region. The crystal structure suggests that the

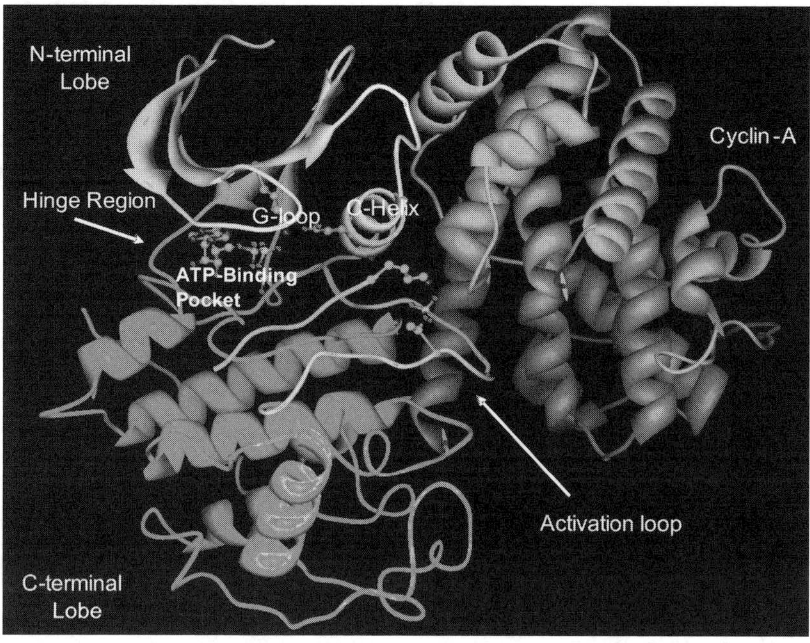

FIGURE 18.1 (See color insert following page 142.) Ribbon representation of the crystal structure of CDK2/cyclin A/ATP complex. Cyclin is shown by cornflower blue and CDK2 by cyan, yellow, and green ribbon. The substrate is colored magenta. The critical residues (K33, E51, and T160 in CDK2 and K in the substrate) involved in the activity are depicted by ball-and-stick model. An ATP analog is also shown.

centrosome interaction domains of cyclin E may be exposed, thereby forming a potential protein recognition site for centrosome localization. The addition of the CDK2/cyclin E complex crystal structure suggests the L14 leucine helix of CDK2 is exposed, allowing interaction with other known regulatory proteins, including CKS1 and the kinase-associated phosphatase KAP, known interacting proteins with CDK2 (Bourne, Watson et al., 1996; Song, Hanlon et al., 2001). p27[KIP1], which is known to inhibit cyclin E/CDK2 and cyclin A/CDK2, involves binding by the p27[KIP1] sequence motif RXLFG, which docks into the cyclin A hydrophobic pocket, both in the α1 helix and α3 helix. These key residues are conserved in sequence in confirmation in cyclin E1, suggesting that the RXL motif of p27[KIP1] can interact with the cyclin E1/CDK2 complex, as predicted by crystal structure (Honda, Lowe et al., 2005).

The usual approach to inhibit CDK activity has been to target the ATP pocket with competitive antagonists of the holoenzyme kinases, although other approaches may also be useful. An homology modeling method was used to generate a structure of the starfish oocyte *Marthasterias glacialis* CDK1/cyclin B complex from the crystal structure of CDK2/cyclin A (McGrath, Pattabiraman et al., 2005), using the program LOOK 3.0 software (GeneMine/Look, Molecular Application Group). This homology model was used to identify a novel set of inhibitors called *paullones*, with nanomolar inhibitory activity against CDK1 (Gussio, Zaharevitz et al., 2000; Kunick, Schultz et al., 2000). Further, a congeneric series of paullones were characterized using a 3-D QSAR (Qualitative Structure–Activity Relationships) with CDK1 inhibition data. Paullones were docked into the ATP-binding site of CDK1/cyclin B models and were optimized with molecular mechanics. Hydropathic analyses of the paullone/CDK1 complexes were performed after the atom types were assigned based on each ligand's electronic properties calculated from quantum mechanics. Hydropathic descriptors formed a significant multiple regression equation that predicts paullone IC_{50} data. Compounds with low affinity for CDK1 were poor charge acceptors and made less than ideal hydrogen-bonding arrangements with the receptor. These considerations led to the prediction that structures such as 9-cyanopaullone would be considerably more potent than the parent compound. The modeling prediction was confirmed by testing these compounds for their inhibitory activity against the purified CDK1 and cyclin B from the starfish oocyte *Marthasterias glacialis*. 9-Nitropaullone emerged as a paullone with similar enzyme inhibition potency and favorable antiproliferative activity in living cells. The discovery and design of paullones strongly suggests that the use of the homology model of the ATP-binding pocket for CDK1 in structure-based inhibitor design could be easily extended to other subtypes of CDKs (Gussio, Zaharevitz et al., 2000).

18.5.1 ATP-Binding Pocket

The ATP-binding pocket of CDKs is a deep cleft formed between the N- and C-terminal lobes of the protein. Apart from bidentate hydrogen bonds formed by the interaction of N-1 and N-6 of the adenine ring of ATP with the backbone carbonyl (81E) and NH (83L) groups in the hinge region of CDK2, other interactions involving the phosphates and sugar moiety render the interactions with ATP lipophilic in nature. Upon analyzing the ATP-binding pocket in the reported crystal structure of the CDK2/cyclin A/ATP complex (Russo, Tong et al., 1998), we identified 22 residues that are important in the binding of ATP (Figure 18.2). The ATP molecule is shown by the ball-and-stick representation. Residues 51E, 80F, 81E, 83L, 84H, 85Q, 86D, 89K, 131N, 137T, and 145D form a "ring" around the ATP molecule, whereas residues 33K and 10I interact with the ATP molecule perpendicular to the ring. From Figure 18.2, it can be seen that the ATP molecule interacts only with a few of the residues at the interface.

18.5.2 ATP-Competitive Inhibitors of CDKs

Progress in producing small molecule inhibitors of CDKs has proceeded principally through the development of agents that target the ATP-binding site and compete with ATP to induce inhibition. Elucidating the basic interactions of ATP and inhibitors with the CDK protein has been facilitated by numerous crystal structure studies. Similar to ATP, most of these inhibitors interact with CDKs on the basis of a donor-acceptor motif. This, in turn, has led to the design and development of new classes of inhibitors with modifications that target specific residues on the protein to improve the potency. Only recently has this effort been directed at achieving selectivity of

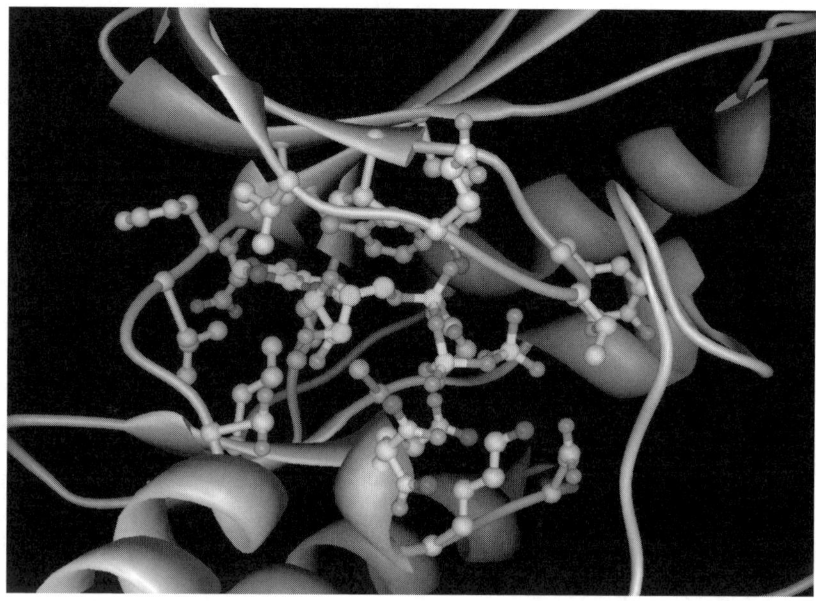

FIGURE 18.2 (See color insert.) The residues of CDK2 in contact with ATP.

the inhibitors toward one particular kinase or kinase subtype. Several classes of inhibitors have been successfully modified to improve the potency of the inhibition of CDKs, and several inhibitors show moderate to significant subtype selectivity.

The first generation of inhibitors relied on the similarity of interactions to known inhibitors of analogous structures. The purine ring system is a component of a number of CDK inhibitors. The initial set of purine analogs is exemplified by olomoucine (Vesely, Havlicek et al., 1994), a compound that exhibits good potency for CDK1, CDK2, and CDK5, but were not active for CDK4 and CDK6 (see Figure 18.3 and Table 18.1). The development of new libraries of analogs with bulky substituents at the 2 and 9 positions showed increased activity and a similar selectivity profile, as illustrated by Roscovitine (Deazevedo, Leclerc et al., 1997). Dose-dependent inhibition of G_1/S phase and $G_2/M/G_1$ phase progression was observed for both olomoucine and roscovitine (Glab, Labidi et al., 1994; Ongkeko, Fergusson et al., 1995).

Staurosporine, a metabolite derived from *Streptomyces sp.*, is a natural ATP-competitive inhibitor of protein kinases (Gadbois, Hamaguchi et al., 1992; Meijer, 1996). It was initially identified as a potent inhibitor of protein kinase C, with an IC_{50} of 1 nM. It was subsequently established that staurosporine was a relatively nonspecific inhibitor that also inhibited the CDK1/cyclin B complex, with an IC_{50} of 3.2 nM (Liu, Marshall et al., 2004). Some of the derivatives of staurosporine, such as UCN-01, K252C, and CGP41251, show excellent anti-tumor activity against different cancer cell lines (Pereira, Belin et al., 1996; Wood, Stoltz et al., 1996). UCN-01 has been shown to reduce the amount of phosphorylated Rb protein, thereby blocking the G_1/S transition, and is in clinical trials (Liu, Marshall et al., 2004; Wu, D'Amico et al., 2005). The lack of specificity of these compounds has incurred the

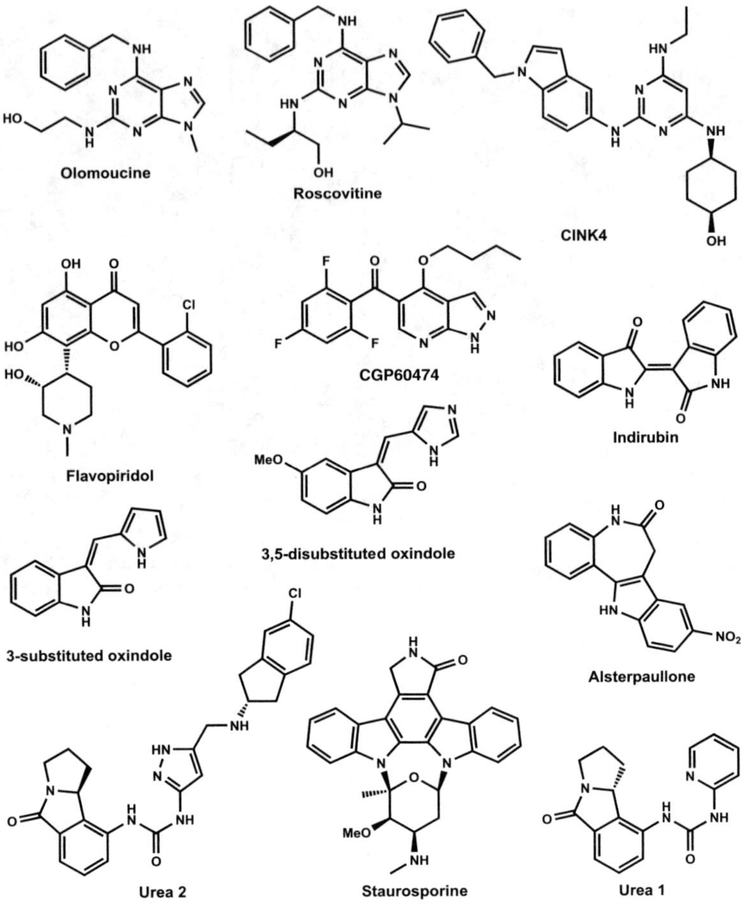

FIGURE 18.3 Chemical structures of selected known CDK inhibitors.

possibility of non-mechanism-dependent side effects in clinical use. The flavones represent another class of compounds that has been studied extensively. Flavopiridol and deschloroflavopiridol are naturally occurring alkaloids showing cytotoxic properties against tumor cell lines (Carlson, Dubay et al., 1996; Sedlacek, Czech et al., 1996). Flavopiridol was the first CDK inhibitor to enter clinical trials. Flavopiridol has its highest inhibitory potency for CDKs and for glycogen synthase kinase-3beta (GSK-3β) (Liu, Marshall et al., 2004). It is more selective for CDK4, CDK6, and CDK1 than for CDK2. Flavopiridol has also been shown to inhibit the CDK7/cyclin H complex and the positive transcription elongation factor (P-TEFb) complex containing CDK9/cyclin T1 (Chao, Fujinaga et al., 2000). This compound has shown promising results in combination with the chemotherapy drug paclitaxel in human trials (Liu, Marshall et al., 2004).

The pyrimidines and pyrido[2,3-d]pyrimidinones series of inhibitors were designed on the basis of purine-based derivatives (Boschelli, Dobrusin et al., 1998;

TABLE 18.1
Subtype Inhibitory Activities of Some Known CDK Inhibitors

Inhibitor	CDK1 IC_{50}	CDK2 IC_{50}	CDK5 IC_{50}	CDK4 IC_{50}	CDK6 IC_{50}
Olomoucine	7 µM	7 µM	3 µM	>100 µM	
Roscovitine	0.45 µM	0.7 µM	0.16 µM	>100 µM	
CINK4	>50 µM	>50 µM	25 µM	1.5 µM	5.6 µM
Flavopiridol	0.4 µM	0.1 µM		0.4 µM	
CGP60474	0.02 µM	0.05 µM		~10 µM	
Indirubin	10 µM	2.2 µM	5.5 µM	12 µM	
3-Substituted indirubin	10.2 µM	10 µM		4.9 µM	
3,5-Disubstituted indirubin	0.04 µM	0.022 µM		0.2 µM	
Alsterpaullone	0.035 µM	0.015 µM	0.04 µM	>10 µM	
Urea 1	0.12 µM	0.078 µM		0.042 µM	
Urea 2	1.8 µM	0.44 µM		0.002 µM	

Arris, Boyle et al., 2000). CINK4, a pyrimidine analog, was identified in a high-throughput screen as a CDK4 ($IC_{50} = 1.5$ µM) and CDK6 ($IC_{50} = 5.6$ µM) inhibitor that also inhibits CDK5 ($IC_{50} = 25$ µM) to a lesser extent (Soni, O'Reilly et al., 2001). Inhibition of CDK1 and CDK2 was observed only at concentrations above 50 µM. The phenylpyrimidine CGP60474 is a potent inhibitor of CDK1 and CDK2, with weaker inhibition of CDK4 (VanderWel, Harvey et al., 2005). Pyrido[2,3-d]pyrimidinones showed potency against receptor tyrosine kinases as well as CDKs (VanderWel, Harvey et al., 2005). Modifications of the core structures have led to compounds that have different selectivity profiles and increased potency.

Oxindoles, including indirubin, have been used in traditional Chinese medicine for the treatment of chronic diseases, such as leukemia (Kent, Hull-Campbell et al., 1999; Eisenbrand, Hippe et al., 2004). Several derivatives of oxindoles have been synthesized to improve their bioavailability and potency. Analogs such as indirubin sulfate and indirubin monoxime have shown good activity and moderate selectivity for CDK1, CDK2, and CDK5, as compared to CDK4 (Eisenbrand, Hippe et al., 2004). 3-Substituted oxindole inhibited CDK4 with moderately more selectivity over CDK1 and CDK2. The 3,5-disubstituted oxindole derivative, SU9516, showed greater selectivity for CDK2 and CDK1 than for CDK4 (Lane, Yu et al., 2001). Some hydrazone analogs exhibit potent inhibition of CDK2 and CDK1 (Luzzio, Bramson et al., 1999). Paullones are another class of compounds that exhibit good potency of CDK inhibition; and most show selectivity for CDK1, CDK2, and CDK5 (Zaharevitz, Kunick et al., 1999). The nitro derivative alsterpaullone and the cyano derivative show further improved potency (Schultz, Link et al., 1999). Diarylurea derivatives were designed on the basis of structural information derived through homology modeling of CDK4 (Honma, Hayashi et al., 2001). A diverse set of compounds were synthesized, some of which show very high potency and good selectivity for CDK4.

18.6 DESIGNING BETTER SPECIFIC INHIBITORS OF CDKS

Two classes of cyclins become successively active during the G_1 phase of the cell cycle, D-type (cyclins D1, D2, and D3; (Wang, Li et al., 2004) and E-type (cyclins E1 and E2; Payton and Coats, 2002) cyclins. Expression of D-type cyclins represents a fundamental link between mitogen/nutrient stimulation and the cell cycle machinery and is often an underlying cause of proliferative diseases such as cancer (Hanahan and Weinberg, 2000). Cyclin D1 binds to CDK4 or CDK6, thereby stimulating their kinase activity (Morgan, 1995).

Recent x-ray crystallographic studies on CDK2/cyclin A complexes (Lowe, Tews et al., 2002) have elucidated the structural basis of one mechanism by which the cyclin subunit enhances CDK substrate phosphorylation. Through a comparison of sequences of proteins — namely, E2F1, p107, p130, Rb, and members of CIP/KIP CDK inhibitor family — a "cyclin recruitment motif" or "cyclin-binding motif (CBM)," ZRXL (where Z and X are typically basic), has been identified. In a recent study, a series of peptides, ZRXLYY′ (where Y and Y′ are hydrophobic residues), containing non-natural amino acids, based on the CBM present in the tumor suppressor proteins p21[CIP1] and p27[KIP1] as a template, were prepared (Kontopidis, Andrews et al., 2003). These peptides were shown to have nanomolar to micromolar CDK2 inhibitory activity by binding to cyclin A. The cyclin A binding site comprises a groove resulting from the exposed surface of the cyclin A α1, α3, and α4 helices. A number of subsites lining the hydrophobic pocket (Met210, Ile213, Leu214, Trp217, Leu253, Glu220, Val221, and Ile281) contribute toward the hydrogen bonding of the ligand.

Benzeno and colleagues recently discovered that Kruppel-like factor (KLF6), a zinc finger tumor suppressor protein, mediates growth inhibition through an interaction with cyclin D1, leading to reduced phosphorylation of Rb at Ser795 (Benzeno, Narla et al., 2004). Furthermore, the growth inhibitory activity of KLF6 was linked to the p53-independent transactivation of p21[CIP1], a key CDK inhibitor. KLF6 disrupts CDK4/cyclin D1 complexes and forces the redistribution of p21[CIP1] onto CDK2, thereby promoting G_1 cell cycle arrest (Hu, Bryington et al., 2002). The KLF6 amino acid sequence contains consensus motifs for a CDK phosphorylation site at Ser171 and a proximal cyclin consensus-binding site at the COOH terminus (ZRXL motif, where Z is a basic residue, at amino acids 279–283). Relative to the KLF6–cyclin D1 protein interaction, which defines a novel mechanism of KLF6-mediated growth suppression, the comparison of the sequence alignment of cyclins D1, D2, and D3 reveals that even though many of the residues at the binding region are similar, certain critical residues are different (Benzeno, Narla et al., 2004).

18.6.1 POTENCY AND CDK SUBTYPE SELECTIVITY

Using the alignment of CDK sequences and the crystal structure of the CDK2/cyclin A/ATP complex, we have identified three groups of residues in the ATP-binding pocket for the binding of inhibitors. The first group of residues is involved in the formation of two to three hydrogen bonds with the inhibitors. The second group of residues is important for increasing the potency of the inhibitors. The third group of residues increases the CDK subtype selectivity of inhibitors. The residues that

---

TABLE 18.2
Residues to Be Targeted for Increasing the
Potency of CDK Inhibitors

Residue	Side Chain[a]	Backbone [a]
E8	Yes	
K9		>C=O
I10	Yes	>C=O
K33	Yes	
E51	Yes	
F80	Yes	
E81		>C=O
L83		−NH and >C=O
H84		>C=O
Q85	Yes	
D86	Yes	
K89	Yes	
E131		>C=O
N132	Yes	
D145	Yes	−NH

Note: The sequence number is based on CDK2.
[a] Interactions

---

increase the potency of CDK inhibitors are listed in Table 18.2. The side chain or
backbone to which the inhibitor might make favorable interactions and the region
to which these residues belong are listed in the table. For residues K9, E81, L83,
H84, and E131, only backbone atoms are identified. For residues E8, A85, D86,
K29, and N132, only the side-chain interactions are identified. For residues I10 and
D145, both backbone and side-chain atoms are identified as potential interacting
atoms. The residues that were identified as potential sites for increasing the subtype
selectivity are shown in Table 18.3. This selectivity could be achieved by interaction
with the C-terminal residues because of the difference in lengths as well as sequences

---

TABLE 18.3
Residues to Be Targeted for Subtype Specificity

CDK1	CDK2	CDK4	CDK5
E8	E8	A10	E8
M85	Q85	Q98	Q85
K89	K89	T102	K89
Q132	Q131	E144	Q130
D138	T137	S150	R136
291-DNQIKKM-297	291-TKPVPHLRL-299	299-EGNPE-303	289-FCPP-292

among the CDKs. As shown in Table 18.3, only CDK2 has the longest C-terminal among the four CDKs. In fact, in the crystal structures of CDK2, the residues at the C-terminus are in close proximity to the hydrogen-bonding residues of CDK2, making it possible to design subtype specific compounds by adding appropriate groups to interact with the C-terminal residues.

18.7 FUTURE DIRECTIONS

The rationale for designing inhibitors to CDK holoenzymes is based on the findings that kinase activities are increased in tumors and expression of genes encoding CDK inhibitors induces cell cycle arrest. Molecular genetic studies have demonstrated complex relationships between kinase inhibitors and the endogenous holoenzymes, with multiple distinct interacting proteins, allele- or oncogene-specific interactions, and important kinase-independent functions of cyclins in tumorigenesis. In this regard, although deletion of either cyclin E or cyclin D confers resistance to tumorigenesis in mice, overexpression of cyclin E or cyclin D1 mutants that are defective in binding Rb and stimulating CDK activity can still collaborate in oncogenesis, suggesting additional kinase-independent functions of the cyclins may contribute to tumor progression. With these caveats, CDK inhibitors can mediate effective antiproliferative and anti-tumor actions *in vitro* and *in vivo*.

The finding that cyclins can also mediate functions not directly related to their roles in cell cycle regulation (e.g., promoting cell migration and angiogenesis, regulating telomerase activity, inhibiting cell differentiation and tumor suppressors such as BRCA1, cellular senescence, and epigenetic functions (Holnthoner, Pillinger et al., 2002; Neumeister, Pixley et al., 2003; Fu, Wang et al., 2004; Fu, Rao et al., 2005; Wang, Fan et al., 2005)) may provide an opportunity for identifying alternate targeted cancer therapies. Thus, the finding that cyclins bind to histone deacetylases to regulate cell differentiation suggests that consideration be given to the role of histone deactylase inhibitors as alternate and complementary therapeutic approaches in cancer treatment. The finding that cyclins regulate heterotypic signaling — including macrophage function, angiogenesis, and cytokine secretion (Neumeister, Pixley et al., 2003; Fu, Wang et al., 2004; Fu, Rao et al., 2005) — raises the possibility that CDK inhibitors may also regulate the local tumor microenvironment. The importance of local stromal fibroblasts, infiltrating immune cells, and the extracellular matrix in promoting tumor progression is becoming an increasingly important auxiliary therapeutic target (Joyce, 2005). The finding that cyclins regulate the expression of genes encoding cytokines, growth factors, and their receptors (Fu, Wang et al., 2005) raises the possibility that therapies targeting aberrant cyclin expression within epithelial cells may have the additional effect of reducing heterotypic microenvironmental signals that promote tumor progression.

REFERENCES

Arnold, A., Papanikolaou, A. (2005). Cyclin D1 in breast cancer pathogenesis. *J. Clin. Oncol.* **23**(18): 4215–4224.

Arris, C.E., F.T. Boyle et al. (2000). Identification of novel purine and pyrimidine cyclin-dependent kinase inhibitors with distinct molecular interactions and tumor cell growth inhibition profiles. *J. Med. Chem.* **43**: 2797–2804.

Bearss, D.J., R.J. Lee et al. (2002). Differential effects of p21(WAF1/CIP1) deficiency on MMTV-ras and MMTV-myc mammary tumor properties. *Cancer Res.* **62**(7): 2077–2084.

Benzeno, S., G. Narla et al. (2004). Cyclin-dependent kinase inhibition by the KLF6 tumor suppressor protein through interaction with cyclin D1. *Cancer Res.* **64**(11): 3885–3891.

Berman, H.M., T. Battistuz et al. (2002). The Protein Data Bank. *Acta Crystallogr. D. Biol. Crystallogr.* **58**(Pt. 6 No. 1): 899–907.

Blain, S.W., E. Montalvo et al. (1997). Differential interaction of the cyclin-dependent kinase (Cdk) inhibitor p27^{Kip1} with cyclin A-Cdk2 and cyclin D2-Cdk4. *J. Biol. Chem.* **272**: 25863–25872.

Bortner, D.M. and M.P. Rosenberg (1997). Induction of mammary gland hyperplasia and carcinomas in transgenic mice expressing human cyclin E. *Mol. Cell. Biol.* **17**(1): 453–459.

Boschelli, D.H., E.M. Dobrusin et al. Preparation of Pyrido[2,3-d]pyrimidines and 4-aminopyrimidines as inhibitors of cellular proliferation. *PCT Int. Appl.* WO 9833798 A2 980806.

Bourne, Y., M.H. Watson et al. (1996). Crystal structure and mutational analysis of the human CDK2 kinase complex with cell cycle-regulatory protein CksHs1. *Cell* **84**(6): 863–874.

Brotherton, D.H., V. Dhanaraj et al. (1998). Crystal structure of the complex of the cyclin D-dependent kinase Cdk6 bound to the cell-cycle inhibitor p19INK4d. *Nature* **395**: 244–250.

Brown, N.R., M.E. Noble et al. (1999). The structural basis for specificity of substrate and recruitment peptides for cyclin-dependent kinases. *Nat. Cell Biol.* **1**(7): 438–443.

Byeon, I.-J.L., J. Li et al. (1998). Tumor suppressor p16^{INK4A}: Determination of solution structure and analyses of its interaction with cyclin-dependent kinase 4. *Mol. Cell* **1**(3): 421–431.

Carlson, B.A., M.M. Dubay et al. (1996). Flavopiridol induces g1 arrest with inhibition of cyclin-dependent kinase CDK2 and CDK4 in human breast carcinoma cells. *Cancer Res.* **56**: 2973–2978.

Carrano, A., E. Eytan et al. (1999). SKP2 is required for ubiquitin-mediated degradation of the Cdk inhibitor p27. *Nat. Cell Biol.* **1**(4): 193-199.

Chao, S.H., K. Fujinaga et al. (2000). Flavopiridol inhibits P-TEFb and blocks HIV-1 replication. *J. Biol. Chem.* **275**(37): 28345–28348.

Cheng, M., P. Olivier et al. (1999). The p21^{Cip1} and p27^{Kip1} CDK 'inhibitors' are essential activators of cyclin D-dependent kinases in murine fibroblasts. *EMBO J.* **18**(6): 1571–1583.

Cook, S.A., T. Matsui et al. (2002). Transcriptional effects of chronic Akt activation in the heart. *J. Biol. Chem.* **277**(25): 22528–22533.

De Bondt, H.L., J. Rosenblatt et al. (1993). Crystal structure of cyclin dependent kinase 2. *Nature* **363**: 595–602.

Deazevedo, W.F., S. Leclerc et al. (1997). Inhibition of cyclin-dependent kinases by purine analogues-crystal structure of human CDK2 complexed with roscovitine. *Eur. J. Biochem.* **243**: 518–526.

Donnellan, R. and R. Chetty (1999). Cyclin E in human cancers. *FASEB J* **13**(8): 773–780.

Eisenbrand, G., F. Hippe et al. (2004). Molecular mechanisms of indirubin and its derivatives: novel anticancer molecules with their origin in traditional Chinese phytomedicine. *J. Cancer Res. Clin. Oncol.* **130**(11): 627–635.

Fero, M.L., E. Randel et al. (1998). The murine p27*Kip1* is haplo-insufficient for tumour suppression. *Nature* **396**: 177–180.

Fu, M., M. Rao et al. (2005). Cyclin D1 inhibits PPARgamma-mediated adipogenesis through HDAC recruitment. *J. Biol. Chem.* **280**(17): 16934–16941.

Fu, M., C. Wang et al. (2004). Minireview: cyclin D1: normal and abnormal functions. *Endocrinology* **145**(12): 5439–5447.

Fu, M., C. Wang et al. (2005). Cyclin D1 represses p300 transactivation through a CDK-independent mechanism. *J. Biol. Chem.* **280**(33): 29728–29742.

Fujita, N., S. Sato et al. (2002). Akt-dependent phosphorylation of p27Kip1 promotes binding to 14-3-3 and cytoplasmic localization. *J. Biol. Chem.* **277**(32): 28706–28713.

Gadbois, D., J.R. Hamaguchi et al. (1992). Staurosporine is a Potent Inhibitor of p34cdc2 and p34cdc2-like kinases. *Biochem. Biophys. Res. Commun.* **184**: 80–85.

Geisen, C. and T. Moroy (2002). The oncogenic activity of cyclin E is not confined to Cdk2 activation alone but relies on several other, distinct functions of the protein. *J. Biol. Chem.* **277**(42): 39909–39918.

Glab, N., N.H. Labidi et al. (1994). Olomoucine, an inhibitor of the CDC2/CDK2 kinases activity, blocks plant cells at the G_1 to S and G_2 to M cell cycle transitions. *FEBS Lett.* **353**: 207–211.

Gussio, R., D.W. Zaharevitz et al. (2000). Structure-based design modifications of the paullone molecular scaffold for cyclin-dependent kinase inhibition. *Anticancer Drug Des.* **15**(1): 53–66.

Hanahan, D. and R.A. Weinberg (2000). The hallmarks of cancer. *Cell* **100**(1): 57–70.

Harper, J.W., S.J. Elledge et al. (1995). Inhibition of cyclin-dependent kinases by p21. *Mol. Biol. Cell* **6**(4): 387–400.

Hinds, P.W., S.F. Dowdy et al. (1994). Function of a human cyclin gene as an oncogene. *Proc. Natl. Acad. Sci. USA* **91**(2): 709–713.

Holnthoner, W., M. Pillinger et al. (2002). Fibroblast growth factor-2 induces Lef/Tcf-dependent transcription in human endothelial cells. *J. Biol. Chem.* **277**(48): 45847–45953.

Honda, R., E.D. Lowe et al. (2005). The structure of cyclin E1/CDK2: implications for CDK2 activation and CDK2-independent roles. *EMBO J* **24**(3): 452–463.

Honma, T., K. Hayashi et al. (2001). Structure-based generation of a new class of potent Cdk4 inhibitors: new de novo design strategy and library design. *J. Med. Chem.* **44**: 4615–4627.

Hu, X., M. Bryington et al. (2002). Ubiquitin/proteasome-dependent degradation of D-type cyclins is linked to tumor necrosis factor-induced cell cycle arrest. *J. Biol. Chem.* **277**(19): 16528–16537.

Huang, S., L. Shu et al. (2003). Sustained activation of the JNK cascade and rapamycin-induced apoptosis are suppressed by p53/p21(Cip1). *Mol Cell* **11**(6): 1491–501.

Hulit, J., C. Wang et al. (2004). Cyclin D1 genetic heterozygosity regulates colonic epithelial cell differentiation and tumor number in ApcMin Mice. *Mol. Cell. Biol.* **24**(17): 7598–7611.

Ikuta, M., K. Kamata et al. (2002). Crystallographic approach to identification of cyclin-dependent kinase 4 (Cdk4)-specific inhibitors by using Cdk4 mimic Cdk2 protein. *J. Biol. Chem.* **276**: 27548–27554.

Ishida, N., T. Hara et al. (2002). Phosphorylation of p27Kip1 on serine 10 is required for its binding to CRM1 and nuclear export. *J. Biol. Chem.* **277**(17): 14355–14358.

Jeffrey, P.D., A.A. Russo et al. (1995). Mechanism of CDK activation revealed by the structure of a cyclinA-CDK2 complex. *Nature* **376**: 313–320.

Jeffrey, P.D., L. Tong et al. (2000). Structural Basis of Inhibition of Cdk-cyclin complexes by Ink4 inhibitors. *Genes Dev.* **24**: 3115–3125.

Joyce, J.A. (2005). Therapeutic targeting of the tumor microenvironment. *Cancer Cell* **7**(6): 513–520.

Kamb, A., N.A. Gruis et al. (1994). A cell cycle regulator potentially involved in genesis of many tumor types. *Science* **264**(5157): 436–440.

Kamura, T., T. Hara et al. (2004). Cytoplasmic ubiquitin ligase KPC regulates proteolysis of p27(Kip1) at G_1 phase. *Nat. Cell. Biol.* **6**(12): 1229–1235.

Kent, L.L., N.E. Hull-Campbell et al. (1999). Characterization of novel inhibitors of cyclin-dependent kinases. *Biochem. Biophys. Res. Commun.* **260**: 768–774.

King, D.M., P. Lara et al. (2003). A phase II trial of flavopiridol in patients with metastatic androgen independent prostate cancer. *Proc. Am. Soc. Clin. Oncol.* **22**: 432.

Kiyokawa, H., R.D. Kineman et al. (1996). Enhanced growth of mice lacking the cyclin-dependent kinase inhibitor function of p27*Kip1*. *Cell* **85**: 721–732.

Koff, A., F. Cross et al. (1991). Human cyclin E, a new cyclin that interacts with two members of the CDC2 gene family. *Cell* **66**(6): 1217–1228.

Kontopidis, G., M. Andrews et al. (2003). Insights into cyclin groove recognition. complex crystal structures and inhibitor design through ligand exchange. *Structure* **11**: 1537–1546.

Kriwacki, R.W., L. Hengst et al. (1996). Structural studies of p21Waf1/Cip1/Sdi1 in the free and Cdk2-bound state: conformational disorder mediates binding diversity. *Proc. Natl. Acad. Sci. USA* **93**(21): 11504–11509.

Kunick, C., C. Schultz et al. (2000). 2-Substituted paullones: CDK1/cyclin B-inhibiting property and in vitro antiproliferative activity. *Bioorg. Med. Chem. Lett.* **10**(6): 567–569.

LaBaer, J., M.D. Garrett et al. (1997). New functional activities for the p21 family of CDK inhibitors. *Genes Dev.* **11**: 847–862.

Lacy, E.R., I. Filippov et al. (2004). p27 binds cyclin-CDK complexes through a sequential mechanism involving binding-induced protein folding. *Nat. Struct. Mol. Biol.* **11**(4): 358–364.

Lane, M.E., B. Yu et al. (2001). A novel cdk2-selective inhibitor, SU9516, induces apoptosis in colon carcinoma cells. *Cancer Res.* **61**(16): 6170–6177.

Lauper, N., A.R. Beck et al. (1998). Cyclin E2: a novel CDK2 partner in the late G_1 and S phases of the mammalian cell cycle. *Oncogene* **17**(20): 2637–2643.

Lew, D.J., V. Dulic et al. (1991). Isolation of three novel human cyclins by rescue of G_1 cyclin (Cln) function in yeast. *Cell* **66**: 1197–1206.

Li, J., I.-J. Byeon et al. (1999). Tumor suppressor Ink4: determination of the solution structure of P18Ink4C and demonstration of the functional significance of loops in P18Ink4C and P16Ink4A. *Biochemistry* **38**: 2930–2940.

Liu, M.C., J.L. Marshall et al. (2004). Novel strategies in cancer therapeutics: targeting enzymes involved in cell cycle regulation and cellular proliferation. *Curr. Cancer Drug Targets* **4**(5): 403–424.

Loda, M., B. Cukor et al. (1997). Increased proteasome-dependent degradation of the cyclin-dependent kinase inhibitor p27 in aggressive colorectal carcinomas. *Nat. Med.* **3**(2): 231–234.

Lolli, G., E.D. Lowe et al. (2004). The crystal structure of human Cdk7 and its protein recognition properties. *Structure* **12**: 2067–2079.

Lowe, E., I. Tews et al. (2002). Specificity determinants of recruitment peptides bound to phospho-Cdk2/cyclin A. *Biochemistry* **41**: 15625–15634.

Lu, H.S., D.J. Chang et al. (2005). Crystal structure of a human cyclin-dependent kinase 6 complex with a flavonol inhibitor, fisetin. *J. Med. Chem.* **48**: 737–743.

Luh, F.Y., S.J. Archer et al. (1997). Structure of the cyclin-dependent kinase inhibitor p19Ink4d. *Nature* **389**: 999–1003.

Luzzio, M.J., N. Bramson et al. (1999). Inhibitors of cyclin dependent kinases: design, synthesis and evaluation. *Am. Assoc. Cancer Res.* **Apr**: 10–14.

Malumbres, M., R. Sotillo et al. (2004). Mammalian cells cycle without the D-type cyclin-dependent kinases Cdk4 and Cdk6. *Cell* **118**(4): 493–504.

Mapelli, M., L. Massimilinao et al. (2005). Mechanism of Cdk5/P25 binding by Cdk inhibitors. *J. Med. Chem.* **48**: 671–679.

Martin, A., J. Odajima et al. (2005). Cdk2 is dispensable for cell cycle inhibition and tumor suppression mediated by p27(Kip1) and p21(Cip1). *Cancer Cell* **7**(6): 591–598.

Martin-Caballero, J., J.M. Flores et al. (2001). Tumor susceptibility of p21(Waf1/Cip1)-deficient mice. *Cancer Res.* **61**(16): 6234–6238.

Matsumoto, Y. and J.L. Maller (2004). A centrosomal localization signal in cyclin E required for Cdk2-independent S phase entry. *Science* **306**(5697): 885–888.

Matsushime, H., M.F. Roussel et al. (1991). Colony-stimulating factor 1 regulates novel cyclins during the G_1 phase of the cell cycle. *Cell* **65**: 701–713.

McGrath, C.F., N. Pattabiraman et al. (2005). Homology model of the CDK1/cyclin B complex. *J. Biomol. Struct. Dyn.* **22**(5): 493–502.

Meijer, L. (1996). Chemical inhibitors of cyclin dependent kinases. *Trends Cell Biol.* **6**: 393–397.

Morgan, D.O. (1995). Principles of CDK regulation. *Nature* **374**(6518): 131–134.

Morgan, D.O. (1997). Cyclin-dependent kinases: engines, clocks and microprocessors. *Annu. Rev. Cell. Dev. Biol.* **13**: 261–291.

Moroy, T. and C. Geisen (2004). Cyclin E. *Int. J. Biochem. Cell Biol.* **36**(8): 1424–1439.

Motokura, T., T. Bloom et al. (1991). A novel cyclin encoded by a *bcl1*-linked candidate oncogene. *Nature* **350**: 512–515.

Nakayama, K., H. Nagahama et al. (2000). Targeted disruption of Skp2 results in accumulation of cyclin E and p27(Kip1), polyploidy and centrosome overduplication. *EMBO J.* **19**: 2069–2081.

Neumeister, P., F.J. Pixley et al. (2003). Cyclin D1 governs adhesion and motility of macrophages. *Mol. Biol. Cell* **14**(3): 2005–2015.

Ongkeko, W., J.P. Fergusson et al. (1995). Inactivation of CDC2 increases the level of apoptosis induced by DNA damage. *J. Cell Sci.* **108**: 2897–2904.

Ortega, S., I. Prieto et al. (2003). Cyclin-dependent kinase 2 is essential for meiosis but not for mitotic cell division in mice. *Nat. Genet.* **35**(1): 25–31.

Parisi, T., A.R. Beck et al. (2003). Cyclins E1 and E2 are required for endoreplication in placental trophoblast giant cells. *EMBO J* **22**(18): 4794–4803.

Payton, M. and S. Coats (2002). Cyclin E2, the cycle continues. *Int. J. Biochem. Cell. Biol.* **34**(4): 315–320.

Pereira, E.R., L. Belin et al. (1996). Structure-activity relationships in a series of substituted indolocarbazoles: topoisomerase i and protein kinase C inhibition and antitumoral and antimicrobial properties. *J. Med. Chem.* **39**: 4471–4477.

Porter, P.L., K.E. Malone et al. (1997). Expression of cell-cycle regulators p27[Kip1] and cyclin E, alone and in combination, correlate with survival in young breast cancer patients. *Nat. Med.* **3**(2): 222–225.

Rodier, G., A. Montagnoli et al. (2001). p27 cytoplasmic localization is regulated by phosphorylation on Ser10 and is not a prerequisite for its proteolysis. *EMBO J* **20**(23): 6672–6682.

Rowlands, T.M., I.V. Pechenkina et al. (2003). Dissecting the roles of beta-catenin and cyclin D1 during mammary development and neoplasia. *Proc. Natl. Acad. Sci. USA* **100**: 11400–11405.

Russo, A.A., P.D. Jeffrey et al. (1996). Crystal structure of the p27[Kip1] cyclin-dependent-kinase inhibitor bound to the cyclin A–Cdk2 complex. *Nature* **382**: 325–331.

Russo, A.A., L. Tong et al. (1998). Structural basis for inhibition of the cyclin-dependent kinase Cdk6 by the tumour suppressor p16INK4a. *Nature* **395**: 237–243.

Schulman, B.A., D.L. Lindstrom et al. (1998). Substrate recruitment to cyclin-dependent kinase 2 by a multipurpose docking site on cyclin A. *Proc. Natl. Acad. Sci. USA* **95**(18): 10453–10458.

Schultz, C., A. Link et al. (1999). Paullones, a series of cyclin-dependent kinase inhibitors: synthesis, evaluation of CDK1/Cyclin B inhibition, and in vitro antitumor activity. *J. Med. Chem.* **42**: 2909–2919.

Schulze-Gahmen, U. and S.H. Kim (2002). Structural basis for Cdk6 activation by a virus-encoded cyclin. *Nat. Struct. Biol.* **9**: 177–181.

Sedlacek, H.H., J. Czech et al. (1996). Flavopiridol (L86 8275; NSC 649890), a new kinase inhibitor for tumor therapy. *Int. J. Oncol.* **9**: 1143–1168.

Serrano, M., G.J. Hannon et al. (1993). A new regulatory motif in cell-cycle control causing specific inhibition of cyclin D/CDK4. *Nature* **366**(6456): 704–707.

Shim, J., H. Lee et al. (1996). A non-enzymatic p21 protein inhibitor of stress-activated protein kinases. *Nature* **381**(6585): 804–806.

Song, H., N. Hanlon et al. (2001). Phosphoprotein-protein interactions revealed by the crystal structure of kinase-associated phosphatase in complex with phosphoCDK2. *Mol. Cell* **7**(3): 615–626.

Soni, R., T. O'Reilly et al. (2001). Selective in vivo and in vitro effects of a small molecule inhibitor of cyclin-dependent kinase 4. *J. Natl. Cancer. Inst* **93**: 436–446.

Spruck, C., K. Won et al. (1999). Deregulated cyclin E induces chromosome instability. *Nature* **401**(6750): 297–300.

Sutterluty, H., E. Chatelain et al. (1999). p45SKP2 promotes p27Kip1 degradation and induces S phase in quiescent cells. *Nature Cell Biol.* **1**: 207–214.

Suzuki, A., H. Kawano et al. (2000). Procaspase 3/p21 complex formation to resist fas-mediated cell death is initiated as a result of the phosphorylation of p21 by protein kinase A. *Cell Death Differ.* **7**(8): 721–728.

Suzuki, A., Y. Tsutomi et al. (1998). Resistance to Fas-mediated apoptosis: activation of caspase 3 is regulated by cell cycle regulator p21WAF1 and IAP gene family ILP. *Oncogene* **17**(8): 931–939.

Suzuki, A., Y. Tsutomi et al. (1999). Caspase 3 inactivation to suppress Fas-mediated apoptosis: identification of binding domain with p21 and ILP and inactivation machinery by p21. *Oncogene* **18**(5): 1239–1244.

Tanaka, H., I. Matsumura et al. (2002). E2F1 and c-Myc potentiate apoptosis through inhibition of NF-kappaB activity that facilitates MnSOD-mediated ROS elimination. *Mol. Cell* **9**: 1017–1029.

Tarricone, C., R. Dhavan et al. (2001). Structure and regulation of the Cdk5-P25(Nck5A) complex. *Mol. Cell* **8**: 657–669.

Tetsu, O. and F. McCormick (2003). Proliferation of cancer cells despite CDK2 inhibition. *Cancer Cell* **3**(3): 233–245.

Tomoda, K., Y. Kubota et al. (1999). Degradation of the cyclin-dependent-kinase inhibitor p27Kip1 is instigated by Jab1. *Nature* **398**(6723): 160–165.

Tsvetkov, L.M., K.H. Yeh et al. (1999). p27(Kip1) ubiquitination and degradation is regulated by the SCF(Skp2) complex through phosphorylated Thr187 in p27. *Curr. Biol.* **9**: 661–664.

VanderWel, S.N., P.J. Harvey et al. (2005). Pyrido[2,3-d]pyrimidin-7-ones as specific inhibitors of cyclin-dependent kinase 4. *J. Med. Chem.* **48**(7): 2371–2387.

Vesely, J., L. Havlicek et al. (1994). Inhibition of cyclin dependent kinases by purine analogues. *Eur. J. Biochem.* **224**: 771–786.

Wang, C., S. Fan et al. (2005). Cyclin D1 antagonizes BRCA1 repression of estrogen receptor A activity. *Cancer Res.* **65**(15): in press.

Wang, C., Z. Li et al. (2004). Signal transduction mediated by cyclin D1: from mitogens to cell proliferation: a molecular target with therapeutic potential. *Cancer Treat. Res.* **119**: 217–237.

Wang, T.C., R.D. Cardiff et al. (1994). Mammary hyperplasia and carcinoma in MMTV-cyclin D1 transgenic mice. *Nature* **369**: 669–671.

Wood, L., B.M. Stoltz et al. (1996). Total synthesis of (+)-RK-286c, (+)-MLR-52, (+)-staurosporine, and (+)-K252a. *J. Am. Chem. Soc.* **118**: 10656–10657.

Wu, K., M. D'Amico et al. (2005). A study of cytotoxic synergy of UCN-01 and flavopiridol in syngeneic pair of cell lines. *Invest. New Drugs* **23**(4): 299–309.

Xiong, Y., T. Connolly et al. (1991). Human D-type cyclin. *Cell* **65**(4): 691–699.

Yu, Q., Y. Geng et al. (2001). Specific protection against breast cancers by cyclin D1 ablation. *Nature* **411**(28 June): 1017–1021.

Yuan, C., J. Li et al. (1999). Tumor suppressor Ink4: comparisons of conformational properties between P16Ink4A and P18Ink4C. *J. Mol. Biol.* **294**: 201–211.

Yuan, C., T.L. Selby et al. (2000). Tumor suppressor Ink4: refinement of P16Ink4A structure and determination of P15Ink4B structure by comparative modeling and NMR data. *Protein Sci.* **9**: 1120–1128.

Zaharevitz, D., C. Kunick et al. (1999). The paullones, a novel class of small molecule inhibitors of cyclin-dependent kinases. *Am. Assoc. Cancer Res.* **Apr**: 10–14.

Zwicker, J., S. Brusselbach et al. (1999). Functional domains in cyclin D1: pRb-kinase activity is not essential for transformation. *Oncogene* **18**(1): 19–25.

Index